Feeding the World

THE PRINCETON ECONOMIC HISTORY
OF THE WESTERN WORLD

Joel Mokyr, Editor

RECENT TITLES INCLUDE:

Farm to Factory: A Reinterpretation of the Soviet Industrial Revolution,
by Robert C. Allen

Quarter Notes and Bank Notes: The Economics of Music Composition in the Eighteenth and Nineteenth Centuries,
by F. M. Scherer

The Strictures of Inheritance: The Dutch Economy in the Long Nineteenth-Century,
by Jan Luiten van Zanden and Arthur van Riel

Understanding the Process of Economic Change,
by Douglass C. North

Feeding the World: An Economic History of Agriculture, 1800–2000,
by Giovanni Federico

Cultures Merging: A Historical and Economic Critique of Culture,
by Eric L. Jones

The European Economy since 1945: Coordinated Capitalism and Beyond,
by Barry Eichengreen

War, Wine, and Taxes: The Political Economy of Anglo-French Trade, 1689–1900,
by John V. C. Nye

A Farewell to Alms: A Brief Economic History of the World,
by Gregory Clark

Power and Plenty: Trade, War, and the World Economy in the Second Millennium,
by Ronald Findlay and Kevin O'Rourke

Feeding the World

AN ECONOMIC HISTORY OF AGRICULTURE, 1800–2000

Giovanni Federico

PRINCETON UNIVERSITY PRESS

PRINCETON AND OXFORD

Published by Princeton University Press, 41 William Street, Princeton, New Jersey 08540
In the United Kingdom: Princeton University Press, 6 Oxford Street, Woodstock, Oxfordshire OX20 1TW

Second printing, and first paperback printing, 2009
Paperback ISBN: 978-0-691-13853-4

The Library of Congress has cataloged the cloth edition of this book as follows
Federico, Giovanni, 1954–
Feeding the world : an economic history of agriculture, 1800–2000 / Giovanni Federico.
p. cm. (The Princeton economic history of the Western World)
Includes bibliographical references and index.
ISBN-13: 978-0-691-12051-5 (cl. : alk. paper)
ISBN-10: 0-691-12051-X (cl. : alk. paper)
1. Agriculture—Economic aspects—History—19th century. 2. Agriculture—Economic aspects—History—20th century. I. Title.
HD1411.F43 2005
338.1'09'034–dc22
2004058634

British Library Cataloging-in-Publication Data is available

This book has been composed in Garamond

Printed on acid-free paper. ∞

press.princeton.edu

Printed in the United States of America

3 5 7 9 10 8 6 4 2

CONTENTS

LIST OF ILLUSTRATIONS

Maps

Graphs

Tables

PREFACE

I STARTED TO WORK on this book six or seven years ago, when my colleague Paolo Malanima asked me to write about a hundred pages on agriculture for a textbook on the economic history of the nineteenth and twentieth centuries, which he planned to edit. The task was both intriguing and challenging at the same time. I had written extensively on Italian agriculture (the attentive reader of this book will surely spot my pet issues) and I had already had a taste of long-range comparative work in my previous book on the history of world silk production (Federico 1997). I liked the opportunity to learn something on the history of other countries and relished the challenge to squeeze such a wide subject into a simple and clear story. The gist of this story is that agriculture has been an outstanding, and somewhat neglected, success story. In the past two centuries, it has succeeded in feeding a much greater population a greater variety of products at falling prices, while releasing a growing number of workers to the rest of the economy. Thus, agriculture has contributed in an essential way to modern economic growth.

This book seeks to explain how this outstanding feat has been achieved. There is no simple answer: agricultural performance has differed greatly among countries and periods as the result of many factors, from the environment to agricultural policies, from technologies to contracts, just to name a few. Focusing on some cases or issues would have yielded a partial and potentially misleading picture. Therefore, I have tried to be as comprehensive and systematic as possible, dealing with all the relevant topics. I have tackled such a wide and potentially unwieldy subject through a combination of modern development economics and intensive data-mining. To be sure, historical data (especially before 1914, if not 1950) are less abundant and less reliable than the current (post-1950) ones. However, special care has been taken to present the data in a comparable way, with no major methodological difference in dealing with "history" or "the present." These features should give this book an appeal that goes beyond the realm of economic history, and make it interesting to rural economists, sociologists, experts in development, and even to theorists in search of stylized facts to explain. It might even convince some of these people that economic history has something interesting to tell them. At least, this is the author's hope.

The research has greatly benefited from the nine months spent at the Department of Economics at UCLA (1999–2000) as Distinguished Visiting Scholar of the All-UC Economic history group. I would like to thank Jeff Williamson, Donald Larson, Pierre van der Eng, and Steven Broadberry for having shared unpublished data, and Alan Olmstead and Peter Lindert for having invited me to the workshop "Agricultural Productivity Changes over the Centuries" (UC-Davis, December 6–7, 2002), where I was able to discuss a very early draft of this work in a friendly and challenging environment. I have also to thank Chris Engert and Linda Truill for having edited an earlier version of the manuscript. Joel Mokyr was taken by surprise when I first contacted him with my proposal, but he recovered quickly, and during these years he has always been extremely supportive and helpful: without him, the project might well have been shelved. The book is dedicated to the memory of my father.

Pisa, September 2004

Feeding the World

Chapter One

INTRODUCTION

Agriculture has always been absolutely necessary for the very survival of humankind. For centuries, it has provided people with food, clothing, and heating, and it has employed most of the total active population. Nowadays, we dress mainly in artificial and synthetic fibers and heat themselves with fossil fuels, but the primary sector still supplies all the food we need. The available projections suggest that the world population will grow further in the next decades, while the nutritional status of the world poor must improve. Thus, agricultural production has to rise, and it has to rise with little or no further environmental damage: modern agriculture has, in fact, the reputation, largely deserved, of being environment-unfriendly.

The challenges ahead, however, should not let people forget the past achievements.[1] From 1800 to 2000, the world population has risen about six- to sevenfold, from less than one billion to six billion.[2] Yet, world agricultural production has increased substantially faster—at the very least, tenfold in the same period. Nowadays, people are better fed than in the past: each person in the world has, in theory, 2,800 calories available, with a minimum of some 2,200 in sub-Saharan Africa.[3] Famines, which haunted preindustrial times, have disappeared from most of the world. The latest survey by the Food and Agricultural Organization (FAO) estimates that 800 million people (i.e., some 10–15% of the world population) are still undernourished—but this may be an overestimation, and the proportion has drastically fallen by about a quarter since 1970.[4] Furthermore, undernourishment and famine are caused much more by the skewed distribution of income (poor entitlements in Sen's definition) and by political events (international wars, civil wars, terrorism), than by sheer lack of food.[5] Actually, many OECD countries have, since the 1950s, been struggling with an overproduction of food. The achievements of agriculture appear even more remarkable if one looks at employment. Agriculture employed more than 75 percent of the total workforce in traditional agrarian societies, and, as late as 1950, about two-thirds throughout the world. Nowadays, in the advanced countries, the share is about 2.5 percent—eleven million people out of 430.[6] In the rest of the world, agricultural workers still account for almost half the labor force, with a world total of some 1.3 billion workers (775 million in China and India alone). Such a massive transfer of labor, one of the key features of modern economic growth in the past two centuries, was made possible by a dramatic increase in product per worker. In short, agriculture is an outstanding success story. Its achievements have been outshone by the even faster growth of industry and services, but the latter would have been almost impossible if the workers had not had sufficient food to eat.

The aim of this book is to describe this success, and to understand its causes. Chapter 2 illustrates the peculiarities of agriculture. Its production depends on the environment: soil, climate, and the availability of water have always determined what peasants could grow, how much they had to work, and how much they could obtain from their efforts. These constraints have been relaxed in recent times, without totally disappearing. The factor endowment, and notably the amount of land per agricultural worker, determines the intensity of cultivation. The combined effects of the environment and the factor endowment have created long-lasting and area-specific patterns of land use, crop mix, and techniques ("agricultural systems"). The next three chapters present the main statistical evidence, loosely arranged in a production-function framework. Chapter 3 deals with the long-term trends in output (which has always been growing), relative prices (increasing in the first half of the nineteenth century, then roughly constant or slowly declining), and world trade in agricultural products (increasing quite fast before 1913 and again after 1950). The focus then shifts to the proximate causes of this growth, the increase in the use of factors (chapter 4) and productivity growth (chapter 5). Historians have a fairly clear idea about the long-run change in factors. The total agricultural work force remained roughly constant all over the world, with the notable exception of Western settlement countries (North America, Australia, Argentina, and so on) during settlement process—that is, until the beginning of the twentieth century. The stock of capital grew fast beginning in the late nineteenth century, as machines substituted labor. Although this conventional wisdom is not exactly wrong, it is, however, inspired a bit too much by the experience of the Western world. The growth of land stock has been much more geographically widespread and has lasted for longer than is commonly assumed. Agricultural capital consists mainly of building, irrigation works, and the like, and thus it increased slowly but steadily throughout the period. The real process of mechanization started only in the 1950s, and the agricultural work force has gone on growing in absolute terms. Thus, the growth of inputs (extensive growth) was the major cause of worldwide growth in agricultural production until the 1930s, but after World War II, it slowed down. Consequently, most of the big increase in total output in the past half-century has been achieved thanks to the growth in total factor productivity. The available estimates, surveyed in chapter 5, suggest that its growth has been increasing over time and that it has been faster in "advanced" countries than in LDCs. In the "advanced" countries, productivity growth has accounted for the whole of the increase in agricultural output. Contrary to a common view, productivity growth has been faster in agriculture than in the rest of the economy, including manufacturing. Chapter 6 focuses on the main source of this great achievement, technical progress. It starts by describing the main innovations, and then focuses on the process of their adoption. As in the rest of the economy, innovations are adopted when profitable, and profitability ultimately depends on the expected productivity gains and on factor endowment and factor prices. However, as the chapter argues, a standard neoclassical model cannot explain all the features of technical progress in agriculture. Agricultural innovations depend on the environment and entail a high level of risk, and many of them yield little or no financial rewards to the inventor. These features call for a greater role of the state, both in the production and the diffusion of innovations. Chapters 7, 8, and 9 deal with the institutional framework of agricultural production. "Institutions" is a fairly vague word, which resists all attempts at a general definition. Chapters 7 and 8 deal with property rights on labor and land, markets for goods and inputs (labor, land, capital), and agricultural co-operatives. Chapter 7 is, to some extent, a general introduction to these issues and to the approaches of economists and historians to institutions. It discusses how institutions work and how they might affect the performance of agriculture. Chapter 8 describes the main changes—the creation of property rights on labor and land, the trends in the average size of farms, in landownership, and in contracts, and the development of markets for goods and factors. It also puts forward some tentative hypotheses on the likely causes of these changes and on their effects on agricultural performance—although, it is fair to say, the discussion on these issues is surprisingly thin when compared to the attention they have received in the theoretical literature. Chapter 9 focuses on the effects of agricultural policies. It argues that state intervention has only really affected agricultural development since the 1930s, and that, by and large, it has reduced the aggregate welfare of the whole population. The tenth, and last, chapter shifts the focus from agriculture to the whole economy. How did the growth of agricultural output and the change in input use affect modern economic growth? This issue has been the subject of much discussion in historical perspective, and it still looms large in the debates about the optimal development strategy for less developed countries. The chapter has no ambition to solve such a controversial issue. It sketches out the prevailing theories and deals very briefly with three case studies. The book closes with some very general remarks about the future of agriculture.

The summary makes it clear that this is quite an ambitious book. It deals with many issues, and covers two centuries of agricultural history in the whole world, from Monsoon Asia to Midwest prairies. Any attempt to be comprehensive would be foolish. The potentially relevant literature spans dozens of languages, and many disciplines, from "traditional" agricultural economics and history to more "trendy" social and environmental history. Just to quote an example, the fourth volume of *A Survey of Agricultural Economics Literature, Agriculture in Economic Development,* contains more than two hundred pages of references.[7] Assuming (conservatively) that there are twenty entries per page, the total sums up to almost four thousand entries. Some of these works may be purely theoretical, and thus outside the scope of this book, but the majority should still be considered. The survey refers only to the less developed countries, deals (almost) exclusively with the post–World War II period, lists only works in English, French, Spanish, and Portuguese published before 1990, and is probably, as with all surveys, not complete. A simple proportion suggests that there are thousands of potentially relevant references. Clearly, no one in the world (certainly not this author) can reasonably claim to master all the literature. And even if this miracle were possible, it would be impossible to review it thoroughly and keep the book to a reasonable size. Selective reading is an imperative. Thus, I have decided to focus on more general contributions, and to favor works that frame their views in economic theory and buttress their statements with data.

This approach has some clear and often rehearsed shortcomings. Mainstream economic theory may appear too abstract to be relevant. Agriculture is a highly local activity, and specialists in agrarian history always warn against broad generalizations, which, they claim, cannot capture the peculiarities of the area that they are dealing with. Many data are missing, unreliable, or sometimes plainly wrong. Reliable "historical" (pre-1950) data are available only for some "advanced" countries (those of Western Europe, USA, Japan, etc). International organizations such as the UN, FAO, World Bank, and the OECD have made a magnificent effort to extract comparable data for all countries from the information provided by national statistical offices, which are sometimes incomplete and/or of dubious quality.[8] However, there are some reasons for hope. Modern development economics, with its emphasis on institutions, transaction costs, information, and so on, provides powerful tools for understanding rural societies, which can also be employed to explore societies of the past. Economic historians have unearthed a great deal of new data, which, in spite of all their shortcomings, do throw light on many key issues. And, last but not least, I feel that there is no real alternative. A history of agriculture based on anecdotal evidence from local case studies would be a boundless and largely meaningless list of details. But details are sometimes fascinating and are useful for illustrating general points—to put some flesh on the bare bones of quantitative analysis, so to speak. The reader may find the selection of these examples somewhat haphazard (why—for example—discuss tenure in China during the 1930s instead of that in Guatemala during the 1970s?). It is, however, guided, whenever possible, by two principles: first, to deal with "large" countries (China, India, Russia, the USA) and, second, to focus on controversial cases. The interest of "big" countries is self-evident, while focusing on controversial issues makes it possible to give the reader a flavor of the current research and debates.

Chapter Two

WHY IS AGRICULTURE DIFFERENT?

2.1 Introduction

The relationship with environment has always been a distinctive feature of agriculture, but its nature has changed deeply in the recent decades.[1] The current worries about the impact of agriculture on the environment would have greatly surprised a nineteenth-century American farmer, and probably an Indian one in the present day, too. The experience led them to consider nature as an enemy: he had to fight for survival against pests, weather, and diseases with inadequate tools and poor knowledge. Agricultural land itself is a product of human ingenuity: generations have toiled to convert marshes, forests, and prairies into fields and meadows. Indeed, most land is unsuitable for agriculture, as map 1 shows. Even nowadays, at the end of a long period of expansion, agriculture uses some 1.7 billion hectares for arable and tree crops, plus 3.5 billion hectares for pasture—out of 13 billion hectares total landmass.[2] The stock could indeed be augmented with suitable (huge) investments and/or the sacrifice of most remaining forests.[3] However, even in most optimistic assessments, about 50 to 60 percent of the total landmass will remain unfit for agricultural purposes, barring some spectacular and so far unpredictable technological breakthrough. This is, in itself, a powerful environmental constraint. But the environment constrains not only where but also what and when it is possible to cultivate, and thereby determines output. The next section will outline these constraints, while section 3 will focus on the relationship between land endowment and the characteristics of agriculture.

2.2 Agriculture and the Environment: An Uneasy Relationship

The environment, in the widest meaning of the word, affects agriculture in four different ways:

1. The environment determines what is possible to cultivate. Each plant or animal has an ideal habitat, which depends on several conditions that interact in a very complex way. We will focus on three key features—the quality of the soil, the temperature, and the amount of available water.

The quality of the soil is determined by its physical characteristics, which include texture, structure, exchange capacity (i.e., the capacity to release nutrients

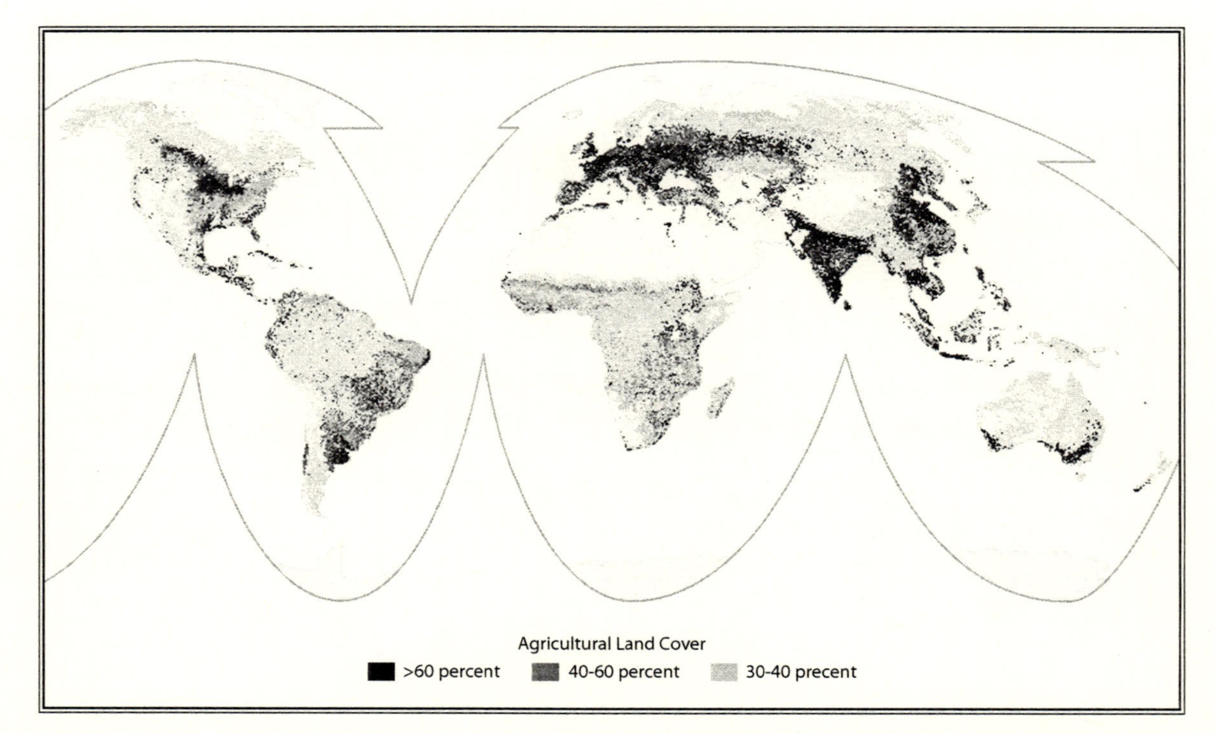

Map 1 adapted from: Stanley Wood, Kate Sebastian, and Sara J Scherer (2000), *Pilot analysis of global ecosystems, Agroecosystems.* International Food Policy Research Institute and World Resource Institute; based on data from Global Land Cover Characteristics Database Version 1.2 (Loveland et al. 2000); NOAA-NGDC (1998); WWF (1999).

to plants), alkalinity or acidity (pH), and quantity of organic matter.[4] A good-quality soil is often assumed to be essential for agricultural performance. Some soils, such as the "black earth" of the Ukraine or Russia, have the reputation of being exceptionally fertile; while the African ones, according to many authors, are qualitatively poor and prone to leaching and erosion, so that "land" in Africa is much less abundant in economic terms than it would appear by simply looking at the available acreage.[5] Some statistical analyses do find a statistically significant relation between soil quality and crop mix or yields.[6] However, the inference is a bit tricky, as the variable soil may capture the influence of other features of the environment and of human endeavor (i.e., the techniques and the amount of input, including skills). In a very careful analysis, Lindert (2000, 132–37) finds that, in China in the 1980s, only extreme acidity or alkalinity really affected land productivity once other factors, such as the intake of fertilizers and other inputs, are taken into account.[7]

The temperature is a constraint on the portfolio of crops in temperate climates because most plants and animals cannot resist a temperature below a minimum threshold. In Northern Europe, for example, Mediterranean plants, such as vines or olive trees, cannot grow, and cattle cannot be left in the open air during the winter, as they would not find grass to graze. Thus, in wintertime, animals have to be stabled or moved to milder climates. Similarly, some crops cannot grow beyond a certain altitude: for instance, in the highland zones of east Africa, where the temperature falls by about 1.8 degrees Celsius every 300 meters (3.3 degrees Fahrenheit every 1,000 feet), bananas and cassava are cultivated only below 1,500 meters, coffee between 1,200 and 1,900, potatoes between 1,500 and 2,500, tea between 1,500 and 2,300, and pyrethrum up to 2,500.[8]

All plants and animals need some water—from about 500 liters for a kilo of potatoes, to 2,000 liters and upward for rice, bananas, and sugarcane.[9] Water is usually abundant in temperate climates, but not in all semitropical and tropical areas.[10] Even very dry areas can be cultivated with careful water management. For instance, the peasants from Bechuanaland, a semidesert area, employed the technique of planting at different intervals of time (staggering) to maximize the chances that at least some plants could get the minimum amount of water during their life cycle.[11] However, a lack of water is clearly a major constraint for agriculture in tropical areas, as much as temperature is in temperate climates.[12]

It is important to stress that these constraints are not immutable. First, the climate itself can, and indeed does, change. Nowadays, the attention focuses on human-induced global warming, but history has featured a succession of relatively warm and relatively cold periods, which have changed the range of possible crops. In the late Middle Ages, the European climate was decidedly warm, even warmer than today, so that vines were routinely cultivated in the United Kingdom. The climate cooled noticeably in the sixteenth and seventeenth centuries, a period known as the Little Ice Age, and was relatively mild in the past two centuries.[13] Second, environmental constraints are not an on-off switch. The further conditions are from ideal, the less productive and the more sensitive plants and animals become to small environmental changes, although there is still a long way to go before cultivation or breeding become totally unfeasible. For instance, potatoes can be cultivated on the Peruvian highlands (ca. 3000 meters above the sea), but they take eight months, instead of two at sea-level, to grow. Last but not least, farmers can adjust to these environmental constraints by modifying the environment to suit the needs of plants and animals, or by modifying plants to suit the environment. Farmers have long known how to correct the defects of the soil, such as by adding lime to soil with a high pH (acidity), or by adding clay and marl to soil with too light a "structure."[14] Over the millennia, peasants have selected varieties of plants that can stand conditions that are very different from those in which the plants were originally cultivated. For instance, wheat, originally from the hot Middle East, is cultivated in the cold open air of the North American prairies. This is a great achievement, but it is also a mixed blessing. In fact, environment-specific varieties are more sensitive to deviations from their ideal conditions—as shown by the comparison of yields of local and imported varieties in some controlled experiments.[15]

2. Plants grow by taking nutrients—potassium, phosphorus, and, above all, nitrogen—away from the soil.[16] These nutrients have to be reintegrated because otherwise, sooner or later, the plants would not find them in sufficient quantities, and yields would fall. Leaving the soil uncultivated restores the level of the nutrients—above all, nitrogen—and thus staves off the decline in yields. The length of this rest has always been one of the defining features of traditional agriculture.[17] In the so-called swidden (or slash-and-burn) agriculture, once widely used in tropical Africa, after two or three years of cultivation the land was laid to rest for up to twenty years—long enough to allow for the regrowth of full-size trees, which were then burnt to start a new cycle.[18] Elsewhere in Africa, and in some areas of North America, Australia, and in the Balkans, the rest was shorter—four to five years—but still sufficient to allow for the growth of small trees, bush, and tall grass. In most European countries, the land was laid to rest one year out of three, or out of two (fallow).[19] In the most intensive traditional agricultural system, in east and southeast Asia, land was never allowed to rest. Peasants routinely grew two crops every two years, with peaks of up to three crops every year in some subtropical areas—and this feat deeply puzzled Western visitors.[20] The different intensity can be expressed in a compact way with the so-called cropping ratio—that is, the ratio between cropped acreage and total acreage. It can vary from 0.05 in slash-and-burn cultivation to 0.5 or 0.67 in traditional European agriculture with fallow to about 1.5 (or even more) in Asia.

The intensity of traditional Asian agriculture shows that rest is not the only way to restore the soil's fertility. Indeed, the stock of nutrients can be restored from four different natural sources, (1) irrigation water, (2) animal manure or night soil, (3) plants and grass, either spontaneous or purposefully cultivated, and (4) other natural substance, from mud and bones to more exotic items such as oil-cakes, birds' plumes, or herrings. All types of traditional agriculture once used a different combination of these sources, according to the demand (i.e., the crop mix) and the local supply (clearly water could be a major source of nutrients only in irrigated culture). The extensive African-style system used the ashes from burnt trees and plants, and some night soil in garden plots near houses.[21] In European-style agriculture, the main sources of nitrogen were the spontaneous grasses and plants from fallow (sometimes buried in the ground or burnt to minimize losses of fertilizing matter), and, above all, animal manure.[22] City waste and night soil were used only in some areas, such as the Netherlands, which, according to Van Zanden, "during these years probably approached the ideal of a 'recycling economy' more than it ever had or would again."[23] Asian agriculture relied mainly on irrigation water, night soil, and poultry and pig manure, supplemented by a careful collection of all available fertilizing material, also from cities.[24] Cattle manure was once a minor source of nitrogen both in China, where land was too scarce to be used to produce feedstuffs, and in India, where the dung had to be used as fuel because wood was extremely scarce.[25] Thus, all agricultural societies had found a way to cope with the problem of restoring fertility. However, all the traditional solutions implied some sort of constraint on the amount of nutrients available. Irrigation often needs expensive works (section 4.3) and the amount of nutrients depends on the quality of available water (Egypt was famously named "a gift from the Nile" because the water was rich in silt). The production of animal manure is constrained by the need to feed them—that is, to leave enough land for the production of feedstuff. Consequently, manure was once precious: a rich French farmer in the 1830s "received the dung from his horse in his own hat, placed the dung in a small container on the horse, put the hat back on his head and resumed conversation with his friends"(Clout 1980, 149). All these constraints were lifted for good only by artificial fertilizers, which were made available to farmers in the 1840s and 1850s (sections 4.3 and 6.2).

3. Plants grow according to a biologically determined life-cycle, which (a) takes a certain amount of time, (b) must start at a given period of the year to maximize the chance of receiving the right amount of rain, sunshine, etc., and (c) needs a well-defined set of cultivation practices that had to be performed in sequence at the "right" time. In the northern hemisphere, wheat must ripen in late spring or early summer to get as much sunshine as possible. To grow it, one has first to plough the ground, then sow, harrow (to get rid of weeds), and eventually harvest. Any anticipation or delay of any of these operations entails losses, which can become serious. For example, ripe wheat, if not cut, would fall on the ground and soon become worthless. In Burkina Faso, land has to be ploughed within three days of the first rain—otherwise, the soil will become too dry.[26] In many cases, peasants have tried to loosen these constraints. For instance, nineteenth-century Sicilian peasants discovered that withholding the supply of water and then suddenly

increasing it retarded the ripening of lemons by about ten to fifteen days (Lupo 1990, 61). For them, this was an important achievement, because late (or early) products fetched higher prices. However, it did not alter the basic seasonality of the production. Such an achievement exceeded the resources and skills of traditional peasants by far.

Livestock-raising also has, or used to have, a distinct seasonal pattern, although each animal needs roughly the same amount of care throughout the year. In fact, the number of animals depends on demand for products and above all on the supply of feed, which, in traditional agriculture, was heavily affected by the seasonal pattern in the production of cereals, hay, and grass.[27] Thus, in the northern hemisphere, farmers arranged births and slaughtering in order to have a minimum number of large animals during the winter, when food was scarcer, and to raise pigs (which can be fed with household waste) in order to slaughter them just before Christmas, when the demand for pork products reached its peak.[28] However, in livestock-breeding, unlike in field cultivation, the seasonal pattern was deeply affected by modernization. The invention of silage for cereals (section 4.3) and the increasing use of industrial feedstuffs drastically reduced this seasonal constraint, transforming livestock breeding into a quasi-industrial activity (see section 7.6).

Seasonality is important as it affects the demand for inputs, capital, and labor and also the level of specialization (Allen and Lueck 2002, 167–71). Agriculture is arguably the most seasonal economic activity, jointly (maybe) with construction and tourism. Each crop needs very little work, if any at all, for most of the year, with short periods of very intense activity (usually peaking at harvest time). The distribution over time and the total number of hours per unit of land, or of output, depends on techniques: hoeing is more labor intensive than plowing with a team of oxen, which is more labor intensive than using a tractor. Technical progress has reduced these peaks, without eliminating them altogether. Traditional agriculture coped with seasonality by trying to obtain as many different crops (with different seasonal distributions of work) or as many repetitions of the same crop as possible, given the environmental constraints and the endowment of land and capital.[29] For instance, Gallman (1970, 22–23) argued, cotton plantations in the antebellum U.S. South could produce corn for slave consumption because land was abundant and because corn demanded labor when cotton did not, and thus the opportunity cost of slave labor was almost nil.[30] This mixed farming strategy is useful, although it forces workers to develop multiple skills and thus to lose potential gains from specialization. But it is almost never sufficient to keep workers employed throughout the year, as at the height of winter, when there is almost no work, except for tending cattle. Thus, the number of workdays for full-time employees is structurally lower in agriculture than in the rest of the economy. Some extremely tentative estimates suggest a minimum of 500 to 1500 hours/worker/year in tropical Africa and a maximum of 2500 in the Netherlands at the beginning of the nineteenth century (Eicher and Baker-Doyle 1992, 83; Van Zanden 2000).[31] The slack time could be employed in off-farm jobs: in the 1950s and 1960s, it was widely believed that this "disguised unemployment" could be mobilized for development purposes (e.g., to build infrastructure) or that agriculture could release labor to the rest of the economy with no effect on production ("unlimited supply of labor").[32] The hope of finding a simple shortcut to development proved, unfortunately, to be vastly exaggerated.[33] However, there is no doubt that the crop mix has had far-reaching implications on the overall labor market. In a recent paper, Sokoloff and Dollar (1997) argue that, in the early nineteenth century, proto-industry did not develop in New England because mixed farming left comparatively little time for off-farm jobs. Thus, in these areas, the factory system was adopted earlier than in southern England, where specialization in cereals left workers plenty of time for off-farm part-time household work.

4. Last but not least, the environment determines output via the effect of weather, pests, and diseases.[34] Ceteris paribus—that is, for any level of inputs and techniques, each product has its own maximum biological output which can be attained if the weather is "perfect" (the definition of which may be different for each crop) and no pests or diseases affect the crop. Clearly, this lucky coincidence is extremely rare: the weather is never perfect, and pests and diseases are endemic. Olmstead and Rhode (2002, 948) estimate that rust, a fungus, reduced the American wheat crop by 5 to 10 percent on average in the late nineteenth and early twentieth centuries.[35] These "normal" fluctuations of output below the maximum were expected and not necessarily harmful to the peasants' income. Under certain conditions, a fall in output would cause local prices to rise, which might even more than compensate for the fall in output.[36] In some cases, however, the deviation from the norm was so great as to have devastating effects. For instance, the ashes from the eruption of the volcano Tomboro in 1815 lowered the temperature in Europe, causing a string of poor harvests, high prices, and a temporary increase in mortality (Del Vita et al. 1998).[37] The whole of Indian agriculture depends on the timely arrival of the tropical rains (monsoon): its failure caused catastrophic famines in 1876– 78 (8 million deaths), 1896–1897 (2.4 million), and 1899–1900 (3.4 million).[38] Weather-caused famines claimed between 9 and 13 million lives in China in 1876–78 and 1.8 in Rwanda and Burundi in 1928 (perhaps a third of the population—the worst famine in recent history).[39] Drought killed about half the Australian flock in the mid-1890s, a blow from which it took about twenty years to recover.[40] Plant and animal diseases, too, took a heavy toll in the nineteenth century. The Great Irish Famine, caused by the potato blight (*Phytosphora Infestans*), killed about one million people in 1845–1848, forcing many more to migrate to the United States (O'Grada 1995, 173—87).[41] In the past two centuries, serious diseases hit cotton (the boll weevil, endemic in the United States from the early 1900s to the 1940s), vines (*oidium*, or powdery mildew, in the 1840s and 1850s; *phylloxera* in the 1860s; and *peronosphora*, or downy mildew, in 1878), bananas (the so-called Panama disaster, in the 1920s), silkworm cocoons (*pebrine* in the

1850s and 1860s), cattle (the rinderpest, which swept Africa in the 1890s), pigs (swine fever in Hungary in the late 1890s and 1900s), and so on.[42] Each of these diseases reduced production in the affected areas, sometimes with far-reaching consequences on the world market: for instance, *pebrine* greatly boosted Asian silk exports, and allowed Japan its first foray on a market that it was to dominate in the 1920s and 1930s. Luckily, none of these diseases hit the production of essential food crops, such as wheat or rice: however, the rinderpest caused a famine in Ethiopia by reducing the number of oxen available for plowing.

Most of the examples quoted so far refer to the period before 1950. One might think that technical progress helped "advanced" agriculture to reduce the effect of weather and diseases, by providing new and more resistant varieties of plants, insecticides to kill pests, medicines to prevent diseases, machinery to shorten the length of key operations such as harvesting, better weather forecasts, etc. This inference is not necessarily true. Some diseases or weather conditions might hit modern varieties of plants and animals more than traditional ones because many of the former share a great portion of their genetic code. The modern varieties of cereals, the so-called HYV (high-yielding varieties) can develop their potential only with a specific set of inputs and cultivation practices (section 6.4) and failure to comply could cause serious losses in bad years. Thus, in principle, their adoption (section 6.3) might have increased yield variability.[43] Some authors have tested this hypothesis for the postwar years, without finding much support for it.[44] Total yield variance declined for wheat and perhaps rice but it increased somewhat for minor cereals. In the long run, the variability of wheat yields has increased in the United States (the coefficient of variation of the residuals of a trend series was 0.70 in 1860–1910 and 0.91 in 1960–2000) and it has remained stable in the United Kingdom.[45] However, the changes seem to depend greatly on the location of cultivation: the extension on unsuitable areas (e.g., areas that are too dry) would increase variability, ceteris paribus. A similar reasoning may hold true for animals as well. Thoroughbred animals need stables, a constant amount of nutritionally balanced feed, and much more care than traditional stock, whereas the product of centuries of random mating could stand harsh environment, poor feeding, and overall mistreatment (for instance, in Italy, cows were quite often used for plowing).[46] Furthermore, agriculture as a whole remains subject to weather vagaries: as late as 1956–57, a frost destroyed about 40 percent and damaged 45 percent of the prized Bordeaux vineyards.[47] But the effects of production fluctuations are, however, much smaller nowadays than they used to be. Trade can smooth the impact on consumption, while insurance and local disaster relief provisions (section 8.8) reduce the negative effect of falls in output on the incomes of farmers.

The features discussed so far (the natural constraints to the portfolio of crops, the effect of cultivation on the soil fertility, the seasonality of work, the output-related risk) differentiate agriculture from any other sector. An industrial plant can be located (almost) anywhere, can operate for as long as the manager and workers deem profitable and convenient (and the laws allow them), and its output is a constant and predictable function of the amount of inputs. In theory, modern technology could reduce this difference almost to nil. Nowadays, almost everything is technically feasible in agriculture, including growing tropical crops in a greenhouse in Alaska. However, with the partial exception of the industrialization of livestock-breeding, these technological opportunities have not been exploited, for one very simple reason: greenhouse production may be rational and profitable for some high-price products, such as flowers or out-of-season vegetables, but it would be economically insane and environmentally devastating to extend it to staple crops. Relative factor prices are not likely to change sufficiently in the foreseeable future to modify this fact, and thus agriculture is bound to remain different.

2.3 Factor Endowment and the Characteristics of Agriculture

Agriculture produces a relatively narrow range of commodities. The statistical database of the FAO lists 156 crops and 39 livestock products, a very small number compared to the thousands of industrial products. Yet, agriculture is highly diversified as well. In any given area and time, it features a specific combination of crop mix, level of specialization, intensity of cultivation, techniques, pattern of settlement (clustering in villages or scattering in the fields), distribution of landownership (or other claims to land use), level of development of markets, and so on. These "areas" can be quite small. Just as an example, let us consider the case of early-nineteenth-century Northern Tuscany, along the course of the river Arno.[48] In the high hills close to the springs of the river, a relatively thin population scraped a meager crop of cereals from their fields, reared sheep and picked chestnuts on common land, and integrated their poor income with emigration and off-farm work. Down the river, in the hills around Florence, the population was quite dense, and the land belonged mainly to city dwellers, who rented it to peasants for a half of the main crops (*mezzadria*). The peasants produced wheat, maize, wine, and oil with typical Tuscan intensive mixed farming. Further west, toward the sea, the original marshes had been or were still being drained. The culture was more extensive, with a wider resort to day-laborers, a higher share of cereals, and extensive cattle-breeding. The Tuscan case is exceptional because three widely different systems coexisted within less than 150 kilometers (just for a comparison, the American prairies extend for thousands of kilometers). Yet, the diversity of systems worldwide, or even in any large country, is astonishing.

Many scholars have tried to frame this huge diversity in elaborated taxonomies: Grigg (1974) describes nine major "agricultural systems," Pearson (1991) describes seventeen "field-crop ecosystems," the recent PAGE survey (Wood et al. 2000) describes sixteen "agro-ecosystems," and so on.[49] These classifications aim at being "scientific," and thus focus on such measurable criteria as the amount of rainfall, temperature, etc., neglecting or ignoring altogether the property rights, the distribution

TABLE 2.1
Land per Worker, 2000 (Hectares)

Belgium	18.9	Pakistan	1.0	South Africa	61.5
Denmark	23.6	Philippines	0.9	Sudan	18.0
Netherlands	7.7	Vietnam	0.3	Zimbabwe	5.9
France	31.9	India	0.7	*Africa*	5.5
Germany, West	16.0	Indonesia (Java)	0.9	Canada	184.4
Greece	11.5	Thailand	1.0	USA	137.6
Spain	22.7	Turkey	2.8	Mexico	12.3
Italy	11.2	*Asia*	1.3	Jamaica	1.5
UK	32.9	Argentina	115.4	Haiti	0.6
Europe	12.2	Brasil	18.6	*North and Central America*	30.8
Former USSR	26.4	Colombia	12.1		
Bangladesh	0.2	*South America*	23.0	Australia	1035.2
Burma	0.6	Algeria	17.1	New Zealand	95.3
Korea, South	0.8	Egypt	0.4	*Oceania*	166.6
China	1.0	Ethiopia	1.3	*World*	3.8

Source: FAO Statistical Database.

of landownership, and the contracts, which feature so prominently in the descriptions of historians. Economists have a compromise position, in that they assume that the features of agricultural systems are determined by "nature" (climate and other environmental characteristics) and by factor endowment. The former has been already dealt with extensively. Thus table 2.1 presents some data on land-labor ratios by continent and country (unfortunately there are not enough data on agricultural capital).

These figures are hardly accurate: land is measured as the sum of arable, permanent tree crops and pasture, and labor as the total of agricultural workers, without distinction by sex or skill or by number of hours (the possible distortions will be discussed in more detail in chapter 4). Yet, the figures clearly highlight the big differences in factor endowment among continents and among countries on the same continent.

In the very long run, arguably, only "nature" is really exogenous. In fact, the land-labor ratio is determined by the population growth and pattern of settlement, which are undoubtedly affected by the overall economic situation. Mortality depends also on income (via nutrition and sanitation), while people decide the number of desired children, the pattern of migration, and the investment in land according to expectations about income, wages, and so on. However, in the short and medium run, land-labor ratios can be assumed to be exogenous from the point of view of agriculture. If so, they determine crop mix and techniques. As any introductory textbook of microeconomics suggests, it is rational to use the more abundant factor intensively. For the same production function, the more intensively a factor is used, the lower its productivity—ceteris paribus. In other words,

TABLE 2.2
Productivity of Land and Labor, 1984–86 (1975 International Dollars)

	Land	*Labor*		*Land*	*Labor*		*Land*	*Labor*
Japan	1327	1590	Belgium	1062	15143	Portugal	253	869
Korea	1354	915	Denmark	770	14265	Spain	219	3802
Taiwan	1871	1430	Netherlands	1858	15193	EU total	470	6128
Indonesia	372	425	France	473	10074	Australia	15	18403
Philippines	544	579	Germany (FR)	709	7251	USA	146	18773
Thailand	344	461	Ireland	247	8326	Canada	86	13132
India	308	277	UK	543	10558	Argentina	57	7653
Pakistan	338	708	Greece	317	3063	Brazil	109	2102
Sweden	433	8047	Italy	608	4805	Mexico	84	1109

Source: van der Meer and Yamada 1990, tables 3.1 and 3.3.

there should be an inverse relationship between the productivity of land and labor (table 2.2). The correlation between land and labor productivity is, as predicted, negative, but extremely weak (–0.02). The Netherlands, Belgium, and Denmark were close to the top of group for the productivity of both labor and land: omitting them, the coefficient would fall to –0.35.[50] The absence of a strong negative correlation is not so surprising. Productivity of land and labor depend on the quantity of capital, which is omitted from table 2.2 for lack of data (cf. section 4.3). Furthermore, the assumption that all countries share the same production function is in conflict with sensitivity to environment.

Chapter Three

TRENDS IN THE LONG RUN

3.1 Introduction

The starting point of the analysis must be a review of long-term trends. How much did total and per capita agricultural production grow? Did the composition of output change? Did the prices of agricultural product rise or fall relative to the prices of other goods and services? How much did trade in agricultural product grow in the long run? The key to answering these questions must be provided by national data series. Price data, especially for staples as rice in Asia or wheat in Europe, are fairly abundant from the early modern period, as the political authorities closely monitored the markets, fearing that shortages of food could trigger political turmoil. It is thus fairly easy to construct long-run series of agricultural prices, stretching, in some cases, back to the Middle Ages. The first statistics of foreign trade were published by England at the end of the seventeenth century and fairly reliable estimates of world trade are available from 1850. Countries such as France, Denmark, and the Netherlands published output statistics from the early nineteenth century, while most "advanced" countries started some decades later, around the middle of the century (UK in 1856, USA in 1869), and LDCs after World War II. Nowadays, all countries provide data on production and trade, which are collected and published by international organizations such as the FAO, the United Nations, and the WTO. These data will be used to discuss the main trends in total output (section 3.2), in prices (section 3.3), in the composition of agricultural output (section 3.4), and in trade of agricultural products (section 3.5).

3.2 Output

The analysis of long-term growth in agricultural output is possible thanks to the painstaking work of scholars all over the world, who have reconstructed yearly series of national accounts and/or agricultural production from all available sources.[1] Before the 1850s, the information is relatively meager. Data are available for only about a dozen countries, almost all in Europe (plus the USA and Australia), and some of the series are based on poor data.[2] The results, however, are quite consistent. In almost all countries agricultural production increased, with rates ranging from a minimum of 0.50 to 0.60 percent in Austria, Belgium, and Spain, to 3 percent in the United States, and 8.4 percent in Australia (which started from a very low level). Production fell only in Portugal. The most unexpected result is the relatively poor performance of British agriculture, which contrasts with its reputation for agricultural innovation.[3] Bairoch suggests that production in the twenty-four OECD countries increased 2.4 times from 1800 to 1870—that is, as much as the population.[4] Actually, this statement seems somewhat pessimistic: per capita production rose in these years in all the countries in table 1 (except Portugal, Austria, and perhaps the UK). The increase was minimal in Australia and in the United States, where agricultural production grew roughly parallel with both settlement and the growth of the agricultural work force, but was quite substantial in some European countries, such as Belgium and Germany. On the other hand, in the 1830s and 1840s, the average heights declined in the United States and in many European countries, and this is strong evidence of a worsening nutritional status, possibly related to insufficient production (Steckel 1995; Komlos 1998).[5]

There are no data on agricultural production outside the Atlantic economy before 1870, with the solitary exception of Indonesia (another success story, with rising per capita output). One can surmise that, in the long run, production must have increased roughly as much as population. In fact, food consumption was low, and GDP per capita, according to Maddison's very tentative estimates (2001, tables B-18, B-12), remained roughly constant; and in most countries there was plenty of land available to settle new farmers (section 4.2). The available figures suggest that population grew steadily in Japan, and that in China it first rose and then plunged.[6] The total population of the LDCs (including China) might have increased by about 0.3 to 0.40 percent yearly in the first half of the nineteenth century—that is, by a quarter or a third.[7] As already stated, production per capita in "advanced" countries was rising, so that one could conclude, very tentatively, that before 1870 world output per capita certainly did not fall and probably increased somewhat.

Beginning in 1870, the country data become much more abundant, and from this year onward it is possible to construct fairly representative yearly indices of world agricultural gross output and Value Added.[8] They are based on series for twenty-five countries at their 1913 boundaries—covering the whole of Europe (except the Balkans), the countries of western settlement, and a substantial part of Asia and South America. Table 3.1 gives the rates of change in gross output by continent, computed (as are all other rates in the book if not otherwise specified) as the coefficient of a log-linear regression with a time trend.

The table shows two important facts.[9] First, agricultural production grew faster during the *belle époque* than in 1913–38. The difference is not surprising. The former period was almost uneventful: the so-called Great Crisis (1873–96) caused a very modest slow-down in southern and north-western Europe, but it apparently did not affect other continents. In contrast, the second period featured major shocks, such as the World War I (in which world production fell by some 10%), the Great Depression, and the Soviet collectivization. Second, the performance by continent differed greatly, and the ranking does not tally with the conventional wisdom.

TABLE 3.1
Rates of Change in Gross Output, 1870–1938, Twenty-five Countries

	1870–1938	*1870–1913*	*1913–1938*
Europe	1.18	1.34	0.76°
North Western Europe	0.97	1.02	1.50
Southern Europe	0.89	0.81	1.19
Eastern Europe	1.67	2.13	0.36°
Asia	0.97	1.11	0.58
South America	3.80	4.43	3.05
Western Settlement	1.37	2.20	0.74
World	1.31	1.56	0.67

Notes: **Northwestern Europe:** UK, France, Sweden, Denmark, Belgium, the Netherlands, Germany, Finland, Switzerland. **Southern Europe:** Italy, Greece, Spain, Portugal. **Eastern Europe:** Austria-Hungary and Russia. **Asia:** Japan, India, Indonesia. **Western Settlement:** Canada, Australia, USA. **South America:** Argentina, Uruguay, Chile.

° not significantly different from zero.

Source: Federico 2004a.

Only Latin America stands out for its high growth in both periods. In contrast, the performance of agriculture in the countries of Western settlement (i.e., mainly the USA) was not so good. In the long run, the growth rate barely exceeded the world rate. Before 1914, Western settlement countries were clearly outperformed by eastern Europe: Imperial Russia achieved one of the highest growth rates in the world. Even the allegedly backward and stagnant Asia managed to increase output substantially (a 1% increase over seventy years corresponds to a doubling of output). However, this aggregate growth is the outcome of widely different country trends, with rates ranging from a paltry 0.7 percent in India to almost 2 percent in Indonesia.

What about the rest of the world? Some additional figures (short series or point-to-point growth rates) for countries outside the sample, including major ones such as Brazil, Mexico, and China, confirm that output was also growing outside the "Atlantic economy." However, the exact rate of growth is highly uncertain. The estimates for China range from a minimum of 0.4 percent (i.e., less than the population growth) to a maximum of 1.2 to 1.3 percent.[10] As before, one can assume that production grew as much as the population. It is thus possible to compute the rate of growth of world production (see table 3.2) by assuming that the rest of the world accounted for 45 percent of world output in 1913 (its share on world population). In seventy years, the total world production increased by 85 percent, and increased per capita by about a tenth, despite the decline in interwar years. These are by no means negligible achievements. On top of this, these figures are in all likelihood a lower bound, for two reasons. (1) Using population as weight underestimates the rate of growth because production per caput in 1913 was in all

TABLE 3.2
Rates of Change in World Gross Output, 1870–1938

	Twenty-five Countries	*Rest of the World*	*Total World*
Total			
1870–1913	1.54	0.58	1.06
1913–38	0.71	0.73	0.72
1870–1938	1.24	0.64	0.94
Per Capita			
1870–1913	0.55	0.00	0.26
1913–38	–0.08	0.00	–0.05
1870–1938	0.32	0.00	0.15

Source: Federico 2004a.

likelihood higher in the twenty-five countries. For instance, if they accounted for two-thirds of world output (as in 1970), the long-run rate of growth in per capita output would be 0.2 percent.[11] (2) Production per capita in the rest of the world is likely to have increased. Maddison (2001) estimates that the GDP per capita increased by almost 60 percent from 1870 to 1950 in all LDCs (China included).[12] Such an increase must have augmented the demand for food—at least by a third (hypothesizing a 0.50 income elasticity and no major change in prices) and thus also of agricultural output. Taking both biases into account, one can, very tentatively, surmise that world output per capita grew at some 0.2 to 0.3 percent per annum from 1870 to 1938 (i.e., by a quarter over the whole period) and that total world output more than doubled.

The performance of agriculture in these years was undoubtedly quite good if compared with the probable stagnation in the previous centuries, but it pales somewhat if compared with the post-1950 growth. According to estimates by the WTO, from 1950 to 2000, world agricultural production grew 2.3 percent yearly—in other words, it more than tripled—while world population increased 2.4 times.[13] Table 3.3 shows the growth rates by continent according to the FAO. By the late 1940s, the production had recovered to the prewar level in both Asia and western Europe, and had exceeded it, by a huge margin, in the Americas and Africa (if African data are accurate). Since then, total and per capita world production have been growing quite fast. By and large, LDCs, notably Asian ones, have outperformed the industrially "advanced" countries. In the 1990s, the rate of growth in western Europe fell from the 2.0 percent per year of the previous three decades to 0.5 percent. The world growth rate in the 1990s was also lowered by the veritable collapse of agriculture in the "transition economies" (the former Soviet Union and other Eastern European countries). They moved from a positive, although not outstanding, growth (1.5% per year) to a precipitous decline (a 40% fall in

TABLE 3.3
Rate of Change in Gross Output, 1938–2000

	(a)	*(b)*	*(c)*
Africa	1.72	3.10	2.25
North Central America	2.63	1.40	1.77
South America	1.68	3.13	2.92
Asia (excluding China)	0.31	3.64	
Asia (including China)			3.54
Western Europe	0.56	2.55	0.91
Europe			0.00°
Oceania	0.81	2.85	1.68
USSR			−1.96°
World			2.27
World (excluding Socialist countries)	1.34	2.69	
World, per capita			0.56
World, per capita (excluding Socialist countries)	0.22#	1.19@	

Notes: Column (a): from prewar (1936–38) to 1948–52—compound rates (# from prewar to 1952–54).Column (b): from 1948–52 to 1958–60—compound rates (@ from 1952–54 to 1958–60). Column (c): 1961–2000.

° not significantly different from zero.

Sources: FAO Yearbook ad annum and FAO Statistical Database.

a decade). This collapse contrasts with the boom in China, where agricultural output stagnated from the 1930s to the early 1960s, grew slowly in the 1960s and 1970s, and zoomed up to 5.2 percent per year after 1979.[14] Sub-Saharan Africa is the other bleak spot. Its total production grew rather fast in the long run (2.24% for the period 1960–2000), but rather slowly in the 1970s and early 1980s (1.3%). The population grew much faster, so that in these decades output per capita declined by about a sixth, and it has not recovered since then.[15]

With the data presented so far, it is possible to piece together a long-run series of world gross output from 1800 to 2000, assuming that (1) production per capita remained constant from 1800 to 1870 all over the world, that (2) it remained constant from 1870 to 1939 in the "rest of the world," and that (3) between 1938 and 1948–52 the production of socialist countries grew as much as that of the other countries. Gross output increased roughly by 10.5 times in two centuries, at a quickening pace—0.5 percent yearly from 1800 to 1870 (as much as population, by construction), 0.9 percent from 1870 to 1938 (0.15% per capita) and 2.2 percent from 1938 to present (0.56% per capita). These figures are, clearly, very tentative. The three previous assumptions are obviously questionable: if anything, they underestimate the overall growth up to 1938. Furthermore, even some country series may be questionable.[16] Indeed, estimating agricultural production is a very daunting task, which requires well-organized statistical offices with substantial resources and considerable good faith. These resources were simply not available in some industrially advanced countries in the past and are not available now in many LDCs. Many country figures are thus only educated guesses. Furthermore, good faith may also be in short supply. The figures for the Soviet Union and other socialist countries were manipulated for political purposes, so that there are several alternative estimates of production in the 1930s, which differ widely. Last but not least, farmers may have an interest in not providing accurate information. Traditionally, they tended to conceal some of their product for fear of the taxman, causing the overall statistics to underestimate the growth of output. The incentive to conceal output was even greater when faced with policies of compulsory delivery to marketing boards, as occurred in many LDCs in the 1960s and 1970s (section 9.5). On the other hand, the massive adoption of price subsidies in many OECD countries since the 1930s (section 9.4) has created the opposite incentive to over-report output. It is impossible to assess the extent to which these conflicting biases affect the data. However, one can remark that even serious errors in the series of individual countries would hardly affect global production, as no country exceeded a sixth of the world output.[17] However, as stated in the introduction, imperfect data are better than no data, or than unsupported anecdotal evidence.

3.3 Prices

In theory, agricultural prices should be measured at farm gate, relative to a composite index of goods and services. Unfortunately, most agricultural price series refer to urban markets and price series for the "rest of economy" (manufacturing and services) are simply not available.[18] They are routinely proxied with general price indices (or implicit deflator of GDP series) and/or with the index of the price of manufactures, obtaining "real" agricultural prices and "domestic terms of trade" respectively. It is important for the following discussion to stress the implicit bias relative to the "ideal" index. Real prices include agricultural prices in both the denominator and numerator, and thus are bound to change less than the ideal index. In contrast, the omission of the prices of services causes the "terms of trade" to overshoot relative to the ideal index (rise more or fall less) if they are less volatile than those of industrial products.[19] It is thus likely that movements of the ideal price index are somewhere in between the two others.

Before 1870 price data, in either definition, are available for only a handful of countries, and most of them are computed as implicit deflators from somewhat unreliable historical estimates of national accounts (table 3.4).

In spite of their shortcomings, these series highlight two important facts. First, as expected, terms of trade changed more than real prices. Second, agricultural prices rose in all countries except the United Kingdom. Most of the increase is concentrated in the "hungry forties": In all countries, agricultural prices rose by about a third from 1842–44 to some time in the mid-1850s.[20] After 1870, the country coverage increases considerably (table 3.5): there are data on terms of

TABLE 3.4
Rates of Change in Prices of Agricultural Products, 1800–70

	Period	*Terms of Trade*	*Real Prices*
USA	1800–70		0.43
France	1815–70	1.14	0.65
Netherlands	1815–70	1.52	1.02
Denmark	1818–70	2.10	0.80
Germany	1850–70		–0.27°
Spain	1850–70	0.12°	–0.28°
UK	1800–70	–0.22°	–0.12°

Note: ° not significantly different from zero.

Sources: **USA:** Historical Statistics 1975, Series E23, E25, E52, E53. **UK:** Mitchell 1988 (chained price indices by Rousseaux, Gayer, Rostow and Schwartz, Sauerbeck, and the Board of Trade). **France:** Toutain 1997, Series V5, V16 and V43 (implicit GDP deflators). **Germany:** Hoffman 1965, ii, Tables 137 and 148 (implicit GDP deflators). **Netherlands**: Van Zanden 2000 (implicit GDP deflators). **Denmark:** Hansen 1974, tables 3 and 4 (implicit GDP deflators). **Spain:** Prados de la Escosura 2004, cuadro A.4 (implicit price deflators).

TABLE 3.5
Rate of Change in Prices of Agricultural Products, 1870–1938

	1870–1913	*1913–1938*	*1870–1938*
Terms of trade			
Western Settlement°	0.76	−1.48	0.01
Europe*	0.42	−0.32	0.10
LDCs#	1.10	0.44	0.30
World	0.70	−0.28	0.15
Lewis (a)	−0.27	−0.52	−0.42
Lewis (b)	−0.18	−2.34	−0.78
Grilli and Yang		−2.23	
Real prices@	0.07	−0.84	−0.16

Notes: ° Argentina (1885–1938), Uruguay, Australia, Canada (1890–1938), USA.

* Great Britain, Denmark, Sweden, Ireland, Italy, Netherlands, France (no data 1914–19), Germany (no data 1914–25), Austria-Hungary (1870–1909), Spain.

\# Egypt, Thailand, India, China (1876–1936), Punjab, Burma, Taiwan (1903–38), Korea (1910–38), Japan (1874–1938).

@ Canada, Denmark, France (no data 1914–19), Germany (no data 1914–25), Italy, Netherlands (1870–1913), Japan (1885–1938), Spain, USA, UK.

Sources: Grilli and Yang 1988, Appendix 1 (the agricultural prices index is computed as a weighted average of prices of food and nonfood agricultural raw materials with weights 0.85/0.15); Lewis 1979, table A11: (a) cereals/manufactures and (b) tropical food/manufactures; others, Statistical appendix, table II.

trade for twenty-five countries and on relative prices for ten of them (plus two series of "world" terms of trade).[21]

On average, agricultural prices did not change very much in the long run, but this result is the outcome of quite different trends by country and period (all the coefficients of variation of area-averages exceed 0.5).[22] Before 1914, real prices remained roughly constant in all countries except the United States, while terms of trade improved. They increased in seventeen countries, declined in Australia, and remained trend-less (co-efficient not significantly different from zero) in the other five. Most of the increase was concentrated in the last twenty years before the war, a veritable golden age for world farmers. In fact, the average world yearly rate of change in terms of trade rose from 0.2 percent in 1870–95 to 0.7 percent in 1895–1913 and the post-1895 rate exceeds the pre-1895 one in two thirds of the countries.

Price trends are much less clear after 1913 than before. The world averages (table 3.5) for the whole period declined somewhat, but these averages conceal quite different country trends. Real prices actually declined in three countries and remained (statistically) constant in five others; terms of trade declined in seven countries, increased in four, and remained trendless in eleven, exactly half the total sample. These long-term variations, however, were swamped by short-term fluctuations, which had a deep impact on the welfare of farmers and thus on the adoption of agricultural policies (section 9.4). It is widely believed that, in the 1920s, overproduction caused a fall in prices, and indeed prices, which had peaked on the eve of the war and declined somewhat during the war in most countries, plunged immediately after the war. Terms of trade hit a trough sometime in the early 1920s, on average about 15 to 20 percent down from the prewar peak. The fall was impressive, but terms of trade remained still higher than in the mid-1890s in the majority of countries.[23] In other words, the postwar crisis caused farmers to lose the gains of the 1900s. Furthermore, prices rebounded in the 1920s, and, in two-thirds of the countries, the peak exceeded the 1911–13 level.[24] Unfortunately, the respite was short-lived, as the Great Depression hit agriculture very hard: on average, terms of trade fell from the peak to the trough (which differed among countries) by 22 percent.[25]

The discussion so far has considered all countries as undifferentiated units of observation, but table 3.5 shows a distinctive clustering in terms of trade. Before 1914, they grew more in Western settlement countries than in Western Europe. These trends can be interpreted as a consequence of globalization: price convergence caused agricultural prices to grow more in land-abundant countries (exporters of agricultural products) than in land-scarce ones.[26] The backlash during interwar years caused prices to diverge, and, in fact, prices fell less in Europe than in the countries of Western settlement. Prices in LDCs (notably in China) grew more than in "advanced" countries in both periods. One can, very tentatively, hypothesize that world rice prices were not affected by technical progress and competition from land-abundant countries as much as wheat prices were. Western settlement countries were not yet competitors on the rice market, while the productive capacity

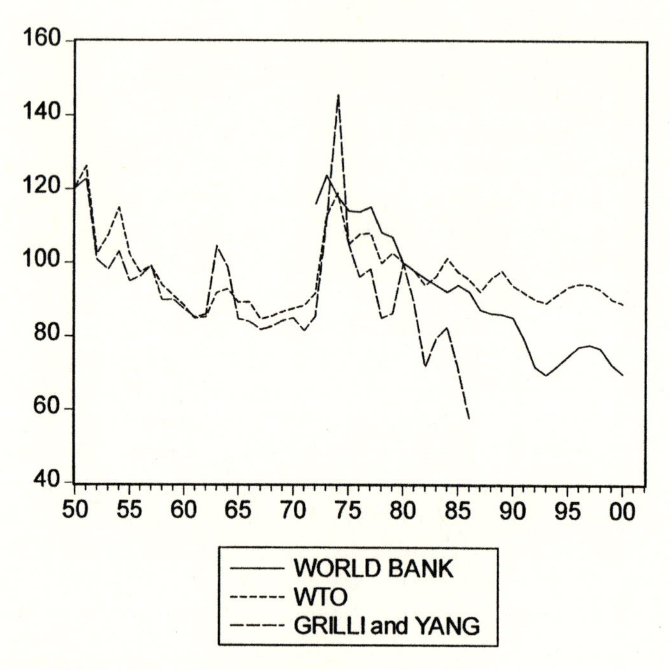

Graph 3.1 Terms of Trade 1950–2000

in land-abundant Asian countries was small relative to the potential demand in the "core" countries of the area. These hypotheses are plausible, but clearly they will need much more solid evidence to be accepted as a fact.

After 1950, agricultural prices remained stable or decreased. Mundlak, Larson, and Crego (1997), on the basis of an extensive collection of data for some 130 countries for the period 1967–92, argue that real prices declined by about 10 to 15 percent. Graph 3.1 shows three different series of terms of trade, by Grilli and Yang, the World Bank and the WTO.[27] The interpretation of long-run price trends depends on the representativeness of the external terms of trade. In the long run, they remained almost constant (the rate of change is –0.03% and is not significantly different from zero) in spite of the peaks during the Korea boom and the first oil crisis. They did fall from 1973, but much more slowly than the domestic prices, according to the World Bank—at a mere 0.72 instead of 2.08 percent.[28] Thus, on the whole, the latter convey a much bleaker outlook than international prices. One can speculate on the causes of the difference—for instance, by hypothesizing that the steeper fall in domestic terms of trade in the 1990s had something to do with the changes in agricultural policies (sections 9.4

and 9.5) and the ensuing reductions in support prices.[29] However, testing this hypothesis would entail a substantial research project.

To sum up, terms of trade increased somewhat in the long nineteenth century (especially in the 1890s and 1900s) and remained stable or fell somewhat in the twentieth century, with a collapse during the Great Depression. The real prices of agricultural goods grew less than the terms of trade: one can surmise that the (unavailable) ideal index—that is, the terms of trade relative to the rest of the economy—remained constant or declined somewhat. Needless to say, these conclusions are very tentative, as the evidence is incomplete and price movements show an alarming variety by country.

In spite of all the necessary caveats, price trends allow us to draw some interesting inferences about the performance of the agricultural sector. In theory, in a purely competitive market, trends in terms of trade are determined by the rates of shift of supply and demand in the two sectors (in the case at hand, agriculture and manufacturing). There is no doubt that the relative demand for agricultural products has been falling. The demand shift depends on income per capital and Engel's law, one of the most widely supported stylized economic facts, states that the income elasticity is lower for *food* than for manufactures and services.[30] Income elasticity for *agricultural products* is even lower, as marketing and processing account for a growing share of the final price of food. Thus, the agricultural terms of trade are bound to worsen if the rate of technical progress in the two sectors is the same, or is faster in agriculture than in manufacturing. The extent of the fall depends on the price elasticies of the demand and supply.[31] Such a fall in the terms of trade has come to be regarded in U.S. political discourse as a state of nature, often referred to as a "farm problem" (see Gardner 1992, 62; section 9.4 of this book). Although very convenient for agricultural lobbies seeking state support, this view is one-sided. The agricultural terms of trade are bound to *improve* if technical progress in agriculture is nil or much slower than in manufacturing (unless agricultural demand collapses). In other words, they can remain constant or fall *only if* agriculture manages to match the technical progress in the rest of the economy. The model is admittedly stylized, but it implies that, contrary to the traditional view, technical progress in agriculture was substantial, especially in the twentieth century.

The price analysis buttresses the conclusion from the production data: agriculture has successfully accomplished its basic task, to feed the population and provide raw materials for industry. In fact, apart from any short-term setback, the average height has risen in most countries (Fogel 2004). On the other hand, being fed is only the minimum for welfare. People can survive and grow tall on a poor and monotonous diet based on high-caloric staple foods such as potatoes, as the Irish did before the famine. However, a wide range of foods (and attractive presentation and cooking, which are outside the scope of this book) undoubtedly enhances their welfare. The range of available food can be widened by changing the composition of agricultural output and/or by trading, the subject of the next two sections.

3.4 The Composition of Agricultural Output

There are dozens of possible ways to aggregate agricultural products, from the basic distinctions according to final destination (e.g., food/nonfood) or method of production (e.g., arable and tree crops/cattle-raising) to quite detailed subcategories. The FAO website provides data for twelve groups of field crops, plus eight groups of livestock products. Unfortunately, the historical sources do not allow such a detailed disaggregation. It is possible to compute shares by destination (food and raw materials) for eight countries only.[32] Raw materials accounted for a major share of gross output only in Australia, while in the other seven countries they did not exceed a sixth of the gross output. In some cases, the shares fluctuated widely, according to trends in the world market for key commodities. For instance, the boom of cotton production and exports caused raw materials to increase from 6 percent of American gross output in 1800 to 15 percent in 1850. In the United Kingdom, in contrast, the share halved from the 1880s to the eve of World War II because the main product, wool, was hit by competition of imports from Australia and elsewhere. But on the whole, shares have remained fairly constant and thus output of raw materials has grown quite steadily. This implies that before 1938 the demand for raw materials of natural origin was not yet seriously affected by the competition of artificial substitutes. Rayon and other textile fibers were marketed as "artificial silk," but, before the invention of nylon, they did not dent the consumption of natural silk, especially for women's stockings (Federico 1997, 43). The competition was felt after the war. From 1961 to the present, according to the FAO, the production of nonfood has been growing at a "mere" 1.24 percent per annum, that is, half the rate of growth of the total agricultural output.[33]

The historical data on crops and livestock are more abundant, and it is possible to estimate, with some approximation, indices of production and share on total output for the twenty-five countries (table 3.6).

Table 3.6
Production, Crops, and Livestock, 1870–1938

	(a)	*(b)*	*(c)*
1870–72	55.5	45.7	38.3
1889–91	68.9	63.0	41.6
1911–13	98.2	97.2	43.4
1920–22	93.1	93.3	44.1
1936–38	117.2	122.1	44.7

Notes: (a) Index of world production of crops (1911 = 100). (b) Index of world output of livestock products (1911 = 100). (c) Share of livestock products on world gross output (measured at current prices).
Source: Federico 2004a.

In the long run, in the twenty-five countries, production of livestock (column b) increased somewhat faster than that of crops (column a). In fact, the share of livestock products on total gross output (column c) increased substantially until World War I, and, more slowly, even in interwar years. In the rest of the world, the share of cattle-breeding was undoubtedly lower than in these twenty-five countries.[34] Thus, the figures of column c overstate the share of livestock on total world gross output at every benchmark year. Trends in world share depend on the composition of output in the rest of the world. Assuming that livestock always accounted for a quarter of the total world gross output, its is possible to compute that its share increased from about 30 percent in 1870 to slightly more than 35 percent in 1938.[35] From 1961, according to the FAO database, crop and livestock production grew roughly at the same rate (2.10% for livestock and 2.22% for crops), and livestock products accounted for about 35 to 40 percent of world output in 1970 and 1990.[36] Thus the best guess suggests that the production of livestock increased proportionally more than gross output until World War I and as much as the total or a little faster since then.

This hypothesis can be supported by the price trends, which are available for a small group of "advanced" countries (table 3.7).

TABLE 3.7
Rates of Change in Relative Prices, Crop/Livestock, 1870–1938

	1800–70	*1870–1913*	*1913–38*
Canada		−0.69	−0.07°
Netherlands		−0.75	−0.50°
Germany	−2.45	−0.36	1.37
Australia		−0.78	1.28°
Japan		0.26	1.41
Austria		−0.73	
USA		−0.94	−1.22
Uruguay		1.53	−3.57
France	−0.80	−1.02	−0.22
Belgium		−0.56	−0.97°
Italy		−0.39	0.55°
UK	−0.89	−0.62	−0.79°

Note: ° not significantly different from zero.

Sources: **Australia:** (1861–1938) Butlin 1962, table 267. **Canada:** (1867–1938): Urquhart 1993, series k35/k36. **Austria-Hungary:** (1867–1909): von Jankovich 1912, 156. **Netherlands:** (1851–1938) Knibbe 1993, addendum (Paasche indices). **Germany:** (1850–1938) Hoffman 1965, ii, tables 56, 58, 59, and 60 (implicit deflators). **Japan:** (1874–1938): Okhawa et al.1967, tables 5, 6. **Uruguay:** (1870–1938): Bértola 1998. **USA:** (1867–1936): Strauss and Bean 1940 (weighted averages of series 84, 85, 86, 90, and 92). **France:** (1809–1925): Levy-Leboyer 1970, table 3 (ratio cereals/livestock). **Belgium:** (1877–1938): Blomme 1992, tables 46, 47. **Italy:** (1861–1938): Barberi 1961, 135–37 (implicit deflators). **UK:** (1800–1938) Mitchell and Deane 1962, 471–75; (1800–1845): Rousseaux index; (1846–1938), Sauerbeck-Statist price index.

The relative prices of livestock products rose (negative sign) before 1913, and they remained roughly constant, with widely different country trends, in the interwar years. The combination of rising prices and a growing share of output suggests a demand-driven growth. Indeed, livestock products were "rich," income-elastic goods, in most traditional, land-scarce, peasant societies (although not necessarily in land-abundant pastoral countries, nor even in Northern Europe, where vegetables and fruits were scarcer than meat or milk). It also suggests that relative productivity growth in cattle-raising did not match the relative shift in demand.

After World War II, world income was undoubtedly growing faster than before, and yet the share of livestock products seems not to have increased. This trend may reflect faster productivity growth in cattle-breeding, but also a composition effect from the demand side. In the past fifty years, the population living in LDCs has risen from 70 to 85 percent of the world total. This increase has depressed the average consumption of luxury goods, possibly balancing the demand-driven growth in the consumption of livestock products in both high- and low-income countries. The equilibrium is likely to be broken in the near future in that per capita consumption of livestock products in LDCs is going to rise with the growing income.[37] This change is likely to put a serious stress on the factor endowment, as cattle-breeding notoriously needs much more land than crop production to produce the same amount of calories.

3.5 Trade

World trade is much better documented than output, at least from 1850. The differences between the available series (table 3.8) can be attributed to different product coverage and the method of construction of the series. Both Lewis and the UN/WTO series cover—in theory—the whole world, while Vidal uses the imports of the four major countries as a proxy for world trade. The WTO series refer to all agricultural products, while Lewis considers all primary products (including minerals), and Vidal considers only foodstuffs (thus excluding agricultural raw materials). In spite of these differences, it is possible to piece together a plausible series of trade in agricultural products in the past 150 years (graph 3.2).[38]

In the long run, trade in "agricultural" products has grown seventy-five times, much less than the total trade (280 times) but far more than output. With some daring assumptions, it can be estimated that the trade/gross output ratio rose from about 5 percent in 1870, to about 15 percent in 1913, and from about 16–17 percent in 1938 and possibly to a quarter or a third in the late 1990s.[39] As is clear from the graph, all the difference in the rate of growth of trade was cumulated after 1950. In fact, during the first globalization, up to 1913, the trade in agricultural products grew as fast as total trade and, if anything, during the Great Depression agricultural trade decreased less. It accounted for exactly half of the total trade in 1913, 1929, and 1937, and fell to about 10 percent at the end of the 1990s.[40]

TABLE 3.8
Rate of Change in World Trade of Agricultural Products, 1850–2000

	Lewis	*Vidal*	*WTO*
Agricultural Products			
1850–90	3.59		
1890–1913	3.35	2.34	
1850–1913	3.44		
1919–38		–0.75°	
1950–2000			3.22
Total world exports			
1850–90	3.63		
1890–1913	3.62	3.23	
1850–1913	3.46		
1925–38		–0.81°	
1950–2000			5.25

Note: ° Not significantly different from zero.
Sources: Lewis 1981, tables 3, 4; Vidal 1990; WTO Statistical Database.

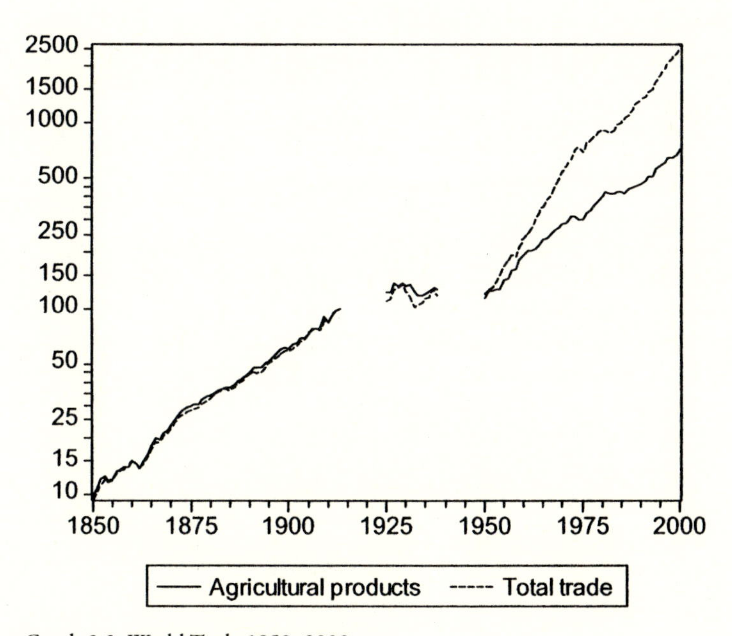

Graph 3.2 World Trade 1850–2000

Table 3.9
Composition of Agricultural Trade by Product, 1913–99

	1913	*1937*	*1929*	*1980–82*	*1997–99*
Tropical Products*	11.9	17.5	14.5	14.4	11.1
Temperate Products	67.2	55.1	59.3	61.1	61.5
Mixed Products°	21.0	27.4	26.2	24.6	27.4
Total	100	100	100	100	100

Notes: * Tea, coffee, cocoa, oilseeds, rubber.

° Fruit and vegetables, sugar, wood, vegetable oils.

Sources: 1913–37: Yates 1959, table A 17. 1980–82 and 1997–99: UN data (World Trade Analyzer)

To some extent, this decline was unavoidable: agriculture, as a whole, grew less than the whole economy as modern economic growth spread, and the scope for specialization and development of intra-industry trade was greater in industry, with its thousands of different products, than in agriculture. However, the relative decline is accounted for also by agricultural protection, which, as we will discuss in chapter 9, has proven to be far more resilient than protection to manufactures.[41]

Here it is impossible to deal with trade flows in any detail, as a comprehensive historical reconstruction would entail a major research project in itself. To sketch out the main trends, table 3.9 shows the essential data on the composition of world trade by product in five benchmark years. Most trade consist in products of temperate agriculture—wheat and cereals, cotton, livestock products, and so on. Until World War I, they had flowed from the countries of Western settlement and Russia (the main exporter of wheat) into Western Europe. The Soviet Revolution halted Russian exports, while, after World War II, Western Europe became self-sufficient and later started to export. From the 1960s, the socialist countries (later, transition economies) have become the main outlet for agricultural exports. The table shows that tropical products accounted for a minor share of world trade. Thus, the situation used to be, and still is, deeply asymmetric: exports of "tropical" crops were essential for the development of LDCs, but they mattered relatively little for the growth of advanced countries.

Chapter Four

PATTERNS OF GROWTH: THE INPUTS

4.1 Introduction

Measuring the growth of inputs is far from easy because the data are incomplete. The available sources refer mostly to stocks (number of workers, acreage, capital, etc.), while one would need data on flows of services. This causes potential biases, which should be considered whenever possible. Furthermore, the coverage varies by country, factor, and period. The data on capital are particularly scarce: there are no aggregate data even for recent years, and thus it is necessary to present data by item. The data on the stock of land and labor are more abundant, since the FAO provides a comprehensive set of data for all the countries of the world after 1950. For the earlier periods, one has to resort to country data, which, from 1910, have been collected by the International Institute of Agriculture (established in 1905 as the forerunner of the FAO) and published in its yearbook. Thus, the country coverage for land and labor increases from a few countries before 1910 to a bigger sample in the interwar years (but still consisting almost exclusively of European and Western settlement countries) to the whole world after 1950. These data are, for obvious reasons, reported in separate tables that show also the most recent data for the same countries in order to give a rough idea of long-term changes.[1]

4.2 Land

In 1800, nomadic or semi-nomadic tribes peopled the Americas (25% of agricultural land now in use), Africa (13%), Oceania (4%), and a vast extension of the land in the former Soviet Union (15%) and Asia (34%). They fed themselves by fishing and hunting, extremely extensive forms of swidden agriculture, or by nomadic cattle-raising. In theory, a Native American roaming on the prairies hunting bison had the same rightful claim to the productive use of land that a modern farmer does (section 7.2). According to the standard definition of the term "productive factor," all the land should be included in the available stock, irrespective of its use (arable, pasture, woodland, etc.). The stock would remain almost constant in time.[2] This interpretation would, however, be too narrowly legalistic: the arrival of European settlers marked a quantum leap in the intensity of the use of land and in the nature of the property rights on it (section 8.2). In the following, we will include in the stock of considered land only the acreage subject to some agricultural use, albeit extensive. In this more realistic, although less elegant, definition, the stock of land greatly increased all over the world in the past two centuries.

The European colonization of overseas territories was a long process. It started with the arrival of the first immigrants in the American colonies in the early seventeenth century, in Australia in the 1790s, in New Zealand in the 1830s, and so on. Initially settlement was very slow. American farmers first crossed the Appalachian Mountains around 1790, settled in the prairies during the 1840s and in California during the 1850s and concluded their westward march to the Pacific Northwest in the 1890s.[3] In 1850, when the available series begins, "land in farms" (i.e., legally owned) extended up to 118 million hectares out of the total U.S. land mass of 762 million hectares.[4] Forty years later, when the frontier was officially closed, the total land in farms had risen to about 250 million hectares. However, much land of poor quality, or poorly located, was still in the public domain. It was distributed during the next half-century, so that the total land in farms peaked at 469 million hectares in 1950. All Western settlement countries experienced a similar process, although it was, perhaps, slightly later.[5] In Australia, the distribution of public land lasted until the 1930s. The Canadian frontier was officially closed in the early 1900s, but the total land in farm went on growing until 1931. The conquest of the American West has become a worldwide myth, thanks to Hollywood movies, but large-scale migration and settlement was not a phenomenon exclusive to countries of Western settlement. Farmers poured out from the core area in Central Russia in two big waves: in the first half of the nineteenth century toward Ukraine and the Volga provinces; and from the 1890s toward Siberia, both independently and with the Russian government's help.[6] Twenty years later, in 1916, there were in Siberia some 8.5 million people, who cultivated 10.7 million hectares. From 1860 to 1940 (and especially in the 1930s), about 8 million Han Chinese settled in Manchuria, causing the population to rise twelve times and the cultivated land nine times.[7] Settlement is still going on in Brazil and other South American countries, although at a reduced pace, and sometimes with government support.[8]

There is no doubt that in the long run, migrations were propelled by population growth, but the timing of the process depended on many factors, including, quite often, political events. The settlement of the Argentinian pampas was made possible by military campaigns in 1833 and 1880, which exterminated the natives and dissolved the threat of a Chilean invasion.[9] The massive migration of Chinese to Manchuria had to wait until the lifting of the ban, which had been imposed by Manchu dynasty to preserve the ethnic purity of its homeland and the military prowess of its warriors (who nonetheless preferred to live in Bejing). The Canadian prairies were open to colonization after they were sold to the government by the Hudson Bay Company (1869).[10] In many cases, settlement was made possible by the improvement in transportation means, notably the building of railways. The construction of the Transiberian Railway in the 1890s greatly enhanced the appeal of colonization both in Western Siberia and in Manchuria. In a famous paper, Harley has argued that the settlement of the American frontier followed a twenty-year (Kuznets) cycle, driven by railroad investment.[11]

The allocation of land to a would-be farmer is only a first step. The land needs large investment of capital to be used productively: it has to be cleared and fenced, trees have to be planted, buildings have to be constructed, and so on. These investments could take several years, and during this time the land would be used less intensively than wanted. In the United States, the agricultural acreage (labeled "improved land" in statistics) accounted for 38 percent of total land in farms in 1850, for 77 percent in 1900 and for 85 percent in 1940.[12] In the following tables, we will consider separately cropland (arable and treecrops) and pastures, for theoretical and practical reasons.[13] In theory, pastures need a different service from land from that required by arable land. In fact, grass can grow without depleting the soil of nutrients, and indeed leaving it to grow is a way to restore fertility (section 2.2).[14] In practice, the data are much less abundant for pasture than for cropland, and, above all, they refer to widely different definitions of "pasture" across countries and time. Sometimes they include only lush natural prairies, while in other cases they refer to all grazing land, including poor rocky soils suitable only for goats.

Table 4.1 collects the available data on the stock of arable land for the period up to 1910 from a variety of country sources—not all equally reliable or homogeneous.

TABLE 4.1
Acreage in Arable and Tree Crops, 1800–1910 (Million Hectares)

	ca. 1800	*ca. 1850*	*ca. 1880*	*ca. 1910*	*2000*
Austria	8.6	10.1	10.3	10.6	18.4°
Hungary	5.0	10.0	15.3	16.1	
Belgium			1.6	1.5	0.8
Denmark		2.0	2.4	2.9	2.4
France	33.5	34.3	32.7	29.6	19.5
Greece		0.85	0.9	0.9	3.9
Germany		24.4	26.2	26.2	12.1
Italy		13.5	15.4	14.8	11.0
Ireland		2.1	2.1	1.9	1.4
Netherlands	1.8	1.9	2.0	2.1	0.9
Portugal		1.9		3.2	2.6
Sweden			3.4	3.4	2.8
Spain	12.5	16.0	15.8	19.1	19.1
England	4.7	5.9			
UK and Ireland		8.0	8.0	6.7	6.4
Russia	80	82.5	103.8	113.4	219
Romania		2.5		5.0	10
Bulgaria			1.8	2.5	4.5

TABLE 4.1 (*continued*)

	ca. 1800	*ca. 1850*	*ca. 1880*	*ca. 1910*	*2000*
Austria	8.6	10.1	10.3	10.6	18.4°
Canada		3.5	6.1	15.6	45.7
USA		45.7	76.1	140.4	179
New Zealand		0.01	0.4	0.8	3.3
Indonesia			7.3	12.0	31.0
Japan			4.7	5.6	4.3
Burma			4.4	6.9	10.1
Egypt	1.3	1.7	1.9	2.3	3.3
Thailand		0.9	1.1	2.0	20.4
Taiwan			0.55	0.7	n.a.
Philippines			1.3	2.4	9.9
Colombia		0.4	0.4	0.9	4
Europe	110	136.9		151.8	119.7
Western Settl.		50.5		165.8	285.2

Note: ° Austria and Hungary.

Sources: 2000 FAO Statistical Database: **Austria:** (1789, 1830–50, 1876-85, 1904–13): Sandgruber 1978, table 8. **Belgium:** (1880 and 1910): Blomme 1992, table 40. **France:** (1781–90, 1845–54, 1875–84, 1905–14) Toutain 1961, table 146. **Indonesia:** (1880 and 1910) van der Eng 1992, table A4. **Canada:** (1851, 1881, and 1911) Statistics Canada 1965, series L8–L10 (1881 and 1911, land under crops and summer fallow) and L351–354 (1851, land under culture). **Denmark:** (1861, 1881, and 1910) Johansen 1985, table 2.2. **England:** Allen 1994, table 5.2. **UK and Ireland:** (1866, 1880, and 1910) Mitchell 1988, Agriculture tables 1 and 2. **Germany:** (1862, 1878, and 1913) Hoffman 1965, ii, tables 46 and 47. **Hungary:** ("end 18th century"; 1860s, 1883, and 1913) Gunst 1996, 23 and 25; **Italy:** (ca. 1855) Correnti and Maestri 1864, 392; (1893–95) ASI 1898, 128; and (1909–14) NPSA 1911–14. **Russia:** (1800 and 1850) Grigg 1974, 266; (1880) Lyaschenko 1939, 451; and (1910) Institut Internationale de Agriculture 1909–21. **Romania:** (1860 and "1900s") Berend and Ranki 1974, 70. **Bulgaria:** (1899 and 1912) Palairet 1997, 316; **USA:** (1850, 1880, and 1910) Historical Statistics 1975, cropland (J52). **Spain:** (1860, 1890, and 1910) personal communications by L. Prados de la Escosura;[15] **Sweden:** (1877–81 and 1910–13) Holgersson 1974, table 12. **the Netherlands:** (1810, 1850, 1880, and 1910) Van Zanden 1994, table 1.8. **Japan:** (1880 and 1910) Hayami and Yamada 1991, table A6. **Burma:** (1896 and 1910) Saito and Kiong 1999, table II.1. **Taiwan:** (1895 and 1911) Lee and Chen 1979, 61, 82. **Thailand:** Manarungsan 1989, tables 2.7 and 3.4. **the Philippines:** (1902 and 1918) Holley and Ruttan 1969, table 7.4. **New Zealand:** (1858, 1880, and 1911) Bloomfield 1984, table V 8.[16] **Egypt:** (1813, 1852, and 1880) O'Brien 1968, table III; and (1907) Owen 1986, table 1.[17] **Greece:** (1860, 1887, and 1911) Petmezas 2000, table 1 (Southern Greece only). **Portugal:** (1867, 1902; excludes fallow) Lains 2003. **Colombia:** McGreevey 1971, table 17; **Europe:** (1800, 1850, 1910, 1995) Bairoch 1999, table 5.1. **Western Settlement Countries:** (1850, 1910, 1995) Bairoch 1999, table 5.1.

The sample of countries is not large, but it may be considered to be fairly representative: the cumulated acreage accounts for almost two-fifths of the total land in use in 2000. The table shows that the stock of land has grown almost everywhere, with the exception of Western Europe. Such a growth is by no means surprising for the countries of western settlement, South America, or even for Russia and

Eastern Europe.[18] It is, however, remarkable how peasants were able to extend acreage also in long-settled Asian countries. In Japan, land increased over the whole Togukawa period, roughly doubling, and the growth rate seems to have peaked during the first half of the nineteenth century.[19] In China, according to Perkins (1969), the acreage increased from 63 to 80 million hectares from 1766 to 1873 (thanks notably to expansion in the south) and to 90 in 1913 (thanks to the colonization of Manchuria).[20] Other authors suggest slightly different figures, but they all confirm that there was growth. In India, the acreage increased by 60 percent from 1860 to 1920.[21] But cultivated acreage expanded all over Asia, from Anatolia to Thailand—and Ishikawa (1967, 61) argues that the "frontier" in rice-growing areas was closed only as late as the 1950s.[22] The great incognita is, as usual, Africa. As argued in the previous chapter, output must have increased at least as much as population—roughly by 70 to 80 percent in the nineteenth century (Biraben 1979, table 1; McEvedy and Jones 1978). Part of this increase may have been achieved by a more intensive use of the already-settled land (section 6.3), but it seems unlikely that total acreage did not increase, as land was quite abundant.

Before 1910, there are data on land in pasture (see table 4.2) for very few countries. One could conclude, very tentatively, that land in pasture remained roughly constant in most countries, including the United States (in spite of the big change in the legal status of grazing land).

Thus, in the "long" nineteenth century (until 1913), the total stock of land was growing almost everywhere, except in Western Europe and in the core areas of China. Land was abundant then. How much did this abundance last? Was it another unique feature of the gilded age of the first globalization? Table 4.3 reports data for the interwar years from the Annuaire of the Institute Internationale

TABLE 4.2
Acreage in Pasture 1800–1910 (Million Hectares)

	1800	*1850*	*1880*	*1910*	*2000*
Denmark		0.4*	0.4	0.3	0.34
France	10.8	6.2	6.5	10.0	10.39
England	7.1	6.5			
UK and Ireland			5.8	7.1	11.25
Hungary		4.1°	3.7	3.3	
Spain		8.8*	8.3	7.7	11.45
USA(a)			49.3	114.9	239.2
USA(b)			357.3	299	
New Zealand			2.6	5.8	13.3

Notes: * 1860 ° 1873

Sources: **USA:** Historical Statistics 1975: (a) grassland pasture in farms series J 52; and (b) grazing land not in farms series J 62. **All others:** see table 4.1.

TABLE 4.3
Acreage in Arable and Tree Crops, 1910–38 (Million Hectares)

	ca. 1910	*1938*	*2000*
Africa	13.9	35.4	42.0
Europe	149.9	149.5	128.3
North and Central America	141.9	165.1	224.7
South America	20.7	33.3	27.2
Asia	152.1	167.8	205.4
Oceania	9.0	14.2	56.4
Former-USSR	113.4	245.1	219
World	600.9	810.3	903

Notes: **Africa:** Algeria, Egypt, Morocco, South Africa, Tunisia. **Europe:** Austria, Hungary, Yugoslavia (in 1938 and 2000), Czechoslovakia (in 1938 and 2000), Belgium-Luxembourg, Bulgaria, Denmark, Finland, France, Germany, Greece, Italy, Ireland (in 1938 and 2000), Netherlands, Norway, Poland (in 1938 and 2000), Portugal, Romania, Serbia, Spain, Sweden, Switzerland, UK (and Ireland). **North and Central America:** Canada and USA. **South America:** Argentina. **Oceania:** Australia and New Zealand. **Asia:** India, Indonesia, Japan. **Former USSR** includes Estonia, Latvia, and Lithuania.

Sources: **Ca. 1910:** Institut Internationale d'Agriculture 1909–21. **Ca. 1938:** Institut Internationale d'Agriculture 1939–40. **2000:** FAO Statistical Database.

d'Agriculture. As indicated in the table, the country coverage is somewhat larger than that of table 4.1: they account for about 60 percent of present-day world acreage.

According to the official data, the world acreage increased further by 35 percent. This figure is somewhat implausible. In fact, about half the total increase is accounted for by an alleged doubling of land in the USSR, a country that had lost some 20–25 million hectares on its western boundaries. In fact, according to the most recent official Soviet figures, the acreage increased only by 15 percent, from 118.4 million hectares in 1913 to 136.9 million hectares in 1938, mainly due to the colonization of the steppes.[23] The cumulative acreage of the other countries in the table increased by some 80 million hectares—that is, by 16 percent. Country sources show that cropland increased also in Asian countries, such as China, Burma, and Thailand, and one can surmise, even without data, that it went on growing also in sub-Saharan Africa, where the population was growing quite fast.[24] Thus, one can tentatively conclude that the prewar growth in cropland continued until the 1930s, albeit at a decidedly slower rate.

At the same time, according to the data from the Institut Internationale d'Agriculture (table 4.4), the acreage under pastures doubled. Such an increase beggars belief: a tenfold increase in the extension of pasture in the Soviet Union can be explained only by a change in definition. The case is surely an extreme one, but all figures conceal huge problems of definition, and thus of comparability across time and among countries. For instance, the 1938 figures refer

TABLE 4.4
Acreage in Pasture, 1910–38 (Million Hectares)

	1911	*1938*	*2000*
Europe	82.1	81.7	77.9
South America	172.9	139.1	154.9
Oceania	14.5	6.8	13.3
USSR	38.7	401.9	356.4
'World'	308.3	629.4	619.6

Notes: **Europe:** Austria, Hungary, Yugoslavia (in 1938 and 2000), Czechoslovakia (in 1938 and 2000), Belgium-Luxembourg, Denmark, Finland, France, Germany. Greece, Italy, Ireland (in 1938 and 2000), Netherlands, Norway, Poland (in 1938 and 2000), Romania, Serbia, Spain, Sweden, Switzerland, UK and Ireland. **South America:** Argentina and Chile. **Oceania:** New Zealand.

Sources: Same as for table 4.3.

TABLE 4.5
Acreage in Arable and Tree Crops, 1940–2000 (Million Hectares)

	1940s	*1961*	*1970*	*1980*	*1990*	*2000*
Africa	173.0	154.9	165.8	178.0	191.2	201.8
Europe	149.0	151.4	145.7	140.7	138.6	133.2
North and Central America	240.0	259.6	268.9	273.9	274.6	268.1
South America	83.0	68.6	82.5	101.0	108.8	116.1
Oceania	18.0	34.8	46.4	48.8	53.3	53.0
Asia	329.0	437.1	448.9	457.6	507.1	511.7
USSR	225.0	239.8	233.2	231.5	228.9	217.5
World	1217.0	1346.2	1390.9	1431.6	1502.4	1501.5

Sources: **1940s:** FAO, Yearbook, 1949, vol. I table.1 (most data refer to 1945/48; the figure for USSR is labeled "estimate"). **1960–2000:** FAO Statistical Database.

to "cultivable waste" in India and to "permanent, artificially sown grasses" in Australia. Furthermore, the table omits the United States, where, in 1940, "grassland pasture" in farm comprised nearly 185 million hectares, and "grazing land not in farm" another 200 million hectares—that is, 5 percent less than in 1910.[25]

Table 4.5 reproduces the data on worldwide agricultural acreage, which the FAO started to publish in the 1940s. This collection marks a great leap forward in our knowledge, thanks to the systematic effort to make definition comparable, and, above all, to be comprehensive.[26] Some of these data are, however, somewhat dubious. Trends in China, which accounts for a large proportion of Asian land, are still uncertain.[27] The 10 percent fall in Africa from the 1940s to 1961 contrasts with the 63 percent population growth during the same period, while the Soviet

TABLE 4.6
Acreage in Pasture, 1945–2000 (Million Hectares)

	1940s	*1960*	*1970*	*1980*	*1990*	*2000*
Africa	509	898.8	894.2	894.8	903.7	891.8
Europe	99	89.8	88.9	86.8	81.2	80.6
North and Central America	422	373.0	357.2	357.6	365.4	367.3
South America	343	418.1	452.4	479.4	499.9	503.1
Oceania	370	444.5	453.2	453.5	430.5	419.4
Asia	266	618.5	650.9	694.5	799.7	841.2
USSR	124	302.0	314.6	321.8	327.3	356.4
World	2133	3144.7	3211.4	3288.0	3409.8	3459.8

Sources: **1940s:** FAO Yearbook 1949, vol. I table.1. **1960–2000:** FAO Statistical Database.

Union figures seem not to show the impact from the "virgin land" campaign (the agricultural settlement of Central Asia).[28] In spite of these reservations, the story is clear: the expansion in acreage continued after World War II, but more slowly than before. The overall growth is the sum of fast growth in some areas—such as Oceania, Asia (with the big question mark about trends in China), and South America (i.e., in Brazil)—and stagnation or decline in the "advanced" countries. In Europe and Northern America, agricultural acreage diminished because of the combined effects of migration from the countryside, losses to urbanization, and policies designed to curb overproduction by setting aside land.

As table 4.6 shows, the long-term increase was not achieved at the expense of pasture land. Even discarding the leap from the 1940s to 1960 as implausible, the FAO data show a steady growth in pasture acreage.

The analysis so far, in spite of the incomplete and sometimes dubious data, boils down to a clear conclusion: the total stock of land for agricultural purposes rose fast in the nineteenth century, while, in the twentieth century, the growth slowed down without stopping. Did the flow of services increase as much as the stock?

1. As previously argued, crops use the productive power of the soil more intensively than natural grass. An increase (decrease) in the proportion of cropland on total acreage would therefore increase (decrease) the amount of land services. This was the case, according to a large body of anecdotal evidence, in most countries of western settlement—the American Wild West, Australia, Argentina, and so on—where land was first used as pasture for extensive ranching and only later settled as farms.[29] In this case, the share of pasture would increase at the beginning of settlement, remain roughly constant throughout the process (as new pastures were being added at the extensive margin, while former meadows were being converted to arable) and eventually decline with the exhaustion of the stock of new land.

Unfortunately, the data are not precise enough to substantiate this hypothesis, as official series of pasture do not cover the early process of settlement. In the United States, the total acreage under pasture seems to have remained roughly constant from 1880 to 1960, but a growing proportion of it was incorporated into farms.[30] Worldwide, the share of pasture on total agricultural land has remained quite stable since 1960 (tables 4.5 and 4.6). Thus, the change in destination seems not to have been an important source of divergence between stock and flow of measures of land in the twentieth century, although one cannot rule out an increase in services (from the conversion of pastures into arable land) in the nineteenth century.

2. The stock and flow measures of land could differ also if, ceteris paribus, the best land was settled first, as stated by Ricardo. If so, the average "quality" of land (section 2.2) would decrease, and thus the raw figures of stock (acreage) would overvalue the actual flow of services. This may have been the case in India, where the poor quality of new land is conventionally reckoned to have been one of the causes of the (alleged) fall in food-grain yields in the twentieth century.[31] However, Ricardo's statement is less compelling than it may seem, as it refers to the choice of a single farmer (or of a community) within the stock of accessible land. But most of the "new" land was added to the stock in large swathes as soon as the political situation and the conditions of means of transportation made it accessible. Thus, its quality was not necessarily worse than that of the already settled areas. Indeed, Manchuria or the Isle of Hokkaido in Japan are cooler and drier than the "core" areas of Central China or Honshu, and are, therefore, unsuited to rice production. In the United States the prairies are drier than the East Coast, and, in a famous study, Parker and Klein estimate that shifting wheat-growing westward reduced yields, ceteris paribus, by 5 percent.[32] However, in other cases, "new" land was better than the "old." For instance, the *pampa humida* (wet pampas) of southern Argentina are decidedly better than the areas around Buenos Aires, which were settled first. It would, therefore, be impossible to assume a priori that the series of land biases the actual flow of service systematically downward.

3. Last but not least, the quality of land can be negatively affected by human action. Excessive use of chemicals can damage land, over-exploitation can cause the loss of topsoil (erosion), especially in the tropics, and irrigation can cause degradation. The evaporation of water can leave a high concentration of salts (salinization), and/or the leaking of superficial water into the ground can cause a rise in the underground water table (waterlogging). Solutions (such as drainage against waterlogging) are available but quite expensive. These problems are attracting a great deal of attention nowadays, but they are hardly new.[33] Salinization was detected in Egypt in the 1880s, waterlogging in China and India roughly at the same time.[34] The estimates of losses vary widely also because authors use different definitions of "degraded" land. Some include all land that has lost some productive power, while

others include only the seriously compromised acreage. A fortiori, it is difficult to assess to what extent degradation is a recent trend or not (let alone to discuss its remedies). In the major historical work on the issue, Lindert (2000, 152–62, 196–206) finds no evidence of growing large-scale degradation in China and Indonesia from the 1930s to present.[35] One would like to have his pioneering work replicated for other areas.

The available evidence is thus too scarce to draw any definite conclusion about any divergence between the flow of services from land and stock. The implicit ratio might have increased somewhat in the nineteenth century and remained constant thereafter—possibly with a decrease in the past decades because of environmental damages. However, the changes seem not to have been large, and thus one could use the series of stock as a proxy for flows, with a reasonable margin of error.

4.3 Capital

Agricultural capital consists of five items: (1) improvements to land (fencing, terracing, planting tree crops, etc.); (2) buildings for agricultural purposes (stables, sheds, granaries, etc.); (3) tools and machinery; (4) livestock; and (5) working capital, in other words, the sums for the purchase of feed and other inputs outside the sector plus the economic value of standing crops and stocks of products.[36] The task of collecting reliable data on all five items is quite a difficult one, and indeed most of the available estimates (table 4.7) omit one or more items.

Total capital stock increased in all countries—except, perhaps, the United Kingdom, and the country-ranking tallies well with the expectation. The highest rates are registered in the United States and Canada during the process of settlement, the lowest in the United Kingdom. The figures of table 4.7, however, tend to underestimate the gap. In fact, as detailed in the sources for the table, most estimates for European countries omit improvements to land and buildings, which, as table 4.8 shows, grew more slowly than livestock or machinery. Thus, before 1940, the main source of growth in agricultural capital was the process of settlement.

There are no worldwide data on agricultural capital after 1950. The FAO database provides data on specific items, such as irrigated land, machinery, and livestock—but not aggregate figures. The most comprehensive data collection reports annual series for fixed capital, orchards, and livestock (thus excluding machinery and working capital) for fifty-seven countries from 1967 to 1992 (table 4.9). Even if it only reports four benchmark years, table 4.9 highlights the main changes well, because 1980 happens to be a break point in the series. Agricultural capital had been growing quite fast in the 1970s, in all likelihood continuing the previous trend, while it fell in all continents (except Asia) in the 1980s. This

TABLE 4.7
Rates of Growth in Capital Stock, by Country, 1880–1938

	Prewar		*Interwar*	
Belgium	1880–1913	0.96	1910–38	0.86
Canada	1871–1911	3.64	1911–21	2.76
France	1870–1914	0.03		
Germany	1850–1913	1.2	1913–38	0.65
Italy(a)	1881–1913	1.12	1913–38	2.32
Italy(b)	1860–1913	1.20	1913–38	0.63
Japan	1880–1915	1.10	1915–40	0.83
Russia	1885–1913	2.31		
USA(a)	1870–1900	2.39		
USA(b)	1870–1913	2.11	1913–38	0.06
USA(c)	1870–1910	2.26	1910–40	0.06
UK(a)	1878–1913	−0.25		
UK(b)	1850–1910	0.32		
UK(c)	1873–1913	−1.00	1924–37	−0.10
India			1920–35	0.13

Sources: **Belgium:** Blomme 1992, tables 63 and 64 (livestock only). **Canada:** McInnis 1986, table 14.7 (buildings, tools, and machinery; livestock and working capital). **France:** Grantham 1996, table 3 (improvements to land, buildings, tools and machinery, and livestock).[37] **Germany:** Hoffman 1965, ii, table 29 (buildings, tools and machinery; livestock and working capital). **Italy:** (a) Vitali 1969, table XII.3.3 (improvements to land, buildings, and tools, and machinery); (b) Federico 2002 (buildings, tools, and machinery; livestock and working capital). **Japan:** Okhawa and Shinohara 1979, table A18 (buildings, tools and machinery; livestock and working capital). **Russia:** elaboration from Gregory 1982 (table H1.C livestock, table J.1 "rural structures", table I.1 D equipment) and Kahan 1978, tables 45 and 46 (transportation equipment). **USA:** (a) Gallman 1986, table 4.A.1 (improvements to land, buildings, tools, and machinery); (b) Kendrick 1961, table B-1 (improvements to land, buildings, tools and machinery, livestock;) (c) Tostlebe 1957, table 9 (buildings, tools, and machinery, livestock and crop inventories). **UK:** (a) O'Grada 1994, table 5.4 ("structure," implement, livestock); (b) Feinstein 1988a, table 10.2, and Feinstein 1988b, table 18.2 (buildings, tools and machinery, livestock, and crop inventories); (c) Matthews et al. 1982 table 8.3 (buildings, tools and machinery). **India:** Shukla 1965, table V.2, Series X.1 (improvements to land, buildings, tools and machinery, and livestock).

result is somewhat surprising, as it contrasts not only with the conventional wisdom, but also with the increase in irrigated land (table 4.10) and livestock (table 4.14). One must therefore infer that in the 1980s the depreciation of existing fixed capital exceeded net investment. On the other hand, more than half of the world decrease in capital from 1980 to 1992 is accounted for by the fall in American capital stock, which may be overestimated.[38] Last but not least, the series omits two fast-growing items: tractors (table 4.11) and the working capital for the purchases of input, such as fertilizers (table 4.15). To sum up, the available aggregate data suggest an increase of agricultural capital in the long run—possibly peaking

TABLE 4.8
Rates of Growth in Capital Stock, by Country and Type, 1850–1940

		Buildings	*Implements*	*Livestock*	*Total*
UK	1850–1910	0.13	1.09	0.56	0.43
Japan	1880–1935	0.69	1.48	1.30	1.03
Germany	1850–1913	1.17	1.57	1.13	1.22
	1923–37	0.48	2.64	2.04	1.22
Canada	1871–1921	2.82	4.53	2.50	2.92
Russia	1885–1913	3.30	2.46	1.65	2.21
France	1789–1870	0.66	0.66	0.80	0.73
	1870–1914	0.02	0.46	0.48	0.03
Italy	1861–1938	0.80	1.26	0.60	1.01
USA	1870–1910	2.25	4.07	1.87	2.16
	1910–40	–0.05	1.40	–0.40	–0.06

Sources: Same as for table 4.7 (**UK** series (b), **USA** series (c), and **Italy** series (b))—except **Japan:** Hayami and Yamada 1991, table A.8.

TABLE 4.9
Total Fixed Agricultural Capital, by Continent, 1967–92
(Billion 1990 US $)

	1967	*1970*	*1980*	*1992*
Africa	55.9	56.3	81.0	63.4
Europe	347.6	332.9	691.1	620.6
North and Central America	382.9	399.0	614.8	411.8
South America	107.7	104.2	130.3	99.0
Asia	318.5	325.4	707.6	1038.5
Oceania	50.1	49.6	78.2	59.6
World	1262.8	1267.4	2302.9	2292.8

Notes: **Countries: Africa (10):** Egypt, Kenya, Magadascar, Malawi, Mauritius, Morocco, South Africa, Tanzania, Tunisia, Zimbabwe. **Europe (16):** Austria, Belgium-Luxembourg, Czechoslovakia, Denmark, Finland, France, UK, Greece, Ireland, Italy, Malta, the Netherlands, Norway, Portugal, Sweden. **North and Central America (9):** Canada, Costa Rica, the Dominican Republic, Guatemala, Honduras, Jamaica, El Salvador, Trinidad, Tobago, and USA. **South America (6):** Argentina, Chile, Colombia, Peru, Uruguay, and Venezuela. **Oceania (2):** Australia and New Zealand. **Asia (14):** Cyprus, Indonesia, India, Iran, Iraq, Israel, Japan, Pakistan, the Philippines, South Korea, Sri Lanka, Syria, Turkey, and Taiwan.

Source: Larson et al. 2000.

around 1980. However, these aggregate data are really incomplete and possibly biased. Thus, they have to be supplemented with additional information, item by item.

1. *Improvements to Land*

The transformation of virgin land into a viable farm entails huge investments—largely, through the back-breaking work of peasants.[39] In the United States, capital in land has always been the largest single item of agricultural capital by far, accounting for 55 to 60 percent of the total in the nineteenth century, and for 61 percent in 1992–94.[40] The cost of land was only the first item, and the least important, in the list of expenditures (many farmers received land for free, as will be discussed in section 8.2). The new farmer also had to clear the land. In forested areas, he had to fell the trees (or possibly burn them, losing valuable wood) and remove the stumps. In the prairies, the task was "simpler," as the farmer had only to plow the soil, even if breaking it needed a lot of draught power.[41] Then, he had to build fences to keep animals away from his crops and a hut for himself, his family, and animals; he also had to live on his own money while waiting for the new crops. Primack (1966) estimates that in the period 1850–1910 the settlement of the United States absorbed some 30 million man-hours (20 million for improvements and the rest for building)—that is, about 5 percent of the adult male work force over the whole period—and over 10 percent in the 1850s.[42] Furthermore, new farmers had to set aside additional money (i.e., capital) for any unforeseen circumstances, which were quite likely, in that many farmers came from very different environments and had to adjust their skills to the new environment. Thus, the total cost could be high. Setting up a new farm in the American North on the eve of the Civil War cost around $1000–1500, equivalent to seven to ten times the yearly income of a yeoman (the figures vary greatly, and some of them seem suspiciously low, so as not to scare new farmers away).[43] American agriculture in the nineteenth century was quite advanced, and new farmers wanted a minimum standard of comfort in the cold climate. In "backward" tropical countries, setting up a farm cost decidedly less—two to four times the yearly wages for workers in cocoa production in Ghana (Ingham 1979, 30) and one year in Thailand (Ingram 1971, 64–65).

In long-settled countries, such as western Europe or the core areas of China, most investments in buildings and land had already been made centuries before. Thus, they may have been already amortized by 1800, and needed only maintenance. In many of these countries, however, substantial investments to improve the land already in use were made in the nineteenth century. In the United Kingdom, the transformation of some five million acres of open fields into "modern" farms from 1760 to 1860 cost 18.5 million (1860) pounds—about a tenth of the gross agricultural output in the 1860s.[44] The acreage under vineyard and tree crops greatly increased in France during the first half of the nineteenth century and in Italy during the 1880s, and literally boomed in California from the 1880s onward.[45]

Both the magnitude of investments and the costs of maintenance are much greater if water was involved—if there is too much of it and it has to be drained, or if there is too little and it has to be brought to the fields. In the nineteenth and twentieth centuries, farmers and public authorities invested massively in water management. For instance, British farmers used underground drainage to help heavy clay soils absorb water. The practice was first adopted in the mid-eighteenth century in the London area. From then to 1900, landowners spent some thirty million (1860) pounds to drain about 1.65 million hectares (roughly half of the 3.6 million estimated by experts to need drainage).[46] Marshes were reclaimed mainly in land-scarce countries, such as the Netherlands and Italy. The drainage of the marshes on the Italian coastal plains increased the total available acreage from the mid-nineteenth century to the 1930s by two million hectares—or by about 6 to 7 percent.[47] In the United States, from 1846 to 1956 some fifty-six million hectares of marshes out of a total of 100 million hectares were drained, mainly during the prosperous 1900s and 1910s.[48]

Irrigation is essential for the cultivation of (paddy) rice in Asian rice-producing countries.[49] In China, major irrigation works had been completed long before 1800: the country underwent a succession of "hydraulic cycles"—a rise and then a fall in the extension and quality of irrigation services. The period from 1800 to the 1950s marked the downward phase of the last of these cycles, with stagnant irrigated acreage and increasingly frequent disasters.[50] In contrast, irrigated acreage doubled in India, and grew extensively in many other Asian countries, such as Japan (and its colonies), Indonesia, and the Philippines.[51] Massive irrigation projects were begun by Europeans in their colonies: the Dutch in Indonesia (as early as the 1850s, but with substantial investments from the 1880s), the British in India (from the mid-1850s) and Egypt (from the 1880s), and the French in both South Vietnam and the Tonkin Gulf (from 1906).[52] Feeny (1982, 64–79) strongly criticizes the Thai government for having invested in railways instead of in irrigation, which would have yielded a higher rate of return and would have staved off the decrease in rice yields.[53] Thus, capital accumulation was a major dynamic force in Asian agriculture.

The big change in the nineteenth century was the extension of irrigated areas in hot, dry regions of Mediterranean agriculture. By 1800, throughout the Mediterranean basin, with the conspicuous exception of Egypt, irrigation was limited to small pockets of land, such as the "gardens" (i.e., citrus fruit plantations). Irrigated areas expanded in the late nineteenth and twentieth centuries in the Mediterranean basin proper (Spain, Italy, etc.), in South America (Mexico, Chile), and in Australia and the United States.[54] The first recorded irrigation scheme in the United States was built in 1848 (by Mormons in Utah), but massive projects started in the late nineteenth century. In 1896, there were already 1.5 million hectares of irrigated land, which rose to 4.7 million in 1910, to 5.9 million in 1930. The acreage jumped to 8.03 million hectares in 1940 and increased quite quickly thereafter.[55]

TABLE 4.10
Irrigated Acreage, by Continent, 1900–2000 (Million Hectares)

								%	%
	1900s	*1930s*	*1961*	*1970*	*1980*	*1990*	*2000*	*1961*	*2000*
Africa	2.5	4	7.4	8.5	9.5	11.2	12.5	4.8	6.2
Europe	3.5	6	8.3	10.4	14.0	16.7	16.9	5.5	12.7
North Central America	4	11	17.9	20.9	27.6	28.9	31.4	6.9	11.7
South America	0.5	3	4.7	5.7	7.4	9.5	10.3	6.8	8.9
Oceania	0	0.5	1.1	1.6	1.7	2.1	2.5	3.1	4.8
Asia	30	57	90.2	109.7	132.4	155.0	180.5	20.6	35.3
USSR			9.4	11.1	17.2	20.8	19.9	3.9	9.2
World	40	81	139.0	167.8	209.7	244.3	274.2	10.3	18.3

Note: % ratio to total acreage (arable and tree crops).

Sources: **1900**: Framji et al. 1981, table ix. **1930s**: Clark 1970, table 25. **1961–2000**: FAO, Statistical Database.

After World War II, the irrigated acreage more than doubled (table 4.10): nowadays, it exceeds the total agricultural acreage on the entire North American continent. Nearly half of the increase from 1960 to 2000 is accounted for by the growth in three countries: India, USSR (where it doubled), and China, where it increased "only" by a quarter. The two last columns of the table show how much irrigated land has increased as a proportion of the total. Its ratio is now approaching 100 percent in the former Soviet republics of central Asia (Turkmenistan, Ubzekistan, Tajikistan) as well as in Egypt. This impressive growth has been essential for the increase in agricultural production: worldwide, irrigated land provides 30 to 40 percent of world gross output (including two-thirds of rice and wheat) and uses about 70 percent of the water extracted for human use.[56]

Clearly, the diffusion of irrigation entailed huge investments. The unit cost can differ widely according to the system employed. Land can be irrigated by simple wells, tanks, or reservoirs, or by river water diverted by contour canals (possibly from afar), or by creeks in river delta areas (where water has to be raised to the fields). The choice of a system depends mainly on the supply of water (amount of rainfall, distribution throughout the year, etc.) and is also affected by the demand for water (i.e., the crop mix and techniques employed). The choice of the system determines the cost of the work. In Tanzania, the traditional gravity system cost less than a tenth of a diesel pump.[57] It is likely that the unit cost of irrigation has been rising in the long run. The real wages of construction workers have been growing, and the simplest and cheapest projects were undertaken first.[58] In fact, the share of technical irrigation (from canals) on the total irrigated acreage has risen all over Asia.[59] Most of the cost was borne by public funds. Private funding of major irrigation schemes (as opposed to digging village wells) was comparatively rare, even in nineteenth-century India, in spite of the strong preference for this arrangement by the British colonial administration.[60] Some governments tried to save money by drafting peasants to work on building sites. This was a centuries-old system in most of Asia, which was still very much employed in the nineteenth century by colonial governments. Peasants were obliged to work for free also in the Soviet Union and in China during the Cultural Revolution (Zweig 1989, 147–67; Huang 1990, 231–36; Becker 1996, 76–79). Clearly, peasant labor is free in appearance only. The peasants lose income from alternative occupations, if any, and also forego leisure: drafting is equivalent to taxation. Most of the investments the nineteenth and twentieth centuries were undertaken with paid laborers, and thus involved huge monetary expenditure. Governments funded most of these investments, either entirely or under some sort of partnership with private capital. For instance, the United States government subsidized the drainage of marshes from 1846, and irrigation, beginning with the Reclamation of New Land Act in 1902 (Hurt 1994, 239–44; McCorvie and Lant 1993).[61] In Japan, the government started to subsidize the building of irrigation works at the end of the nineteenth century (forcing reluctant peasants to join in, if necessary) and increased its participation in the expenses, from the original 15 percent to some two-thirds in the 1950s and 1960s.[62] In Indonesia, public expenditure on irrigation (including maintenance) grew almost ten times from the early 1880s to the peak in 1929–33, from 0.15 to 0.55 percent of the output.[63] The increasing involvement of governments has reflected the growing complexity of irrigation projects, and the political demand of farmers. In fact, government-managed irrigation works have often ended up in the subsidizing of water consumption, with severe distortion of relative price of inputs.

2. *Buildings*

Stables, warehouses, and other buildings are necessary to protect animals and stocks from rain and cold weather, and thus account for a sizeable share of total agricultural capital—in the United States, for example, it accounted for 17 percent in 1992–94, the second largest item after land (USDA 2004, table 1). Unfortunately, the data are scarce, and this is not surprising. Most agricultural buildings are annexes of farmers' houses, and thus their value is not estimated separately, even though it should be (in national accounts, the houses for the rural population belong to the residential housing stock). The missing data have to be substituted with some educated guesses. The amount of capital in building is mainly determined by demand. The land to build on is, by definition, abundant everywhere, and the cost of construction remained relatively low throughout this period as long as peasants could work during the slack season. As a first approximation, one would expect capital in building to have grown proportionally with gross output. The modernization of agriculture, however, affects this relation in quite a complex way. Mechanization increased the need for warehouses, and the pure breeds of animals needed stables and more storage space for winter fodder. On the other hand, the number of animals grew less than their output (section 5.1) and the layout of buildings improved. For instance, the invention of ensilage drastically reduced the need for space to store winter fodder.[64] Last but not least, the commercialization of agriculture (section 8.5) has also reduced capital in buildings per unit of output. In fact, it shifted a growing share of the total stocks away from farmers' premises (i.e., agricultural capital) into the warehouses of processing companies (industrial capital), which belong to industrial capital. It is difficult to predict a priori the outcome, and the data are scarce.[65] A straightforward comparison between the rate of growth of capital in buildings (table 4.8) and the rate of growth of output suggests that, in most cases, the former grew less than the latter. For instance, from 1870 to 1910, the stock remained constant in France and the United Kingdom, where new investments barely compensated for the depreciation of the existing stock. After 1960, total fixed capital increased much less than output (Larson et al. 2000). On the other hand, capital in buildings grew more than output in the United States and Canada in 1870–1910, and probably also in other countries during settlement.

3. Tools and Machinery

Farmers used, and still use, a wide range of tools, from sticks to the modern combine harvesters (harvesters with threshers). Traditional tools were simple and inexpensive, and the total stock was roughly proportional to the number of workers and/or to the land to be worked, as long as there was no technical progress. This was the case for China: most scholars, following Perkins, rule out any major improvement in tools from the end of fifteenth century to the beginning of the twentieth.[66] But elsewhere, the range of tools in use widened and their quality improved, and thus the stock per worker also rose. For instance, French peasants started to use shears, instead of knives, to prune vines in the early nineteenth century (Price 1983, 381; Loubère 1978, 82–83). African peasants substituted their traditional hoes with Western-style plows: the movement started in the late nineteenth century in South Africa, under the direct influence of Western technology, and plowing spread throughout the whole continent South of the Sahara from the 1920s and 1930s (even though hoeing is still largely practiced).[67] In eastern Europe and also, to some extent, India, the traditional wooden plows were substituted by iron plows in the second half of the nineteenth century.[68] In "advanced" Western countries, the quality of plows improved, with better shapes and more resistant material—from wood, still diffused even in Colonial America, to iron, cast-iron, and eventually steel (McClelland 1997, 16–63). Until the end of the nineteenth century, farmers used few machines, and almost exclusively for processing, with the exception of the reaper for harvesting wheat (section 6.3). Massive investments for mechanization for both processing and fieldwork started only in the twentieth century. In the United States, the number of farms with

TABLE 4.11
Number of Tractors, by Continent, 1920–2000 (Thousands)

	1920	*1930*	*1939*	*1950*	*1961*	*1970*	*1980*	*1990*	*2000*
Africa			6	95	235	334	439	532	528
Europe		130	270	990	3,698	6,077	8,454	10,356	9,650
North and Central America	294	1,030	1,576	4220	5,326	6,038	5,606	5,841	5,808
South America			17	70	297	465	880	1186	1293
Oceania				142	351	428	427	403	401
Asia				35	200	783	3,475	5,599	6,884
USSR		78	440*	430°	1,212	1,978	2,646	2,609	1,861
World				5552	11,318	16,102	21,932	26,526	26,424

Notes: * 1935. ° 1948.

Sources: 1920–1939: **North and Central America:** sum of USA (Historical Statistics 1985, K184) and Canada (Historical Statistics Canada 1965, series L 318). **Europe:** Svennilson 1954, table A19.[70] **USSR:** (1930) Clarke and Matko 1984, table 59 (cumulated total of deliveries from 1924–25 to 1935). **Africa:** (South Africa only) Duggan 1986, 158. **South America:** (Argentina only) Diaz Alejandro 1970, table 3.11; 1950 FAO Yearbook. **All others:** FAO Statistical Databases.

milking machines grew from some 12,000 in 1910 to 175,000 in 1940, and jumped to 636,000 by 1950.[69] In Western Europe, the first machines were introduced in the late 1930s and spread after World War II. Table 4.11 shows the available data on the number of tractors, the most important machine for fieldwork. Before World War II, tractors were used massively only on the cereal farms of the Great Plains and the Pacific region of North America, and in the Soviet Union, where motorization happened to be one of the priorities of the five-year plans.[71] The use of tractors spread in other "advanced" countries of Western Europe and Oceania in the 1950s and 1960s, in Asia and South America in the 1970s and 1980s, and in Africa in the 1980s.

Estimating a series of capital in tools and machinery from these scattered data is quite difficult. Probably, the United States set the upper bound of capital intensity in traditional agriculture. According to textbooks for prospective settlers, in the 1860s–1870s, tools for a new farm cost $200–300 (about a fourth of the total cost), but this sum included "luxuries" such as wagons, although no machines (Bogue 1963, 170; Atack and Passel 1977, 277–78). The cost must have been much lower in the rest of the world, where only the richest peasants owned carts and wagons, and the average quality of tools was surely inferior. Throughout the nineteenth century, capital in tools was the fastest growing item in total stock (table 4.8), but the absolute cost was still low. The reaper was a simple horse-drawn machine. Massive investments in machinery started in the twentieth century. The number of machines such as tractors (table 4.11) tends to overstate the increase in capital. In fact, the unit price of machinery fell—and this decrease contributed to mechanization.

Hayami and Ruttan (1985, table c.2) estimate that the price of machinery relative to agricultural products halved in the United States; and it fell by one-third in Japan from the 1880s to the 1920s (Hayami and Yamada 1991, tables A10, A11).[72] From the 1920s onwards, prices increased in the United States and fell further in Japan. However, these unit prices are not an accurate measure of the cost of capital services: in fact, they do not take the quality of improvement into account. The real price of comparable tractors in the United States fell by two-thirds in the 1910s, and then crawled back somewhat to remain roughly stable from the 1930s to the 1960s.[73] On the whole, it is likely that the number of machines undervalues the (quality-adjusted) stock, and thus the flow of potential services from this category of capital.

4. *Livestock*

The historical data on the number of animals are quite abundant, as many countries held periodical censuses from the nineteenth century. Table 4.12 summarizes the results for some fifty countries accounting for 60 percent of the world stock of cattle and 80 percent of pigs in 2000.[74] Before 1937, the number of cattle and pigs grew faster in countries of Western settlement than in Europe or Asia. This pattern is consistent with the overall expansion of agriculture in these areas. It is, however, worth stressing that the difference is smaller for cattle than for land or agricultural work force. The number of sheep and goats, in contrast, grew mainly in Oceania and in LDCs, while it has remained stable in "advanced" countries.

From the 1950s the FAO data (table 4.13) show that the overall growth in stock has been decidedly slower in "advanced" countries (excluding Oceania) than in the rest of the world. From the 1930s to 2000, in Europe and North America the total number of cattle increased by a third, that of pigs increased by 80 percent, and the number of sheep fell slightly. In the rest of the world, however, the number of cattle and sheep/goats more than doubled, and the number of pigs increased 4.3 times. Asia (including mainland China) accounted for three-quarters of the increase in the number of "small" animals (pigs, sheep, goats) and only for a third of the growth in the cattle stock.[75] The overall increase in the number of large animals greatly increased the supply of manure, which still accounts for a substantial proportion of total fertilizers.[76]

The number of animals may not be a good proxy of the value of the stock, which depends on the unit value of animals and on their composition by race, sex, and age. The unit value of animals has augmented, especially in "advanced" countries, because of the diffusion of pure-breed races. For instance, Friesian cows rose from 3 to 85 percent of the British stock from 1919 to 1985.[77] Better breeds are more productive, and thus more expensive. In the United States, real prices per head increased by two to three times between the 1860s and 1910–14, then remained roughly constant during the interwar years, before doubling again in the

Table 4.12
Number of Animals, Selected Countries, 1850–1940 (Millions)

	ca. 1850	*ca. 1880*	*1913*	*1937*	*2000*
Cattle and buffalo					
Europe	57.5	73.4	93.3	104.3	93.6
USSR		31.8	53.0	36.5	62.2
Africa			16.6	29.2	37.7
Asia			222.8	259.2	372.9
North America	25.4	43.2	63.4	75.0	111.5
South America			40.4	54.0	116.3
Oceania	1.9	7.8	13.8	17.9	35.6
Pigs					
Europe	28.0	34.1	65.9	79.9	150.8
USSR		11.3	14.2	25.7	35.4
Africa			1.8	1.7	3.6
Asia			83.4	73.5	482
North America		51.0	57.4	46.6	74.6
South America			4.4	6.6	15.8
Oceania		0.8	1.1	1.9	3.0
Sheep and goats					
Europe	132.0	121.3	112.5	138.7	147.6
USSR		60.9	75.0	57.3	55.7
Africa			66.6	85.5	108.1
Asia			99.0	122.4	533.7
North America		45.2	42.9	48.3	15.2
South America			74.8	84.1	61.7
Oceania	16.3	67.0	111.3	145.8	161.5

Countries: **Africa:** Algeria, Egypt, Lybia, Magadascar, Morocco, South Africa, Tunisia, Zimbabwe. **Europe:** Austria, Hungary, Yugoslavia (in 1937 and 1999), Czechoslovakia (in 1937 and 1999), Belgium-Luxembourg, Denmark, France, Germany. Greece, Italy, the Netherlands, Norway, Poland (in 1937 and 1999), Portugal, Romania, Spain, Sweden, Switzerland, the UK and Ireland. **North and Central America:** Canada and USA. **South America:** Argentina, Chile, Colombia, Peru, Uruguay, Venezuela. **Oceania:** Australia and New Zealand. **Asia:** China, India, Indochina, Indonesia, Japan, Korea, the Philippines, and Thailand. **USSR:** includes Estonia, Latvia, and Lithuania.

Sources: **Indonesia:** van der Eng 1996, table A5. **USA:** Historical Statistics 1975, series K564–582. *Russia:* Gregory 1985, table H. **Japan:** Historical Statistics 1987–88, Vol. 2, table 4-9. **Others:** Mitchell 1998a–c, with interpolations by the author when necessary.

1970s (except for hogs).[78] Thus, the number of animals undervalues the growth in the stock. On the other hand, this undervaluation might be balanced by composition effects. First, the increase in the share of animals in LDCs countries has lowered the worldwide average quality of animals—and thus their unit value. Second, the composition of the stock has changed. All over the world, tractors have substituted

TABLE 4.13
Number of Animals, by Continent, 1940–2000 (Millions)

	Ca.1938	*1948–52*	*1961*	*1970*	*1980*	*1990*	*2000*
Cattle and buffalo							
Africa	84	104	122	151	174	191	231
North America	95	114	140	166	172	159	160
South America	107	135	145	178	243	274	308
Asia	302	295	405	446	465	545	615
Europe	104	100	117	123	134	125	102
Oceania	18	20	27	31	35	32	38
USSR	60	56	76	96	115	119	60
Total	769	822	1,030	1,189	1,338	1,444	1,576
Pigs							
Africa	3	4	6	7	10	17	19
North America	64	77	72	81	101	86	94
South America	31	35	38	45	52	52	55
Asia	85	85	119	222	382	437	535
Europe	80	71	109	130	175	183	168
Oceania	2	2	3	4	4	5	5
USSR	32	27	59	56	74	79	36
Total	296	301	406	547	798	858	912
Sheep and goats							
Africa	176	186	229	271	321	378	457
North America	72	51	55	42	34	34	28
South America	114	144	137	134	121	127	99
Asia	270	251	431	477	593	702	819
Europe	158	144	149	137	135	174	150
Oceania	140	145	201	241	205	230	165
USSR	75	103	140	136	149	145	57
Total	1,005	1,024	1,343	1,437	1,559	1,790	1,775

Sources: ca. **1938** and **1948–52**: FAO Yearbook. **1961–2000**: FAO Statistical Database.

draft animals—oxen or he-buffalo in Asia and in Mediterranean Europe, horses and mules in the rest of Europe and in countries of western settlement. In many countries (especially "advanced" ones), the average age of the rest of the stock has decreased. In traditional agriculture, animals were slaughtered at an old age, so that the unit cost of the production of meat was lower. In modern livestock-raising, only dairy cows are kept throughout all their productive life, while oxen and calves are slaughtered while relatively young to produce meat. The age at slaughterhouse also depends on the relative prices of milk and meat and on the taste of consumers. Ceteris paribus, cows cost less than oxen, and young animals cost less than old ones. Thus the change in composition might have reduced the value of stock.

Given all these effects, it is difficult to draw any firm conclusion about the difference between the number of animals and the value of the stock.

5. *Working Capital*

The working capital includes the value of crops (both as standing crops and as the post-harvest inventories) and the amount of outlays for productive expenditures. The value of standing crops rose as much as output, while that of on-farm inventories, as already stated, is likely to have declined as a proportion of output because of commercialization. In fact, harvested and growing crops accounted for about a third of total capital of the United Kingdom in the nineteenth century, while "inventories" accounted for a mere 8 to 10 percent in post-1950 United States.[79] On the other hand, expenditures for purchases outside the sector have grown greatly. As said earlier, they were negligible in traditional agriculture, with the exception of small areas of intensive cultivation. Probably they did not exceed 2 percent of gross output, as in the United States in 1800 or in Africa in the early 1960s.[80] The modernization of agriculture and technical progress, which will be described in more detail in the chapter 6, has greatly increased the expenditure for "industrial" inputs, such as fertilizers, insecticides, and fuel for machinery, as well as the purchase of feedstuffs (including processed maize and industrial products such as bran or whey), selected seeds, and so on. Thompson (1968, table 5) estimates that total expenditures increased thirty-fold in Britain from the 1810s to the 1850s.[81] In the United States, they grew by five times from 1870 to 1910 (at 4.1% yearly). The growth slowed a bit during interwar years (about 1.5% yearly), resumed in earnest in the 1950s and 1960s (about 3%), and has slowed down again, quite drastically, in the past thirty years (about 0.55%).[82]

This increase boosted the ratio of off-sector purchases to gross output (table 4.14). These data are not perfectly comparable, as they have been collected from different sources that may use different definitions of expenditures. However, the upward trend in the technically advanced countries is clear, and there is very little evidence of an inversion of the trend. Unfortunately, there are no recent data about LDCs: the data on the use of fertilizers (table 4.15) and on mechanization (table 4.11) imply that purchases must have risen, especially in Asia.

The most important item in the shopping list has always been, and probably still is, the purchase of fertilizers. As already said (section 2.2), Chinese and Japanese farmers had always resorted to purchased fertilizers, such as bean-cake, and the amount of these purchases seems to have risen in the eighteenth century.[83] In contrast, these expenditures remained low in Europe until well into the nineteenth century. In the 1830s–1840s, Europe started to import nitrates from Chile and guano (the excrement of certain birds) from Peru, and Thompson (1968) claims that purchased fertilizers triggered a "second agricultural revolution" in the United Kingdom.[84] Although growing rapidly, the quantity remained small and the real boom started only later, with the production of artificial fertilizers (section 6.2).

Table 4.14
Purchases as Share of Gross Output, by Country, 1870–2000

	1870–72	*1911–13*	*1920–22*	*1936–38*	*1950–52*	*1963*	*1985*	*1998–00*
"Advanced" Countries								
France	0.116	0.12	0.135	0.18	0.185	0.274	0.400	0.55
Greece					0.084	0.331	0.415	0.32
Germany	0.078	0.117	0.239	0.235	0.272	0.425	0.461	0.65
Italy	0.069	0.072	0.052	0.113	0.132	0.197	0.314	0.34
Ireland						0.331	0.415	0.51
Belgium	0.237	0.367	0.24	0.347	0.216	0.36	0.476	0.63
Portugal	0.06	0.079	0.083	0.09		0.524	0.497	0.56
UK	0.242	0.336	0.33	0.432		0.542	0.449	0.65
Denmark						0.523	0.47	0.64
Netherlands	0.238	0.436	0.303	0.299	0.328	0.482	0.463	0.54
Spain	0.006	0.057	0.064	0.119	0.073	0.353	0.457	0.37
Sweden	0.083	0.179	0.17	0.216	0.28	0.533	0.481	0.74
Australia	0.17	0.18	0.182	0.177		0.368	0.332	0.48
USA	0.08	0.157	0.198	0.186		0.501	0.446	0.52
Canada	0.108	0.134	0.177	0.171		0.364	0.528	0.64
New Zealand			0.32	0.32		0.42a		0.43
Japan	0.156	0.134	0.148	0.173		0.284	0.45	0.45
South Korea						0.113	0.26	0.34
Taiwan						0.247	0.413	
LDC Countries								
South America*						0.08	0.141	
Asia°						0.214	0.199	

Notes: *Argentina, Brazil, Mexico. Indonesia, India, Philippines, Pakistan, and Thailand.

Sources: **Columns 1870–72 to 1936–38: Australia:** Butlin 1962, tables 40, 41, 53, 54. **Italy:** ISTAT 1957, tables 2, 3, and 4. **Belgium:** Blomme 1992, tables 22, 42. **Netherlands:** Knibbe 1993. **Japan:** Okhawa and Shinohara 1979, table A16. **Canada:** Urquhart 1993, tables 1.1, and 1.9. **France:** Toutain 1997. **Germany:** Hoffman 1965, II, tables 58, 64; **Spain:** Prados, de la Escosura personal communication. **Sweden:** Schon 1995, table J1, J11. **New Zealand:** Hawke 1985, 240. **UK:** (1870–72, 1911–13) Afton and Turner 2000, table 38.8, and 1920–22, (1936–39) Ojala 1952, table XX (5 year average). **USA**: gross output: Federico 2004a; expenditures 1870–72: Towne and Rasmussen 1960; expenditures (1911–13, 1920–22, and 1936–38): Historical Statistics 1975, series K276–K278. **Column 1950–52:** Johnson 1973, table 4.3. **Columns 1963–85:** van der Meer and Yamada 1992, table B.1. **Column 1998–2000:** OECD Statistical Database.

Bairoch (1999, table 5.5) estimates that the consumption in technically advanced countries grew from 0.25 million tons (in fertilizing content) in 1880 to 1.38 in 1913, and to 7.54 in 1938.[85] The same author (Bairoch 1999, 89) surmises that in 1938 the LDC's all together consumed 1.47 million tons (half of which was consumed in Japan), but the estimate seems somewhat too optimistic.[86]

Since World War II, the world consumption of fertilizers has boomed (graph 1), increasing at 8.2 percent per annum in the 1950s, and 4.4 percent in the 1960s and 1970s. World consumption declined substantially in the early 1990s. It recovered in the second half of the decade, but the current consumption is still lower

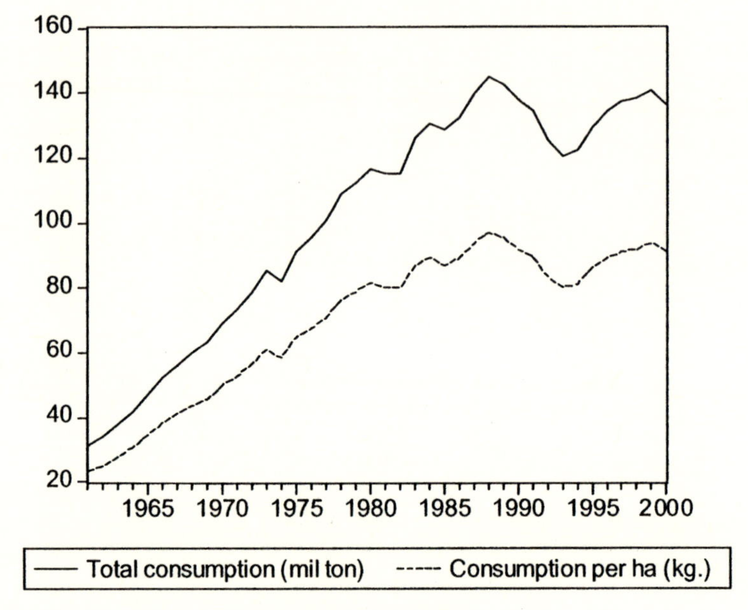

Graph 4.1 Consumption of fertilizers, 1960–2000

than the 1988 peak. As table 4.15 shows, the worldwide change is the outcome of widely different trends by area. Before 1980, the consumption of fertilizers increased all over the world, albeit more slowly in Europe or North America than in Asia, Africa, or South America. Since then, the consumption of fertilizers boomed in Asia and South America, while it stagnated in North America and fell in Western Europe. This reduction may reflect the growing awareness of the negative consequences of the intensive use of fertilizers and the ensuing search for a more environmentally friendly use (Smil 2001, 209–10).[87] However, this fall would not have prevented the growth of worldwide consumption had it not been for the veritable collapse in consumption of fertilizers in the "transition" economies (section 9.6). If their consumption had remained constant at their 1990 level, the world total would have been 14 percent higher in 2000 than it had been ten years before.

The consumption of other chemical products—notably herbicides—is much less documented, as the FAO does not provide data before 1990. It is certain that purchases boomed only after World War II, although farmers had been using some chemicals since the nineteenth century (section 6.2). The rate of growth in consumption was extremely high in the 1950s and 1970s—around 8 to 9 percent in the United States (USDA 2004).[88] It then slowed down, possibly due to health concerns, which are quite serious (Naylor 1994, 68–69; Pingali et al. 1997, 110–14, 258–60; Pretty 1995, 65–69). According to an estimate, the worldwide

TABLE 4.15
Use of Fertilizers, 1950–2000 (Thousand of Tons)

	1950/51	*1961*	*1970*	*1980*	*1990*	*2000*
Africa	256	716	1,615	3,242	3,629	3,881
Europe	6,990	13,955	24,883	31,196	26,414	19,472
North and Central America	4,798	8,469	17,614	25,636	23,605	22,868
South America	179	558	1,633	5,291	4,999	10,284
Oceania	525	964	1,428	1,651	1,545	3,166
Asia	1,044	3,809	11,823	30,948	55,993	71,898
USSR		2,710	10,312	18,756	21,634	3,866
World	13,792*	31,182	69,308	116,720	137,819	136,435

Note: * Omitting USSR.
Sources: **1950/51**: FAO 1952. **1961–2000**: FAO Statistical Database.

consumption of pesticides (mainly herbicides) grew at 3.2 percent per annum from 1982 to 1998—roughly as much as fertilizers during the same time (Wood et al. 2000, 36).[89]

The big increase in the agricultural consumption of chemicals was undoubtedly fostered by falling prices, especially in the nineteenth and early twentieth centuries. The price of fertilizers halved in the United States, from 1880 to 1910, and fell by a further fifth until the 1930s (Hayami and Ruttan 1985, table c.2, series U23, U13). The real price of fertilizers also fell in other countries, such as Japan, the United Kingdom, the Netherlands, Spain, Italy, and Indonesia.[90] The 1930s probably marked a historical trough in the United States. From 1948 to 1994, the price of fertilizers relative to output increased 2.6 times and those of pesticides increased 3.2 times (USDA 1994). However, the American prices are not necessarily representative of farm-gate prices all over the world. In fact, many LDCs started to produce fertilizers after World War II, and, in spite of the inefficiency of many new factories, savings on transportation costs did cut domestic prices. Furthermore, governments heavily subsidized the consumption of fertilizers, at least until the 1990s (section 9.5). On the whole, therefore, it seems likely that, just as for machinery, the quantity data overvalued the growth of total expenditures, although the bias seems smaller.

It is time to take stock. The data on capital are scarce relative to those on land and labor, and somewhat imprecise, but there is no doubt that, on the whole, modern agriculture is more capital-intensive than the traditional one. All over the world the capital stock has grown in the past two hundred years, along quite distinctive area patterns:

a. In 1800 the total capital stock in the Western settlement countries was very small and geographically concentrated in few areas, such as the East Coast of the United States. It grew very fast during settlement, until the early twentieth century. The growth slowed down in interwar years, to resume, as a consequence of mechanization and technical progress, in the thirty years after World War II. Nowadays, agriculture is very capital-intensive.

b. The "advanced" long-settled countries of Western Europe had traditionally a very substantial capital stock. Thus it grew until World War II decidedly slower than in the western settlement countries. The postwar spurt was similar, and nowadays their agriculture is only slightly less capital-intensive than the American one.

c. In less "advanced" long-settled countries such as China, the capital stock was quite large around 1800, possibly even greater than in Europe because of the irrigation works. It grew very slowly or did not grow at all until quite recently and then boomed, with the intensive use of fertilizers and mechanization.

d. One can hypothesize that in less "advanced" but not entirely settled countries (i.e., Africa), the capital stock grew as much as population, probably until World War II. Since 1950, the per capita stock of capital has increased, but much less than in Asia.

The series of the stock of capital should be a good proxy for the flow of services, if correctly measured, taking into account depreciation and changes in quality. The measurement is seldom perfect, although in the previous discussion it has, to some extent, been possible to take the quality change into account by means of a mixture of data and reasonable hypothesis. For some categories of capital, such as tools and machinery, the flow of services also depends on the number of hours of use. The data on stock might overvalue (undervalue) the flow of services if the capital is used for a shorter (longer) time during the year. This issue can be discussed when dealing with the change in the numbers of work hours.

4.4 Labor

The data on the agricultural work force are quite abundant, as population censuses routinely reported the data on employment. It is possible to collect data for a fairly representative group of countries, which includes almost all the "advanced" countries during the initial stage of their modern economic growth (table 4.16).[91] The table shows that the total agricultural work force did not start to decline until a highly advanced stage of modern economic growth. In fact, before 1910, it diminished in absolute terms only in the United Kingdom, Belgium, and France, while remaining roughly constant or even increasing in quickly industrializing countries such as Germany and the United States. The interwar years may have marked a change in regime, especially in Western settlement countries. The rise in agricultural population slowed down markedly (as in Argentina) or turned into a decline (as in the United States).[95] Many of these changes, however, remain within the limits of the (wide) measurement errors. One has to conclude that before 1940

TABLE 4.16
Agricultural Work Force, by Country, 1800–2000 (Millions)

	1800	*1850*	*1880*	*1910*	*1938*	*2000*
Males and Females						
Austria-Hungary		13.8	14.9	15.5	NA	NA
Belgium		1.0	1.1	0.8	0.65	0.08
Bulgaria				0.9	1.3	
Denmark	0.7		0.95	0.5	0.56	0.11
Finland		0.40	0.45	0.6	0.55	0.14
France	8.4	9.1	8.6	7.7	6.0	0.90
Germany		8.3	9.6	10.5	9.0	1.01
Italy			9.4	10.5	10.6	1.35
Netherlands	0.37	0.49	0.62	0.68	0.66	0.25
Norway			0.22	0.37	0.41	0.11
Portugal			1.1	1.1	1.3	0.65
Romania				1.6	1.9	
Spain	3.5	4.5	4.9	4.4	4.1	1.29
Sweden		0.7	1.1	1.0	1.0	0.15
UK	1.7	2.0	1.6	1.5	1.4	0.54
Canada			0.68	0.96	1.1	0.39
USA	1.3	4.9	8.9	11.8	9.6	3.0
New Zealand		0.004	0.05	0.11	0.15	0.17
Australia				0.48	0.5	0.45
Argentina				1.1	1.8	1.46
Chile	0.12	0.24	0.4	0.5	0.65	1.00
Brazil			3.3	6.5		
Indonesia			7.4	13.6	19.5	49.6
India	60	67	74	92	91	263.7
Japan			14.6	14.0	13.7	2.77
Egypt				2.4	4.3	8.60
Males Only						
Europe	31.2	36.2	39.2	41.4		
Russia	12.8	17.6	25.7	37.0		
Western Settlement	1.2	5.1	8.9	11.0		
Developed Countries	45.3	59.0	74.0	89.5		

Sources: **2000** FAO Statistical Database. **Males only** (last 4 rows): Bairoch 1999, table 2.1. **Others: Argentina:** (1910–14, 1935–39) Diaz Alejandro 1970, table 5. **Austria-Hungary:** (1869, 1890, 1910) personal communication by M. S. Schultze. **France:** (1806, 1856, 1881, 1911, 1936) Marchand and Thelot 1991, Table 3h. **India:** (1800 and 1850) estimated from Roy 2000, table 9.1, assuming a constant ratio to total population, (1881, 1910, and 1938) Roy 2000, table 9.5.[92] **Japan:** (1880, 1910, and 1935) Hayami and Yamada 1991, table A.5; **Canada:** (1881, 1911 and 1931) Urquhart 1993, table 1.5. **Germany:** (1852, 1880, 1910, and 1938) Hoffmann 1965, ii, table 20.[93] **Finland:** (1880, 1910, and 1938) Hjerrpe 1989, table 11 A. **Italy:** (1881, 1911, and 1936) Vitali 1968, table V.2. **Netherlands:** (1807, 1849, 1889, and 1909) Van Zanden 2000; and (1930) Mitchell 1998c, table B1. **Spain:** males (all years) personal communications by L. Prados de la Escosura, females (1877, 1910, and 1940) Nicolau 1989, cuadro 2.15. **USA:** (1800 and 1850) Weiss 1992, table 1.1. (1880, 1910,

very few "core" farmers were lured away by the attraction of industry or urban life, although many of their sons and daughters were (or were forced to). If the agricultural work force did not decline in "advanced" countries during their industrialization, a fortiori, one could not expect a decline when modern economic growth was only beginning (as in northwestern Europe and the United States in the first half of the nineteenth century) or had not yet started at all (as in the LDCs). For instance, in Imperial Russia, the rural population, including nonagricultural workers and nonworkers, rose almost as much as the total population, from 41 million in 1811 (93% of the population) to 66.2 million in 1866, and to 136 million in 1914 (85%).[96] After the war, and the loss of huge amounts of territory, the rural population remained constant at some 115 million, according to Clarke and Matko (1984, table 1), while, according to official sources, the total agricultural work force declined by 20 percent from 1926 to 1939.[97] It is reasonable to assume that, in LDCs, the share of the agricultural work force out of the total active population, and the proportion of the latter on the total population, remained fairly constant. From these assumptions, the agricultural work force in the LDCs (in Asia, Africa, and South America) increased from about 300 million in 1820, to 350 in 1870, to 450 in 1913, and to 680 million in 1950.[98] The estimate is extremely crude, but the fact that the worldwide agricultural work force rose before World War II seems certain.

The FAO Database provides data on the agricultural work force beginning in 1950, which can be extrapolated to the 1930s by assuming that the ratio of agricultural workers to the rural population was constant at the 1950 level (table 4.17). The past half-century featured a spectacular divergence between the "advanced" countries and the rest of the world. The number of agricultural workers plummeted in western Europe and Japan (falling by 80%), and in the United States and Canada (by 60%), while it remained constant in Australia and New Zealand, where acreage increased by three-quarters (table 4.5), and where the crop mix shifted toward more labor-intensive products, such as wine. In Eastern Europe and the USSR, the decline started later, but it was massive as well. In contrast, the agricultural work force kept growing in the LDCs, especially the Asian ones: in China alone, it increased by 230 million—or 20 percent more that the total work force in Europe, the USSR, North America, and Oceania in 1950. So far,

1940) Lebergott 1984, table 7.3. **UK:** (1801) Deane and Cole 1969, table 31; (1851, 1911, 1931) Mitchell 1988, Labor force table 2 (1921 boundaries).[94] **Indonesia:** (1870, 1910, and 1938) van der Eng, personal communication. **New Zealand:** (1858, 1878, 1911, and 1936) Bloomfield 1984, tables IV.2 and IV.3. **Bulgaria:** (1910 and 1930) and **Romania:** (old kingdom, 1913 and 1930) Lampe and Jackson 1982, 104. **Chile:** (1810, 1850, 1880, 1910 and 1938) Braun 2000, table 7.2. **Australia:** (1911 and 1941), **Belgium:** (1846,1880, 1910, and 1930), **Brazil:** (1872 and 1920), **Egypt:** (1907 and 1937), **Denmark:** (1805, 1880, 1911, and 1940), **Norway:** (1875, 1910 and 1930); **Portugal:** (1890, 1910, and 1940), **Sweden:** (1860, 1880, 1910, and 1930) Mitchell 1998a, b, and c, table B1; and Bairoch et al. 1968, table A2 (includes forestry and fishing).

TABLE 4.17
Agricultural Work Force, by Continent, 1930–2000 (Millions)

	1930s	*1950*	*1960*	*1970*	*1980*	*1990*	*2000*
Africa	61.6	86.8	100.2	117.9	137.4	165.6	197.1
Europe	63.1	66.2	54.3	40.8	31.3	24.2	17.6
North and Central America	23.2	21.0	18.6	17.4	20.1	20.6	20.7
South America	18.5	21.1	23.8	26.7	29.3	28,2	26.9
Oceania	1.9	1.7	1.8	1.9	2.1	2.5	2.8
Asia	436.6	571.6	605.8	693.7	817.1	953.4	1,031.8
USSR	44.1	41.0	38.6	29.6	29.8	26.7	21.7
World	648.9	809.5	843.0	928.7	1,067.1	1,221.2	1318.6

Sources: **1930s**: FAO Yearbook. **1950–2000** FAO Statistical Database.

the great boom in the urban population in the Third World has been fueled by population growth, as in nineteenth-century Europe. It is likely, however, that sooner than later the agricultural work force will begin to decline in these countries as well.

The number of workers is a handy proxy for the input of labor—but is it a good one as well? Unfortunately, the reply would not be positive, even in the (implausible) hypothesis that the data were totally accurate. In fact, the number of registered workers can differ from the flow of service that they provide for at least four major reasons:

1. Agriculture in the nineteenth century resorted to nonagricultural workers during seasonal peaks of labor.[99] For instance, in Canada, harvesters were recruited in eastern cities, while in Argentina, in the late nineteenth to early twentieth centuries, the *brazeros* came from Italy (the so-called *golondrinas* or swallows).[100] Marchand and Thelot (1991, 153) venture to put forward a figure for nineteenth-century France: people not permanently employed in agriculture supplied between 7.5 and 11 percent of total agricultural labor in 1862–66 and less than 2 percent in the 1890s.[101] This contribution fell because mechanization aimed first at curbing the seasonal peak of demand (section 6.3). The effects on agricultural employment depend on the definition of "worker" used in national censuses. Most of them classified workers according to their prevalent occupation, and thus omitted part-time workers from the agricultural work force.[102] Thus, the data of table 4.16 would undervalue the labor in the nineteenth century. The gap is likely to shrink with technical progress.

2. The quality of agricultural labor differs according to the skills and physical strength of workers. In theory, if the labor market works well, any such difference should show up in wage differentials, and, thus, it would be possible to estimate an "adjusted" series of the work force by weighting the total number of workers

with their wages. Unfortunately, this method not only entails strong and questionable theoretical assumptions about the functioning of the labor market, but it also needs detailed data on both wages and the composition of the work force, which are hardly available in the historical perspective. Some censuses did record data on age of manpower by sector of activity or (very seldom) on skills, but unfortunately the available collections do not report these data.

The only available data refer to the distribution by sex (table 4.18), which is undoubtedly relevant, as average female earnings were substantially lower than the comparable ones for males.[103] Two features stand out. Females accounted for a small proportion of the "official" work force in most countries, and the gender ratio did not change in most countries.[104] This reflected a long-lasting division of labor by gender in traditional Western agriculture: the core agricultural tasks, such as harvesting or plowing needed mainly strength and thus were typically considered "male" jobs, while women traditionally tended cattle and poultry. The division of labor was traditionally different in Sub-Saharan Africa: women worked in the fields, while men cleared them and tended the cattle–and, indeed, women have accounted for more than half the agricultural work force after 1950 (table 4.19). From 1950, the proportion of female workers rose slowly but steadily throughout the world, with the conspicuous exception of the Soviet Union, where the initial share of women was exceptionally high after the huge migration of male workers to cities during the 1930s and the large war-time losses.[105]

Even if historical data on the composition by age and skills are extremely scarce, one can reasonably argue that the changes were modest until the twentieth century and the start of the great migration out of agriculture. The age composition of the agricultural work force did not differ from that of the overall work force, and all workers were endowed with roughly the same skills, gained from a hard on-the-job training: "A lad was never shown how to do a thing; to show him how was to spoil him. The only way to learn either plowing, thatching, stacking or any other skilled work was to watch how other people did it and then earn your skill by trial and error" (Brassley 2000a, 622). The human capital was thus substantial if measured with the number of years of training; but this training was highly specific and thus not very useful for acquiring new skills and managing technical progress.

Migrations changed the composition of manpower in all "advanced" countries because the young and the educated were the first to leave, causing the average age of farmers to drift upward—up to fifty-five years in the United States according to the 2002 census.[106] The effect of emigration on (formal) skills was balanced by the spread of literacy and of institutions for agricultural training (section 6.5; Huffman 2001, 348–52). In the United States, the proportion of farm operators with a college degree rose, from 5 percent in 1949 to 37 percent in 1991, and the average (wage-weighted) skill of all workers increased by a fifth (Craig and Pardey 2001, 47).[107] However, the agricultural work force still remains comparatively poorly educated even in OECD countries.[108]

TABLE 4.18
Share of Females in Agricultural Work Force, by Country, 1850–1938

	ca.1850	*ca.1880*	*ca.1910*	*ca.1938*		*ca.1938*
Canada		0.015	0.016	0.016		
Mexico			0.009			
Argentina		0.170	0.079	0.054	Peru	0.696
Chile			0.102	0.064	Venezuela	0.064
Egypt			0.043	0.163		
South Africa			0.506	0.284		
India			0.322	0.280	Philippines	0.285
Thailand				0.499	Turkey	0.332
Japan		0.465	0.458	0.493	Taiwan	0.335
Belgium		0.365	0.271	0.218		
Bulgaria			0.490	0.509		
Denmark		0.501	0.214	0.226	Czechoslovakia	0.417
Finland		0.274	0.367	0.457		
France	0.30	0.30	0.37			
Hungary		0.271	0.227	0.232	Greece	0.317
Ireland			0.076	0.165		
Netherlands	0.341	0.268	0.252	0.167		
Norway		0.227	0.142	0.099		
Portugal		0.312	0.229	0.151		
Romania			0.499	0.507	Poland	0.443
Spain		0.185	0.085	0.055	USSR	0.500
Sweden		0.310	0.254	0.232	Yugoslavia	0.366
Switzerland			0.212	0.072		
Germany		0.308	0.465	0.723		
UK	0.114	0.071	0.040	0.052		
Italy	0.380	0.374	0.394	0.386		
USA			0.101	0.056		
Australia			0.033	0.050		
New Zealand			0.069	0.040		

Sources: **France:** Caron 1979, table 1.10. **Netherlands:** Van Zanden 2000. **Italy:** ca. 1850: estimate of the author; other years: Vitali 1968, tables 21, and 46. **Japan:** Hayami and Yamada 1991, table A.5; **USA:** Historical Statistics, 1975, series D196, D230. **UK:** Mitchell 1988, table 2. **All others:** Mitchell 1998a, b, and c, table B1; Bairoch et al. 1968, table A2.

How did these changes affect the flow of labor input? The increase in formal education did, in all likelihood, improve the quality of the manpower, but what about the increase in the proportion of women and elderly workers? In traditional agriculture, which relied on physical strength, these changes were bound to reduce the overall productivity. However, modern mechanized agriculture needs much

TABLE 4.19
Share of Females in Agricultural Work Force, by Continent 1950–2000

	1950	*1960*	*1970*	*1980*	*1990*	*2000*
Africa	42.8	43.4	44.5	46.0	46.5	47.5
Europe	37.1	38.6	41.0	42.9	41.1	39.6
North and Central America	11.1	12.1	11.1	17.3	16.4	17.2
South America	11.7	13.3	14.5	17.6	18.1	21.8
Oceania	27.1	30.0	34.7	37.6	39.8	43.2
Asia	39.7	40.3	41.8	43.1	43.5	44.3
USSR	53.3	54.3	50.2	45.6	38.3	35.0
World	39.0	39.8	41.0	42.3	42.7	43.6

Sources: FAO Yearbook; FAO Statistical Database.

less brute strength, and thus the efficiency losses have, in all likelihood, been more than compensated for by the increase in human capital. Therefore, the overall effect of the changes in the composition of manpower on the flow of labor is likely to have been modest but positive, especially in the "advanced" countries in the past decades.

3. The labor input depends also on the number of hours per "core" worker. As previously said (section 2.2), the number of hours at each moment in time depends on the crop mix and technology. Did these changes follow a consistent pattern in the long run? In a famous article, Jan de Vries (1994) hypothesized that the number of hours of work in Europe had increased in the seventeenth and eighteenth centuries (a process he termed the "industrious revolution").[109] Historians have tried to prove de Vries's conjecture with imaginative use of the available evidence, but with conflicting results. Using wage data, Clark and van der Werf (1998) argue that neither the length of the workday nor the number of days changed noticeably in the United Kingdom. In contrast, Voth (2001), using judicial sources, finds an increase from 2,311 hours per year per (male) agricultural worker in 1750, to 3,431 in 1800, followed by a decrease to 2,762 hours per year in 1830—however implausible such a change may seem.

In the late nineteenth and twentieth centuries, the average number of hours per worker in advanced countries decreased quite substantially (Maddison 1995, table J-4 Huberman 2004). Table 4.20 shows some estimates for agriculture only. These estimates are obtained by an array of methods, from sophisticated computations of demand for labor by crop, which take changes in the crop mix and in the productivity by crop into account (and, at least in principle, include changes in the number of hours by nonagricultural workers), to simple assumptions on the number of hours by category (owner-occupiers, workers, etc.), which simply reflect the change in the composition of the work force. In spite of these differences, they are rather consistent. Before 1910, the number of hours per worker either

TABLE 4.20
Rates of Change in the Number of Hours and of Workers, by Country, 1850–1950

Country	*Years*	*Hours*	*Workers*	*Hours/Worker*
France	1862–66 to 1892–96	−2.24	−4.05	1.81
	1896 to 1911			0.20
	1911–38			−0.50
Ireland	1851–1911			1.00
Belgium	1846–1910	−0.38	−0.19	0.19
	1910–50	−1.58	−0.66	0.92
Netherlands	1807–1909	0.61	0.67	0.06
	1913–38	0.29	0.21	0.08
Germany	1913–29			−0.63
UK	1924–37	−1.15	−1.43	0.28
USA	1840–1900			0.47
	1910–12 to 1937–39	−0.56	−0.31	0.25
Japan	1880–1915	0.92	−0.13	1.05
	1915–40	−0.81	−0.21	−0.60

Sources: **France:** Marchand and Thelot 1991, 128–30, 153, and Table 3h. **Belgium:** hours: Goosens 1992, table 53, and Blomme 1992, table 38; workers: Mitchell 1998c, table B1. **Ireland:** Turner 1996, 184. **UK:** hours: Matthews et al. 1982, table D.2; and workers: Feinstein 1972, table 59. **Germany:** Wunderlich quoted by Bairoch 1999, 47. **Netherlands:** (1807–1909) Van Zanden 2000; and (1913–38) van der Meer and Yamada 1990, table C.1. **Japan:** Hayami and Yamada 1991, table A.5. **USA:** (1840–1900) Craig and Weiss 2000; and (1910–40) Historical Statistics 1975, series K174 and K183.

increased or remained constant, whereas afterward it started to decrease (with the somewhat odd exception of Belgium). This trend continued after World War II, at least, according to Bairoch (1999, 48), until 1980.[110] Agriculture had to adjust to the economy-wide trend toward a reduction of the hours of work even for its core workers—although full-time farmers still worked more than industrial workers.[111] This effect was reinforced by the widespread diffusion of part-time work in all advanced countries. The proportion of part-time farmers in the United States rose from 6 percent in 1929 to about 40 percent in 2002, and the average farming household drew only a sixth of its income from agriculture (Gardner 2002, 71; USDA 2002).[112] In the European Union and Japan, most rural households work part-time.[113] Thus, the number of workers (table 4.16) may undervalue labor input before 1940, and is likely to overvalue it afterward.

4. Last but not least, even the best measure of the number of working hours, properly weighted with some measure of skill, might not be good enough if the intensity of work differs. A worker might be physically unable to work very intensively, or simply might decide not to. The perverse effect of malnutrition on work performance is widely recognized in the literature about present-day LDCs, and it must have also been widespread in the pre-industrial age (Behrman and Deolikar

1988, 683–90; Strauss and Duncan 1995, 1905–17). Indeed, according to Bekaert's (1991) computations, the caloric intake was insufficient even in mid-nineteenth-century Belgium, which, by then, was not exactly a poor country. In addition, well-fed workers might also want to work less intensively than they would if they were working for someone else (an incentive problem, which will be discussed at length in sections 7.3 and 7.6) or because they value leisure more than the monetary income that they would gain from working harder. In a well-known paper, Clark (1987) has argued that, at the beginning of the nineteenth century, labor productivity in similar tasks (harvesting and threshing wheat) was lower in Eastern Europe than in the western part of the continent or in the United States simply because the work ethic of Eastern workers was different. Komlos (1988) replied that east European workers were less productive simply because they had to use inferior tools. The discussion highlights an important point and shows the ingenuity of economic historians in finding sources to deal with apparently impossible issues. However, it does not settle the issue in the specific case and, a fortiori, cannot be generalized.

It is now time to return to the initial question: to what extent are the census figures (tables 4.16 and 4.17) representative of the total labor flow? And if not, was the gap positive (labor input grew more than the number of workers) or negative? So far, we have argued that, ceteris paribus, (1) the census data undervalued the actual number of workers, but the gap shrank because fewer and fewer nonagricultural workers were employed part-time in agriculture; (2) the changes in the composition of the work force may have increased the quality-adjusted input of labor, especially in the "advanced" countries and in the last half-century; (3) the number of hours increased (positive gap) until the interwar period and fell (negative gap) after 1950 because of the diffusion of part-time work; (4) the intensity of work may have increased somewhat, especially at very low levels of development, but the issue is extremely uncertain. In brief, some of the effects have been positive and some negative, and, as argued earlier, many of them were rather modest, especially before World War II. Afterward, the changes mainly affected the agriculture of the "advanced" countries, which by then accounted for a tiny proportion of the worldwide agricultural work force. Thus, one could conclude that the series of the number of workers are not very biased, either upward or downward: the flow of labor does not differ hugely from the stock. Needless to say, this conclusion is tentative, as it relies on incomplete evidence. Unfortunately, historical evidence on some issues is very difficult to find.

4.5 Conclusion: Factor Endowment and Factor Prices in the Long Run

The data discussed so far, despite all their limitations, do suggest a first conclusion. The quantity of agricultural inputs did grow quite fast, especially before

TABLE 4.21
Land per Worker, by Country, 1800–1938 (Hectares)

	1800	*1850*	*1880*	*1910*	*1938*
Austria-Hungary		1.42	1.68	1.72	n.a.
Belgium			1.45	1.88	1.69
Denmark			2.53	5.80	4.82
France	3.99	3.77	3.80	3.84	3.80
Germany		2.94	2.73	2.50	2.22
Italy			1.53	1.41	1.44
Netherlands	4.86	3.88	3.23	3.09	1.67
Portugal				2.91	2.46
Sweden			3.09	3.40	3.80
Spain	5.00	3.56	3.22	4.34	4.10
UK and Ireland		4.00	5.00	4.47	5.21
Canada			8.97	16.25	21.45
USA			5.13	11.90	14.73
New Zealand		2.50	8.00	7.27	5.33
Indonesia			0.99	0.88	0.78
Japan			0.32	0.40	0.49
Egypt				0.96	0.51

Source: tables 4.1 and 4.16.

1913, and not only in the countries of Western settlement or in the Russian steppes, as everyone would expect, but also in supposedly "overcrowded" areas such as Java and China. Clearly, this conclusion holds true only as a very broad generalization, and one can easily find counterexamples—such as the quasi-stagnation of the stock of land and work force in Western Europe or of capital in China. The growth slowed down remarkably, almost down to nil in some cases, in the interwar period. After World War II, the pattern diverged between the "advanced" countries and the rest of the world. In the "advanced" countries, the stock of capital went on growing (until the 1980s), while that of land remained constant and the work force fell. In the rest of the world, notably in Asia, all factors continued to grow, albeit at different rates. In the long run, the growth of inputs (or extensive growth) played a key role for the overall increase in production.

By definition, different rates of growth in inputs caused a change in factor endowment, which is usually measured with the land-labor ratio (table 4.21). Land per worker increased in the Western settlement countries, declined in some (already overcrowded) countries, such as Egypt and Indonesia, and remained roughly constant in most others (especially European ones).

After 1950, it is possible to compute the same ratio by continent (table 4.22). In the 1950s and 1960s, the land endowment per worker increased slightly in all continents (excluding Africa). In the two next decades, it continued growing in

TABLE 4.22
Land per Worker, by Continent, 1950–2000 (Hectares)

	Arable			*Total*		
	1950	*1970*	*2000*	*1950*	*1970*	*2000*
Africa	2.0	1.7	1.1	7.9	10.6	5.8
Europe	2.3	2.7	7.1	3.7	4.3	11.4
North and Central America	11.4	14.5	13.0	31.5	33.7	30.7
South America	3.9	3.5	4.3	20.2	22.5	22.8
Oceania	10.3	26.2	20.4	222.6	282.1	181.7
Asia	0.6	0.7	0.5	1.0	1.8	1.3
USSR	5.5	6.0	9.7	8.5	14.2	25.6
World	1.5	1.6	1.1	4.1	5.5	3.8

Source: FAO Statistical Database.

Europe, South America, and the USSR, and it fell in Asia and Africa. Needless to say, all these data are very crude, as they compound all the errors in measurement that have been discussed so far.

The change in factor endowment is important, to the extent that it affects factor prices and thus the choice of techniques (chapter 6). In a purely agricultural closed economy, with no technical progress, an exogenous increase (decrease) in the land-labor ratio would immediately cause the wage-rent ratio to rise (fall), and make it more profitable to use labor-saving techniques for each product and to switch to land-intensive products. This simple causation no longer holds true in an open economy with an urban sector and technical progress both in agriculture and in the rest of the economy. In this case, wages and interest would be determined in the domestic or international market for labor and capital. In contrast, the price of land, as a specific factor, should equal the discounted value of the expected earnings (i.e., rents). In the long run, the latter depend on (1) the supply of land; (2) land productivity (i.e., the techniques and the crop mix); (3) terms of trade of agricultural products (any improvement increases returns, given physical productivity); and (4) any agricultural policy that affects the farmers' income (e.g., a subsidy). However, in the short and medium run, expected (ex-ante) returns can diverge from actual (ex-post) ones. Lindert (1988) shows that in the United States, land prices overshot implicit rents—that is, they rose more when the latter were rising (as in the 1920s) and vice-versa, as in the 1940s).[114] Speculative bubbles have been quite frequent during the process of settlement, when information on the supply of land was highly imprecise, as in Argentina during the 1870s and 1880s, or in Canada during the 1900s (Adelman 1994, 40, 80; Solberg 1980, 56). In other words, in even moderately complex economies, it is impossible to predict a priori changes in factor prices from those in factor endowment.

In this case, one can turn to actual data. There is no evidence for the first half of the nineteenth century, while, for the period 1870–1938, it is possible to rely

TABLE 4.23
Rate of Change in Relative Factor Prices, 1870–1938

	1870–1913	*1913–38*	*1870–1938*
Full Sample			
Wages/Rent			
Western Settlement	−3.78	1.76	−0.82
Europe	1.60	1.37	1.53
LDCs	−3.27	−0.06	−0.77
"World"	−1.27	1.10	0.10
Rent/Capital World	−0.12	3.37	−0.05
Wages/Capital World	1.19	4.92	1.66
Only Significant Rates			
Wages/Rent	−1.43	0.87	−0.09
Rent/Capital	0.45	1.36	0.05
Wages/Capital	1.26	4.92	1.66

Notes: **Wages/Rent: Western Settlement:** Argentina (1885–1938), Uruguay, Australia, Canada (1901–38), and the USA. **Europe:** Great Britain, Denmark (1870–1914), Sweden (1870–1930), Ireland, France (no data 1914–19), Germany (1870–1914), and Spain (1870–1933). **LDCs:** Egypt, India, Siam, Burma (1890–1923), Taiwan (1903–38), Korea (1910–38), and Japan (1885–1938). **Rent/Interest and Wages/ Interest:** Canada (1901–38), USA, Great Britain, Sweden (1870–1930), France (no data 1914–19), Germany (1870–1914), and Japan (discount rate 1885–1938).

Source: Statistical Appendix, table III.

on Williamson's database of wages and land price series, and to proxy, quite crudely, the cost of capital with the interest rates on state bonds (on commercial paper for the United States). Needless to say, the series are far from accurate in measuring the cost of inputs to farmers. Many land price series refer to a specific area, and thus may be poorly representative of prices in large countries. The urban wages and the rates of interests on state bonds may not proxy rural wages and the cost of capital to farmers (section 7.4). Last but not least, for all investment in land improvements and/or buildings, the opportunity cost of peasant labor—the real wage—may be as relevant as the cost of capital. In spite of all these caveats and shortcomings, the results are interesting (table 4.23).

Throughout the world and in all periods (with the exception of the war decade 1911–20), capital has become cheaper relative to labor. Capital became cheaper relative to land in interwar years, and possibly also before 1910: unfortunately, the number of countries is small and the trends are sufficiently divergent to make any conclusion difficult.[115] The data on wage-rent ratios are much more abundant, and show a clear differentiation by area/period. Before 1910, the ratio rose in western Europe and fell in the Western settlement countries and in the LDCs.[116] During the interwar years, it grew in most countries, with little difference among

Europe, the LDCs, and the countries of Western settlement. However, in ten out of fourteen cases, the country trends are not significantly different from zero. Thus, the wage-rent ratios converged among the Atlantic economy countries before the war, but not during interwar years. Williamson (2002) interprets these trends as a case of factor-price equalization driven by globalization (interrupted by the postwar backlash).[117] However, it is not possible to pursue the issue further here. Suffice it to say that, from the point of view of agriculture, capital has become less expensive relative to other factors, while trends in wage-rent ratios have differed.

One would surmise that these trends continued after World War II, but, unfortunately, data are extremely scarce. Mundlak et al. (1997) can find data on land prices in only four countries.[118] Land prices rose, especially in the 1970s—but the wage-rent ratio may have fallen, as wages rose even faster. However, as we will discuss more thoroughly in chapter 9, overall input price ratios became much less relevant after the war because investment choices were increasingly determined by specific agricultural policies.

Chapter Five

THE CAUSES OF GROWTH: THE INCREASE IN PRODUCTIVITY

5.1 Introduction

Chapter 3 has shown that world agricultural production increased quite considerably, especially after World War II. In contrast, the growth of inputs, although quite fast until 1914, slowed down remarkably afterward. In many countries, the quantity of labor fell. Thus, the overall productivity must have risen. This increase is often proxied with the production per unit of land (or per unit of seed) or per unit of labor.[1] These partial measures are quite popular among historians and economists because they are relatively easy to compute. Furthermore, output per worker has a straightforward economic interpretation: if the activity rate does not change, it moves in parallel with GDP per capita, which is the most common measure for welfare. The available evidence will be discussed briefly in section 5.2. However, partial productivity measures will not be the main focus of the chapter. In fact, they can be misleading, because the productivity of a factor (e.g., land) depends on the quantity of other inputs (capital and labor). Labor productivity can be relatively high in backward agricultures, if blessed by a large endowment of land (Boserup 1951, 28–34; Eicher and Baker-Doyle 1992, 91). At the beginning of the twentieth century, labor productivity was roughly double in land-abundant Thailand and Cambodia than in technically advanced but densely populated Japan (van der Eng 2004, table 3).[2] In the 1950s in Thailand, swidden rice cultivation was 15 percent more productive than settled cultivation using the same tools (Feony 1982, 36). Vice versa, land productivity was high in densely populated countries. O'Brien and Prados de la Escosura (1992, table 1), in a classical article, show that in 1911 land productivity in Italy was 65 percent higher than in the United Kingdom, and some recent (and controversial) work show that the gap in land productivity was even greater in the early nineteenth century with China.[3] These particular gaps are partially accounted for by structural differences—such as double cropping in China and the diffusion of tree crops in Italy—but land-scarce countries often achieve higher yields for the same product. According to the latest FAO data, the average yield for wheat in 1998–2000 was 8.4 tons per hectare in Ireland (the highest in the world), 3.7 in China, and only 2.9 tons in the United States. It would be difficult to argue that China's agriculture is more advanced than America's. Thus, agricultural productivity must be measured by Total (or Multi-Factor)

Productivity, which takes all inputs into account. Indeed, there are literally hundreds of estimates of TFP: section 5.3 collects them and discusses the main trends. As discussed briefly in the last section, this growth is the joint effect of technical progress (chapter 6) and increases in efficiency from a better allocation of resources.

5.2 The Productivity of Land and Labor

The data on yields, especially of cereals, are relatively abundant from the Middle Ages onward, and are often the only surviving evidence of agricultural production. They are also apparently easy to interpret: the higher the yield, the more technically advanced the agriculture. According to this principle, the performance of Western agriculture was decidedly good. Bairoch (1997, table 4.6; 1999, 99–115) surmises that wheat yields grew in Europe (excluding Russia) from 0.86 tons per hectare in 1800 to 1.26 in 1910 and to 1.48 in 1950, while in North America they fell from 0.96 tons per hectare in 1800 to less than 0.8 in the mid-1870s, and they rose thereafter (e.g., to 1.16 tons per hectare in 1950).[4] British wheat yields in the eighteenth and nineteenth centuries have attracted a lot of attention: in their recent book, Turner, Beckett, and Afton (2001, 129–40) argue that the yields stagnated during the eighteenth century, rose remarkably in the first half of the nineteenth century, and remained constant until World War I.[5] Yields also rose during the nineteenth century elsewhere in Europe. For instance, they doubled in Spain, a country not particularly reputed for its technical prowess, from 0.49 tons per hectare in 1818–20 to 9 in 1903–12 (Bringas Gutierrez 2000, table 1.1).[6] Trends in Asia were mixed, but the evidence rules out a massive increase. According to the Indian official statistics, yields of food-grain fell by 15 percent in India (notably in Bengal) from the 1890s to the 1940s. This decrease is controversial, but even Heston (1983), the most sanguine among historians, rules out an increase.[7] Yields also decreased in other Asian countries, such as Indonesia and Thailand, but not in others—notably Japan).[8] There are no official statistics for China, but informed opinions range from total stagnation (corresponding to a fall in per-capita availability) to a small increase.[9] World yields can be computed as a weighted average between the growth in technically advanced countries and the stagnation or slight decrease in the LDCs. Given the world distribution of the cultivation of the two main cereals, one would surmise that, until World War II, worldwide yields rose somewhat for wheat and remained stable for rice. This hypothesis contrasts markedly with postwar trends: from 1961–63 to 1998–2000, rice yields doubled, and wheat yields more than doubled (up to more than 2.7 tons per hectare; see FAO Statistical Database). Furthermore, yields in LDCs increased as much as or even more than in the "advanced" countries: Asia outperformed all other continents from this point of view.

There is no simple way of measuring the productivity of tree crops or cattle-breeding. The production of tree crops per unit of land depends on the distribution

TABLE 5.1
Production of Milk per Cow, by Country, 1810–1940 (1910 = 100)

	Austria	*Belgium*	*Denmark*	*UK*	*USA*	*New Z.*	*India*
1810		55.3					
1850		59.1			66.9		
1870	60.5		56.3	81.6	72.2		
1880	62.1	70.9	62.6	93.2	78.9		
1890	76.4	84.5	77.4	95.2	86.0		
1900	84.1	94.4	82.1	94.8	94.6		100.4
1910	100.0	100.0	100.0	100.0	100.0	100.0	100.0
1930		112.7	106.0	93.8	134.6	115.6	102.0
1940		118.2		88.1	126.7	140.2	107.5

Sources: **Austria:** (1869, 1880, 1890, 1900, and 1910) Sandgruber 1978, table 32. **Belgium:** (1812 and 1846) Goosens 1992, 128; and (1880, 1890, 1900, 1910, 1930, and 1938) Blomme 1992, tables 15 and 36. **Denmark:** (1871, 1881, 1888, 1898, 1909, and 1929) Jensen 1937. **UK:** (1868, 1879, 1890) Turner 2000, table 3.30; and (1900, 1910, 1930, and 1938) Mitchell 1988, Agriculture tables 6.7 and 9.[12] **USA:** (1850–1910) Bateman 1968, table 1; and (1930 and 1940) Historical Statistics, tables K595 and K597. **New Zealand:** Bloomfield 1984, table V.14. **India:** Sivasubramonian 2000, table 3.8, appendix table 3(h).

of plants. Vines can be planted in vineyards, with or without intercropping, but can also be planted on the edge of fields or can be allowed to grow on trees (the so-called *piantata*, widely used in northern Italy): in the latter cases, the very concept of yield is meaningless. The yield of a vineyard depends on the density of plants and on the methods of cultivation, especially of pruning. Indeed, the quality of the product (and thus its unit price) can be inversely correlated to production per plant—so that the total revenue per unit of land may not be proportional to physical productivity. The vineyards produce much less wine per unit of land in Champagne than in the south of France ones, but the higher price of the former easily compensates for the difference (Loubère 1978, 15–20).[10]

The production per animal is not a good proxy for productivity in cattle-raising. The production of meat (the average dressed weight at the slaughterhouse) depends, ceteris paribus, on the breed of the animal and its age when slaughtered. Both features are strongly affected by consumers' tastes. Western consumers, for example, have increasingly preferred younger and leaner animals, and thus the average weight may well have fallen.[11] The production of milk, a fairly homogeneous good, is a somewhat more meaningful figure. Bairoch (1997; 1999, table 6.7) reckons that the production of milk per cow in western Europe doubled from 950 liters in 1800, to 1800 in 1910, and then increased by a further 15 percent to 2,090 liters as of 1950. The available country series confirm that production of milk per lactating cow (i.e., those aged—approximately—three years and over) has been rising, with the exceptions of India and, unexpectedly, the United Kingdom during interwar years.

After World War II, the average world production per cow rose by a quarter from 1,760 liters in 1961 to 2,180 liters in 2000. The rate of growth (0.5%) is deceptively low. In fact, the productivity grew at 1.5 percent per annum in developed countries (from 2,450 to 4,300 liters) and at 1.3 percent in developing countries (from 630 to 1,020 liters). The overall rate of growth comes out as much lower because low-productivity LDCs accounted for a growing proportion of the world stock (table 4.13). However, even a perfectly measured production per animal would be a poor proxy for *land* productivity. In fact, the growth in their unit output may reflect an increase in the amount and quality of feed they received, and thus, *ceteris paribus*, in the quantity of the land used to produce it.[13] To sum up, the evidence on the growth in land productivity is much less abundant and compelling than one might assume looking at cereals only.

The evidence on labor productivity by crop is scarce because measuring the labor input for each crop exactly is very difficult. The available data confirm the trends in yields. The production per worker in the cultivation of cereals roughly doubled both in Japan (from 1874 to 1940) and in the United States (from 1840 to 1940), while it remained stable in Indonesia.[14] In contrast, productivity in American cattle-breeding remained constant before 1900 and rose by only a tenth in the first half of the twentieth century.[15]

The previous analysis highlights wide differences in the rates of productivity change, by product as well as by country. Thus, the aggregate change also depends on the composition of output, and productivity must be measured for the whole agricultural sector. Bairoch (1999) has recently put forward estimates of changes in labor productivity, measured in calories per worker, from 1800 onward.[16] He sharply contrasts the growth in Europe and countries of Western settlement with the decline in the LDCs. He admits that his figures are simple guesses and that the production of calories is not, in any way, a good proxy of output in the long run. In fact, ceteris paribus, the increase in the share of livestock products (section 3.4) reduces the efficiency in the production of calories per unit of land. Thus, calories bias downward growth in output and therefore in productivity.[17] The conventional measures of output do not suffer from this bias—and, indeed, they show very rapid growth in productivity. The available estimate of labor productivity for the United Kingdom during the first half of the nineteenth century ranges from 0.33 to 1.6 percent.[18] Hayami and Ruttan (1985), in their classic book, estimate productivity growth for five "advanced" countries (table 5.2). These data highlight three stylized facts: (1) both productivity of land and labor increased in the long run, but labor grew decidedly faster; (2) productivity growth accelerated over time, with some evidence of slowdown in the 1990s; (3) in 1880, land productivity was much higher in Europe than in the United States, and was growing faster, while the United States led in labor productivity, although the gap was smaller. As predictable, we know very little about the trends in countries outside the "core" group. Tentative estimates suggest some growth even in the allegedly "backward" countries, such as Russia, the Philippines and Indonesia.[19]

TABLE 5.2
Land and Labor Productivity, Five Advanced Countries, 1880–1993
(UK 1880 = 100)

	USA	*Japan*	*France*	*Denmark*	*UK*
a) Land					
1880	45.5	300	100	109.1	100
1910	45.5	427.3	109.1	181.8	100
1940	54.5	545.5	136.4	254.5	127.3
1980	109.1	1263.6	409.1	581.8	272.7
1993	166.3	n.a.	510.8	622.6	297.9
Rates					
1880–1910	0.00	1.18	0.29	1.70	0.00
1910–40	0.60	0.81	0.74	1.12	0.80
1940–80	1.74	2.10	2.75	2.07	1.90
1980–93	3.24	n.a.	1.71	0.52	0.68
b) Labor					
1880	82.8	14.6	47.1	66.9	100
1910	104.5	24.8	59.2	107.6	108.3
1940	161.8	45.9	80.3	146.5	157.3
1980	1833.1	222.9	691.1	964.3	742.7
1993	3315.4	n.a.	1370.0	1524.7	995.7
Rates					
1880–1910	0.78	1.77	0.76	1.58	0.27
1910–40	1.46	2.05	1.02	1.03	1.24
1940–80	6.07	3.95	5.38	4.71	3.88
1980–93	4.56		5.26	4.52	2.26

Sources: Hayami and Ruttan 1985, tables B.1–B.5, extrapolated to 1993 with the data from Ball et al. 2001, tables A.2, A6, A8.

Others, just as tentatively, rule out growth in the Balkans before World War I, and in Spain for most of the nineteenth and the early twentieth centuries.[20] As usual, opinions on China diverge quite widely—from the pessimistic views of Perkins (1969) and Pomeranz (2000); (no growth at all throughout the whole period) to the optimistic assessment of Rawski, based on wage trends (a 15% increase in the first thirty years of the twentieth century).[21] To sum up, before 1940, labor productivity grew in the "advanced" countries and stagnated or increased very slowly elsewhere.

After 1950, partial factor productivity can be computed as a simple ratio of gross output to the stock of land and the number of workers (table 5.3).[22] The data may not be as accurate as they seem, but they do show an impressive growth. This result is not unexpected, given the evidence about production growth (section 3.2) and the change in labor and land (sections 4.2 and 4.4). The sharp fall in

TABLE 5.3
Land and Labor Productivity, by Continent, 1950–2000 (1950 = 100)

	1960	*1970*	*1980*	*1990*	*2000*	*Yearly Change*
(a) Labor						
Africa	131.1	139.7	138.2	162.8	159.8	0.40
Europe	178.7	266	342.1	392.4	277.3	0.88
North and Central America	131.7	166.2	187.1	199.1	240.3	1.20
South America	128.6	140.8	175.9	276.5	331	1.89
Oceania	145.3	166.8	170.8	208.1	183	0.46
Asia	138.8	154.9	175.9	247.7	290.8	1.48
USSR	185.3	305.2	303.2	606.2	317.5	1.08
World	134.1	149.5	175.6	200.2	202	0.82
(b) Land						
Africa	169	198.1	212.6	260.6	311	1.22
Europe	144.1	175.8	223.3	246.7	254.2	1.14
North and Central America	107.6	122.8	156.9	169.7	211.3	1.35
South America	175.4	179.4	200.7	241.2	302	1.09
Oceania	76.4	70.8	76.9	80.6	101.1	0.56
Asia	110.7	137.8	180.8	242.1	337.5	2.23
USSR	163.7	212.6	214.2	271.8	173.8	0.12
World	126.2	150.1	196.7	220.6	266.7	1.50

Source: FAO Statistical Database.

productivity in the "transition" economies during the 1990s shows that the collapse in production was not compensated by a comparable out-migration of workers and/or by a massive reduction (setting aside) of land.

5.3 The Total Factor Productivity

In a famous article, the Nobel laureate Robert Solow (1957) suggested measuring the growth in Total Factor Productivity (TFP) as a residual, that is, as the difference between the rate of growth in output and aggregate inputs, weighting the rates of change in inputs with the respective shares on production (growth accounting).[23] As an alternative, one can use price instead of quantities, but this ("dual") approach is less common in empirical work because input prices are more difficult to find. Both measures have two serious shortcomings.[24] The comparison between benchmark years might cause the loss of valuable information on short- and medium-term changes, or even, in the worst case, be positively misleading. If the initial year is a bumper one, and the end-year is a poor one, the "primal" measure of TFP would be biased downward, and vice versa.[25] Second, they embody some strong, and possibly not realistic, assumptions such as perfect competition, constant returns to scale, and constant factor shares (as implied by neutral technical progress with a Cobb-Douglas production function). The two former assumptions are plausible, at least in agriculture. The third one may not hold if differences in relative prices of factors steer technical progress toward saving the scarce factor, as argued by Hayami and Ruttan (section 6.1). This shortcoming can be tackled by other methods, such as the Tornqvist-Theil index or the estimation of profit or production functions.[26] However, most recent comparative work uses a different measure, the so called Malmquist "distance."[27] It measures, with linear programming, the difference between the actual output and the maximum feasible production, given the available inputs (the "best practice"). The latter is assumed to be the same for all units of observation (farms, countries)—that is, the Malmquist index assumes the existence of a common set of production possibilities. It also assumes neutral technical progress, or, more precisely, it can misinterpret factor-biased technical progress as regress (cf. Nin et al. 2003a). Both assumptions are questionable, in that agricultural technical progress is often biased and always location-specific (section 6.4) and quality of inputs can vary greatly.[28] In short, there is not an ideal measure of TFP growth—and, by implication, no compelling theoretical reason for discarding theoretically "inferior" estimates.

Many scholars have put forward estimates of TFP in England before 1850, as a key information in the debate about the role of agriculture in the English industrial revolution (section 10.4). Unfortunately, the figures differ quite widely (table 5.4).

TABLE 5.4
Rates of Change in Total Factor Productivity, England, 1700–1850

	1700–1800	*1800–1850*
Deane-Cole	0.2	1.4
McCloskey		0.45*
Crafts	0.2	1.1
Jackson	0.1	
Allen	0.6	0.50
Hueckel		0.34°
Harley		0.70*
Brunt#	0.10	−0.01
Clark	0.0	0.60

Notes: ° 1800–1870. * 1780–1860. # Arable only.

Sources: McCloskey 1981, table 6.2; Harley 1993, table 3.6; Hueckel 1981, table 10.3; Overton 1996, table 3.9; Brunt, **table 5.4**, personal communication; Clark 2002b, table 8. All others: Allen 1994, table 5.5.

Extracting a consensus view from these figures is not so easy. However, it seems fair to stress how productivity growth was slow (or perhaps nil) in the eighteenth century and picked up in the first half of the nineteenth, somewhat exceeding the average in Continental Europe in the same period (table 5.5). Clark (1993, table 4.1) estimates that by 1850 Britain was the most productive country in Europe, but, in his view, this lead had been cumulated in centuries of progress.[29]

Table IV of the Statistical Appendix collects all the estimates of TFP that the author is aware of. The collection is likely to be incomplete, but the number is, however, impressive. There are ninety-one "historical" estimates (i.e., for the period up to 1940) for twenty-seven "countries" (including Ireland and Poland before 1910) and 545 "current" (post-1950) ones for 131 countries (plus the European Community and sub-Saharan Africa).[30] Table 5.5 reports the corresponding average rates.[31] These figures have to be taken with caution. First, as is clear from the column "coefficient of variation," the underlying dispersion of country estimates is quite large: their averages may give excessive weight to minor countries and/or to outlandish estimates. For instance, the decline in Asian TFP in interwar years reflects the fall in productivity in the Philippines: productivity grew quite fast in Japan and Taiwan and remained constant in India and China. However, in most cases, the median is not very far from the average, showing that the distribution is not too greatly skewed. But estimates differ markedly even for the same country/period: those for South Korea since 1961 range from a disappointing −2.76 percent to an excellent 3.3 percent. Some of the figures are so extreme as to be hardly credible: in four cases, all in Sub-Saharian Africa, TFP is said to have fallen at an yearly rate over 10 percent, which corresponds to a 95 percent cumulated change, relative to the best practice over thirty years. Some of these differences are accounted for by different set of inputs (most authors do not take into account human capital, and some even omit land) or different data.[32] For instance, Perkins's (1969) estimates of TFP change in pre-revolutionary China range from modest growth to a slow decline according to the assumptions about production growth.[33] But in a substantial number of cases (including South Korea) the difference depends on the techniques of estimation: rates are lower if computed with nonparametric techniques (the Malmquist "distance") than if computed with any other method (see Federico 2004c).[34] Last but not least, nationwide figures can conceal fairly wide differences by area or sector. State-level estimates of TFP growth in the United States from 1949 to 1991 range from a minimum of 1.36 percent per annum in Vermont, to a maximum of 4.11 percent per annum in North Dakota[35] (Craig and Pardey 2001, table 4.2).

All these caveats should be kept in mind when comparing point estimates across time and space, but as a whole the database suggests four important stylized facts.

1. Agricultural productivity has grown remarkably in the large majority of countries and periods. Almost 70 percent of estimates (438 out of 636) are positive,

TABLE 5.5
Rates of Change in Total Factor Productivity, Average of Country Estimates, 1800–2000

	Year	*Number*	*Rates of Change*		*Coefficient*
		Countries	*Average*	*Median*	*Variation*
Before 1870					
Europe	various	5	0.45	0.56	0.38
Africa	1821–75	1	3.41		
1870–1910					
Europe, Van Zanden	1870–1910	15	0.78	0.82	0.50
Europe, Other Estimates	various	10	0.71	0.75	0.25
Western Settlement Countries	various	2	0.70		0.42
Asia	various	3	1.24		0.88
Africa	1875–1912	1	0.83		
1910–40					
Europe	various	7	1.16	1.01	0.42
Western Settlement Countries	various	2	0.53		0.49
Asia	various	6	0.08	0.36	10.92
Africa	1912–36	1	−0.21		
After 1945					
OECD Countries					
Henrichsmeyer and Ostermeyer-Schloder	1965–85	8	1.74	1.65	0.16
OECD	1973–89	8	1.78	1.70	0.24
Trueblood and Coggins	1961–91	22	1.63	1.69	1.01
Bernard and Jones	1970–87	14	2.60	2.15	0.53
Martin and Mitra	1967–92	15	3.29	2.89	0.43
Pryor	1970–87	8	2.18	2.16	0.35
Ball et al.	1973–93	10	1.89	1.86	0.23
Kawagoe and Hayami	1960–80	22	2.49	2.59	0.32
Barkaoui et al.	1973–93	11	1.51	1.40	0.58
Nin et al.	1965–94	9	0.55	0.86	2.45
Arnade	1962–92	21	1.29	1.42	1.20
Other	various	6	2.10	1.81	0.59
Less Developed Countries					
Asia					
Pingali and Heisey	1971–86	6	1.36	1.11	0.61
Martin and Mitra	1967–92	6	2.48	2.51	0.24
Fulginiti and Perren	1961–85	6	−1.17	−0.30	−2.43
Kawagoe and Hayami	1960–80	8	−1.04	−0.96	−1.53
Suharlyanto, Lusigi, Thirtle	1965–96	10	0.26	−0.22	4.91
Nin et al.	1965–94	4	−0.53	−0.57	−0.70
Arnade	1962–92	4	−1.15	−1.29	−0.88
Trueblood and Coggins	1961–91	17	−0.77	−0.90	−1.29

TABLE 5.5 (*continued*)

	Year	*Number*	*Rates of Change*		*Coefficient*
		Countries	*Average*	*Median*	*Variation*
Central and South America					
Pingali and Heisey	1971–86	2	2.26		0.24
Martin and Mitra	1967–92	10	1.00	1.51	1.96
Fulginiti and Perren	1961–85	5	−0.64	0.00	−3.38
Kawagoe and Hayami	1960–80	8	0.72	0.71	0.98
Nin et al.	1965–94	11	0.44	0.64	1.56
Arnade	1962–92	18	0.30	−0.51	6.94
Trueblood and Coggins	1961–91	22	−0.06	0.10	−23.43
Africa					
Pingali and Heisey	1971–86	21	0.89	0.74	1.25
Martin and Mitra	1967–92	9	1.87	1.86	0.93
Suharlyanto, Lusigi, Thirtle	1965–96	47	1.28	1.16	1.74
Fulginiti and Perren	1961–85	5	−2.16	−0.10	−1.39
Kawagoe and Hayami	1960–80	4	0.93	1.06	1.64
Nin et al.	1965–94	16	−0.04	0.06	−19.45
Arnade	1962–92	8	−0.83	−1.12	−2.15
Trueblood and Coggins	1961–91	41	−1.68	−0.40	−2.43
Others	various	2	0.77		
All LDC countries					
Pingali and Heisey	1971–86	29	1.08	1.01	1.01
Martin and Mitra	1967–92	25	1.67	1.90	1.05
Suharlyanto, Lusigi, Thirtle	1965–96	47	1.28	1.16	1.74
Fulginiti and Perren	1961–85	16	−1.31	−0.10	−2.11
Kawagoe and Hayami	1960–80	21	0.06	0.44	27.09
Nin et al.	1965–94	31	0.07	0.05	10.66
Arnade	1962–92	39	−0.42	−0.83	−4.44
Trueblood and Coggins	1961–91	119	−0.84	−0.55	−3.74
Others	Various	3	0.77	0.78	0.45
Socialist Countries					
Pryor	1970–87	12	0.75	0.84	2.03
Wong	1960–80	8	−1.44	−1.18	−0.72
Lazarcik	1970–86	6	0.68	0.45	1.93
Nin et al.	1965–94	4	0.68	0.16	1.57
Arnade	1962–92	7	0.36	−0.74	6.07
Trueblood and Coggins	1961–91	10	0.55	0.40	2.90
China					
Pre-Reform (1979)	Various	8	−1.02	−0.54	−2.20
Post-reform	Various	13	3.64	3.30	0.54

Source: Statistical Appendix, table IV.

and the average TFP growth for the whole database is 0.58 percent per annum. No estimate of the database goes beyond the mid-1990s, but the available data for the United States and the United Kingdom suggest that growth has continued, although at a slower pace, also in the second half of the decade.[36] As said, the database includes all available estimates without a priori selection. Therefore the differences for the same country and period must be settled according to the consensus view: a country will be classified as a "good" ("poor") performer if a majority of the estimates is positive (negative). Out of 175 country/period observations, there are 109 "good" performers, 53 "poor" performers, and 13 undecided (i.e., countries with as many negative as positive estimates). The distribution of these countries is, however, heavily skewed by period and location, with a clear distinction between the "advanced" (OECD) countries and the rest.

2. The performance of OECD countries has been quite good. TFP has grown in almost all cases, with the exceptions of "border" cases such as Portugal and (perhaps) South Korea after 1950 and, notably, of the United States from 1900 to 1920. This latter decline is, however, balanced by growth in previous decades, so that over the whole period 1870–1913 the change is still positive. Furthermore, the growth in TFP has been accelerating in almost all cases. The rates are higher (up to three and even four times) after 1950 than before for all countries (except Portugal). In most cases, including the United States, the rates are higher in interwar years than in 1870–1913.[37]

3. The performance of LDCs has been decidedly mixed, to say the least. Before 1950, there are very few, and possibly not so accurate, estimates for these countries: productivity declined in Egypt and the Philippines, stagnated in India, and increased in Argentina (where it had been declining in the prewar decade 1902–10), Mexico, Taiwan, and Korea. After 1950, the database coverage is almost complete. The main feature is the large number of "poor performers"—forty-five out of a total ninety-four LDC countries. They thus outnumber the thirty-six "good" performers, with twelve undecided cases. As expected, Sub-Saharan African countries performed quite poorly, but also Asian ones, in spite of the successes of the Green Revolution did not fare that much better. The majority of estimates for Pakistan, Bangladesh, and India after 1960 are negative.

4. As hypothesized by looking at prices (section 3.3), the agricultural sector moved from laggard to leader, at least from a purely statistical point of view. In the United States, the TFP grew in agriculture slightly less than in the non-farm sector from 1870 to 1913, matched almost perfectly the performance of the rest of the economy in 1913–50, and dramatically outperformed it after 1950.[38] The same happened in the United Kingdom.[39] Unfortunately, a precise comparison is impossible for other countries before 1950. The leadership of agriculture in TFP growth is, in contrast, fairly well established for the postwar period. Martin and

Mitra (2001, tables 2, 3) show that, in 1967–92, the rate of growth in TFP was higher in agriculture than in manufacturing both on average (2.29% vs. 1.74%) and for individual countries (22 out of a total of 36).[40] However, this statement may not hold true for socialist countries. The rates of economy-wide TFP growth

TABLE 5.6
Contribution of Total Factor Productivity Growth to Production Growth, Average of Country Estimates, 1800–2000

	Year	*Number*			
		Countries	*Average*	*Median*	*CV*
Before 1870					
Europe	Various	3	0.49	0.58	2.36
Africa	1821–75	1	0.74		
1870–1910					
Europe, Van Zanden	1870–1910	15	0.90	0.86	0.28
Europe, Other Estimates	various	9	1.08	0.61	0.65
Western Settlement Countries	various	2	0.30		1.75
Asia	various	3	0.35		0.56
Africa	1875–1912	1	0.39		
1910–40					
Europe	various	5	0.73	0.66	0.31
Western Settlement Countries	various	2	0.72		0.86
Asia	various	6	−0.07	0.16	−12.54
Africa	1912–36				
After 1945					
OECD countries					
Henrichsmeyer and Ostermeyer-Schloder	1965–85	8	0.83	0.80	0.34
OECD	1973–89	8	1.06	1.17	0.29
Other	various	4	1.14	1.16	0.10
Less developed countries					
Asia					
Pingali and Hesey	1971–86	3	0.38	0.30	0.33
Pryor	1967–92	1	−0.61		
Central and South America					
Pryor	1967	2	0.56		
Africa					
Pingali and Hesey	1971–86	19	0.39	0.42	1.65
Pryor	1967–92	4	0.66	0.48	1.41
Others	1947–97	2	0.28		

TABLE 5.6 (*continued*)

	Year	*Number Countries*	*Average*	*Median*	*CV*
All LDC Countries					
Pingali and Hesey	1971–86	24	0.40	0.43	1.44
Pryor	1967–92	8	0.44	0.48	2.70
Others	Various	3	0.30	0.33	0.42
Socialist countries					
Pryor	1970–87	11	0.53	0.25	0.84
China					
Pre-reform (1979)	various	2	0.17	0.17	
Post-reform	various	6	0.46	0.45	

Source: Statistical Appendix, table V.

exceeded all estimates by far for agriculture only—but the comparison may not be accurate as the figures are taken from different sources (cf. Campos and Coricelli 2002, tables 1, 2).

The data on the change in factor endowment (sections 4.2 to 4.4) suggest that the contribution of TFP growth to the increase in production must have differed by area and period. Van Zanden (1991, 228), commenting on his results for Europe before World War I, contrasts the intensive growth in Western Europe, based on TFP growth, with the extensive growth in the east, based on increase in inputs.[41] The contribution can be measured as the ratio of rates of growth in TFP and in output. This ratio can be computed only for the Solow residual estimates, and not for all of them, as many authors do not report the rates of growth of inputs. Results are nevertheless interesting, especially in historical perspective (table 5.6). The contribution comes out to have been positively related to the level of development, with the conspicuous (but expected) exception of countries of Western settlement before 1914. TFP growth accounted for almost all the production growth in both Western Europe and in the countries of Western settlement after 1914, while it accounted for between a third and a half in the LDCs.

Summing up, agriculture does not, by and large, deserve its reputation of having been stagnant and backward. The statistical evidence, albeit imperfect and incomplete, shows that Total Factor Productivity grew throughout the period (at an accelerating pace) in nowadays "advanced" countries and has grown, albeit more slowly, in the LDCs after World War II. Furthermore, in the past fifty years, TFP has grown faster in agriculture than in the rest of the economy.

5.4 Conclusion: On the Interpretation of Total Factor Productivity Growth

Total Factor Productivity can, in theory, grow because a new and more efficient technique is invented or because farmers decide to adopt already available, more productive, techniques. In the latter case, they move from a point inside the frontier of possibility of production (the set of most productive techniques), while innovations shift the whole frontier outward. The very possibility that the economy is within the frontier of possibility of production requires an explanation—as it may seem to contradict one of the basic principles of rational behavior. In theory, under the spur of competition, all firms should adopt suitable innovations as soon as they are made available. However, real world conditions may differ for at least four major reasons: (1) some innovations might not be adopted because they are perceived as being too risky, or being unsuitable to the specific environment (section 6.4) or to characteristics of the farm, such as size (section 7.6); (2) imperfections in the markets for credit may prevent farmers from funding investments (section 7.4); (3) government policies may hamper the adoption of suitable technology—for example, by subsidizing alternative, inefficient ones by preventing the adjustment of farm size (sections 9.4 and 9.5); (4) poor access to markets for goods may prevent a technically feasible specialization of farm production. Changes in all these parameters can boost (or reduce) productivity growth. In particular, better access to markets would enable farmers to specialize according to their comparative advantages ("Smithian growth"). As pointed out by Farrell (1957) in the same year as Solow's seminal paper, changes in allocative efficiency are potentially a major component of TFP growth.[42] This insight has long been neglected, due to a lack of a measure of efficiency changes as simple as the Solow residual for changes in time.[43] All changes in TFP have routinely been attributed, more or less consciously, to technical progress.

From this point of view, the recent advances in nonparametric estimation are a great leap forward. In fact the Malmquist index allows us to distinguish between the "pure" technical progress (the shift in the maximum feasible production or "frontier") and changes in allocative efficiency (the movements of actual output relative to the "frontier"). Many recent comparative works report separate estimates of these two components.[44] For instance, according to Fulginiti and Perrin (1997), growth in allocative efficiency has (partially) rescued a dismal technical performance, while according to Suharlyanto et al. (2001), efficiency has decreased, dragging down total TFP. Trueblood and Coggins (2004) find both widespread technical regress and decline in productivity efficiency, which contribute roughly to the same extent to the quite poor aggregate performance. These differences, which are only partially accounted for by different country coverage, show that nonparametric techniques still need improving, and probably better data are needed as well. Anyway, some sort of decomposition of TFP growth is essential to understand the role of technical progress.

Chapter Six

TECHNICAL PROGRESS IN AGRICULTURE

6.1 Introduction: Productivity Growth and Technical Progress

Technical progress has always attracted the attention of economists and economic historians as the key to sustained economic growth. As a result, theories on its causes abound, but none of them really seems suited to explain the whole process, nor can they take the specificity of agriculture into account. It would be pointless to discuss the economic theory of technical progress here. We will focus on two competing interpretations of technical change in agriculture: a loosely defined "modernization" or "diffusion" approach, and the "neoclassical" models of Boserup (1951) and of Hayami and Ruttan (1985).

The "modernization" approach was very popular among agricultural "experts" in the nineteenth and early twentieth centuries and inspired much of the technical advice given to LDCs up until the 1970s.[1] It assumes that alternative techniques could be ranked according to technical criteria, and that the "best" (most productive) technique must be adopted. Any refusal or delay in doing so is evidence of the peasants' so-called conservatism, an allegedly irrational fear of change, which had to be overcome by the enlightened shove of the landowners and/or the state. Arthur Young, the most famous eighteenth-century expert, harshly criticized Irish and French farmers for not imitating British-style continuous rotation, even in contrast with his own data about yields.[2] In the early twentieth century, experts blamed West African natives for neglecting cocoa trees and extolled the careful tree management of Western plantations (see Clarence-Smith 1995, 164–65. Agricultural historians widely used these expert opinions to assess the innovative performance of agriculture—usually finding it poor.[3] Yet, this modernization approach is basically flawed, and its fallacy was exposed as early as 1840 by the German economist von Thunen.[4] In fact, it contrasts with a basic principle of rational economic behavior—that agents choose the most profitable techniques according to the expected profits, which depend on the relative prices of factors. It would be economically irrational to adopt a given innovation, although technically feasible, if it were not economically profitable.

E. Boserup (1951) is known for having stated that population growth causes technical progress, but this summary of her work is both somewhat unfair (as her book is much richer, and includes some seminal insights on institutions) and inaccurate. In fact, in her view, population growth causes an "intensification" of agricultural production—that is, a "gradual change toward patterns of land use which make it possible to crop a given area more frequently than before" (1951, 43).[5] This change can be achieved by a more intense use of labor, and thus, as Boserup explicitly states, intensification causes the productivity of land to rise and the productivity of labor (and thus its returns—or wages) to fall. In economists' jargon, "intensification" reflects a choice of techniques along the known production function. It thus differs from technical *progress*, which, as stated at the end of the previous chapter, consists in the shift of the whole frontier or in a movement of the production point toward the frontier itself.

Hayami and Ruttan (1985) share Boserup's view about factor cost as the main determinant of technical change, but they focus on technical progress and, above all, add two further ideas, which greatly increase the reach of their analysis. First, they argue that the *production*, and not just the adoption, of new techniques, depends on factor prices.[6] The available scientific knowledge determines a general set of feasible innovations ("meta-production function") and each country tends to develop the technology that best suits its own relative prices ("induced innovation"). A land-scarce country would invest more on R&D into land-saving technology, a labor-scarce one in labor-saving innovations. Second, the needs of technical progress also determine changes in institutions, such as the development of suitable property rights or the establishment of research facilities ("induced institutional innovation"). This hypothesis is the least developed and the most controversial part of the Hayami-Ruttan model (1985, 110–12), and it is almost impossible to devise a proper test for it.[7] As the authors themselves admit, the supply of institutional innovation also depends on "cultural endowment," a loose concept that may include everything from political institutions to the mentality of the peasants.

This basic, "neoclassical" model will be used to interpret technical progress in section 6.3, after a description of the main innovations (section 6.2). It will be argued that, by and large, this model works, especially if one takes physical and human capital and not just land and labor into account. However, the neoclassical model cannot account entirely for technical progress in agriculture (section 6.4). Section 6.5 deals with the production and transmission of innovations, and on the institutions needed for these tasks (research institutes, experimental stations, etc.). This is the first part of a wider discussion of "institutions," which will continue in the three next chapters. Section 6.6 concludes by addressing the contribution of these institutions to technical progress.

6.2 The Major Innovations

For ease of exposition, innovations can be grouped in four categories: "new" plants and animals (or biological innovations), practices of cultivation, chemical products, and machinery.

Biological Innovations

In the strictest meaning of the word, only natural selection (or biological engineering) can produce new plants and animals, and humans only have to discover or domesticate them. The whole history of agriculture can be construed as a long-term process of the diffusion of plants and animals from one or several "original" locations (the Middle East for wheat, South Eastern Asia for rice, America for maize, potatoes, and tomatoes) to the whole world (Diamond 1997, 93–191; Grigg 1974, 24–27; Bray 1986, 9–10). As late as 1862, a Mr. Fulz discovered by chance a new variety of wheat (later named after him), which the United States Department of Agriculture (USDA) distributed (see Olmstead and Rhode 2002, 941–43).[8] The importance of pure discovery was reduced by its own success. The quest for new plants or varieties reduced the "raw material" (i.e., the number of unknown plants, species, or varieties to be discovered) faster than natural evolution could produce them. Thus, in the past few centuries, there has been no domestication of any species of animals, and only one discovery of an entirely new plant, the sugar-beet. Its sugar content had been first noticed in 1747, but it remained a scientific curiosity until the Napoleonic Wars, which cut Continental Europe off from the supply of sugar cane from the Western Indies (Bill and Graves 1984, 18; Tonizzi 2001, 23–31).

Even without any major entry, biological innovations did play a major role in agricultural progress during the nineteenth century, as improvements in transportation and the increase in transatlantic migration boosted the worldwide transfer of known varieties.[9] European colonists tried to re-create the familiar agriculture of their motherland wherever they settled, bringing with them the animals and the seeds that they were used to. In some cases, this approach did not work, and thus the settlers resorted to long-range imports, often with the financial support of their homeland institutions. For instance, the first settlers in the Canadian and northern American prairies discovered that the wheat varieties cultivated in the eastern region of the continent were unsuited to the short growing season, and so they imported early ripening varieties of spring wheat from Russia.[10] The Merinos sheep, especially bred for producing wool, were imported from Spain to South Africa and Australia as early as the end of the eighteenth century.[11] These transfers also played a key role in the development of agriculture outside the Atlantic economy. Imported plants—such as sweet potatoes in Papua New Guinea and in some areas of China, cassava in Africa, and maize in Africa and southern Europe—became staple foods for native populations.[12] Western administrators in Africa introduced plants and trees that became main cash crops, such as cocoa in Ghana (perhaps the greatest success-story), cotton in Uganda, coffee in Kenya, and so on.[13] The process was, of course, far from smooth. There were countless failures, some of which very expensive: the pioneer of Californian wine industry, A. Haraszthy, changed five locations (investing huge sums in each of them), before finding the ideal environment in the Sonoma Valley (Olmstead and Rhode 2000b).[14] In some cases, transfers were positively harmful: new colonists in the U.S. prairies also brought with them pests, such as the Hessian fly, or weeds, such as the Russian thistle, which were to haunt wheat-growing for a long time.

Biological innovation via worldwide transfer is bound to reach a limit, when all known plants and animals have been tested in all the reasonably suitable environments. This limit was reached toward the end of the nineteenth century. Since then, almost all "new" varieties of existing plants have been produced with systematic hybridization, and later, genetic engineering (new plants need genetic engineering and, so far, they are in the realm of laboratory science). The first recorded experiments of hybridization date back to the late eighteenth century (Perkins 1997, 40–43), and the first hybrid wheat was produced in the 1870s and possibly as early as the 1840s (Olmstead and Rhode 2002, 940). In the 1890s, hybridization scored its first huge success when the grafting of American vines onto native ones saved European vineyards from phylloxera, after chemicals had failed and/or had proved to be too expensive (Agulhon et al. 1976, 389; Unwin 1991, 290–92; Loubère 1978, 161–65).[15] These early attempts were however totally unscientific, and their results were consequently unpredictable. The research work was greatly boosted by the (re)discovery of Mendel's laws of genetics toward the end of the nineteenth century and, later, by the development of techniques for mass production of hybrids in the early twentieth century.[16] The first great breakthrough of this new science-based approach was the production of hybrid corn in the United States in the 1930s.[17] The new seed augmented yields by 20 percent, ceteris paribus, and was thus an instant success story. This success proved the potential of scientific hybridization and stimulated research on both hybrids and open-air pollinated varieties. The results were particularly impressive in the LDCs, where traditional varieties still prevailed. The work started in 1941–43 in Mexico for wheat, and in the 1950s for rice, using semi-dwarf varieties of Japanese origin.[18] Researchers focused on increasing yields of wheat and rice in "core" areas (e.g., irrigated lowlands for rice), with great success. The IR-8 rice, released in 1966, could produce, under experimental conditions, up to eight times the amount produced by traditional varieties. Researchers soon discovered, however, that the initial achievements had to be sustained with a continuous flow of new investments (Byerlee 1996, 697–702; Biggs and Clay 1981, 323; Gill 1991, 104–11). Hybrids tend to lose their high-yielding properties after some years of use and to become vulnerable to pests, so that the original releases have to be replaced by new ones. The initial batch of "high-yield varieties" (HYVs) was hardly suited to "marginal" areas (e.g., rain-fed highlands for rice), was quite vulnerable to pests, and sometimes did not satisfy the tastes of consumers. Furthermore, the intensification of cultivation created the need for new varieties with a shorter growing season, which could be cultivated two or three times in the same year. Last but not least, the research had to be extended to other cereals, such as maize and sorghum, which had been neglected in the first wave. Research has tried to address all these needs, with varying degrees of success. In recent times, the number and the quality of new releases has slowed down somewhat, and thus experts on the field have called for a renewed research effort.[19] Most research efforts have concentrated on cereals and, subordinately, on tropical cash crops (section 6.5). Years of research and experimentations transformed rubber trees from semi-wild plants of the Brazilian rain forest into plantation trees, allowing for the production boom in South Asia (Headrick 1988, 243–48). The production of sugar per unit of land greatly increased thanks to high-yielding varieties of cane sugar such as the POJ2878, developed in Indonesia, and to the increase in sugar content of sugar beets.[20] In contrast, the research on roots such as cassava, a key staple in many LDCs, has started only quite recently, albeit with substantial funding.[21]

We know relatively little about progress in animal husbandry, which seems to have attracted much less funding than crops.[22] Clearly, hybridization is simpler for animals than for plants, and "natural" hybrids—such as mules—have been known for millennia. If anything, the existing stock of animals has suffered from an excessive level of hybridization, as the outcome of millennia of almost random mating—hence the research aimed at producing "pure" breeds with stable and desirable features (pedigree). To this end, the mating was limited to pure animals of the same breed, which thus had to be listed (herd-book).[23] The pioneer in this work was Robert Bakewell, who, in the 1760s and 1770s, selected a new breed of (Leicester) sheep especially for the production of meat. In Australia, where selection aimed to increase the production of wool, a new cross between the hairy sheep from Bengal with fine-wool Merinos from the Cape colony was a great success, jumping from 10 percent of all the clip in 1805 to 100 percent five years later. The first herd-book for horses was established in the United Kingdom in 1822. Systematic scientific work to improve races started much later, well into the twentieth century, and the work was greatly eased by the adoption of artificial insemination in the late 1930s (Huffman and Evenson 1993, 195; Kostrowicki 1980, 469–70).[24] A substantial part of the research funds for cattle, however, has been used to fight animal diseases, which often are a health hazard for humans as well.[25]

The potential for biological innovations has been greatly increased by the recent development of genetic engineering. The production of genetically engineered varieties started in the late 1980s, and the first commercial one, the Flavr-Savr tomato, was approved for release in 1994.[26] As is well known, biotechnologies are extremely controversial and no one knows whether or not they will be allowed to develop their potential.

Agricultural Practices

The expression "agricultural practices" refers to all systems that farmers use to meet the basic needs of plants and animals (water, seeds, nutrients) in order to maximize output. Thus, it includes the skills to perform simple operations (such as plowing), the knowledge of optimal timing of operations, and the knowledge of the best succession of different crops (rotation). Most of this information had been discovered during the process of the domestication of plants and animals, and thus the proper credit for the key innovations will never be attributed (How did Chinese peasants succeed in cultivating rice twice in a year? Who was the first African peasant to abandon the traditional slash-and-burn system?). The past two centuries, however, have featured a massive change, which the literature often contrasts with an alleged stagnation in the previous period. Some of these changes were forced, as practices had to adapt to the needs of new plants and animals and to the environment of newly settled areas. It is difficult to describe the process, which consisted in micro-innovations and local adaptations. Some innovations, however, were big enough to capture the attention of contemporaries, and thus of historians. One of them was dry farming—a sort of preventive deep plowing to capture the moisture of the soil—which was advertised as the key to settling the semi-arid prairies of Canada and the northern United States in the 1900s. In the first years, yields (helped by a string of unusually wet years) were high enough to make small-scale farming profitable, but in the long run they fell, and thus dry farming had to be abandoned in favor of more extensive methods.[27] Most innovations aimed to increase the number of crops per unit of time. Japanese peasants succeeded in adding a second or a third crop of cocoons in late summer and/or early autumn to the traditional spring one, doubling the production of raw silk.[28] Egyptian peasants moved from the traditional three-year rotation, already quite an intensive one, to a two-year rotation in the 1890s and 1900s (Richards 1982, 70–72).[29] European peasants succeeded in making do without fallow ("continuous rotation"). The advantage was clear: fallow produced only a meager pasture, and thus any substitute was bound to boost gross output, provided, of course, that it could reintegrate the nutrients that cereals extracted from the soil. The range of suitable substitutes included maize, roots (turnips, potatoes), legumes (peas, beans, vetches), and dozens of various grasses (lucerne, sainfoin, *etc.*).

The historians' attention has focused on one specific set of substitutes, the grasses and roots (mainly turnips) used to feed cattle. They are regarded as a major technical breakthrough because they solved a problem that had always haunted European agriculture: the conflict between cattle-breeding and wheat-growing over scarce land. In fact, the increase in available fodder augmented the number of animals, and their manure increased cereal yields. According to the conventional wisdom, this practice ("new husbandry") was first adopted in eighteenth-century Britain, and the ensuing "agricultural revolution" was a key prerequisite for the industrial Revolution.[30] This view is now very controversial and scholars such as Robert Allen (1992, 1999) and Gregory Clark (1993) deem the very name "revolution" to be misleading.[31] Continuous rotation was known in Britain well before the eighteenth century and had boosted land productivity already in the seventeenth century (the "yeoman revolution" in Allen's words). Elsewhere in Europe—notably in the Netherlands and in the Po Valley in Italy—it had been adopted since the Middle Ages (Shiel 1991, 54–58; de Vries and van der Woude 1997,

203–5).[32] On the other hand, the reputation of Britain in the nineteenth century as a beacon of technical progress greatly helped the spread of "new husbandry" all over Europe (albeit not fast enough in the eyes of agricultural "experts") and beyond. For instance, the practice of sowing alfa-alfa (a kind of grass) after two or three years of wheat spread in Argentina in the early 1900s, fostering the integration between agriculture and ranching.[33] Throughout the nineteenth century, rotation in Europe became more and more complex, involving periods of up to eight or nine years.[34]

In spite of the reputation of new husbandry, other substitutes for fallow had a deep impact on agriculture and the welfare of rural population. Maize and potatoes, both of American origin, spread in Europe beginning in the late seventeenth century.[35] They both yielded more calories per unit of land than wheat or rye, making for a substantial increase in rural population (with catastrophic results for its health) possible.[36] But arguably, maize has had the most lasting impact on world agriculture. It is still a major staple food in some parts of Latin America and Africa, and it has become the main feedstuff for large cattle and hogs. This use of maize was pioneered in the American Corn Belt in the nineteenth century, and then spread in Western Europe and the Soviet Union in the second half of the twentieth century.[37] Nowadays, the total world output of maize is about 600 million metric tons, as much as wheat, but, according to one estimate, about two-thirds of it (three-quarters in developed countries) is used as fodder.[38]

Chemical Products

The history of modern fertilizers started in the 1840s. In 1840, the German chemist J. Liebig published the first theory of fertilization in his book *Die organische chemie in ihrer anwendung auf agricultur und physiologie*.[39] The next year, Lawes patented the first method to produce phosphates by dissolving bones in sulphuric acid, which was used in commercial production from the early 1840s.[40] Quite soon bones were substituted as a raw material with mineral phosphate, which was widely available in many areas. Potash fertilizers were made available by the discovery of a natural deposit around Stassfurt, Germany, in 1856 (exploited from 1861 onward). Nitrogen was first provided by "natural" sources, guano and Chilean nitrates, while the first artificial nitrogen fertilizers, ammonium sulfate, a by-product of coke, became available in the 1860s–1870s. Calcium-cyanamide was invented in the 1900s, but the real breakthrough was the Haber-Bosch method of producing ammonium sulfate—invented in 1909. The industrial production started immediately after the war, and, a few years later, it was supplemented by urea, another nitrogen fertilizer, which could be delivered as a liquid. In 1926, BASF marketed the first composed fertilizers. Fertilizers provided a much more flexible way of reintegrating natural fertility than complex rotations, which were progressively abandoned beginning at the end of the nineteenth century

(Perkins 1981, 90). Nowadays, a crop can be cultivated on the same plot of land for many years in a row without affecting yields.

The quest for chemical products to fight pests and weeds started in the first half of the nineteenth century with two remarkable successes, both against vine parasites—the use of sulphur against oidium and of the so-called Bordeaux mixture (lime and copper sulphate) against peronosphora (Unwin 1991, 283, 292).[41] In the early twentieth century, British farmers used nicotine and other insecticides, which were, however, very expensive and highly dangerous (Brassky 2000e, 560). Most attempts to fight insects or pests failed, however, because chemical science was not sufficiently developed. The breakthrough came in the 1940s with the discovery of DDT and of the first modern herbicide, the 2.4-D (Gardner 2002, 24; Bairoch 1999, 96).[42]

Machinery

Throughout the nineteenth century, traditional hand-powered tools were being continuously changed and improved. By their nature, these incremental innovations are poorly documented. However, McClelland (1997, esp. 49–63, 129–50) succeeds in meticulously listing dozens of these improvements in wheat-growing on the American East Coast, although he is unable to provide much evidence on their adoption.[43] He argues that, as a whole, all these improvements are so impressive that they deserve the label of "the first agricultural revolution." His focus leads him to miss the two major innovations of the pre-mechanical era, the cotton gin and barbed wire. The cotton gin, invented by Eli Whitney in 1793, mechanized the stripping of lint from the cotton flower.[44] Although initially hand powered, it saved a lot of labor, because alternatives (such as manual stripping) were extremely labor intensive. Without the gin, it would have been impossible to increase cotton output to match the needs of British factories. Barbed wire also solved a major problem—how to enclose fields in the American prairies, where lumber, imported from the east, was extremely expensive.[45] The idea of using iron wire dated back to the 1860s, but the key feature, the addition of barbs, was patented separately in 1874 by Glidden and Haish. The yearly production of wire jumped to 40,000 tons in 1880 and to 200,000 in 1907. Barbed wire was a quintessential labor-saving innovation: the work of building and maintaining fences fell from 4.5 percent of the total labor input of American agriculture in 1850 to a mere 0.5 percent in 1910. Furthermore, the possibility of building cheap corrals changed the whole business of cattle-breeding. Animals could be fed better, monitored more easily by fewer cowboys, and mated according to the owners' wishes, thereby easing selective breeding.

Arguably, the first modern agricultural machine was the wheat thresher, invented in 1786 by a Scotsman, Meikle (Hurt 1982, 67–76; Bairoch 1999, 55–59; McClelland 1997, 170–202; Federico 2003). Like the cotton gin, it stripped the wheat kernel from its external protection, but, unlike the cotton gin, it was not an instant success. In fact, the early threshers were either not powerful enough (if horse-powered) or cumbersome (if fixed steam-powered ones), while the need for a labor-saving innovation was not so pressing: wheat was usually threshed during winter, when labor was fairly cheap. Mechanical threshing was to spread after the mid-nineteenth century, thanks to new portable steam-powered models. Many other processing operations were mechanized in the nineteenth century.[46] The crushing of cane-sugar was thoroughly mechanized in Cuba as early as 1860, the processing of coffee in Brazil started in the 1880s, the butter separator for milk was invented in 1878–79 (separately by Nielsen and de Laval), and presses to extract wine and oil appeared in the 1920s.[47]

The mechanization of fieldwork developed in the twentieth century, with one major exception, harvesting, the most labor-intensive task in traditional agriculture. After decades of failures, in 1833–34 O. Hussey and C. McCormick patented two (slightly different) models of horse-powered wheat reapers, which were later improved and adapted also to harvest other cereals and to mow grass.[48] This success stimulated research into combining harvesting and threshing in a single machine (the "combine").[49] Working combines were built during the 1880s, but they had to be pulled by a large number of horses (usually 25–30, and sometimes up to 50). This prevented their adoption, with the notable exception of California: combines were to gain popularity in the 1920s, after the development of tractors. The research for the mechanization of other agricultural operations started in the second half of the nineteenth century, but with disappointing results over a long period of time. In fact, the lag between the patents embodying the correct principles and the marketing of reliable, cost-effective machines was quite long. The first milking machine was patented in 1872, the first corn picker in 1900, and the first cotton harvesters in the period 1907–12 (all in the USA), but their use started to spread in the 1900s, 1920s, and during World War II respectively.[50] Different types of machines for harvesting cane sugar, suited to different types of cane, were invented in the 1920s, 1930s, and 1950s.[51] Machines for harvesting and transplanting rice were built in the 1950s in Italy, but they proved too big and unwieldy for the Asian landscape, and thus smaller ones were built, but only in the late 1960s.[52] Mechanizing the harvest of tree crops proved even more difficult. The first machines were built in Brazil for coffee trees in the 1900s, but they proved unsuitable.[53] A massive research effort started only after World War II, and it is arguably not yet over. Picking machines are indeed available for many products (the first tomato picker was patented in 1959–62, grape-picking machines in the 1970s), but they are not suited to all needs, and fruits and vegetables are still picked by hand in many areas.[54]

The "delay" in the production of suitable machines can be explained, at least in part, by specific technical problems, such as the need to work in deep water and on soft ground (which made the mechanization of rice-growing so difficult) or on irregular terrain (as on hillsides for vineyards), and to handle irregularly sized, delicate objects, such as fruits or cow udders).[55] However, the mechanization of fieldwork was also hampered by the lack of a suitable source of power. Agriculture needs a small-scale and movable source of power, and for a long period of time the only available sources were animals, horses, or mules. A single animal, however, can provide a limited amount of power, and augmenting the number of animals to increase power created awkward problems of coordination. Furthermore, animals have to be fed, and thus a massive increase in their total number would "waste" precious land for the production of fodder. The obvious solution, with nineteenth-century technology, was to use steam. It worked in processing, which could be concentrated in one location only, although later electrification greatly helped to downsize processing equipment.[56] Steam was not a solution, however, for mechanizing fieldwork, with the exception of wheat threshing. Steam plows were built from the 1830s, but they were quite cumbersome, semi-fixed machines, which had to be placed at the border of a field and worked with a complex set of belts. They had a very limited success, almost exclusively in the United Kingdom (Collins 1983, 92; Bridgen 2000, 508–10; Turner et al. 2001, 93). Steam tractors were built in the 1890s and 1900s, but they were too big and needed too much fuel and water to be useful in most circumstances (Sargen 1979, 3; Wik 1953, 200–205; Hurt 1982, 110–16). Furthermore, they could not be used to pull a reaper or power a combine, because sparks from steam boilers could ignite ripe wheat, with disastrous consequences. These defects could have been overcome with sufficient research, if a simple alternative, the internal combustion engine, had not been available. The first oil-powered tractor was built in the United States in 1889, and commercial sales started in 1902.[57] However, the initial diffusion was quite slow: in the whole country there were 600 tractors in 1906 and only 2,000 in 1909. They were still quite expensive, and their uses remained quite limited until the early 1920s. The appeal of gasoline tractors was substantially increased by the invention of the power-take-off shaft (or PTO), which transformed the pulling power of the engine into a rotatory movement, and by the successful downsizing of machines. The PTO greatly widened the range of possible uses of tractors, while smaller machines could cultivate crops in rows, such as legumes (Burnell 2001, 533–34; Olmstead and Rhode 2001, 666; White 2001, 495–96). The first tractor fitted with a PTO, the McCormick Farmhall, was marketed in 1924. The downsizing of machines was essential in Japan and in the whole of Asia, where tractors had to adjust to the small size of the rice fields. The first suitable tractor (known as a power tiller) was built in 1926 by Nishizaki Hiroshi, a farmer living near Tokyo (Francks 1996, 785–88). At the other end of the range, the progress in engines allowed the production of increasingly powerful machines for large-scale cultivation. In the 1920s and 1930s, agriculture also benefited from the development of general-purpose technology, such as the (oil or electricity-powered) water pump and the truck.[58]

As said in the introduction, the "neoclassical" model of technical progress explains the production and adoption of innovations as a consequence of factor endowment. From this point of view, the key feature is the factor content of each innovation. Hayami and Ruttan (1985, 74–75) assume that mechanical innovations are labor saving, while all others are prevalently land saving. New rotations increase the number of crops per unit of land and new seeds and chemical products increase the unit production of each crop. This view is only a very first approximation: the factor intensity of the innovations depends on the alternative techniques available ceteris paribus (i.e., for the same flow of services). Fertilizers and pesticides save labor relative to traditional systems, such as collecting and spreading manure or extirpating weeds by hand. Tractors save the land needed to produce feed for horses: Olmstead and Rhode (2001, 665) estimate that, in 1915, at the historical peak of the stock of draught animals, the production of fodder needed 79 million acres—or, about a sixth of the combined extension of arable land and pasture.[59] Furthermore, arguably, focusing on two factors only causes us to miss the key feature of modern technology, its intensity in human and physical capital. It is important to stress that biological innovations are capital intensive. Purchased seeds are, by definition, more expensive than home-grown ones, and, in most cases, they need a greater amount of fertilizer or irrigation to develop their potential (section 6.4).[60] All innovations need some additional human capital. Farmers have to learn how to use a machine, how much fertilizer to apply, and so on. They can learn faster if they are literate and technically knowledgeable.

6.3 The Macroeconomics of Innovations: Factor Prices and Technical Progress

Given the factor intensity (section 4.5), the extended induced innovation model would suggest that

1. Europe and Japan led in land-saving technical progress, and the Western settlement countries (i.e., mainly the USA) led in labor-saving ones.
2. In "advanced" countries, labor-saving innovations become increasingly appealing as wages increased relative to rents and interest rates. In other words, one would expect a convergence in the pattern of adoption toward the "American" model.
3. The LDCs were unlikely to import most innovations from the "advanced" countries as they were too capital intensive for them. LDCs would have needed low-cost land-saving innovations (in Asia) or labor-saving innovations (in Africa), but they were too technically backward to produce them.

The hypothesis refers to both the production and the adoption of innovations, but testing the factor bias in R&D (research and development) is extremely difficult. In theory, one would need data on input for the research process—that is, on the total amount of resources allocated to R&D in land-saving or labor-saving innovations. Unfortunately, these data are scarce even nowadays and are simply not available in historical perspective. Until quite recently, most research was conducted by farmers, landowners, and other enterprising individuals. At best one can muster some indirect evidence from technical literature (technical pamphlets, journals, and so on). Relying on these sources, Olmstead and Rhode (2000b) argue forcefully, against the Hayami-Ruttan hypothesis, that in the nineteenth-century United States "biological" innovations were given much more attention than mechanical ones. However, their evidence is indirect and their category of "biological" innovations is all encompassing (everything but machinery). Indeed, a quick perusal of the country of origin of innovations confirms the basic insight of the model. All major mechanical innovations (the reaper, the tractor, the cotton and corn pickers, and so on) were invented or developed in the United States, while Europe led in the development of fertilizers and chemical products as well as in new practices of cultivation. This difference largely disappeared in the second half of the twentieth century: nowadays, companies on both sides of the Atlantic invest in R&D on machinery and on biotechnologies for the same market. This fact would support, as a first approximation, Hayami and Ruttan's hypothesis of induced innovation. On the other hand, the production of relevant innovations depends on many other factors—including genius, sheer luck, progress in related industries, and, above all, technological capabilities. Most LDCs would undoubtedly need land-saving and capital-saving innovations, but they have nowadays (and, *a fortiori*, had in the past) few scientific and technological capabilities. Thus, they have produced comparatively few innovations, and only in specific cash crops, such as rice and cocoons in Japan, coffee in Brazil, and so on. Most research on HYV was initiated and funded by the advanced countries.

The factor content of R&D is very important to understand long-run agricultural performance, but it is not strictly necessary to account for technical progress. In theory, a (3- or 4-factor) neoclassical model could explain technical progress even if the location of invention were totally random. Farmers would adopt the innovation or not according to local factor prices and thus, ultimately, to local factor endowment (although the relation between prices and endowment is not so straightforward, as was argued in section 4.5). Consequently, factor shares would be inversely proportional to factor prices, *and* changes in factor shares would be inversely proportional to changes in prices.[61] This inverse relation would hold, however, even without any technical progress, as a consequence of factor substitution along the same production function (section 6.1). In other words, changes in factor ratios can be evidence of an induced innovation only if they exceed the pure substitution effect. To test this hypothesis, Hayami and Ruttan (1985) compute a counterfactual (no progress) series of factor shares in the United States and Japan from 1880 to 1980: the difference with the actual shares is negatively related to factor prices and thus, they argue, must have been caused by technical progress.[62] Olmstead and Rhode (1993b) have contested this view for the United States. They stress the inconsistency between the fall in wages/rent ratio before 1930 and the alleged labor-saving bias in technical change (i.e., the diffusion of mechanical innovations).[63] In another paper (1998), they discover a positive relationship between factor intensity by state and long-run trends in factor prices.[64] In a very recent paper, Thirtle et al. (2002) rehabilitate the induced innovation hypothesis, showing that long-term factor price movements account for about 70 percent of changes in the aggregate machinery/labor, and about 60 percent of those in fertilizers-acreage ratios from 1880 to 1990.[65] On top of this, with few exceptions, the long-term estimates of aggregate production functions for the United States (and Canada) show that technical progress has been labor-saving, a stylized fact consistent with prior assumptions about factor endowment.[66] Thus, by and large, the econometric works do support the induced innovation hypothesis. However, they cannot be considered conclusive evidence—if anything because they refer almost exclusively to the United States during the past one hundred years. Extending their conclusions to the whole world and to the two past centuries would be excessive.

As an alternative, one can focus on specific innovations and explore whether the country pattern of their adoption fits with the a priori assumptions about the factor endowment.

1. As expected, there is very little evidence on the adoption of new practices, a microeconomic process that, by its nature, escapes measurement. The only available proxy is the cropping ratio, the ratio of cultivated crops to total arable land (table 6.1).[67] The table confirms that Asia led by far in the intensification of land use. According to Chao (1986), the cropping ratio in China rose from about 0.5 around the first year A.D. to 1 in the Sung period (12th century), to about 1.4 at the beginning of the nineteenth century.[68] Further intensification without technical progress was difficult, but some Asian peasants did succeed in increasing the cropping ratio, possibly losing something in terms of yields.[69] The increase in the cropping ratio in European countries reflects the reduction of fallow, which is, however, still practiced in some areas of the western United States (Gardner 2002, 43). The table omits Africa, for lack of data. Yet there is no doubt that African agriculture has become more intensive, and that the traditional swidden system is disappearing.[70] The worldwide intensification process is still going on: the FAO (2003) predict that, in the LDCs, the cropping ratio is going to rise from 0.93 at the end of the 1990s to 0.99 in 2030.[71] However, this intensification is powerful, albeit indirect, evidence of the adoption of "new" practices.

2. The historical evidence on seed use is scarce. It is often assumed that traditional farmers were diffident toward new varieties and used only local seed, if possible, from their own farm.[72] Indeed, even nowadays in "advanced" countries noncommercial seed (i.e., not produced by seed companies) accounts for a sizeable proportion of the total. However, the long-range trade of seed, produced by specialized farmers and seed companies, started as early as the late eighteenth century in the United States.

TABLE 6.1
Cropping Ratios, by Country, 1800–1938

	1800	*1850*	*1880*	*1910*	*1938*
China		1.4	1.4	1.4	1.3
Java			0.73	0.89	1.04
India				0.95	0.97
Japan			1.33	1.42	1.32
Taiwan				1.19	1.34
Korea				1.37	1.53
Burma			0.78	0.77	0.82
Egypt			1.35	1.44	1.60
Austria	0.70		0.86	0.97	
Belgium		0.94	0.98	0.99	1.00
Denmark		0.91	0.89	0.99	0.99
France	0.76	0.84	0.88	0.91	0.93
UK			0.96	0.98	0.97
Hungary		0.82	0.86	0.92	
Spain	0.46		0.53	0.54	0.60
Greece					0.72
Italy					0.84
Bulgaria			0.70	0.80	0.81
Yugoslavia				0.85*	0.94
Russia			0.60	0.70	
Europe				0.84	0.91
USA				0.93	0.91

Note: *ca 1920.

Sources: **Europe:** (1910) Institute Internationale d' Agriculture 1909–21. **Greece and "Europe":** (1930) Moore 1935, table 9; **Java:** (1880, 1910, and 1938) van der Eng 1996, table A.7. **China:** Chao 1986, 196–197. **India:** (1906–16 and 1936–45) Blyn 1966, table 6.1. **Burma:** (1896, 1910–12, and 1935–36) Saito and Kiong 1999, table II.1. **Japan:** (1878–82, 1908–12, and 1933–38) Hayami and Yamada 1991, table A6. **Taiwan:** (1911–14 and 1936–39) Lee and Chen 1979, table T-4. **Korea:** (1918 and 1938) Ban 1979, table K 4. **Egypt:** (1897–98, 1909–12, and 1936–38) Richards 1982, tables 3.6 and 4.1. **UK:** (1879–81, 1909–11, and 1936–38) Mitchell 1988, Agriculture table 1. **France:** (1781–90, 1845–54, 1875–84, and 1905–14) Toutain 1961, table 146. **Denmark:** (1861 and 1881) Jensen 1931, table II; and (1910 and 1938) Johansen 1985, table 2.2. **Belgium:** (1846) Goosens 1992, table 47; and (1880, 1910, and 1938) Blomme 1992, table 42. **Austria:** (1789, 1876–85, and 1904–13) Sandgruber 1978, table 8. **Hungary:** (1870, 1890, and 1910) Eddie 1971, 306. **Spain:** (1818–20, 1886–90, 1903–12, and 1930–35) Bringas Gutierrez 2000, cuadro I.3. **Bulgaria:** (1896 and 1910) Palairet 1997, 317; and (1938) Berend 1985, 188. **Yugoslavia:** (ca. 1920 and 1938) Berend 1985, 188. **Italy:** (1923–28) ISTAT 1939. **Russia:** (1870s and 1900s) Antsiferov 1930, 16. **USA:** (1910 and 1940) Historical Statistics 1975, series J53 and J54.

Purchased seeds accounted for two-thirds of the total seeds used in the United Kingdom on the eve of World War I (Brassley 2002c, 531). These companies must have promoted new varieties, if anything, because the farmers would otherwise have had little incentive to purchase seeds instead of using the local ones. Indeed, the available data show that varieties changed greatly, at least in the "advanced" countries. In Japan, improved varieties of rice rose from 36 percent in 1910 to 62 percent in 1939 of the total seed. In the United States, the Marquis and Durum varieties of wheat jumped from 14 to 87 percent of the acreage in just seven years, from 1914–21, and hybrid corn from 2 to 90 percent of total corn acreage in ten years, 1936–45.[73] The big change, however, was the diffusion of HYV varieties in LDCs (table 6.2). The table highlights huge differences by product and continent. To some extent, these differences reflect the factor endowment—lack of capital, for example, may have delayed the adoption of HYVs in Sub-Saharan Africa. But clearly the pattern has been determined by the availability of suitable seeds and thus by the initial choice of allocating most research funding to wheat and rice, and to varieties suitable for Asia and South America (section 6.2). On the other hand, Hayami and Ruttan (1985) would interpret the allocation of research funds as a rational concentration of scarce resources to meet the most immediate needs of land-scarce, over-populated countries such as Mexico or the Philippines.

3. As stated in section 4.3, artificial fertilizers—the quintessential yield-enhancing, land-saving innovation—started to be used toward the end of nineteenth century, and their worldwide consumption boomed after 1950. The pattern of the diffusion in "advanced" countries tallies well with their relative endowment of land (table 6.3).[74] The United States used very little fertilizers—roughly as much as Spain and little more than the allegedly backward Russia.[75] The difference among the "advanced" countries still persists today, and it is reproduced in the pattern of consumption of pesticides.[76] As anticipated (see table 4.15) the big change has been the boom in the use of fertilizers in LDCs (table 6.4).

Only Africa still lags clearly behind, in spite of a twelve-fold increase. South America uses as many fertilizers as the traditional land-intensive countries of Western settlement, while Asia (as a whole) matches Europe (as a whole). The so-called Asian rice bowls, such as Jiangnan in China or the central provinces of the Honshu island in Japan, attain the highest level of consumption in the world per unit of land (see Smil 2001, 164–173 and Appendices S and R). The consumption in Western Europe is still high, although it is declining, while the countries of Western settlement and South America trail behind. The consumption of fertilizers is low in Africa and in the transition economies, after the collapse of the Soviet system. By and large, this pattern is consistent with predictions 1 and 3: land-saving innovations were widely adopted in land-scarce countries as soon as farmers were able to afford them.

Table 6.2

Share of Land Planted with High Yielding Varieties on Total, by Continent, 1970–2000

	Latin America				*Asia*				*North Africa Middle East*				*Sub-Saharan Africa*			
	1970	*1980*	*1990*	*1998*	*1970*	*1980*	*1990*	*1998*	*1970*	*1980*	*1990*	*1998*	*1970*	*1980*	*1990*	*1998*
Wheat	11	46	83	90	19	49	74	86	5	16	38	66	5	22	32	52
Rice	2	22	52	65	10	35	55	65					0	2	15	40
Maize	10	20	30	46	10	25	45	70					1	4	15	17
Sorghum					4	20	54	70					0	8	15	26
Millet					5	30	50	78					0	0	5	14
Cassava	0	1	2	7	0	0	2	12					0	0	2	18

Source: UN Development Report 2001, 41.

TABLE 6.3
Fertilizer Consumption per Hectare, by Country, 1890–2000 (kg. of nutrient)

	Germany	*UK*	*Denmark*	*France*	*Italy*	*Spain*	*Netherlands*	*Belgium*	*Japan*	*USA*
1880	3.1									1.4
1890										1.9
1900	15.6							15.3	0.1	2.9
1913	42.0	28.2	17.9	19.7	13.3	4.2	163.7	35.5	3.4	5.8
1922	72.7	21.1	16.2	25.3	11.7	1.5	188.6	85.9	11.8	7.0
1937–38	143.9	60.1	53.8	40.6	26.0		299.2		29.7	8.7
1957–58	150.9	82.9	118.0	69.3	42.4	24.9	204.2	196.1	131	23.4
1998–00	240.9	299.4	162.7	235.0	158.6	119.5	494.5	369.4	290.4	108.3

Sources: 1890–1957: **Germany:** (1880, 1900, and 1913) Achilles 1993, table 22. **United States:** (1880, 1890, 1900, 1910, 1920, 1940, and 1960) Historical Statistics 1975, series K193 and J53 (assuming an average content of nutrients 17.5%); **Japan:** (1898–1902, 1908–12, 1918–22, 1933–37, and 1958–62) Hayami and Yamada 1991, table A5. **Belgium:** (1895, 1910, 1929, and 1960–61) Blomme 1992, table 48. **Others:** Pezzati 1993, table D; 1998–2000 FAO Statistical Database.

TABLE 6.4
Fertilizer Consumption per Hectare, by Continent, 1950–2000
(kg. of nutrient)

	1950	*1970*	*2000*
Africa	1.5	9.7	19.2
Europe	46.9	170.8	146.2
North and Central America	20.0	65.5	89.0
South America	2.2	19.8	88.6
Oceania	29.2	30.8	59.8
Asia	3.2	26.3	140.5
USSR		44.2	17.8
World	11.3	49.8	90.9
USA and Canada	22.3	71.4	93.4
Japan	131.0	387.3	293.8
Western Europe		182.0	182.1

Sources: **1950:** FAO 1952. **1970 and 2000:** FAO, Statistical Database.

4. Mechanical innovations also fits fairly well with the neoclassical framework. The evidence on the diffusion of improved tools is scarce, but it strongly suggests that technical change depended on factor endowment. A good example is the choice of harvesting tools before the invention of the reaper. Farmers could choose among the sickle (short), the scythe (long), and the grain cradle (a sort of improved

scythe).[77] The scythe is difficult to handle, especially for women, and its straw is unsuitable for thatched roofs as well as for winter fodder. Yet, in the nineteenth century, it spread all over northwestern Europe just because it saved 25 to 35 percent of the work at a time of rising wages. In the meantime, American farmers preferred the grain cradle, an even more efficient labor-saving technique. Another interesting case is the adoption of the plow instead of the hoe in Sub-Saharan Africa.[78] The innovation was undoubtedly capital intensive (a plow was very expensive by the poor African standards, and it needed expensive oxen), but also land saving (it increased yields, and the manure from oxen fertilized the fields). The effect on total labor demand is complex: plowing saved time per unit of land, but it needed preliminary work to prepare the ground, clearing it of rocks and stumps, and it entailed the additional work of tending the oxen. It is thus likely that the innovation was labor intensive, and thus consistent with the process of intensification.

As expected, Western Europe lagged behind the United States and other countries of western settlement in the diffusion of all key machines. The use of the reaper spread in the United States during the 1850s: on the eve of the Civil War, about half of U.S. wheat was harvested by machine, and, at the end of the war decade, the share had risen to 80 percent.[79] In contrast, reapers started to spread in the United Kingdom during the 1880s and on the Continent some twenty years later. Tractors accounted for 10.8 percent of total American draft power in 1920, 40 percent in 1930, 64.8 percent in 1940, and 100 percent in 1960.[80] On the eve of World War II, tractors accounted for about perhaps a third of total power in the United Kingdom and perhaps a fifth in Germany (and for less than 1% in Japan).[81] After World War II, the number of tractors rose quite fast in all the advanced countries (see table 4.11). Yet, as the figures of table 6.5 make clear, even nowadays, Europe and Japan have not yet caught up with the United States in terms of tractors per worker. The gap would appear even wider if one could measure the total draft power of tractors. American machines are bigger than their European counterparts, and both are bigger than the Japanese ones, which continue in the tradition of power tillers, suited to small Asian fields. As expected, the (capital-scarce) LDCs use fewer tractors than the "advanced" countries and, among them, tractors are more diffused in labor-scarce South America than in Asia or Africa.

Factor costs may account for other features of the process of mechanization. Reapers and, above all, threshers, a real worldwide success story, spread quite early in labor-scarce, albeit "backward" and capital-scarce, areas such as Russia and eastern Europe, Anatolia (from the 1880s), and Mexico (from the 1900s).[82] Bateman (1969) argues that the development and adoption of milking machines, even in the United States, were comparatively slow because the opportunity cost of female labor for milking was low.[83] Last but not least, the scarcity of manpower during both world wars greatly stimulated the adoption of labor-saving machinery, from tractors in Europe during World War I (with government support) to power tillers

TABLE 6.5
The Diffusion of Tractors, by Continent, 1950–2000

	Tractors/hectare			*Tractors/Worker*		
	1950	*1970*	*2000*	*1950*	*1970*	*2000*
Africa	0.001	0.002	0.003	0.001	0.003	0.003
Europe	0.007	0.042	0.072	0.015	0.149	0.549
North and Central America	0.018	0.022	0.019	0.201	0.347	0.281
South America	0.001	0.006	0.005	0.003	0.017	0.048
Oceania	0.008	0.009	0.003	0.081	0.225	0.142
Asia	0.000	0.002	0.130	0.000	0.001	0.007
USSR	0.002	0.008	0.009	0.010	0.067	0.086
World	0.005	0.012	0.018	0.002	0.017	0.020
USA and Canada	0.020	0.026	0.025	0.528	1.332	1.600
Japan	0.003	0.055	0.447	0.001	0.027	0.762
Western Europe	0.009	0.034	0.080	0.024	0.177	0.857

Sources: **1950** FAO Yearbook; **1970 and 2000** FAO Statistical Database.

in Japan and cane harvesters and cotton pickers in the United States during World War II.[84]

To sum up, both the econometric estimates and the anecdotal evidence show that a neo-classical model in its augmented, four-factor version can explain a good deal. It cannot, however, account for all technical progress in agriculture. To understand this, one has to take the peculiarities of the sector into account.

6.4 The Microeconomics of Agricultural Innovation: Appropriability, Complementarity, Environment, and Risk

Most agricultural innovations share three main features. They are not fully appropriable, they are complementary (interrelated) to each other, and they are highly location specific. An innovation is said to be nonappropriable (or to be a public good) if the inventor does not receive the whole return for his endeavors.[85] Two innovations are said to be complementary (or interrelated) if one or both cannot deliver its or their full benefits if not adopted jointly. Finally, an innovation is location specific if it can work only in a given environment. None of these features is exclusive to agriculture—but arguably they are more widespread and relevant in agriculture than in other sectors.

1. It is commonly believed that most agricultural innovations are not appropriable and that this feature reduces the amount of private R&D below the socially optimal level.[86] This belief provided a rationale for the establishment of public

research institutions (section 6.5) long before the development of the modern theory of the public good. Before endorsing this argument, however, one should carefully distinguish among different categories of innovations. There is no doubt that most agricultural practices are public goods: anyone can imitate a rotation, or a way of pruning, or a mixture of fertilizers. On the other hand, machinery and chemicals can be patented like any other industrial product. Mechanical companies might find it difficult to enforce their patents, in extreme conditions, such as in the American West (Atack et al. 2000, 271). Fertilizers are subject to fraud (such as mixing them with other substances), which seriously hampered their diffusion at the beginning.[87] These problems are bound to disappear, however, when the industrial producers grow big enough to build a reputation, and rich enough to afford to defend themselves in courts. Thus, for these products, public R&D has no clear advantage over private R&D. Actually, it may be positively harmful if the state endorses, with its full authority, a wrong solution. In the 1920s, the Japanese Ministry of Agriculture imported Western tractors, much too big for Japanese fields, and spent large sums on a fixed-engine cable-pulled plough, which proved useless.[88]

The case of "biological" innovations is perhaps the most interesting, as it features a far-reaching change in the degree of appropriability. Traditional and modern open-pollinated varieties of plants are almost impossible to protect, as a farmer could simply replant the best seeds from his own crop and/or mate the best animals of his herd. In contrast, hybrid seeds (or, a fortiori, genetically modified ones) have to be produced every year in a laboratory by skilled workers. As a result, they are out of the reach of a farmer, but not of a competing company. Discoverers of new varieties and producers of seeds have always tried to appropriate as many benefits as they could from their work (or luck). The American tree nurseries and Italian cocoon-egg producers in late nineteenth century tried to protect their markets by building "brand awareness" among their customers, but this strategy was slow, expensive, and uncertain.[89] Thus, very many "inventors" received recognition, and maybe the odd prize from a farmers' association, but little else. Their conditions changed for the better when the right to patenting was extended to living beings. The American producers of seeds and plants had requested this extension from the beginning of the twentieth century, but it was granted, after a long political battle, only in 1930 and only for new varieties of tree crops.[90] The European Community allowed the patenting of seeds in 1961 (if "manifestly superior" to existing ones) and the United States followed suit in 1970–71. As of 2001, 48 out of 207 independent countries have signed the UPOV (International Union for the Protection of New Plant Varieties) Convention. It creates a legal framework for the worldwide recognition of patents, which obliges companies to file for a patent in each country. The moral right to ownership of living beings is still heatedly disputed, to say the least, and the effects of the diffusion of new seeds and varieties and the possible constraints from patents to R&D in LDCs are also quite controversial.[91] On the one hand, patents increase the cost of new seeds or varieties, and thus they may hamper their diffusion. On the other, they increase the returns to investment in R&D and also act as a powerful signal of the quality of the seeds and thus increase their appeal.[92] Furthermore, enforcing the patents of reproducible plants against thousands of small farmers can be difficult and politically costly.

2. The concept of complementarity (or interrelatedness) is intuitive, although somewhat hard to pin down in practice. In an all-encompassing version, one could extend it to the provision of credit and to other "institutional" innovations (which will be dealt with in the next chapters), or argue that all techniques, new and old, form a coherent whole, and thus they are all necessary to make agriculture feasible and economically profitable. However, the concept will be used here in a more narrow meaning: two innovations are defined as complementary if joint adoption reduces the average costs per unit of output and/or the risks of failure relative to a separate adoption.

This concept was first used in agricultural history by Paul David to explain the British "delay" in adopting the reaper in the 1850s and 1870s. He argued that the machine could work only if the fields were suitably prepared—that is, if they were cleared of rocks or of deep ditches and were drained.[93] The cost of this additional work was so high as to make the innovation too expensive for marginal fields. The proportion of wheat harvested by machine rose when the fall in wheat prices forced farmers to stop growing wheat in these fields. The size, shape, and physical features of fields hampered the adoption of other machines, such as the steam plow in the nineteenth century, or forced the development of specific ones, such as the straddle tractor in the French vineyards or the already mentioned power tiller in Japan.[94] But complementarities were essential in the adoption of many innovations. Plows needed animals, and thus often needed a radical reorganization in agricultural practices to produce feed. Furthermore, animals had to be strong to pull modern implements and allegedly the small size of many Russian horses hampered the adoption of iron plows in the nineteenth century (Moon 1999, 128–29). Cotton and tomato pickers needed varieties of plants that all ripen at the same time.[95] Arguably, the most important case of complementarity involves HYVs, chemical fertilizers, and, in many cases, irrigation—the package popularly known as the Green Revolution. HYVs can deliver high yields only if they receive sufficient fertilizing matter and water. But they can exploit fertilizers better than traditional varieties because, they are semi-dwarf and thus less prone to lodging (falling down under the weight of the ear). It has been estimated that the elasticity of the use of fertilizers relative to the acreage under HYV ranges from 1 in India, to 2.4 in the Philippines.[96]

3. The relationship between biological innovations (new seeds, new rotations) and local conditions (e.g., soil, climate) is self-evident. For instance, many

attempts to imitate the nineteenth-century British "new husbandry" in Mediterranean countries failed because most grasses did not survive the long, dry, and warm Mediterranean summers (Galassi 1986; Simpson 1995, 36–38, 100–9). It is more interesting to stress that even chemical and mechanical innovations are location specific and depend on the environment. Each type of soil needs a specific combination of basic nutrients (an "integrated plant nutrient management" in the experts' jargon): a wrong one causes the farmer to waste money and may even harm the soil (and the reputation of fertilizers).[97] The shape of traditional plows and the amount of draft power required by different types of soil depend on their texture. Heavy soils, like the clay ones of northern Europe, need strong plows pulled by powerful animals. This combination would be unsuitable to light, semi-arid Mediterranean soils, as it would strip all their moisture. In fact, the traditional plow of southern Europe, the Middle East, and north Africa ("scratch plow" or "ard") was much lighter and was pulled by fewer animals (McClelland 1997, 16; Valensi 1985, 138; McCann 1996, 45–47; Brunt 2003b, 455–60). The machines for harvesting sugar cane differ according to the weight of the cane to be harvested: light in Louisiana, medium to heavy in Australia, and very heavy in Hawaii (Burrows and Shlomowitz 1992).

It is important to stress that this "circumstantial sensitivity," to use an expression of Evenson and Westphal (1995), affects also R&D.[98] Even today, in spite of all the advances in science, it is extremely difficult to predict a priori, without a field test, whether a given plant would thrive or die, or how it should be cultivated in a new environment. A fortiori, this was true in the past. Some recent work show that the spill-ins of R&D expenditure (i.e., the effects of R&D on the growth of TFP in other areas) are negatively related to the "distance" between the two areas—although the measured effect varies according to the design of the test.[99] Thus, "circumstantial sensitivity" contradicts the assumption of neoclassical growth models that techniques can be transferred at no cost.

Innovation is always risky, but complementarity and circumstantial sensitivity cause it, ceteris paribus, to be more risky in agriculture than in the rest of the economy. An innovation may fail because it does not fit the environment, or because another complementary innovation is not adopted. In some circumstances, it may even be difficult to establish a clear relationship between a specific innovation and changes in output.[100] The potential cost of failure differs according to the type of innovation. If a machine breaks down or proves to be unsuited to a task, the farmer can still revert to the manual system (even if hiring, say, a team of reapers many be more costly at the peak of the season than at its beginning). If fertilizers do not increase the crop as much as hoped, the farmer will lose the sum spent on it, but he still has a crop. In contrast, using a new variety of seed entails the risk of losing all the harvest, should it prove unsuitable to the environment. Ceteris paribus, the adoption of an innovation depends on the amount of risk that the individual is willing to take. Most people do not like risk (they are "risk averse," in jargon), but the degree of this feeling may vary greatly.[101] Clearly, wealthier individuals can cope better with a failure—and thus landlords are supposed to lead technical progress (and are scolded if they fail to comply).[102] Innovation is also positively related to education, which enables individuals better to understand the opportunities afforded by new technologies and to use them properly. Foster and Rosenzweig (1996) show that, in the 1980s, literate Indian farmers were ceteris paribus more ready to adopt HYV seeds and obtained higher profits than the others.[103]

Institutions can greatly help the farmer overcome all these problems, and to smooth technical progress. For example, an efficient credit system can help reduce the cost of modern inputs, and co-operatives can provide technical assistance and share both the risk and the financial cost of innovations. Many historians would argue that the "wrong" size of the farm, and the "wrong" pattern or the "wrong" set of contracts (notably sharecropping) could hamper innovation. We will deal with these claims in the next two chapters, after having surveyed the historical evidence. Here, it is sufficient to anticipate the conclusion: "institutions" may have slowed down the adoption of suitable and profitable innovations for some time (or for a long time, in certain cases), but they have only seldom prevented it. In the long run, institutions have adjusted to agricultural needs.

6.5 The Microeconomics of Agricultural Innovation: Research Institutions and Technical Progress

Public funding of agricultural R&D has a very long history. Already in the late eighteenth century, many universities created chairs in botany and related issues, and several botanic gardens can claim to have started agricultural research. University professors, including J. Liebig, often did applied work. Universities, however, are only one component of an "ideal" system of public research, which includes other types of institutions. In the United States, "universities sponsored the fundamental study of various sciences. Experiment station scientists tailored general ideas to the soil, climate and crops of specific states. Finally, county agents acted as a communication system carrying researchers' results to farmers and farmers' complaints to researchers." (Clarke 1994, 28). This three-tier public research system developed slowly in Western "advanced" countries in the first half of the nineteenth-cenutry. The American federal government started to collect foreign plants and seeds systematically (resorting to the navy, to the consular service, and, from 1839, to the Patent Office) in the early nineteenth century, even though, lacking an experimental farm, it had to rely on the collaboration of farmers for field tests (Juma 1989, 57–60; Huffman and Evenson 1993, 6–10; Fowler 1994, 14–20; Kloppenburg 1994, 53–57). Saxony established the first public experimental station at Mockern, Germany, in 1851, based on the model of two privately owned estates: Bechelbrom in France (established in 1834 by Boussingault) and Rothamstead in the United Kingdom (established in 1843 by Gilbert), which vie for the honor of having been the first modern experimental farms.[104] Public research facilities proliferated in the second half of the nineteenth-century.[105] In 1875, there were already ninety of them in Europe, and the number rose to 233 by 1892.[106] In the United States, the first station was established in 1875 (by the state of Connecticut), and federal funds were allocated to agricultural research twelve years later, through the Hatch Act.[107] They had to be used to set up an experimental station in each state, which had to focus on applied work, leaving the science-oriented research to the Department of Agriculture. Among the great "advanced" countries, the United Kingdom clearly lagged behind in developing a public system. Universities started to offer courses in agricultural technology in the 1890s, specific public funds were allocated only after 1910, and most research institutions were linked to universities instead to the Ministry of Agriculture, as elsewhere. (Brassley 2000a, 613–17; Perren 1995, 28–30; Palladino 1996, 123–25; Koning 1994, 86). Japan explored a different way, after an initial unsuccessful attempt to import Western technology—technology that was clearly unsuitable to the country.[108] The government relied heavily on innovations by prominent farmers (called *rōnō*): the experimental stations (set up from 1885) were supposed to test them and, if the innovations proved suitable, to issue the results throughout the country. Thus, the early Japanese system can be seen as a forerunner of the integration between peasants' knowledge and outside expertise, which nowadays is often touted as the alternative to the traditional model of government-run, science-based R&D.[109] The Japanese model was indeed initially successful: the peasants' varieties of rice such as the Shinri-ku contributed substantially to the increase in yields during the nineteenth century. The backlog of farmers' techniques and innovations was limited, however, and the flow of new discoveries soon proved to be insufficient. Thus, to foster further innovations, Japan in the 1900s adopted the "Western" model, which gave a key role to organized research in agricultural stations.

Research in tropical and semi-tropical LDCs started later than in the "advanced" countries and focused on cash crops for exports. The first experimental stations were established in the 1880s, by the British Royal Niger Company for coffee and cocoa and by a group of sugar producers in Java; many others followed suit.[110] In contrast, research on food crops for the native population lagged behind, with few exceptions. The brightest one was in Java, where the Dutch government collected rice varieties beginning in the 1900s and set up a full-fledged network of stations in the 1920s and 1930s.[111] The overall neglect of research on food crops, Lewis (1972, 202) argued, had dire effects on the growth of agriculture in tropical countries. It is probable that colonial governments did not care much for the welfare of the native population, but one should not underestimate the difficulties of the task. European knowledge was of little use, as the environment was very different and the new techniques were intensive in both financial and human capital, both of which were lacking in the colonies. Furthermore, formally independent countries such as China or Thailand did not shine in their commitment to agricultural R&D. China had a glorious tradition, but the Imperial government did nothing, and the new Republican one invested only small sums in industrial crops such as cotton and silk.[112] Indeed, the research facilities, which discovered the HYVs, such as the CIMMYT in Mexico and the IRRI in the Philippines, were funded by the Rockfeller and Ford foundations, and later by the American and other Western governments as well as by donors. In the 1960s, these national research facilities were upgraded and connected in a network of "international advanced research centers" (IARC), which have been coordinated by an umbrella group, the CGIAR (Consultative Group on International Agricultural Research) since 1971.[113] In theory, the IARC should develop basic germoplast for large areas with broadly similar environments, and distribute it to national experimental stations, which produce the varieties suited to local needs. In other words, IARC provides a "public good" for national networks of experimental stations.

The production of suitable innovations is only a first step. Farmers have to know that an innovation exists, and to be convinced that its benefits are great enough to run the risk of adopting it. For millennia, information has traveled by word of mouth, which is perhaps a trustworthy method, but undoubtedly a very slow and inefficient one.[114] The speed in transmission of information was boosted in the nineteenth century by transoceanic migrations and, above all, by the development of specific institutions. Some of these institutions relied on the willingness of farmers to pay to be informed ("market" solution), but the majority conveyed the relevant information for free.

The market path to knowledge is not exactly a nineteenth-century novelty. Agricultural textbooks have been available since Roman times, but their price put them out of reach of everyone but the very richest landlords. The number of published books increased somewhat in the eighteenth and nineteenth centuries, and, above all, the print runs grew substantially: the most popular British agricultural textbook sold more than 60,000 copies from 1890 to 1913.[115] The real change, however, was brought about by the development of the periodical press at the end of eighteenth century. General-purpose local newspapers in rural areas published news on agricultural issues and technology, and the specialized press flourished. In the United Kingdom, there was not a single agricultural journal in 1751, but the number increased to 15 in the 1800s and to 170 in the 1880s. They informed farmers on many issues (including prices and local politics), but they dealt mainly with technology. In "advanced" countries such as the United States, from the 1920s onward the written word was supplemented by radio and later television, which broadcast special programs for farmers (Gardner 2002, 27). This "market" solution had two serious drawbacks. First, the quality of information varied greatly. Sometimes it was sound, clear, and reliable, sometimes potentially good but insufficient (e.g., it is difficult to understand how a machine works from few drawings), sometimes plainly wrong or crazy. To some extent, these defects were the unavoidable consequence of the imperfection of agronomic science and of the media itself. But, in other cases, the author of a book or the editor of a journal had a financial interest in selling a machine, or in promoting a new variety of seeds, regardless of its actual worth for farmers.[116] Second, the written word could reach only a subset of farmers. They had to be literate, interested in innovation, and "rich" enough to afford to spend some money in professional updating. Only a minority of farmers fulfilled these conditions, even in the "advanced" countries. According to Goddard, in the United Kingdom around 1850, only approximately a half of landowners and tenants of large estates actually read a newspaper (Goddard 1983, 122–24 2000, 673–82; Turner et al. 2001, 33). The diffusion of newspapers was more impressive in the United States, growing from an estimated 0.35 million readers around 1860 to 4 million readers forty years later (out of an agricultural work force of 6 million).[117] Clearly, the potential role of the "market" path was increasing with the spread of literacy and the rise of farmers' income, but until quite recently it has played only a minor role in the dissemination of agricultural knowledge.

The mass of poor, illiterate peasants had to be informed of potentially fruitful innovations in other ways. They could be shown the innovation at a fair, a test exhibition, or a conference. They could be encouraged to adopt it through prizes for the best animal, the highest yield, and so on. They could also receive a regular visit from an extension worker (the so-called training and visit system). All these activities were expensive. Who paid for them? For appropriable innovations, such as a new machine or a new chemical product, the task fell to the balance sheet of the industrial companies producing the innovation.[118] Industrial companies advertised in specialized journals, but they resorted also to specialized shops, salesmen, and a variety of other means. In the 1930s and 1950s, for example, the biggest Italian producer of fertilizers, Montecatini, organized a traveling cinema: the (mostly illiterate) southern peasants could watch the movie only after long commercials extolling the virtues of the fertilizers. The diffusion of non-appropriable innovations had to be entrusted to nonprofit organizations. A popular solution in the eighteenth and nineteenth centuries was the agricultural society, a spontaneous association of (wealthy) farmers to foster technical progress. The earliest of these associations in the United Kingdom, the Society of Improvers in the Knowledge of Agriculture was established in Scotland in 1723, but it disappeared in 1741. The oldest surviving one is the Dublin Society (established in 1731), and the most important is the Royal Agricultural Society of England (established in 1838).[119] The Société Royale d'agriculture de la généralité de Paris, established in 1761, survives as the Académie d'Agriculture de France. In the nineteenth century, these associations proliferated in "advanced" countries, both at national and local level. In the United States, the Society for Promoting Agriculture was established in 1785; there were almost 100 such associations in 1819, and, after a great boom in the 1850s and 1860s, almost 2000, together totaling 400,000 members—or about one farmer out of ten nationwide, and probably more in the Northeast (McClelland 1997, 203–16; Huffman and Evenson 1993, 13; Scott 1970, 10–12). The agricultural societies published journals and books, and funded applied research (such as competitive tests of fertilizers) and extension work. Although at times very useful, they could hardly be a permanent generalized solution, mainly because they faced a serious free-riding problem. If information on, say, the best type of plow could be easily collected by going to a competitive test, why should a farmer invest his own time and money to organize an association to perform the test? Thus, in most cases, the initial enthusiasm wore off and many of these societies started to lose members, sometimes falling prey to internal infighting, and ultimately became dormant.[120]

The information about (nonappropriable) techniques is, therefore, largely a public good, to be provided by the state.[121] The institutions with this aim varied greatly across time and space. Some of them targeted the young generations. Some countries included agricultural topics in the syllabus at general-purpose schools in rural areas or set up specialized agricultural schools or colleges.[122] In the United States, the Morrill Act (1862) allocated a large extension of federal land to the states for free, to set up colleges not only for training but also for research—the first research facilities being Berkeley and Chapel Hill (Huffman and Evenson 1993, 11; Alston and Pardey 1996, 16–19; Hurt 1994, 189–91; Olmstead and Rhode 2000a, 714; Clarke 1994, 28–31; Scott 1970, 26–36). The overall results were not outstanding. Only a small minority of potential pupils (5 percent in the acclaimed Danish system in the 1890s, 1 percent in France in 1913, 0.4 percent in Indonesia in 1939) was exposed to specific agricultural education.[123] The number of college-level students reading agricultural issues had always been very small, and most of them ended up in other jobs (including agricultural research) instead of going back to farms.[124] Experts often criticized this behavior as a sort of betrayal of the mission entrusted to them by the state. However, it appears rather rational: in the nineteenth century, the investment to obtain a higher education degree was substantial, and thus the graduates sought a better status and income than family farming could offer. The big increase in the skills of farm operatives is quite a recent development (section 4.4).

Educating the young generations may be a sensible strategy, but it works only in the long run. In the short run, it is necessary to transmit specific information on new techniques, crops, and so on to adult farmers, and this is the specific task of the extension services. The Japanese government entrusted the task to formally free farmers' associations, which it helped to organize: in 1905, it made subscription to the National Agricultural Association compulsory for all farmers.[125] Most countries, however, relied on the civil service for this purpose, and, even in Japan, the state gradually stepped in, while the farmers' associations increasingly concentrated on lobbying. In Imperial China, reading agricultural textbooks and teaching peasants had always been an important tasks of civil servants (Gang Deng 1973, 120–30; Purdue 1987, 20–25, 131–34). In Western countries during the nineteenth century, experimental stations were often also asked to perform extension services and possibly other tasks, such as fertilizer testing (Grantham 1984, 192). But R&D and extension did not fit well together, in that trained scientists resented being forced to perform these "menial" tasks, while farmers complained that scientists were wasting their time and the public money in "theoretical" work (Marcus 1988, 21–23; Kloppenburg 1994, 75–77). Thus, from the early twentieth century, the spreading of knowledge was entrusted to specialized institutions. In Italy, the government set up a network of *cattedre ambulanti* (travelling lecturers) starting in 1886; in Indonesia, the Dutch colonial administration organized a (small) extension system as early as 1905; and, in the United States, the Smith-Lever Act in 1914 allocated specific federal funds to extension.[126] Extension services greatly expanded after World War II, especially after the introduction of the so-called training and visit system, pioneered in Turkey in the 1970s.[127] The number of extension workers grew from 180,000 in 1959 to 600,000 in 1989, with two-thirds of them working in LDCs (Alexandratos 1995, 344–48).

This anecdotal evidence strongly suggests that expenditures on R&D and extension had greatly increased, both in real terms and as a share of output. The available data, although somewhat imprecise, confirm this impression. From the 1880s to the 1930s, the public expenditure in real terms increased forty times in the United States (i.e., from 0.02% in 1889–91 to 0.81% of gross output at the prewar peak in 1931–33) and twenty-three times in Japan (from 0.02% in 1880 to 0.20% in 1935).[128] Comparable figures for European countries are hard to find, but, according to an estimate in 1900, the European countries spent some 2 million of U.S. dollars—or, in proportion of output, a little less than the United States.[129] The increase in R&D expenditure continued after World War II: at the end of the 1950s, the United States spent a third more than it had twenty years before. Table 6.6 shows some worldwide estimates (excluding "transition" economies), from different sources, of expenditure in R&D and extension in the past forty years.

Table 6.6
Expenditure in Agricultural Research and Extension, 1959–96 (millions 1993 US $)

	R&D (a)	*R&D (b)*	*R&D (c)*	*Extension*	*Private (a)*	*Private (b)*
1959	2,611			1,851		
1970	7,117			3,780		
1971		9,758				
1976			11,837			
1980	10,296			4,702		
1981		15,031			5,353	
1984–86			16,424			
1991		20,034			8,910	
1994–96			21,692			11,501

Sources: columns R&D (a) and extension: Judd et al. 1986, tables 1 and 2. R&D (b) and private (a): Alston et al 1998, table 1. R&D (c) and private (b): Pardey and Beintema 2001, 4.

In spite of the really puzzling difference between estimates for R&D around 1970, these figures highlight some important stylized facts.[130]

1. First, and foremost, the total expenditure in R&D has been growing in real terms, albeit at a declining rate—9.1 percent from 1959 to 1970 (8.7% including the transition economies), 4.3 percent from 1971 to 1981, 2.9 percent in 1981–91, and 2.0 percent in 1991–95. The growth exceeded the increase in world gross output in all periods but the last one.

2. Agricultural R&D has traditionally been a public undertaking, and it still is, but this feature is changing. According to the estimates of table 6.6, state finances still provided two-thirds of funds in the mid-1990s, but this figure probably overstates its role. It is likely that the share of private R&D has grown further since 1994–96, and anyway the data omit the R&D expenditure of chemical and engineering companies, such as Bayer or Ford (including internal spillovers from other research projects).[131] Private R&D spending exceeded its public counterpart in the United States beginning in the 1980s.[132] This increase in the share of private spending reflects structural change in the appropriabilty of biological innovations, but also political choices, such as the privatization of research facilities and the concentration of public funding on "pure" research (Palladino 1996, 123; Juma 1989, 82–85; Alston, Pardey, and Taylor 2001, 8).

3. International research accounts for a small proportion of total expenditure. Its share peaked in the mid-1980s (3.5%), and a decade later it was down to 1.5 percent (Iftikhar and Ruttan 1988, 12; Alston et al. 1998, 1067; Pardey and Beintema 2001, tables 1, 2). These figures may, however, underestimate its effective impact. In fact, international centers provide the germoplasm for the development of new varieties, which is a sort of a public good for national R&D systems.

4. Almost every country, including some very poor ones, undertakes some R&D, but the developed countries still take on most of the expenditure. Their share on public R&D declined from two-thirds in 1959 to about half in 1994–96, but they still monopolize private R&D, accounting to more than 95 percent of total expenditures in the mid-1990s.[133] The LDCs spent much less as proportion of output (0.6% vs. 2.6%) and especially per agricultural worker (8.5 dollars vs. 594). The latter gap increased from fifty-two times less in 1976 to seventy times less in 1994–96. To be sure, researchers in the LDCs cost much less than in advanced countries, and are not necessarily less productive, so that the expenditure in R&D can be more efficient.[134] This potential advantage, however, cannot compensate for the decades of higher investment in R&D by "advanced" countries. Pardey and Beintema (2001, 17) estimate that the discounted value of past (formal) R&D is equivalent to eleven times the gross output in the United States, and to less than one in Africa (Gallup-Sachs 2000, table 3). Furthermore, the gap is bound to grow further in the next decades, as few LDCs have the capabilities for cutting-edge research in biotechnology.

5. Expenditures in extension grew more slowly, at least from 1959 to 1980, than those in R&D (4.4% instead of 6.5%).[135] Proportionally, the LDC countries have spent more on extension than on R&D: they accounted for 42 percent of world expenditures in 1959 and for 53 percent in 1980. Such a focus on extension is rational to the extent that farmers need more support in the LDCs than in the West (and, anyway, they get much less of it, in per capita terms).[136] This strategy assumes, however, that imports of technology can supplement the local supply, and this may not be the case.

How did much expenditures in R&D and extension pay off? The conventional wisdom does not rate the achievements of free associations very highly: as Clout (1980, 37) succinctly puts it, "There is precious little evidence that such information was diffused from the cultivated few to the cultivating masses." However, the issue is still under-researched, and, consequently, this assessment may be too dismissive. In contrast, the effect of public R&D and extension attracted a huge amount of attention, because past returns are supposedly a good guide for further investments. Most work follows the method pioneered by Zvi Grilliches in his famous 1957 article about hybrid corn.[137] He computed the return to expenditure in R&D as the ratio of net social benefits (i.e., the increase in production less direct costs) to the total expenditure on that project, suitably discounted. He found exceedingly high rates (about 700%). Some, more recent, works compute the rates of return from the coefficient of R&D expenditures in a production function. The two methods have applied to literally thousands of projects (table 6.7). The figures may overestimate the actual benefits of agricultural research. First, some estimates are extremely high, and thus they tend to bias the average upward (thus, the median is likely to be more representative than the mean). Second, the overwhelming majority of estimates compare costs and benefits of specific projects in a given country, neglecting spill-overs (to other countries/projects) but also spill-ins (from other countries/projects).[138] On top of this, estimates may overstate the rates of return as they refer to successful projects and neglect the failed ones. Last but not least, the computation is subject to several possible measurement errors.[139] It is, however, unlikely that these mistakes are so large as to lower returns

Table 6.7
Rates of Return to Investment in Agricultural Research

	n.	*Mean*	*Median*
R&D	1144	99.6	48.0
Extension	80	84.6	62.9
R&D + extension	628	47.6	37.0
Total	1852	81.3	44.3

Source: Alston et al. 2001, table 12.

down to a "normal" level—say 5 to 10 percent. In other words, actual investment in R&D and extension paid off handsomely if compared with alternative investments of capital.

This fact does not imply that all investments in R&D have been optimal. In fact, no matter how high returns have been, it is impossible to rule out the possibility of even higher returns in some other lines of inquiry. To some extent, mistakes in the allocation of R&D funds are unavoidable, but in quite a few cases the misallocation was intentional. Science-trained people are likely to prefer pursuing original lines of research instead of testing foreign techniques, even if the latter are more economically efficient (Byerlee 1996, 1710). As previously said, funding of R&D in the LDCs was biased toward export crops in the nineteenth century, and it is still somewhat biased towards food crops for the urban population, neglecting those for the rural population.[140] In South Africa, during apartheid, the state funding for R&D and extension was geared toward labor-saving, capital (or intermediate input) using innovations, which were clearly unsuited to a labor-abundant country.[141] In the United States during the 1980s, farmers' lobbies started to influence the allocation of funds to experimental stations, and Marcus (1988) strongly suggests that this "pork-barrel science" yielded a suboptimal allocation of resources.[142] The R&D in the Soviet Union was severely hampered by Lysenkoism, the theory that the environment could permanently change plants and animals.[143] It lacked any scientific base, but Stalin, and later Mao, believed or pretended to believe in it.

The computed returns on expenditure for extension are almost as high as for R&D (table 6.7), but their effect is more controversial (Boserup 1955, 66; Eicher and Baker-Doyle 1992, 130–34; Anderson and Hoff 1993, 474–77; Antholt 1998, 356–59; Huffman 2001, 359). Extension workers are often accused of behaving as if they were the only repositories of truth and of disregarding the peasants' feelings. Econometric estimates of the impact of extension yield mixed, but overall positive, results. As a whole, macroeconomic estimates find a more positive impact of extension on some measures of productivity than do household-level studies. Some of the latter find no relationship between the efficiency of single farms and the amount of extension services received by the cultivating households.[144] These results may reflect the transfer of information from the contacted households to their neighbors. There is also, however, some troublesome evidence that extension services and R&D are substitutes instead of complements—that is, they duplicate efforts.[145] All these results do not imply that extension services are useless. They do suggest, however, that their worth varies according to the circumstances. They also depend on the quality of the advice they transmit, which can be "wrong"—in other words, economically or environmentally unsound.[146] In this case, the extension workers may waste their time (if peasants do not follow their advice) or the farmer's time and money (if they follow the advice). As a rule, extension services are most effective when farmers are poor, strongly risk-averse, and poorly educated, and thus extension benefits tend to wane with economic development.[147] But designing an effective extension system is not easy.

6.6 Conclusion: On the Causes of Technical Progress

The last three chapters yield one firm conclusion: technical progress has been a major source of production growth, and its role has been increasing. Throughout most of the nineteenth century, technical progress was limited to modest improvements in tools and to the dissemination and adaptation of plant varieties and cultivation practices, and thus output increased, thanks mainly to additional inputs. The first half of the twentieth century featured a boom in the consumption of fertilizers in Europe, and a boom in mechanization in the United States. After World War II, mechanization extended to all the "advanced" countries, the consumption of fertilizers boomed in the LDCs, especially the Asian ones, and the HYVs greatly boosted yields. As we will argue in chapter 10, modern economic growth would have been much more difficult, if not impossible, without technical progress in agriculture.

This chapter also provides some general ideas about the determinants of technical progress. Undoubtedly, change in relative prices of inputs can explain much. The fall in capital cost fostered the adoption of modern innovations first in the capital-abundant "advanced" countries, and later in LDCs. The pattern of adoption was shaped by relative factor prices. By and large, land-scarce countries, such as those in Western Europe, adopted land-saving innovations—for example, fertilizers—earlier than did land-abundant countries, such as the United States (and vice-versa for labor-saving innovations like machinery). The exact factor intensity of each innovation is hard to guess, however, also because of widespread complementarity. Moreover, the diffusion of innovations requires more time and needs more funding in agriculture than in the rest of the economy. In other sectors, a specific technology—say, railroads—can be invented once, and then be adopted, with small changes, all over the world. In agriculture, this is simply not true. Machines and fertilizers have to be adjusted to the environment, and all "biological" innovations are highly location specific. Furthermore, R&D is necessary to stave off environmental challenges and to maintain the level of productivity. Thus, technical progress needs substantial investments in R&D and extension, and rates of returns to these investments are usually high. Publicly funded institutions provided a substantial share of the necessary funding, especially for non appropriable innovations. Clearly, some societies were more efficient than others in marshaling resources, and in setting up the appropriate research institutions.

This summary is couched in deliberately vague terms. How much did technical progress contribute to productivity growth, and how much did formal R&D contribute to technical progress? Was the expenditure in R&D worth the investment? High rates of return are not sufficient evidence, as high returns on a puny sum would not raise overall growth substantially. One has to consider the contribution of total expenditure in R&D and extension to TFP growth, under the questionable but necessary assumption that the output of the research process is related to the inputs. The expenditure can be used as an independent variable in a production function (the so-called "integrated" approach) or in a regression to explain TFP growth (two-stage decomposition).[148] As usual, the research is much more abundant for postwar United States than for any other country. In a pioneering work, Evenson and Huffman (1993) find that public R&D expenditure did indeed contribute to TFP growth, but the elasticity was not high, especially in interwar years.[149] They also estimate that from 1950 to 1982, public research and extension together accounted for some 35 percent of the total production growth for crops, and private research for a 15 percent (the rest being accounted for by growth in input), while for cattle-breeding the proportion was quite different. Public funds accounted for a mere 15 percent of total growth, and private R&D for a full 45 percent. In another contribution, for a slightly different period (1949–91) with a different estimation procedure, Gopinath and Roe (1997) find a rather high elasticity of TFP for public research 1949–91 (0.8) and no significant contribution from private research. The authors surmise that companies were able to appropriate the full returns of their investments in R&D by charging higher prices for their products (seeds, machinery, chemicals, etc.). Schimmelpfenning and Thirtle (1999) stress the role of the transfer of technology, at least among advanced countries: expenditures in RD and patents in other countries affect TFP growth even more than in domestic ones.

The effect of R&D is a part of a broader research agenda on the sources of productivity growth. Until quite recently, this issue has been somewhat neglected, as the attention of scholars focused on factor bias of technical progress. Some recent work, however, deals with the causes of TFP growth. Pingali and Heisey (2001) survey twelve studies on crop production in LDCs.[150] The explicative variables include—besides research and extension, literacy, and other measures of education—the diffusion of HYV and other indices of technological progress (e.g., mechanization), infrastructures, institutional reforms (as in China), and so on. The authors sum up the main results in four points: (1) research and development contributed quite substantially to overall growth; (2) political reforms can be useful to reduce market imperfections, and thus to enhance the benefits of R&D; (3) infrastructures (or related variables such as market density or irrigation) are positively related to TFP growth, but the result may be spurious (i.e., the most advanced areas attract more investment in infrastructures); (4) results on education and extension are mixed, possibly because measures of education are too general to capture the investment in agricultural education. Unfortunately, as Pingali and Heisey point out, as these works are not inspired by a consistent theory on the causes of TFP growth. The choice of variables is somewhat haphazard and, above all, they refer almost exclusively to the causes of technical progress. They neglect the contribution of changes in efficiency (section 5.4), brought about by institutional changes, most notably commercialization (chapter 8), and by agricultural policies (chapter 9). These causes are taken into account in a new econometric analysis, which uses as dependent variables the estimates of TFP change by country of Table IV of the Statistical Appendix (Federico 2004c).[151] Productivity growth comes out to be positively related to latitude (a rough proxy for environment), literacy (as proxy of the endowment of human capital), openness (as a general measure of orientation of the macroeconomic policies) and, somewhat weakly, to political freedom. It is negatively related to a dummy for socialist countries (collective agriculture) but also to the total agricultural production, with a nonlinear effect (possibly reflecting the smaller scope for spill-ins and technology imports).[152] These results suggest that farmers can increase their productivity in almost all circumstances, provided they are exposed to the "right" policies (support to RD, openness, education, and so on). As we will detail in the next chapters, this has often not been the case. This conclusion must be regarded as provisional and tentative, as the analysis needs a lot of improvement, especially in the explicative variables. But the issue is worth an additional effort, as agricultural TFP growth has been essential for agricultural growth and is going to be so also in the future, especially in LDCs.

Chapter Seven

THE MICROECONOMICS OF AGRICULTURAL INSTITUTIONS

7.1 Introduction: What Are the Institutions, and Why Should We Care about Them?

In theory, agricultural households could be completely self-sufficient if they were ready to work very hard and to accept a very low living standard. They can obtain much more by interacting with other households, and pooling or exchanging factors of production and goods. These interactions need a set of formal or informal rules to determine the initial ownership of the goods and factors (property rights) and to regulate the exchanges (contracts, markets, and other forms of distribution). This chapter aims at understanding how these rules ("institutions") work and why they change. Agrarian historians and economists have, or used to have, widely different opinions on this issue. The former assume that institutions are determined by "tradition," and that they can change only as a consequence of decisions by political elites. They also assume that elites are motivated by the quest for power and status more than by economic gains. Thus, institutions are exploitative and more likely than not to jeopardize agricultural growth. Economists, in contrast, view institutions as the outcome of a bargaining process between parties, with a reasonable chance to yield a socially optimal (i.e., production-maximizing) outcome. Furthermore, they argue, any change in the underlying economic conditions (demand, factor endowment, technology, and so on) that affects the interests of the parties would trigger changes in the outcome of the bargain. In other words, institutions can adapt to the needs of the economy, and do not necessarily hinder economic growth. As of late, the gap between the two disciplines has been narrowing. The majority of agricultural historians have abandoned their stereotypes, or have abandoned the field altogether, attracted by other, more exciting, issues. Economists and economic historians have increasingly realized that not all institutions are rational and efficient. People can be motivated by reasons other than potential monetary gains (from the quest for power and social status to altruism), property rights can be poorly defined, and, above all, markets can be imperfect or nonexistent. Most of modern development economics deals with the consequences of these deviations from the neoclassical ideal: in the words of Binswanger and Deininger (1997), "missing or incomplete markets can explain many of the characteristics of rural societies'.[1] Some agricultural markets are bound to be missing or incomplete because of the structural features of agriculture, but, in many other cases, they are missing or incomplete for historical reasons. Markets can work properly only within a suitable institutional framework: not by chance, the 2002 World Development Report (World Bank 2002) is entitled "Building Institutions for Markets."

This chapter, the first of three dealing with institutions, views them as the solutions to specific problems arising from the peculiar features of agriculture. What is the optimal design of property rights (section 7.2)? How are factors (land and savings) to be matched in a viable enterprise via the respective markets (section 7.3)? How can peasants get the capital they need (section 7.4)? What advantages do the free association in co-operatives offer peasants (section 7.5)? Is some category of farm, by size or tenure, superior to others (section 7.6)? The whole chapter thus reads as a sort of "theoretical" introduction to the analysis of long-run changes (chapter 8).

7.2 Property Rights

Modern property rights allow agents (individuals or households) to make freely whatever decisions about production and consumption they feel are best in order to maximize their welfare, given their preferences for consumption and leisure. To this aim, a set of property rights must fulfill four conditions: (1) Individuals have the right to dispose of their time, money, and other assets as they wish, with some well-defined exceptions (e.g., taxation). Landowners have the exclusive right to decide whether and how they cultivate their estates or even to leave the land idle if they want. (2) They can bequeath their estate to their heirs, whom they can choose, at least within some limits. (3) They can trade their own labor, other assets, and goods in markets, where they can pay or get a clear set of prices. (4) They can have their rights enforced by institutions (e.g., courts), which apply the same rules to everyone.

It is doubtful whether these conditions are fully met in "advanced" Western countries nowadays. They were certainly not met in the past. Land and (therefore) most of the capital, belonged to communities, and/or there were multiple claims on it and on its products. Sometimes, even the right to labor belonged to other people—either totally (slavery) or partially (serfdom). Property rights on land and on men were often linked, and jointly were the main sources of social and political power. The markets were not the sole or the main way of allocating resources and redistributing products. Sometimes, they simply did not exist at all: for instance, in many cases land sales were forbidden. Last but not least, property rights, especially on abundant factors, were often poorly defined, and their enforcement was often haphazard and related to the social status of the parties involved.

These property rights are usually defined as "traditional" or "customary," because they are the outcome of the process of institutional change over millennia. Ester

Boserup (1955) has tried to model this process as no historian would probably ever dare to, and thus she has been the forerunner of the modern literature on property rights.[2] In her model, the creation of property rights on land and its products is a key component of the intensification process (section 6.1). The more intense agriculture is, the more long-term commitments it requires from producers; and thus the more defined property rights have to be in order to extract these commitments. In her model, population growth entails an orderly transition from common ownership and use of resources to the private use of collective resources (temporarily allocated to households) and to private rights with some constraints from the group. Actually, in history, such an "orderly" evolution has been the exception, not the rule. In fact, "free" (unsettled) land or militarily weak agricultural societies have always attracted powerful outsiders, such as warrior tribes (like the Moghuls in India), European conquistadores (in Africa and Latin America), or "peaceful" settlers (in North America, Oceania, etc.), who succeeded in securing claims on land or only on its products. Some of these "interventions" long predate the period covered in this book, while others, such as European colonialism, developed in the nineteenth century. The joint effect of intensification and outside intervention, in all their countless local varieties, created an "inordinately complex" web of formal rules, informal obligations, customs, practices.[3] It is doubtful whether the available sources (mainly produced by "outsiders," such as western colonial administrators or anthropologists, rural economists, etc.) can really capture all the complexities and subtleties of these "traditional" rights (Tomlinson 1993, 50). A fortiori, this would be impossible here. Thus, they will be squeezed in four "ideal types": the hunting-gathering tribe, the swidden agricultural tribe, the communal village, and the feudal estate.

1. In hunting and gathering societies, there are no individual property rights on land, although each group (extended household, clan, tribe, etc.) has to protect its own territory against the claims of other groups. All members work together and they share the product according to some simple rules.

2. In swidden agriculture, land is common property, but each household has the right to cultivate a specific plot and to consume its products.[4] It cannot, however, let or sell the plot, and at the end of the cultivation cycle, the land returns to the clan, who can allocate it to another household the next time. Cattle-breeding is clearly distinct from agriculture and is often practiced by specialized households or tribes. They interact with farmers mainly when they have to compete for scarce land (and thus sometimes the competition is not peaceful).[5]

3. In the "commune" system, common ownership is limited to a part, albeit a substantial part, of the land, which cannot be sold or otherwise traded. Furthermore, the members of the commune (who may not coincide with the whole population) have some sort of rights on private land. The extension of the commons and the type of collective rights may vary greatly. At a minimum, the members collectively own pastures or woodland, which they can use

according to well-defined rules. In other cases, members have to let other villagers' cattle graze on their land during fallow (vaine pature), or they are subject to collective decisions about agricultural practices. In the Russian *mir*, perhaps the most invasive case, arable land itself was periodically redistributed among households.

4. The feudal system differs from the other three because a noncultivating outsider (the "lord") has the right to claim part of the production. Such claims differ from "modern" ownership because they are granted by the ruler as a reward for performing specific tasks (military services, tax collection). Thus, at least in principle, they are not saleable and not hereditary. In the classic European feudal system, the landlord had the exclusive right to the product of a part of the estates (demesne) and to part of the product of the peasants' land (manors). The peasants had to work on the demesne (corvées) for a certain number of days per year or pay an equivalent amount of money, and they could not leave the estate without the lord's permission. A variant of this system may have been the so-called debt peonage, widely diffused in Mexico and in other Latin American countries.[6] The workers were legally free to leave the estate, but only after having refunded their loans. This was allegedly almost impossible because the landowner (*haciendero*) used his power to set wages and prices at the estate shop in order to prevent workers from saving enough money and to evict indebted workers from whatever land they still possessed.

The description focuses on property rights on land and capital, but it also has implications for rights on labor: both the commune and the feudal system limited the personal freedom of agents. Serfs and debt peones, however, when they had fulfilled their obligations (including residence on the estate), were free to use the rest of their time as they wished. Thus, they differed from slaves, who, as the personal property of their masters, had, in theory, no freedom at all. Unlike serfdom, slavery was not necessarily related to land, and, indeed, many slaves worked in homes. Its very classification as a "traditional" property right may be questionable. In the United States before the Civil War, slaves had a perfectly defined, although morally repugnant, "modern" legal status as human chattel.[7]

All these traditional property rights have a very poor reputation among economists. Slavery not only deprives human beings of one of their basic rights, but also stifles their entrepreneurship and thus deprives the economy of a pool of potential talents. Modern property rights on land are considered superior to traditional ones because (1) they make it possible to transfer land to the ablest peasants via sales or rent; (2) they make it possible to use land as collateral and thus to access the formal credit market (section 7.4); (3) they prevent over-exploitation of forests and common land, a typical free-rider problem also known as "the tragedy of commons"; (4) they reduce the need to invest in private protection of property rights, which is socially less efficient than state-provided enforcement; and (5) they make it possible to invest without fear of being compelled to share the returns with other claimants (e.g., in case of redistribution).[8] As we will see in section 8.7, historians strongly emphasize this last point, stressing the negative effects of traditional property rights on technical progress. The advantages of modern property rights, it is said, outweigh the losses of any benefit from communal risk management. This idea has been authoritatively endorsed by the World Bank, since the publication of its Land Policy Paper in 1975. It has sponsored campaigns for land registration (titling)—i.e., the setting up of a cadastre—and the allocation of land to cultivators with full ownership.

These arguments are plausible, but the contrast between modern and "customary" rights is more stark on paper than in the field. The former are not fully secure, as owners can be expropriated with compensation (e.g., for public construction) or even without it (as in many cases of land reform, as will be discussed in section 8.3). On the other hand, customary property rights are sometimes quite well defined (at least for natives) and provide a reasonably secure tenure.[9] In these cases, formal, Western-style registration of landownership would add little and not be worth its cost.[10] Titling is an expensive process, which needs resources (specialized staff and funds), and which could be used more productively in other ways, as in the financing of R&D or extension. Land registration may prove a total waste of money if, as sometimes happens, peasants continue to respect traditional rights and ignore the modern ones (Hyden et al. 1993, 418–19). In its most recent documents on titling, the World Bank seems to accept, at least partially, these criticisms: It recommends a gradual strategy, which should build on traditional property rights and possibly award long-term secure rental contracts instead of Western-style ownership (World Bank 2002, 35–37; World Bank 2003, 51–78, 185–87). Anyway, the cost and benefits of titling should be assessed case by case.

7.3 The "Structure": Matching Land and Labor

The potential supply of labor from land-owning households almost always differs from the demand of the farm, given the state of technology and the crop mix.[11] Such a mismatch is to some extent the "natural" consequence of the seasonality of agricultural work and of demographic changes. In the very short run, a work force adequate for the peak season (e.g., the harvest) would exceed the needs of the slack season, and *vice-versa* (the most common case being somewhat in the middle—some excess supply in the slack season and some excess demand in the peak season). In the medium run, the labor supply can no longer match the farm demand if the household work force changes for natural events (the coming of age of a son, or the aging of a man) or for decisions about migrations and off-farm work. But the mismatch between supply and demand of labor is quite often permanent or "structural." Some households ("landlords") own more land than they can cultivate while others ("workers") do not have enough land, or, in some cases, do not have any.

Tackling these imbalances, so as not to leave labor and land idle, requires suitable institutions, notably markets to exchange factors.[12] Short-term disequilibria can be tackled via the market for labor: farmers can seek off-farm jobs during the slack season or hire extra hands at peak season, possibly recruiting them from outside agriculture (section 4.4). In a world of perfect markets, long-term disequilibria could be in theory tackled via the market for land-cum-credit. A land-scarce household could purchase land from a land-abundant one, if necessary borrowing the money.[13] This solution is, however, not very frequently used: even where property rights are perfect, the market for land is sticky and the number of transactions is small. Family farms and estates tend to remain in the possession of the same family for long periods, even for centuries, and to be sold only in distress.[14] For instance, field research on post-independence India shows that only between 0.2 and 1.7 percent of the total land stock was sold each year: thus, a full turnover of land would take between five hundred and sixty years if each plot were sold only once.[15] In the overwhelming majority of cases, land and labor are matched via the market of labor—stipulating a contract between "worker" and "landlord." The former can commit himself to cultivate land in exchange for a fixed wage, or to pay the landlord a given sum (fixed-rent tenancy) or a predetermined share of the output (share-tenancy). Usually, tenancy contracts involve all the labor force of the worker household, which is entrusted to manage a farm. In contrast, most wage contracts refer to individuals who are hired to perform some specific task under the direction of the landlord or of a manager.

The interaction between patterns of ownership and contracts yields three main cases:[16] (1) the "family farm" (i.e., a farm owned and managed by a rural household), (2) the "tenanted estate" (i.e., divided into smaller units and rented out to households with a fixed rate or share tenancy contract), and (3) the "managerial"—or "capitalistic"—estate (i.e., cultivated as a unit, under the management of the owner or a tenant, with a salaried work force). The distinction may be clear on paper, but its application is fraught with difficulties. Many family farms rent in or out some land: the "mixed farms" account for about a quarter of total acreage according to the last World Agricultural Census (FAO 1990). Family farms can also hire permanent workers, and indeed the wealthiest farm households can hardly be distinguished from poor "managerial" landlords who work on their own fields.[17] Similarly blurred is the distinction between managerial and "tenanted" estates. In fact, decisions have often been shared among landlords and tenants. For instance, the large estates of nineteenth-century Tuscany (*fattorie*) were rented out in smaller units (*poderi*) to share-croppers (*mezzadri*), but the landlord managed, directly or through an agent (*fattore*), the processing and sale of products, and maintained the right to oversee the cultivation.[18] The *fattoria* could be thus classified as a tenanted or a managerial farm, according to the circumstances.

Last but not least, contracts show an amazing variety of clauses: a description of existing contracts in Italy in the late 1880s is published in a 813-page volume (MAIC, 1891). A contract could last a few hours or days, as in most of the wage labor contracts (which accounted for the overwhelming majority of the total) or several years or more. The permanent tenancy contracts in the Yang-Tzi basin in China could last for decades.[19] The same farm can be subject to more than one contract among different parties: in many areas, from Ireland to the Mediterranean Basin to southern China, intermediaries rented large estates, and then sublet them in small lots to peasants.[20] Different products of the same farm can be subject to different contracts, as in nineteenth-century Brianza (the area north of Milan), where the tenant paid a fixed rent with the proceeds of the cultivation of cereals, while the cocoons, the staple of the area, were shared between him and the landlord (Serpieri 1910, 121–91). Payments can be in cash or in kind, with widely different criteria to set the price per unit of product. Laborers can also be paid with the right to cultivate a specific plot of land, as in the so-called *ezbah* system in Egypt (Richards 1982, 31–36, 58–69, 118). Fixed-rent contracts can include clauses for a reduction in rents in the event of a very poor crop, as in the Saga plain in Japan (Francks 1964, 120). The division of net product in sharecropping contracts can greatly vary, according to shares by product and to the division of expenditures (inputs, wages, if any, and so on).[21] The recorded tenant's share ranged from a minimum of 20 percent for the traditional *al-varum* contract in Tamilnadu (India) to 75 to 80 percent for Argentina in the 1890s.[22] Last but not least, contracts can change in time, keeping the traditional name but adjusting clauses to the different economic conditions of the parties.[23]

In short, there is a huge variety of solutions for the same problem—how to match land and labor permanently. This variety however, does not trouble most agrarian historians. They take the existing pattern of landownership for granted, and assume that it can change only as a consequence of major political events, such as wars and land reforms. They also assume that contracts are traditional and/or imposed (and changed) by the landlord according to his economic interest or to his social and political aims. Thus, the pattern of ownership and contracts may well be inefficient or harmful from the point of view of the society as a whole. These statements cannot satisfy economists. Ownership and contracts, like any other institution, are endogenous and thus their features have to be explained. Why are land and labor matched via the market for labor than via the market for land-cum-credit? And why do contracts for labor differ so much?

The former question has not received much attention: "the analysis of the land market has been largely ignored in the literature on agrarian economics" (Hayami and Otsuka 199:176). Scholars simply assume as a matter of fact that the market for land is sticky, as a consequence of the high costs of transaction and of the imperfections in the market for credit (section 7.4). Indeed, transaction costs are high because land is a sui generis commodity—highly specific and often hardly divisible. Each plot has its specific features like soil quality, which are better known by the seller than by the buyer. A farm is both a legal and a coherent productive unit. Dividing it according to the needs of the parties entails, at least, legal fees for registration, surveying, and possibly substantial investments for new building, fencing, and so on.[24] Stickiness, however, may reflect two further causes: the existence of legal or quasi-legal restrictions to sales, and of a "land premium." The transfer of land to "inferior" races could be prohibited (as under apartheid) or, even if legal, made difficult by racism, as in the American South after emancipation.[25] Some governments set limits on the sale of peasant land to prevent unwelcome developments, such as the migration of workers to cities (a policy adopted by Nazi Germany in the 1930s), or the re-creation of large estates after a land reform.[26] The land premium is the difference between the market price and the "true" value of land—that is, the stream of future rents, suitably discounted. This premium is said to be higher for small plots and farms than for large estates, and indeed this is often the case, although it is quite difficult to prove that this difference in prices does not reflect other characteristics of the land (the quality of the soil, the amount of capital, and so on).[27] By definition, a land premium reduces land transactions by discouraging the potential buyers looking for "normal" returns. It does not however, discourage the *actual* buyers: Why do they buy land for more than it is worth? The literature suggests four possible explanations: (1) purchasers may deem current interest rates too low relative to expected inflation or may nurture too high expectations because they do not have sufficient information about future returns—the mechanism of speculative bubbles (section 4.3);[28] (2) they might be willing to pay a premium because land is the best or the only feasible collateral for borrowing;[29] (3) ownership may bring economic benefits, such as a low risk of eviction and joblessness, and the right to be helped by the community in case of distress; (4) last but not least, land can be sought after in traditional societies for "noneconomic" reasons.[30] Landowning peasants could have a higher social status, and greater political rights, than tenants or laborers, and working on one's own land may give other satisfactions (the feeling of freedom, the better quality of home-grown food, and so on; see Chavas 2001, 269). In the past, landownership was the main source of social status and political power for elites. Offer (1991) argues that land retained a powerful symbolic value even in the United Kingdom in the nineteenth century, while Clark (1998a) is not so convinced. For him, the price of land was high because it reflected expected capital gains. In theory, these noneconomic motivations would reduce the supply of land, inducing owners not to sell their estates in order to hand them to their heirs, even if a sale were sensible on purely economic grounds. Unfortunately, measuring the "land premium" is very difficult and estimating the impact of noneconomic reasons on it is next to impossible.

The market for agricultural labor, unlike that for land, has attracted huge attention among economists. Alfred Marshall, in a well-known passage of his *Principles of Economics* (1920), argued that sharecropping is not efficient because the workers have an incentive to undersupply labor.[31] This statement has spawned, admittedly with some delay, a huge economic literature on the efficiency of sharecropping and, more broadly, on the choice of contracts. In fact, any inefficiency contrasts with the basic principle of neo-institutional economics—that free contracting between rational individuals must yield an optimal, output-maximizing, solution (the so-called Coase theorem). The parties would choose this solution even if its benefits are unevenly distributed, because the advantaged party could compensate the other and still remain better off than under any other alternative. This theorem, however, assumes perfect markets, perfect information, and perfect foresight, and none of these conditions is likely to hold in the real world. Economists use different concepts (moral hazard, adverse selection, the principal agent problem, asymmetric information, and so on) but arguably all their explanations boil down to different versions of the same problem: an information deficit.[32] In the real world, information is costly and sometimes simply unavailable (infinite costs), while optimal contracting needs a great amount of information. Ex ante, the two parties need to know the conditions of the markets for land and labor and the (unknown) "true" quality of factors—that is, the skill of the prospective tenant and quality of the land. During the contract, each party has to monitor the behavior of the other, who has a clear incentive to undersupply his own specific input and to overexploit the other's input. The worker could shirk—or, work less intensively than he could. The tenant could over-exploit the land to maximize the output throughout the duration of the contract. The landlord could try to save on working capital or on maintenance, provided this behavior did not damage the value of his estate in the long run. Last but not least, each party must try and prevent the other from breaking the contract, or force him to pay for his breach.[33] Workers or tenants might flee before completing their obligations, landlords might not pay his workers, and both parties might not respect the pre-arranged division of the product and so on. In theory, the damaged party could resort to outside power—such as a court—in order to force the other to comply. However this procedure is at the very least expensive and in many cases, unfeasible—as the outside power may not have the necessary information to judge. In short, many agrarian contracts are very difficult to enforce. Clearly this problem is not restricted to agriculture, but the features of agriculture, such as its scattered localization, the existence of multiple tasks and above all the uncertainty of output, makes it more difficult than in manufacturing or in services.

This framework, albeit very general, has three relevant implications:

1. The costs are lower among members of the same household, who all know each other, than among strangers.[34] Household members have few incentives to damage their own farm or to shirk, unless the distribution of work or income among them is extremely contentious. The benefits of ownership may induce household members to forfeit some potential monetary gains from off-farm work or to work harder than hired hands (or to need less supervision to extract the same amount of work).[35] The incentives would be even greater if the market for labor were imperfect or even missing, and thus household members would have no option but to work on the family farm.[36] In this case, "self-exploitation" (to use Chayanov's

well-known expression) would not be limited by returns from off-farm employment, but only by the disutility of work.[37]

2. Imperfect information drastically reduces the scope for insurance against the risks of crop failure.[38] An insurance company can assess the effect of bad weather or of other "natural" factors, but it is powerless against negligence, shirking, and undersupply of input. The state could help by subsidizing insurance schemes, but full cover against all types of risks and events is beyond the reach of even the most generous scheme. In other words, agents have to bear a high proportion of the total risk, and each type of contract implies a different allocation of it. The risk is borne entirely by landlords in managerial estates (and, of course, also by family farmers) and by fixed-rent tenants, while it is divided between the two parties in share contracts.

3. The choice of contract depends, ceteris paribus, on the personal features of the two parties, on the crop mix, and on the social environment. Wealth, education, and skills of the tenant determine his degree of risk-aversion, his access to credit (section 7.4), and the amount of supervision he needs. The wealthier and more skilled a prospective tenant is, the more likely it is that he can assume an entrepreneurial role and/or get credit at favorable terms. Usually, wealth and skills increase with age, and thus a worker can experience a change in contracts during his lifetime. As it is said, he can climb an "agricultural ladder"—starting as a wage laborer, then becoming a sharecropper, a fixed-rent tenant, and eventually a landowner.[39] The wealth of the landowner affects his risk aversion, while his farming skills, personal preferences, and the opportunity cost of his time determine how much and how well he can monitor his workers. An institution, an absentee landlord, or a widow, ceteris paribus, would prefer a less monitoring-intensive contract than an active farmer who is letting a part of his farm. The crop mix can affect the choice of contracts as different crops may require different amount of supervision and/or be more or less variable.[40] Last, but surely not least, both risk-aversion and the amount of information depend on the extent of social relationships. In a small, closely-knit village, it is relatively easy to build a good (or bad) reputation. Workers have less incentives to shirk or cheat and landlords to appear greedy by not helping a tenant in distress.

An ideal theory of agricultural contracts should take both risk and transaction costs into account, allow for long-term relationships and the economic value of reputation, take the personal characteristics of tenants and landlords into consideration, consider the linkage with the market for credit, and, possibly, account for the main features of different types of contracts, such as their length. So far, economic theory has not provided such a "super model."[41] Most contributions focus on one contract alone (with a predilection for sharecropping to solve Marshall's puzzle) and try to explain under which conditions it can be the most efficient. As far as the author knows, there is no comprehensive model of the full range of choices (fixed wage/share tenancy/fixed rate tenancy) under both uncertainty and positive transaction costs. The most ambitious attempt to date is the model by Hayami and Otsuka (1993).[42] They rule out "capitalist farming," as entailing too high monitoring costs, and focus on the choice between share-cropping and fixed-rent tenancy. The two contracts would be equivalent if perfectly enforceable (i.e., if the worker performance were perfectly monitorable). If contracts are not perfectly enforceable (as is more plausible), fixed-rent would be more efficient than sharecropping, as argued by Marshall. But sharecropping could be adopted anyway if both parties are equally risk averse, or if, as is highly likely, the tenant is more risk-averse than the landlord. A risk-based explanation of sharecropping is endorsed by Huffman and Just (2004).[43] They argue that sharecropping is relatively more common in LDCs because the gap in risk-aversion between the parties is much wider than in the advanced countries. As an alternative, one can explain the choice of contract with monitoring costs, without resorting to risk-aversion as deus ex machina. Allen and Lueck (2002) start from the assumption that neither fixed-rent tenancy nor sharecropping is optimal.[44] Sharecroppers have an incentive to under-use labor, as posited by Marshall, and to cheat by underreporting output. All tenants can gain from overexploiting land (i.e., pruning treecrops too much or manuring the fields insufficiently), but the incentive would be weaker for a sharecropper, who would get only half of the additional output, than for a fixed-rent tenant. A rational landlord might prefer sharecropping, forfeiting some rent, to minimize long-term losses on the value of his estate.

The literature on contract choice is largely theoretical. Most empirical works focus on a specific area and try to explain the prevailing contracts with features such as farm size, technology, crop mix, and so on. For instance, Carmona and Simpson (1999) study the so-called *rabassa morta,* a traditional sharecropping contract for viticulture in Catalunia.[45] The peasant committed himself to plant and tend a vine for an indefinite time in exchange for a 70 to 80 percent share of the wine. This share was unusually high because planting new vines was a very labor-intensive operation, which would have been prohibitively expensive with wage labor. The indefinite length of the contract reduced the peasants' incentives to overexploit the vines in the short term (e.g., with heavy pruning). The contract was slowly abandoned since the end of the nineteenth century, because of the combined effect of the phylloxera, which made it necessary to substitute the whole stock of vines, and of the rise in agricultural wages, which raised the opportunity cost of labor. Clearly, such reasoning, although convincing for the case at hand, cannot be generalized. One would need explicit econometric tests of competing models of contract choice and there are very few of them.[46] Alston and Higgs (1982) find that monitoring costs affected the choice of contract in a sample of plantations in Georgia (USA) in 1911, but their regression does not include any risk-related explicative variable.[47] In a more recent paper on the American South, Ransom and Sutch (2001) confirm that the personal characteristics of the tenant mattered, and they find a positive relation between literacy and the diffusion of fixed-rent tenancy, which is consistent with the higher rank of this contract in the agricultural ladder.[48] Allen and Lueck (2002) back their strong theoretical stance for a transaction-cost explanation with a non-nested test against an alternative risk-based model.[49] Their dataset consists of four large samples of contracts in North America in the 1980s and is by far the most detailed one among all the empirical works. Yet they have to resort to (possibly not accurate) proxies. Furthermore, the case is not really representative, as Allen and Lueck themselves admit. In fact, most owners were retired farmers or rural residents, not very different from their tenants in wealth and skills (the personal features of the tenant are not significant). It is thus likely that the gap in absolute risk-aversion between them is much smaller than between, say, liberated slaves and their former masters in the South after the Civil War. Last but not least, all these regressions assume that the explicative variables (crop mix, wealth, etc.) are exogenous. This may not be the case. The crop choice may depend on the contract and, as Ackerberg and Botticini (2002) stress, landlords could choose among prospective tenants those who best match the characteristics of the farm: risk-neutral tenants would end up cultivating riskier crops with fixed-rent contracts.[50] Clearly, the issue need further work, hopefully on additional data-bases. So far, one must conclude that both risk and monitoring costs are likely to be important and must be considered.

Two main points should be clear by now. The economic theory of contracts is a powerful tool to explain the features of actual contracts, but much work still needs to be done to apply it to additional case studies, and, above all, to put forward and test a more comprehensive and realistic model. A really comprehensive one should also be able to explain not only the choice of a contract, given the pattern of landownership, but also to account for the pattern itself—that is, the working of both markets for land and labor. This is, admittedly, a tall order.

7.4 Finding the Money: Formal and Informal Credit

Agriculture needs capital to finance household consumption, pay taxes and other current expenditures (such as wages for laborers or the purchase of fertilizers), invest in machinery or in land reclamation, and, possibly, purchase land. Capital needs are common to the whole economy and to all agents, but agriculture is different on many grounds.[51]

The demand for credit in agriculture, unlike in the rest of the economy, is typically seasonal. Crops can be sold only after harvest, some months after the outlays for the cultivation cycle. Furthermore, in most cases, the same agent (i.e., the household managing a farm) needs capital for both production and consumption purposes and resorts to the same institutions, while, in the rest of the economy, these needs are expressed by different agents and managed by different financial institutions.[52] Agents widely differ in terms of not only observable features, such as wealth or access to land, but also unobservable ones, such as managerial skills. Last but not least, the demand for capital varies substantially, and unpredictably, from one year to another: it falls in years of good harvest or of high price, and rises in years of poor harvest or low prices.

The supply of capital differs as well. Few, if any, agricultural corporations are big enough (or have sufficiently bright prospects for future growth) to be able to raise money by issuing shares or bonds on the stock market.[53] Thus, by default, all farmers have to resort to credit, but the sources differ. Landowners and the richest family farmers could tap the so-called "formal" market—those institutions recognized and regulated by the state as legitimate lenders (banks, insurance companies, etc.).[54] The overwhelming majority of peasants had to resort to other sources, collectively known as the "informal" sector. One major source was the professional moneylenders and pawnshops, often belonging to some group or religion (such as the Jews in Europe, or the Chettyars, a caste of Southern Indian origin, in Burma). They were very common in all areas where independent family farms prevailed, such as Asia. According to conventional wisdom, they provided most credit in India and cumulated large land estates by seizing the property of their debtors.[55] Some recent work on India downplays the share of the moneylenders on total credit and, above all, the extent of their land accumulation. Professional moneylenders were not able to cultivate land or manage estates, and thus preferred to resell the lands seized as soon as possible. Owners of "tenanted estates" were often committed formally, as part of the contract, or informally, to finance their tenants. For instance, in Burma, "it is doubtful whether they [the landlords] could get tenants at all if they were not ready to finance them" (Cheng 1968, 172). Financing (trustworthy) tenants or permanent workers in distress was in the landlord's best interest. If they fled, the landlord would be forced to find a replacement and would incur substantial transaction costs. In many cases, as in nineteenth-century Tuscany, credit entailed a permanent relationship.[56] Landlords anticipated the payment of taxes and farm expenditures by tenants, provided food during the year, and managed the sales on the tenant's behalf. Also, most sales of agricultural products and purchases of manufactures entailed a credit transaction (the so-called interlinkage). In fact, peasants could seldom afford to wait for the sale of their crop to the final purchasers, who, in the case of exported products, could well be thousands of miles away. The merchant would then anticipate the proceeds to them, either in money or in kind (food, industrial products, etc.). Borrowing from merchants was also widespread among relatively well-to-do producers, such as the nineteenth-century French vignerons, or the twentieth-century American farmers. French wine-growers were financed by intermediaries (*negociants*) who sold their wine, while farmers purchased most machinery and other equipment with long-term loans from the sellers.[57]

The prevalence of "informal" credit can be interpreted as a consequence of strongly asymmetrical information about a highly variable crop. An inept or negligent borrower could attribute his failure to repay the loan to natural circumstances, compelling the lender to start a legal procedure to recover his money. To maximize his chances, the lender could ask for collateral to be seized in case of default. The best, or at least the most frequently used, collateral is land, but it is far from being a foolproof one. Its value is subject to fluctuations, as proven by the plight of rural banks in the United States during the early 1930s.[58] However, other forms of collateral, such as crops or livestock, are even worse than land from this point of view. They are more subject to moral hazard (the borrower can steal a part of the crop), more difficult to manage (a repossessed cow has to be fed, stabled, etc.), and the fluctuation in their value could inflict much higher losses to a potential lender (land could recover its value in the long run, but there is no long run for vegetables). In some traditional societies, the lender could expect to be refunded by the extended family, group, or tribe of the borrower, should he fail to honor his commitment. These networks tended to disappear with modernization of property rights, and the resort to common responsibility shifts the problem of reputation from the individual household to the wider group. Actually, the best protection for the lender is to select his borrowers ex ante. As a rule, an "insider" living in a village has much better information on the trustworthiness of potential borrowers than a formal institution. The latter might have access to this information only by paying insiders, and the cost might not be worth it for the small-scale loans that are typical of the agricultural sector. Moreover, borrowers might prefer to resort to insiders than to banks. A peasant might be illiterate or simply unused to legalized procedures, and thus he might have difficulty in applying for a bank loan without the (expensive) help of intermediaries.[59] Furthermore, in the event of trouble, he could expect more forgiveness from a village money lender than from a city bank. Thus, the literature argues, formal institutions are at a loss in financing agriculture. One should not, however neglect their advantages. Banks have more assets and access to other sources of funds (notably the rediscount from central banks), and they can diversify their portfolios—so that they are less exposed to idiosyncratic shocks. Thus, a bank may be, ceteris paribus, more willing to finance risky investments in technical progress. Moreover, a bank is better informed than a village money lender on the trends in the world economy, and thus may be better able to predict the prospects of the export market. Thus, one would expect, ceteris paribus, the market share of the formal sector to be higher in "advanced" countries, and/or for secured loans (especially if the collateral is land and especially if the estate is large), and/or in credit for long-term investments (including the purchase of land).

Scholars emphasize the source of credit because informal sources are assumed to be expensive and exploitative. The first statement is indeed backed by much anecdotal evidence. British colonial administrators in India organized several surveys on the issue and were convinced that high interest rates were the ultimate cause of peasant unrest (including the so-called Deccan riots of the late 1870s).[60] Unfortunately, the data on actual rates on 'informal' loans are scarce. The best source is an

TABLE 7.1
Interest Rates on Informal Credit in LDCs around 1970

	(a)	*(b)*	*(c)*
Asia	32	17	80–200
Latin America	38	28.5	80–300
Middle East	24	9	
Africa	65	24	100–500
All Countries	40	19	

Source: Wai 1977, table 4.
(a) "Usual" nominal interest rate from "informal" institutions.
(b) rate differential between informal and formal credit.
(c) "higher, exceptional" rates in "informal" transactions.

extensive survey of fifteen Third World countries (table 7.1) commissioned by the World Bank in the 1970s. The rates of column(c) are meant to be the real upper boundary of the range, but also the standard ones (column a) exceed the market rates from formal institutions by quite a wide margin (column b).[61] The historical evidence, albeit scattered and hard to check, confirms this fact. Sugarcane growers in South Kanara (India) and Bengal peasants paid up to 50 percent interest for a four-month loan (i.e., 200% yearly) but these rates are likely to be higher than average.[62] In fact, the most common figures for east Asian countries range between 25 and 40 percent.[63] According to a very accurate survey, in the 1930s, a time of distress and deflation, about half of Chinese peasant households were indebted, and two-thirds of them paid rates between 20 and 40 percent, while the unlucky ones could pay up to 100 percent.[64] There also is no doubt that mortgage-backed loans were cheaper than unsecured ones.[65] In Bengal, they cost about 18 to 35 percent, versus up to 200 percent for unsecured ones, whereas in Japan in the early 1910s they cost a "mere" 11.5 percent—a few points above formal loans from government banks (7–8%). In the United States in the twentieth century, the rates on mortgage-based credit were low (5–6%) and, above all, they did not differ between formal institutions and others (i.e., individuals). This was an exception: agriculture was highly developed and the efficient legal system gave all lenders the same fair chance of recovering their money in cases of default. In fact, in the 1830s and 1840s, the new settlers in the Great Plains had to pay 20 to 30 percent (i.e., four to six times the yield on federal bonds) on loans from the speculators who sold them the land.[66] Finally, one could hypothesize that rates could be lower in transaction among people with personal ties, but the evidence on this issue is extremely scarce. Loans from relatives were indeed interest free in Thailand in the nineteenth century, but the case is not necessarily representative.[67]

The high level of interest rates is deemed to be only one of the exploitative features of the informal credit market. It is said that landlords and merchants exploited peasants by gaining fat margins on funds they could borrow from formal institutions. This case was quite common. In Chile in the 1860s and 1870s, landlords borrowed from the *Caja de Crédito Hipotecario* (set up in 1855) at 8 to 9 percent and then lent to tenants and peasant owners at 10 to 20 percent.[68] In Argentina in the nineteenth century, the intermediation was even more complex: the banks financed trading companies at 6 percent, the latter financed local merchants at 12 percent, and, finally, the merchants lent to farmers at 25 to 30 percent.[69] It is also said that interlinked contracts are exploitative because the price for the same good is lower at the moment of financing than at refund. Neither practice, like high interest rates, is by itself sufficient evidence of exploitation. The refund price in interlinked contracts is bound to be higher than the financing price since it includes the return to capital: a point of interest rate on a three-month loan increases the refund price by 0.4%. Credit intermediation may be an efficient way of conveying information: landlords or merchants are more trustworthy than peasants from the point of view of a 'formal' institution (they have land or other real estate to pledge, and it is easier to get information on few rich people than on a multitude of peasants), and, at the same time, they are more knowledgeable about the situation and skills of the final borrowers. Similarly, interlinked transactions reduce transaction costs—if anything, because the merchant already has the product as collateral.

High interest rates in informal lending might simply reflect information costs and the risk premium. Even insiders have far from perfect knowledge of the borrowers, and even highly trustworthy peasants might be hit by random production shocks. Moreover, informal lenders can seldom diversify their portfolios to spread risk like, for example, a city bank. On the other hand, any lender, formal or informal, can squeeze extra-rents if he enjoys some sort of local monopoly for credit. In a famous book, Ransom and Sutch (1977) argued that this was the case in postbellum southern United States. The storekeepers could charge inflated prices (up to 50% higher for corn) for the wares they sold on credit to poor former slaves.[70] Critics such as Claudia Goldin replied that storekeepers could not extract any extra-return because they were always threatened with the possible entry of competitors, as barriers to entry in the market for credit were quite low. At the end of the day, whether high rates are the product of monopoly or of a risky and information-poor environment is an empirical issue, which can be settled only on a case-by-case basis.[71]

Arguably, "exploitation" is not the main problem of credit markets in traditional agriculture, as featured by risky environment and imperfect information. In these conditions, credit is expensive if abundant or, possibly, rationed if reasonably priced.[72] The high cost or shortage of capital reduces investments and makes the land market more sticky. Lowering interest rates is bound to improve long-term agricultural growth, as proven by the great transformation of Californian agriculture in the late nineteenth century.[73] For this reason, many states set up institutions to supply low-cost capital to farmers, but with mixed results (section 8.5).

7.5 The Co-operative: The Best of All Possible Worlds?

A co-operative is an association of farmers, who join in order to perform tasks collectively such as the purchase and management of inputs (fertilizers, machinery, etc.), the processing, packaging, and marketing of products; the provision of credit; and so on.[74] It can perform several or all these tasks together (general purpose co-operative) or specialize in only one (single-purpose co-operatives). In the latter case, the same farmer may join more than one co-operative. By definition, when joining a co-operative, a farmer loses part of his freedom and incurs substantial transaction costs in order to participate in the process of collective decision-making. Indeed, few farmers seem ready to forfeit their freedom entirely: collective farming (such as the Israeli *kibbutz*) is quite rare, and needs very strong ideological motivations.[75] But co-operatives developed quite fast from the end of the nineteenth century, and nowadays they account for a sizeable share of the market for agricultural products in all advanced countries (section 8.6). What advantages could convince farmers to join? Most answers are variants of the same idea: the co-operative is a hybrid institution that can successfully unite the optimal incentives of family farming with the advantages of size.

The advantages of size are minimal, if any, in actual cultivation (section 7.6), but are increasingly substantial in marketing and industrial processing. Therefore, setting up a processing co-operative is an appealing alternative to selling the product to an industrial company, possibly with the intermediation of a merchant. In fact, the first Danish dairy co-operative was set up in Hjedding in 1882, just four years after the invention of the separator, which, as stated earlier, greatly increased the minimum scale of butter production (Henriksen 1992, 171). All members pledged to confer their entire milk produce to the co-operative, which could reject it, if its quality was not good enough. Henriksen (1999) argues that this clause, which was to be imitated by most processing co-operatives all over the world, was the key to its success because it ensured a regular and steady supply of milk, while peer-monitoring ensured its quality.[76] Good milk and careful processing yielded high-quality butter, and Danish co-operatives started to market their butter under the collective brand of "Danish Butter" as early as 1899 (Jensen 1937, 316; Knapp 1969, 265). This branding strategy was adopted by many American co-operatives in the dairy and fruit businesses, with household names such as Sun-Kist, Sun Maid, and Land o' Lakes.[77] The common processing/peer monitoring/branding strategy, however, was not always successful. The co-operatives for wine production are a case in point (Simpson 2000, 117–20; Loubère 1996, 137–53). They developed later than dairy producing co-operatives in the Mediterranean countries (and did not develop at all in the United States). There was no technical breakthrough in processing comparable to the cream separator, and the scale economies were, on the whole, smaller. The (measurable) sugar content was only one dimension of the intrinsic quality of grapes, and thus many peasants preferred to keep their best grapes for domestic wine-making. Thus, the co-operative product did not enjoy a reputation for high quality, and it was often given back to producers for direct marketing.

Joining forces could also be a sound strategy when farmers faced potential monopolists (or fraudsters) in the market for their input and/or a monopsonist in the market for output. As Louise Smith (1961, 3) puts it, at the start of her history of co-operation, "co-operation everywhere has necessity for its mother." In the United States, co-ops were set up to fight local monopsonists, such as the owners of grain elevators in the Great Plains.[78] In the Netherlands, potato producers succeeded in breaking a cartel of producers of potato starch in 1898 by establishing a new factory, and the *Federconsorzi* (a national network of landowners' associations, mainly from the north) became the major producer of phosphatic fertilizers in Italy during the early 1900s.[79] The concentration of firms in food production, increasing the potentially monopsonist power of great food companies (and of large producers of industrial input), might have enhanced this "defensive" role of co-operatives. Many of these companies, however, sign long-term contracts with farmers, and not necessarily on poor terms. Squeezing producers too hard could, in fact, jeopardize the company's reputation and endanger its supply chain in the long run.[80]

The advantages of credit co-operatives for a potential member are clear enough. He can find a reliable outlet for his savings and, above all, borrow with little or no collateral, at a rate that is usually lower than those charged by informal lenders, especially if the latter enjoy a local monopoly.[81] Furthermore, he can reap the profits, either in cash or as a lower interest rate. The risk of credit co-operation is clear as well: a few defaults can wreck the whole initiative. Thus, a credit co-operative is viable only if the access to membership (i.e., to the right to borrow) and the service to loans are strictly monitored with the credible threat of expulsion. An efficient peer-monitoring is thus essential.

The advantages of co-operatives would seem to be big enough to ensure their development all over the world. Indeed, the nineteenth-century "apostles" of the movement dreamt of a world of family farms organized in co-operatives, and many governments actively supported them. Western governments granted privileges to the co-operatives, such as a reduced taxation or, as in the United States, the exemption from anti-trust laws (the Capper-Volstead Act in 1922).[82] Several governments in LDCs went even further, actively promoting the development of co-operatives. In several cases, however, these co-operatives subsequently went bankrupt, showing that state support was not sufficient to create a viable system if the conditions were not ripe. Indeed, co-operatives have not been successful in most LDCs (section 8.6), where, arguably, they are most necessary. Co-operatives have been undoubtedly successful in Western Europe and in the United States, but even there they are far from being the only organization. Thus, by stretching the argument a bit, one can speak of a "failure" of the co-operative movement in comparison to the initial dream. This failure can be explained by six main causes:

1. The diffusion of co-operatives depends on the demand for potential membership, which is the highest among family farmers. Most managerial estates, such as the British ones in the nineteenth century, are big enough to exploit scale economies, while they cannot benefit from better monitoring, as they have to resort to hired manpower anyway (Ilbery 1985, 163; Hunt and Pam 2002, 247–48). In well-organized tenanted estates, the landlord takes care of the main co-operative tasks such as processing, marketing, etc. (Simpson 2000, 117). In other cases, as in Argentina during the nineteenth and early twentieth centuries, the fast turnover of tenants prevented the buildup of mutual knowledge and trust among potential members.[83] Last, but not least, the incentive to set up co-operatives may be reduced if the same services are provided by similar institutions: the competition posed by parish savings banks accounted for scarcity of co-operative banks in nineteenth-century Denmark (Henriksen 1998).

2. The success of production and marketing co-ops seems more likely for perishable products that need quick delivery and processing close to farms, and/or for products directly aimed at the consumer who wants reliable quality, as embodied in a recognized brand. The advantages are smaller for highly standardized and nonperishable products, such as wheat or even meat. In fact, the co-operative movement among bacon producers in Denmark developed more slowly, as pork could be easily transported and thus farmers could shop around for the best price (Henriksen 1999, 73).

3. Co-operatives are highly sensitive to correlated shocks, such as natural disasters or a fall in prices, which affect all members at the same time. This risk is particularly great for credit co-operatives, whose assets can be drained away if all members apply for distress loans at the same time. Thus the repercussions of a failure can be far-reaching and damage the reputation of the institution for a long time and over a wide area. Thus, co-operatives have sought the help of similar institutions, by setting up regional and national networks.[84]

4. Co-operatives can fail if their managers are not up to the task or simply use the co-operatives for their personal welfare (e.g., borrowing money without repayment). Mismanagement and embezzlement are often quoted as causes of failure for co-operatives, but it is difficult to assess how frequent or how damaging this kind of behavior is for the whole co-operative movement.[85] In fact, German credit co-operatives, the model for the whole world, were all audited.[86]

5. Co-operatives can fail to develop or they can decline for political reasons. The fascist regime in Italy regarded independent organizations made up of peasants as a threat to its power, and it adopted tight control measures, which stifled the promising prewar development of the co-operative movement (Galasso 1986, 457–89).

6. The most intriguing and controversial issue is, however, the alleged role of "cultural" factors. Two-thirds of credit co-operatives in Imperial Russia failed because peasants took the (subsidized) loans but refused to repay them, nor

could officials force them to with the threat of foreclosure, because land was collectively owned.[87] Many frustrated would-be organizers of co-operatives attributed their failure to the peasants' stubborn refusal to understand the great advantages of co-operation. Galassi (1998, 2001) argues that the lack of mutual trust among prospective members made it very difficult to establish a *cassa rurale* (credit co-operative) in southern Italy in the early twentieth century, even if the *casse*, once established, were reasonably efficient.[88] The classic case study is, however, late-nineteenth-century Ireland (O'Grada 1977; O'Rourke 2002). Around 1860, Ireland supplied about half of Britain's imports of butter, but in the next decades it steadily lost its market shares to all other suppliers—most notably Denmark (but also Russia). Yet, in spite of the valiant efforts of some pioneers, the co-operative movement developed sluggishly and belatedly in Ireland, allegedly because of the farmers' stubborn individualism and conservatism. Some years ago, in a pioneering analysis, O'Grada (1977) put forward a different explanation, based on the strong positive relationship between the diffusion of creameries and the density of milk cows. Farmers organized co-operatives only if the local milk supply exceeded the minimum profitable threshold (and the lack of co-operatives in some counties dragged down the nationwide average). O'Rourke (2002), in a very recent work, confirms O'Grada's core result but adds that the diffusion of creameries was also inversely related to illiteracy and to the extent of litigation between tenants and landowners. Thus, cultural factors mattered, although success or failure depended mainly on the economic conditions.

7.6 Conclusion: Is There an "Ideal" Farm?

The combination of property rights, patterns of ownership, size of operational units, contracts, and so on yields an apparently infinite number of different cases. Is any of them superior to others—either always or in a set of clearly defined circumstances? The issue has interested generations of scholars for its normative implications. The superior efficiency of (adequately sized) family farms over traditional latifundia has been a powerful argument for land reform all over the world (section 8.3). American farmers often called for state intervention to defend their livelihood against the competition of (allegedly) more efficient, large-scale corporate farms.[89] Unfortunately, the debate has been as confused as it has been lively, because the participants do not agree on the definition of "superiority" and the classification of farms. Economic historians focus on the propensity to innovate, while economists try to measure static efficiency. Some authors contrast "small" (or "family") farms with "large" (or "tenanted" and "marginal") ones; while others, possibly under the influence of the Marxist literature, consider only the "managerial" (or "capitalistic") farms as really modern and innovative, and they bundle family and tenanted farms together.

As argued in section 7.3, family farms enjoy clear advantages in terms of incentives to workers. To what extent are these advantages compensated by disadvantages on other grounds? Family farms are bound to lose the benefits from specialization, which, however, are not so great in agriculture, as the division of labor is limited by the seasonality in the demand, especially for crops (see section 2.2).[90] Furthermore, a family farm may not be able to use some indivisible input or technology (including organization) and may be denied access to "formal" credit (section 7.4). Allen and Lueck (1998; 2000, 167–96), formalize these insights in a model, comparing family farms with corporations (i.e., managerial farms). The latter are more efficient, and thus more likely to prevail, ceteris paribus, the smaller the gains from specialization of manpower, the higher the cost of supervision of the workers, and the greater the variance in output (i.e., the greater the risk). Allen and Lueck do not consider tenanted estates, which can be construed as a reasonable compromise (like the co-operatives). The tenants' incentives are greater than in a managerial farm because the tenants' income is somehow related to their performance, while the scale of the estate is large enough to exploit scale economics and the landlord is rich enough to borrow on more favorable terms.[91] Thus, theory highlights a potential trade-off between size and incentives, and, by definition, cannot settle the issue.

The empirical literature about the static efficiency of different types of farms is quite abundant. The key reference is still the pioneering book by Berry and Cline (1979, 31–37). Using data from the 1970 FAO census of world agriculture, they proved that, throughout the world, large farms utilized land less intensively (i.e., left a large proportion idle) than small ones, and that the bias was larger in land-abundant countries. They also showed that in six countries (Brazil, Colombia, Philippines, Bangladesh, India, and Malaysia), land was more productive in small than in large farms. This inverse relation between land productivity and farm size in LDCs is nowadays routinely quoted as a major stylized fact.[92] However, the historical evidence is rather thin and not so clear-cut. Myers (1970) finds an inverse relationship between size and land productivity in Hebei and Shandong (North China), but neither Huang (1985; for the same region) nor Brandt (1989; for the Yang-Tzi area) agrees with him.[93] This simple comparison, however, may be biased as it does not take the quantity of other factors into account. Land productivity may be higher in family farms (and tenanted estates) simply because they use more labor, and possibly more capital, per unit of land than managerial estates.[94] In other words, efficiency should be measured with the TFP. Berry and Cline (1979, 127–30) do report some estimates, but the results are mixed. The historical evidence is mixed as well. African native small-holders often out-competed Western farmers and planters.[95] Allen (1998, 1991) argues that the late eighteenth-century British managerial farms were more efficient, but his data (based on Arthur Young's work) are strongly contested by Clark (1991; cf. Brunt 2004, 214–15). Brandt (1987, 159–69), in one of the most thorough historical works on the issue, finds that, in the 1930s, small farms in North China were a bit less efficient than large ones, but he attributes the difference to fragmentation, which caused the loss of valuable land and time. The most convincing historical evidence of a TFP differential in favor of large farms refers to the antebellum American South. Fogel and Engerman (1977) argue that the great slave plantations were more efficient than free farms, both in the South and in the North, because they could afford a complex division of labor, within the productive teams and among slave households.[96] After the abolition of slavery, the plantations were divided into small farms and rented out, and the ensuing loss of scale economies caused level and growth rates of agricultural production to fall, with negative consequences on long-term southern development.[97] Ransom and Sutch (1977) contested this interpretation of the relative decline of the South.[98] Productivity in slave plantations was high only because slave-masters could extract an amount of work from the slaves, which they would not have performed had they been free. When freed, slaves returned to their true preferences, but the fall in output could not be considered a welfare loss.

Any point comparisons between efficiency of small and large farms, with either yardstick (land productivity or TFP), must distinguish the two categories. The distinction is rather clear in the case of slave plantations, but in most cases there is no obvious discontinuity in the distribution of farms by size. Thus, authors have resorted to somewhat arbitrary criteria. For instance, Berry and Cline (1979) simply assume that, in each country, large farms accounted for the top 40 percent of the distribution by size (acreage) and small ones for the bottom 20 percent.[99] A theoretically more appealing alternative is the econometric testing of the hypothesis of constant returns to scale in a production function framework. With this method, Lamb (2003) finds an inverse relation between size and production (or profits) in the 2300 Indian farms of the ICRISAT sample. This result would imply diseconomies of scale, but the author shows that it is fully accounted for by the quality of land and by the imperfections in the land and labor market (and also by measurement errors).[100] Other microeconomic studies yield fairly similar results. On the other hand, a comparison between differently sized family farms may not be relevant to the issue of the relative efficiency of different categories of farms. It is almost impossible to find a suitable microeconomic sample, since family farms and managerial estates very seldom co-exist in the same area with the same specialization. Thus, many authors have used macroeconomic data at national or international levels finding, very little or no evidence of scale economies.[101] This comparison is more meaningful than the microeconomic ones for the issue at hand, but, at the same time, it is less accurate, because national data conceal the huge differences among countries and across time in input quality, market imperfections, crop-specific technology, environment, and so on. Thus, neither micro- nor macro-estimates are ideal, but the broad convergence of their results cannot be wholly casual. One should conclude that pure returns to scale in agriculture are nil or very small, possibly with exceptions at the lower (and upper) tails of the distribution.[102] This feature may change, as returns to scale are likely to grow with technological progress. In the Unites States, livestock production is industrializing, and large farms owned by corporations account for a substantial share of total production.[103] For the foreseeable future, however, technical and organizational economies of scale in agriculture will remain negligible if compared with those in manufacturing or some services, and barriers to entry will stay relatively low. Agriculture is bound to remain a highly competitive sector.

The available evidence does not support the idea that farm category and/or size determine static efficiency. What about the effects on technical progress, and thus on long-run agricultural growth? From the economist's point of view, this question belongs to the wider issue of the ultimate causes of technical progress in agriculture, which is still far from being adequately tackled (section 6.6). For most historians, this is *the* issue. Many of them seem to reason backward with a reverse engineering approach: they infer the virtues of the prevailing farm type in a country according to its agricultural performance. For instance, it is quite common to attribute the technological backwardness of Latin American agriculture to the prevalence of large estates (*latifundia*).[104] It is said that landowners, instead of managing their farms and introducing the latest techniques, preferred to live in cities and to use their political and social power to squeeze as much rents as they could from the hapless peasants. The reasoning is not convincing. One can understand why a landlord would prefer to live in the city instead of on his estate, but it seems somewhat implausible that he would forego substantial gains from suitable innovations. It seems more plausible that innovations were not adopted because the gains were smaller than assumed. In fact, some recent research has found ample evidence of innovation in large tradition estates.[105] Mexican *haciendas* were quite prepared to innovate and to seize market opportunities (such as those offered by railways).[106] Alan Taylor (1997) shows that, in Argentina during the 1900s, the rented farms in tenanted estates did not differ in any of the relevant parameters (size, amount of capital in tools and livestock, etc.) from family farms, and that scale did not affect productivity.[107]

Conventional wisdom is exactly the opposite for the "advanced" countries, especially the European ones. Scholars such as Koning (1994) assume that large British-style managerial farms were structurally more innovative than family farms.[108] The prevalence of the latter is considered to be the major cause of the (alleged) poor performance of French agriculture in the nineteenth century (section 10.3). Small size, it is said, hampered technical progress, because it reduced the access to credit, prevented the use of optimal techniques, as in the American Great Plains (Hansen and Libecap 2004b), and, above all, because it reduced the use of indivisible inputs, such as machinery. For this reason, the French government (and many other European ones) adopted the so-called consolidation policies from the 1950s onward (section 8.3). Indeed, most machines can be profitably operated only if the production exceeds a minimum scale or threshold.[109] It is assumed that managers of small-sized farms would prefer not to purchase a machine rather than to use it suboptimally (which may not always be the case).[110] In a seminal paper, P. David (1966) adopted a threshold model to explain the slow adoption of reapers in the American Great Plains.[111] He argued that most farmers did not adopt the reaper during the 1840s and early 1850s because the wheat acreage in most farms did not exceed the minimum threshold for a profitable use of the machine at the prevailing wage for harvesters. The machine spread quite quickly when the rise in agricultural wages caused this threshold to fall. David's paper on reapers sparked a lively debate, which was terminated by Olmstead (1975). He exposed some defects in David's estimates and showed that farmers circumvented the threshold effects with sharing (via collective purchase) or with custom-work (i.e., out-sourcing specific tasks to specialized entrepreneurs who owned the machinery required). He thus argued that the adoption of reapers was delayed by the technical defects of the early reapers. This was not, however, the end of threshold models. They have been used, in a more sophisticated version (e.g., taking into account the expectations on future trends in prices and price instability and the availability of credit) to explain the diffusion of tractors in the United States.[112] Olmstead and Rhode (2001), however, estimate, with a simultaneous-equation econometric model, that farm size affected the diffusion of tractors only marginally.[113] In the short run, farmers resorted to custom work and sharing, while, in the long run, farm size adjusted to the needs of motorization. Although most attention has focused on machines, other inputs, such as animals, are also indivisible. It is argued that sharing animals is more difficult, as their well-being depends on the treatment they receive from all co-owners, which is obviously hard to monitor, but there is evidence of shared animals as well.[114] In short, farm size and type do not seem to have been a key determinant of technical progress and performance in history.

Some of the arguments about farm size and technical progress are reproduced in the parallel debate about contracts. Economists have long tried to measure the extent of Marshallian (static) inefficiency from share-tenancy. Hayami and Otsuka (1993) summing up the literature in a table, find no evidence of lower yields in share-cropped farms.[115] Actually, the test is quite crude, as inefficiency refers to the use of variable factors, and a lower quantity of labor does not necessarily reduce land productivity. In a famous paper using the ICRISAT sample, Shaban estimated that Indian peasants used less variable inputs (household labor, bullocks) on land rented under share-tenancy contracts than on their own land, causing output to be 15 percent lower.[116] Unfortunately, his results may be biased by institutional constraints to the use of fixed-rent contracts. The lack of suitable microeconomic databases has prompted many authors to pursue an econometric short-cut by adding a variable for contracts in regressions to explain production or productivity. Results are mixed. Contracts turn out not to be significant in regressions for Italy in 1911 (Cohen and Galassi 1992) and for some LDCs in the 1970s (Berry and Cline 1979, 128), but they are negative and significant in a regression for Korea

(Yoong-Deok and Young-Yong 2000, 263). This test, however, is a bit too simplistic to be really conclusive.

Historians tend to use anecdotal evidence on (alleged) technical stagnation to blame contracts, with a special predilection for share-tenancy. In a famous passage of his book on Italian agriculture, Sereni (1966) contrasts the thriving capitalist farms of the Po valley in Italy with the estates of Tuscany and Umbria, apparently well cultivated and orderly but actually technically stagnant because sharecropping prevented the introduction of British high farming.[117] On a somewhat different note, Fegerler (1993) blames landlords in the southern United States for having underprovided their croppers with implements, causing their income to trail behind that of other tenants.[118] This reasoning however, is not wholly convincing. It rests on the assumption that landlords preferred to accumulate land (clearly for noneconomic purposes) instead of investing. Furthermore, the (alleged) backwardness can be explained in other ways, which should be discussed and discarded before blaming contracts. For instance, Adelman (1994, 253) attributes the difference in capital use between share-tenants and farmers in Argentina to a selection process of tenants: individuals who entered into a sharecropping contract had a highly speculative frame of mind and thus they were not ready to invest. Some authors put forward much more elegant and theoretically sophisticated models. Whatley (1987) argues that sharecropping delayed the diffusion of tractors in the southern United States as it provided permanent, low-cost manpower throughout the year (as the total work force depended on peaks). He realizes that the argument would apply to all permanent workers, irrespective of contract, but he further adds that low-cost labor discouraged investments and the R&D to build a cotton-picker machine aimed at cutting the seasonal harvest peak.[119] Kauffman (1993) relates sharecropping to the prevalence of the mule over the horse as a draft animal: he argues that mules, a much hardier animals than horses, were more suitable to the situation when monitoring was expensive. Share-tenancy is by far the most controversial contract, but other ones do not escape their share of criticism. For instance, it was argued that in the United Kingdom in the nineteenth century, the short duration of contracts and the failure to compensate tenants for improvements reduced the incentives to long-term investment by the tenant (Perren 1995, 24–26; O'Grada 1994, 156–59). It is difficult to assess how serious this problem was, as it implied quite short-sighted behavior by landlords. Not compensating tenants for monitorable investments (e.g., in farm-building planting trees, etc.), they would forfeit the long-term gains from repeated investments by their tenants for the sake of short-term gains from expropriating them of the first investment they made.[120] However, the principle of compensation for tenants was established in the Agricultural Holding Acts of 1875 and 1883 (Malatesta 1989, 136–55). All these case studies are interesting for their own sake, but are hardly conclusive, especially given the uncertainty about the causes of technical progress in agriculture.

To sum up, neither theory nor the empirical evidence bring really convincing evidence to state that a specific type of farm (small or large, family farm or managerial) is *structurally* superior to others. Each type has its own advantages and disadvantages, and much depends on the local conditions. If anything, the evidence suggests the strong capacity of institutions to adapt themselves to the circumstances.

Chapter Eight

AGRICULTURAL INSTITUTIONS AND GROWTH

8.1 Introduction

Understanding how agricultural institutions work from a "theoretical" point of view is interesting and important for its own sake, but is it also a preliminary step to tackling the really big issues. How did institutions change in the long run? What caused the change? And, last but surely not least, how much did institutional change affect agricultural performance? Did institutions adjust to the need of agriculture or did they evolve independently, possibly slowing down growth?

This chapter addresses these basic issues. The first section describes the establishment of "modern" property rights separately on labor (the abolition of slavery and serfdom) and on land and its products (the diffusion of modern Western-style ownership). Section 8.3 deals with state intervention on land ownership, tenancy, and so on, after the establishment of modern property rights. Section 8.4 marshals the (unfortunately scarce) evidence on changes in the average size of farms, in patterns of ownership, and in contracts, which also affected the development of the market for land and labor. The development of the markets for factors and goods is discussed in section 8.5. Section 8.6 outlines, very briefly, the growth of co-operatives. The last three sections aim at explaining these long-term trends and assessing their effects on agricultural performance. Section 8.7 deals with property rights, section 8.8 with landownership and contracts, and section 8.9 with commercialization.

It is important to stress that the interpretation depends greatly on the nature of the institutional change (section 7.1). If, as is assumed by "historians," institutions are exogenous, then a given feature would be a cause of agricultural transformation (or a lack thereof). If, as is assumed by "economists," institutions are endogenous, then the change itself would be an effect of agricultural transformation. A good example is the relationship between farm size (institution) and mechanization (transformation). If institutions were exogenous, an insufficient farm size would hinder mechanization: an institution causes (a failure of) technical change and (a poor) performance. If institutions were endogenous, the need of mechanization would cause farm size to adjust: thus, technical progress causes institutional change. As discussed in the previous chapter, this was the case in the United States in the twentieth century, according to Olmstead and Rhode (2001). Their conclusions, however, hold true for this specific case only, and it would be hard to replicate their analysis for most other similar cases (e.g., the adoption of the reaper) for lack of data. A fortiori, a statistical analysis is impossible for less well-defined topics—such as the causes and effects of the development of markets. Thus, the last three sections will adopt a pragmatic stance, dealing with institutional change as mainly endogenous or exogenous as it seems to fit the case.

8.2 Prelude: The Establishing of Modern Property Rights

As anticipated in section 7.2, two hundred years ago, "traditional" property rights prevailed all over the world. A large number of workers were deprived of the basic right to personal freedom, albeit to different degrees. According to one estimate, in 1800, there were about 10 million slaves in the world (i.e., 1% of the total population).[1] Slavery was widespread in Africa and Asia, and it was essential for the economy of the Western hemisphere. Slave plantations produced cotton and sugar, by then arguably the most important trade commodities. Serfdom was concentrated in Eastern Europe—some 20–25 million in Russia and perhaps another 5–10 million in other countries (Moon 1999, table 3.1; Lyaschenko 1970, 310). The number of "debt peones" is highly uncertain—as this condition was not legally defined. Walsh-Sanderson (1980, 16) puts forward a figure of half the total of Mexican peasants on the eve of the revolution (i.e., 3–5 million out of a total population of about 15 million), but Katz (1974), in a classical article, strongly downplayed their diffusion.[2] Modern property rights on land did exist in western Europe, in the areas of Western offshoots that had already been settled by European colonists (e.g., the eastern coast of the United States), and in some Asian countries, including, most notably, China.[3] Nowadays, the outlook is totally different. Slavery is officially outlawed all over the world (even though slaves are still numerous in many countries). Serfdom and debt peonage have disappeared and modern property rights on land prevail, even though the traditional ones have by no means disappeared.

To outline this great transformation, it may be helpful to consider the evolution of the four "ideal" types of section 7.2 separately:

1. Hunting and Gathering Societies

Upon the arrival of western colonists, the United States and the other countries of Western settlement were thinly populated by hunter-gatherers or migrant cattle-breeders. It is unclear whether natives had started a "Boserupian" process of intensification, driven by population growth: according to the latest estimates, the native population of North America was shrinking (Thornton 2000, esp. table 2.4). The process, however, was not allowed to continue at its "natural" pace. The new authorities assumed the land to be free from any property rights, and extended their own legal system as part of the settlement process (chapter 4). Natives disagreed, of course, but they were exterminated or otherwise silenced (only in New Zealand were they recognized as rightful owners of the land, at least in principle).[4] Many settlers interpreted the official stance to mean that land was free literally, and grabbed whatever land they wanted, hoping to have their rights recognized afterward (squatting). The anecdotal evidence shows that squatting was widely diffused, from Brazil, where it was the main pathway to landownership, to Australia during its first pastoral boom.[5] By and large, it paid off: sooner or later, the squatters obtained the title to the land that they had seized. Most of the land, however, was first seized by the state and later distributed "lawfully." This distribution entailed two different choices: whether to privilege prospective farmers or to give the land to any bidder (including "speculators" who expected to resell it to farmers, thereby earning some capital gains) and whether to distribute land for free or for a nominal fee, or to sell it at "market" price (e.g., by auction). Thus, a government, in theory, had four alternatives: (i) to sell land to farmers at market prices, (ii) to sell land in large blocks, (iii) to distribute land for free to farmers, or (iv) to give land for free in large blocks. All these methods have been tried somewhere at some time.[6]

The sale of land to farmers at market prices was tried in the United States with the Land Ordinance (1785): the land was to be surveyed, divided in 160-acres lots, and sold at public auction, with a minimum price, but with also the right to delay payment. The Land Act (1820) abolished any limit to the size of lots, and stipulated that the price had to be paid in cash—thereby opening the way to large-scale sales for speculators. Later, in the 1850s and 1860s, the American and Canadian governments sold huge tracts of land to railways companies, which committed to build the intercontinental lines at their expense, hoping to recover the initial outlay and make profit from selling the land, at an enhanced price, to farmers. Jefferson had argued that any "unemployed" citizens had a right to receive public land for free in the late eighteenth century, but this idea was first implemented only some decades later, in 1862. The Homestead Act stipulated that anyone could receive 160 acres of land, provided he paid a small registration fee and lived for five years on the farm (or paid a commutation price). The American Homestead Law was imitated by Canada ten years later in order to distribute the prairie land purchased from the Hudson Bay Company. Free distribution of land also prevailed outside the countries of Western settlement—for instance, in Siberia, Serbia, Thailand, and Burma.[7] In these cases, however, the land formally remained state property (or royal property, as in Thailand): in Siberia, the peasant had to pay a token annual rent (until 1913), and, in Thailand, he could lose his rights if he did not cultivate the land. The concession of large estates almost for free was a traditional Latin American practice (known as *encomienda* in Spanish colonies and *sesmarias* in Brazil). The Argentinian provincial governments followed this tradition to distribute the land of the pampas, in spite of the approbation of federal laws for bulk sales

(1877) and for homestead (1884). Furthermore, in Australia, huge extensions of land were given for a nominal fee to large-scale ranchers at the beginning of colonization.

Hard data on the share of these methods of land distribution are available only for the United States and Canada.[8] Homestead accounted for 40 and 50 percent respectively, sales to railway companies for 30 and 40 percent and (auction) block sales for the balance—30 and 10 percent. The worldwide share of "homestead" (including squatting and some border cases) may have been somewhat higher. A tentative back-of-the envelope computation for the four main Western settlement countries, suggests that, in 1938, about 50 to 60 percent of total acreage and 30 to 40 percent of arable farming had been distributed with some forms of homestead, including the peculiar Australian mixture of squatting and large-scale concession of sheep farms.[9] As stated before, homestead was also the main method of land distribution outside the countries of Western settlement. Thus, one can very tentatively surmise that most of the available land was distributed directly to cultivators and ranchers from the beginning of the great colonization.

2. *Swidden Agriculture*

European conquest did not have the devastating impact on "customary" property rights in areas of swidden cultivation—notably in sub-Saharan Africa—that western settlement had in America.[10] To be sure, with few (but not negligible) exceptions, European colonial authorities were no more respectful of natives' rights than the American federal government was. They assumed that the natives needed only the plots they were currently cultivating, conveniently neglecting the fact that swidden agriculture needed a much larger acreage (section 2.2). From the point of a colonial administrator, these were forests and savannahs, not agricultural land—and as such, they were open to colonization and settlement by white immigrants. The latter settled mainly on the best land in temperate areas, as in South Africa and southern Rhodesia (Zimbabwe).[11] A few Europeans set up American-style plantations in tropical areas, to grow bananas, coffee, and other crops, forcefully recruiting the natives to work.[12] These undertakings raised great hopes for the economic development of colonies, but, at the end of the day, they accounted for a very small proportion of total tropical land. Thus, in most countries of sub-Saharan Africa, the natives maintained most of their customary rights on land.[13] As a rule, the colonial administrations did not meddle further.[14] In the long run, population growth caused the area under swidden cultivation to shrink, and permanent or semi-permanent cultivation brought about a need for better-defined property rights. The colonial administrations started to register native land in the 1940s, but the process was quite slow for a long period.[15] It accelerated from the 1980s by massive campaigns of land-titling sponsored by international organizations such as the World Bank. Communal property still accounted, according to the FAO censuses, for 60 percent of African land in 1970 and for 40 percent in 1990.[16]

3. *The Commune System*

By 1800, "communes" were quite widespread all over the world, from Mexico to the Middle East. In Western Europe, forests were quite often common property and there were also sizeable tracts of common arable land. Almost all this land was divided ("enclosed") in the next decades. In the United Kingdom, from 1800 to 1860, parliamentary acts enclosed some two million hectares—that is, 5 to 6 percent of the total acreage of the country.[17] In Continental Europe, the division of common land and the slow demise of communal institutions continued throughout the nineteenth century, and sometimes into the twentieth century.[18] Common ownership lasted longer in Eastern Europe than in Western Europe: at the beginning of the twentieth century, it still accounted for a quarter of the total arable land in Hungary and for some 55 percent of the total private land in Imperial Russia.[19] In 1906, the Russian prime minister, Stolypin, believing that peasant property would be the best bulwark against the threat of a revolution, allowed communes to disband with a two-thirds majority and each single household to claim full property of the land it was cultivating at that moment under the system of periodical redistribution.[20] This reform was not an instantaneous success: ten years later only a quarter of peasants, owning some 15 percent of the total land, had availed themselves of the right to separate from the *mir*. The share was highest in the more advanced areas—up to a half in the Ukraine. Then, the outbreak of the war stopped the process of dissolution, and, in 1917–18, many independent farmers were forced to return to the commune (Allen 2004, 46).

The British enclosures and the Stolypin reforms extended the areas under modern property rights. In many countries, however, there were no modern property rights, and thus their creation had to begin with the approval of new legislation. The process has been long and sometimes socially inequitable. For instance, in Mexico, the Leyes de Desamortizacion (1856) outlawed communal property, and, in the next forty years, the (so-called *Porfiriato*) *hacienderos* managed to extend their estates by all possible means, including that of debt peonage.[21] Elsewhere, the outcome was more favorable for peasants. The Turkish 1858 Land Code recognized the peasants' permanent and hereditary usufruct to land, provided that they cultivated it, but did not grant them full ownership. The latter was not recognized by the 1926 Land Law, and Turkish peasants had to wait for a series of court decision in the 1940s and 1950s, within the framework of a new 1947 law.[22] In Indonesia, the so-called Agrarian Law (1870) gave peasants an inheritable right to the land that they cultivated and divided the common land, but stopped short of full ownership both in principle (land still officially belonging to the state) and in practice (land could not be sold to money lenders or foreigners).[23] Full ownership was granted only in 1960.

4. The Feudal System

In its historical cradle, northwestern Europe, the "feudal" system had long disappeared by 1800, and its last remnants, some miscellaneous rights and mainly token payments to French lords, had been abolished in August 1789. Feudalism and serfdom survived in some areas of the Mediterranean and East of the Elbe river.[24] Furthermore, all over the world, powerful individuals, from Turkish officers (*spahija*) in the Ottoman empire, to *zamindaris* in northern India, to Japanese landlords, had the right to claim part of the peasants' crop. Scholars have long, heatedly, and somewhat fruitlessly, debated if and to what extent these people could be assimilated to European feudal landlords. Most recent research downplays the similarity. The *spahija* were given, on a temporary basis, the right to collect some taxes and the exclusive right to have a (smallish) holding to farm under sharecropping or, sometimes, with *corvées*.[25] The *zamindaris* were also entitled, on the basis of the 1793 Permanent Settlement with the British government, to claim part of the product of the otherwise independent peasants.[26] Scholars used to consider this right a quasi-feudal rent, but recent research has stressed that the *zamindaris* were mainly tax collectors who obtained their position at auction and could be dismissed if they did not deliver.

The remaining European feudal systems were abolished in the first half of the nineteenth century: in Prussia and southern Italy in 1806–7, in the 1830s in Serbia, in 1848 in Hungary, in 1861 in Russia, and in 1864 in Romania.[27] The serfs were given their personal freedom, and the land was divided between them and the lord. The conditions were quite harsh in Prussia, where peasants got only one-half of the total land, and had to pay compensation to the landlord. In Russia, they received a higher share (losing about one-fifth of their preemancipation holdings, however) and had to pay compensation at above-market land prices, although, in the long run, many of them eschewed their dues. Hungarian peasants received the best deal, as the state paid the compensation. In almost all these cases, with the essential exception of Russia, peasants and former landlords received full modern property rights on their share of land. In Russia, the rights to peasant land were transferred to the *mir*, until the Stolypin reform. Moreover, also most of the "feudal" rights outside Europe disappeared in the nineteenth century. In Japan, they were abolished after the Meiji restoration and peasants received almost all the land that they cultivated, while the landlords were compensated by the state.[28] The Ottoman *spahija* lost their revenue-collecting power after the so-called *Tanzimat* (reform) period of 1838–44, and, after independence, the Muslim right-holders transferred, more or less willingly, their rights back to peasants.[29] The Indian *zamindaris* resisted longer, although they lost a part of their role, especially after the Mutiny (1859).[30] They were severely hit by the peasants' refusal to pay their dues (the no-rent campaign) in the 1930s, and, after independence, the remaining *zamindaris* forfeited all their rights in exchange for some land and monetary compensation.

The first half of the nineteenth century also saw the triumph of the anti-slavery campaign.[31] It had started in the late seventeenth century among British and American religious radicals and had scored its first major success about a century later, with the abolition of slavery in the United Kingdom (1772). The key turning point was, however, the prohibition of the slave trade in 1807. The trade did not disappear overnight, but the increasingly effective repression caused the price of imported slaves from Africa to increase, up to prohibitive levels. The end of the slave trade dealt a very serious blow to the system both in the West Indies and in Brazil, where the net population balance was negative (i.e., imported slaves from Africa were necessary to prevent the total slave population from declining). In the first half of the nineteenth century, slavery was abolished in many South and Central American countries, including the British and French colonies (in 1838 and 1848 respectively). Thus, by 1860, most of the world's slaves lived in the southern United States.[32] There, the system was economically quite viable: the great plantations were highly efficient and profitable (section 7.6), and the slave population was rising, thanks to a very high fertility rate. Slavery was abolished only after the military defeat of the South in the Civil war (1865). Slavery officially survived in a few countries of the western hemisphere, to be abolished later in all of them, the last two being Cuba in 1886 and Brazil in 1888. Elsewhere in the world, events were not so dramatic. Slavery was, however, less strict a regime than on the cotton plantations, and it slowly withered, to be definitively outlawed in the twentieth century (in Thailand in 1915, and as late as 1936 in Northern Nigeria).[33]

To sum up, the transition from traditional to modern property rights entailed a stupendous change, which is not yet over, at least not for land. We will discuss its implications for agriculture in section 8.7, but it is necessary to stress that the process had effects on the economy, society, politics, and everyday life, which goes much beyond the scope of this book.

8.3 Meddling with Property Rights: Land Reform and Other Structural Interventions

In the nineteenth and twentieth centuries, many people were highly critical of the existing pattern of landownership and/or contracts. Most criticism focused on large-scale estates, which were accused of being technically backward and inefficient (section 7.6) and socially inequitable. The concentration of landownership and the exploitative contracts forced the mass of the population into misery for the well-being of a small élite of owners who wielded all the political power. Thus, governments were urged to expropriate owners ("land reform"), with or without compensation, and distribute land to the peasants, or, at the very least, to change the clauses of contracts in favor of the tenants ("tenancy reform").[34]

Although social reformers had campaigned for land reform from the nineteenth century onward, there was very little action until the 1910s. Some very partial measures were implemented in Bulgaria, Ireland, Germany, and New Zealand.[35] The turning point was the start of the Mexican Revolution: as we will document in the next section, the distribution of land at the end of the *Porfiriato* was extremely unequal, and thus the land question was at the top of the radical agenda.[36] A wave of expropriation started at the onset of civil war, and the principle that land belonged to the tiller was enshrined in the 1917 constitution. The process of reform, however, set a world record for duration. In fact, the lawful transfer of land started in 1920–21, and went on until 1964, for a total of 50.5 million hectares—about half the total land (including pastures). The time pattern depended pretty much on the ebbs and flows of the political climate—with a peak during the Cardenas presidency (1936–40). The land was transferred to *eijdos*—village communities, which were supposed to manage it collectively. This rule covered a wide range of cases, from true collective cultivation to household farming with little or no collective overseeing. The *eijdatarios* were not allowed to sell their rights, or to mortgage land. In short, the Mexican reform, like the abolition of serfdom in Russia, entailed a step backward in the evolution of property rights—from modern (albeit very unequally distributed) rights to traditional ones. The upheavals caused by World War I put land reform at the top of political agenda in Europe, too. After the end of the war, all the new Eastern Europe states enacted some reform (Poland in 1919–20, Czechoslovakia in 1919–20 and 1925, Romania and Hungary in 1920, Bulgaria in 1921–22, and Yugoslavia in 1923): on the whole, they transferred some 7 million hectares to the peasants, mainly at the expense of former Russian and German landlords.[37]

After World War II, the trickle became a flood. King's (1977) book, which does not cover the "advanced" countries, discusses twenty-three cases of land reform up until 1975.[38] They ranged from limited measures targeted to specific types of estates in given areas (e.g., the *latifundia* in southern Italy) to blanket expropriation of all noncultivating owners. It would be impossible to deal with all of them here: as an example, let us consider the case of Japan.[39] Before the war, a sizeable extension of land belonged to landlords and was rented to tenants. The Allied administration, regarding landlords as a key component of the prewar elite, pressed for reform. The reform expropriated land beyond one hectare if the farm was owned by an "absentee" landlord, and beyond three hectares if it was managed by the owner—this being considered the maximum size of a rice farm if cultivated by a household. Former tenants were to pay the landlord a compensation, which was, however, quickly eroded by inflation—so that, at end of the day, they got the land almost for free. The reform affected about 40 percent of total acreage and was implemented smoothly and quickly, lasting three years only (1947–50). As a whole, it enjoyed the reputation of having been a great success.

The timing of "tenancy reform" was broadly similar—few antecedents in the nineteenth century, with the bulk of the intervention in the second half of the twentieth century. The forerunners were the British, who approved, almost at the same time, the Deccan Agriculturists' Relief Act (1879) and the Irish Land Act (1881).[40] The former, the first in a long series of area-specific acts, protected tenants against eviction by money lenders (if interest they charged was excessive), while the latter set a fair rent and offered tenants a safer tenure. Tenancy reforms multiplied after World War II, at the same time as land reform (King 1977, 208; Ray 1998, 444–45). For instance, the new Taiwanese government set a 37.5 percent maximum to the landlord's share of product in share contracts. Several Indian states adopted ("unsystematic, uncoordinated, and contradictory") measures to enhance safety of tenure (in some cases, making it permanent or giving peasants a preemption right) and to set ceilings for rent.[41] Many countries banned or severely limited sharecropping, which, as recalled in section 7.3, had always enjoyed the reputation of being usurious, exploitative, and inefficient. It was made illegal (at least officially) in Italy in 1964 and in the Philippines, although this latter reform was implemented beginning only in 1971.[42] In both cases, sharecropping was substituted by fixed rent contracts, at quite favorable terms for the tenant. In the Philippines, the rent could not exceed a quarter of the pre-reform average of production (the traditional share was 40%). In Italy, rents could not exceed a given proportion of the very low cadastral value of the land.

In the twentieth century, would-be reformers strongly argued that the small size of family farms and the fragmentation into too many fields were a major shortcoming of European agriculture, because these features prevented mechanization and technical progress.[43] The evidence for this statement is not so compelling (section 7.3), but many governments followed the advice of experts and adopted "consolidation" policies. These policies were pioneered in the interwar years by the Netherlands (from 1924) and by Nazi Germany.[44] The regime wanted to prevent both the migration of peasants toward the cities and the reduction of the size of family farms, which was the unavoidable consequence of the rise of the farm population under partible inheritance. In 1933, it forbade the hereditary division and the sale of family farms (*Erbhof*), and, in 1936, it gave itself the power to mandate the merging of different farms and the exchange of fields (*Umlegung*). These policies, although innovative and far-reaching on paper, affected only a minority of farms, with less than one million hectares out of some 30 million in the whole country. In France, a law for voluntary consolidation had existed since 1860, but very few villages had used this possibility. Consequently, the government enacted a quasi-compulsory consolidation policy from the 1950s onward, with some success (Grantham 1980, 527–31; Wade 1981, 240–43; Ilbery 1984, 152–54; Barral 1982, 1450–53). The European Union followed this tradition, with the so-called Mansholt plan, which was first proposed in 1968, and was approved in 1972.[45] In its original version, it aimed at increasing the average farm size, at accelerating retirement of older farmers, and at setting aside land. The plan met with strong resistance from farmers, however, and thus the approved version was substantially reduced and proved fairly ineffective. Some LDC countries also adopted consolidation policies: the Indian government subsidized the reduction of fragmentation up to half the total costs.[46] However, these policies had only limited success, and fragmentation is still widespread in LDCs. In the 1990 World Agricultural Census, twenty-one countries provided data about the number of parcels per farm, which is still quite high: on average, the number was 3.3, up to a maximum of 9.6 in Burkina Faso (FAO 1990).

To sum up, the state intervened heavily in land ownership, contracts, farm size, and layout even after the establishment of modern property rights. The effects of these interventions are highly controversial, as we will discuss in section 8.7 after presenting the evidence on long-run changes.

8.4 The "Structural" Change in the Long Run

The macroeconomic evidence on farm size, contracts, and land ownership is surprisingly thin in comparison with the huge theoretical literature on these issues, which relies mostly on microeconomic data. Very few countries collected data before the 1930s, when the Institute Internationale d'Agriculture started to coax them into surveying landownership and tenure with homogeneous criteria. The results were published in four World Censuses of Agriculture.[47] These are really impressive works, but they suffer from five major shortcomings. (1) The country coverage is never complete: the censuses exclude socialist countries and, as is easily predictable, underrepresent the LDCs, especially the African ones.[48] Furthermore, the coverage differs from one census to another—so that the "world" total is not comparable. Therefore, tables 8.3. and 8.5 report data also for comparable ('homogeneous') sets of countries (the list is at the end of each table). (2) The data measure farm size with acreage, which is clearly a very poor proxy of size. Thousands of hectares of barren land or of meager pastures may produce much less than a few hectares of vineyards or greenhouses. If land is more productive on family farms than on large estates, the use of acreage biases the share of the former downward. (3) The unit of survey is the "holding," which is defined as "all land which is used wholly or partly for agricultural production and is operated by one person (the holder) or with the assistance of others, without regard to title, size or location."[49] This definition is clearly more suited for dealing with "technical" issues, such as the relative productivity of small and large farms, than with tenure. Furthermore, in some cases, different people, such as the landlord and the tenant in the nineteenth-century Tuscan *fattorie* (section 7.3) may share management of the same land: the allocation of these mixed cases in a census unit is bound to be arbitrary. (4) The land under traditional property rights is lumped together in a generic category of other forms of tenure in the 1930 and 1950 censuses, while it is considered separately (as "tribal" and "squatted" land) in the censuses of 1970 and 1990. (5) The layout of the published data is not adequate to capture the three-way distinction among family farms, managerial estates, and tenanted estates.

In fact, the censuses report data on ownership, contracts, and size separately but offer no cross-tabulation by ownership/size. It is thus impossible to distinguish, in the category "owner-managed farms," the small family farms from the large managerial estates. It will thus be necessary to consider trends in average size, landownership, and contracts separately.

Farm Size

Nowadays, farm size differs hugely among countries—from a minimum of 0.5 hectares in Bangladesh, to a maximum of 469 hectares in Argentina (FAO 1990). This difference between Asia and the countries of Western settlement date back to the early colonization. On the eve of World War I, the average Argentinian farm had 530 hectares, while the Chinese or Japanese farm barely exceeded one hectare.[50] The difference may have increased in the long-run, as farm size had been decreasing in many LDCs as a consequence of population growth. In China the trend had started, according to Chao (1986), in the seventeenth century and continued at least until World War II, with a temporary reversal in the 1860s to 1870s, after the Tai'Ping Rebellion. From 1870 to 1930, the average size of Chinese farms diminished by a third, from 1.4 hectares to 0.9 hectares.[51] In contrast, farm size in the advanced countries remained roughly constant until the interwar years and increased substantially thereafter. Long-term series from homogenous sources (such as agricultural censuses) are only available for a handful of countries (table 8.1). In the United States (and also in the United Kingdom), growth concentrated in the right-hand tail of the distribution: large farms (above 105 hectares) accounted for 46 percent of the total acreage in 1910, and 87 percent in 1987.[52] The data for other European countries (table 8.2) are scarce and hardly comparable across countries as they refer to different definitions of acreage. It is quite difficult to draw any firm conclusions about changes before 1960, but the later growth in size is evident.

The difference between the growing size in "rich" countries and the declining size in "poor" ones is quite evident after World War II (table 8.3). The decline in farm size North and Central America seems to be an exception among "advanced" countries. It reflects, however, the fall in "poor" Mexico and Central America, while average size did rise in the United States and Canada. Indeed, the coefficient of variation in average farm size by country in 1990 was as high as 1.30—the figures ranging from a minimum of 2 hectares in the Barbados to a maximum of 242 hectares in Canada. The dispersion was smaller in Europe and Asia, but the coefficient of variation was still around 0.7.

It is likely that differences in farm size were smaller if measured, more accurately, by gross output, sales or Value Added, because production per unit of land is often inversely proportional to size (section 7.6). For instance, in 2000, the average Greek and Italian farm produced 55 and 66 percent of the European average gross output, with respectively 25 and 33 percent of the average acreage (European

Table 8.1
Average Size of Farms, by Country, 1850–1990 (Hectares)

	USA	*Canada*	*Belgium*	*England*	*Germany*	*Japan*	*Java*
1850	82						
1860	80.5						
1870	62	40					
1880	56	40	6.3		6.0		
1885				33			
1890	56	45					
1895			6.7	32			1.44
1900	59	50					
1910	56	65		31.5	5.55	1.04	
1920	59	80				1.08	1.1
1924				31.5	ca. 6	1.08	
1930	63.5	91	6.4			1.05	
1940	70						0.84
1950	87	113	6.8	33		0.82	
1975	178			51			0.5
1990	197	242	17.6	70	28.1	1.2	

Sources: 1990: FAO 1990. Other years: **United States:** Stanton 1993, table 4.1. **Canada:** Historical Statistics Canada 1965, tables L2 and L7 (including unimproved land). **Belgium:** Blomme 1992, 229. **England and Wales:** Grigg 1987, table 1. **Germany:** Rolfes 1976a, tables 2 and 3, and 1976b, 763 (unadjusted for boundary changes). **Japan:** Historical Statistics Japan 1987–88 vol. 2, tables 4.1 and 4.4. **Java:** van der Eng 1996, 146, and Booth 1988, table 2.12.

Commission 2000, table T 26). Unfortunately, historical data on gross output by farm are available only for very few advanced countries—notably the United States. They confirm, however, the basic insights: farm size increased after World War II, and the growth was concentrated in large farms. In the United States, the gross output per farm rose by only a half from 1870 to 1940, tripled by 1970, and increased a further 5.5 times by 2002, when farms with more than one million dollars in sales (1.4% of the total) accounted for almost half of the total sales.[53]

Land Ownership

Historians have very strong opinions about the traditional distribution of land, even if they can quote very little or no data at all to back them. They believe that family farms prevailed only in the countries of Western settlement, and possibly in some western Europe countries (notably France), while in the rest of the world most of the land belonged to landlords, who managed their estates directly or rented them out. The unfortunately thin quantitative evidence shows that this conventional wisdom is only partially true. It is by and large true for Latin America and for the countries of Western settlement. Latin America had the least equitable distribution of land: in Mexico on the eve of the revolution about 1 percent of

TABLE 8.2
Average Size of Farms By Country, Europe, 1900–2000 (Hectares)

	ca.1900	*ca.1930*	*1960*	*2000*
England	70a	63b	32.0	69.3
Eire	40		17.1	29.4
Norway	9		5.2	7.1i
Finland	16		10.6	23.7
Sweden	18	16	18.4	34.7
Denmark	40		15.7	42.6
Netherlands	20c		9.9	18.6
Belgium	11		8.2	20.6
France	30d	26	17.1	41.7
Switzerland	10e		10.9	30.3h
W. Germany	13	14	9.3	32.1
E. Germany	35	27		
Austria	65		10.2	16.3
Hungary	180	35		
Greece		9	4.0	4.3
Italy		6.3	6.8	6.4
Portugal				9.2
Spain				21.2
European Russia	37			

Notes: a1875; b1924; c1910; d1892; e1905; g1939; h1992; i1980

Sources: **1900** and **1930:** Dovring 1965, table 17 (except Italy, from ISTAT 1936): Eire, France, Austria, Hungary, and Spain: total area acreage, including privately owned forests and unproductive land; England, Norway, Denmark, Netherlands, Belgium, Switzerland, E. and W. Germany, Greece, Italy, and Russia: arable land, tree crops, and pasture; Finland, Sweden, and Portugal: arable land and tree crops only. **1960:** Hayami and Ruttan 1985, table A.4 (arable land and tree crops). **2000:** Norway, Hayami and Ruttan 1985, table A.4; Switzerland Eurostat 1996, and, all other countries: European Commission 2000, table T 26 (arable land and tree crops).

landowners owned 97 percent of the land, while in the 1950s great estates (with more than 12 workers) accounted for 36.9 percent of land in Argentina, 59.5 percent in Brazil, 81.3 percent in Chile, 49.5 percent in Colombia, and so on.[54] Family farms indeed prevailed in the countries of western settlement, although the myth of a nation of small farmers was much closer to reality in Canada (about 90% of the total acreage from 1871 to 1951) than in the United States (see table 8.4).[55] In the first half of the nineteenth century, tenancy was an option for the impecunious but enterprising American farmer, while after the Civil War, it was almost universally adopted in the former cotton plantations throughout the South (Atack and Passel 1994, 407–411; Atack et al. 2000, 315; Wright 1988). The total share of rented land has remained surprisingly high since then: as shown by the row "part owner" in table 8.4, many family farmers rent land from retired neighbors to increase the size of their farms (Gardner 2002, 54–56).

Table 8.3
Average Size of Farms, by Continent, 1950–90 (Hectares)

	All Countries			*Homogeneous Sample**		
	1950	*1970*	*1990*	*1950*	*1970*	*1990*
Africa	19.4	13.1	5.0			
North and Central America	80.2	130.6	88.1	88.3	141.0	92.0
South America	121.9	53.8	71.8	90.3	56.9	62.5
Asia	3.3	2.3	1.2	2.9	2.3	1.7
Europe	11.1	12.7	16.7	13.8	13.9	16.8
Oceania	1142.6	1647.3				
Total World	21.4	15.6	5.1	17.8	13.3	10.6

Note: * Austria, Czechoslovakia, Denmark, Finland, Ireland, Luxembourg, the Netherlands, Portugal, Switzerland, Canada, the Dominican Republic, Mexico, Panama, USA (including Puerto Rico and Virgin Islands), Brazil, Colombia, Uruguay, Venezuela, India, Japan, and the Philippines.
Sources: FAO 1961a, table 1, FAO 1981, table 3.1, FAO 1990.

Table 8.4
Shares on Total Acreage, by Type of Tenure, United States, 1880–2002

	*1880**	*1900**	*1900*	*1910*	*1920*	*1930*	*1950*	*1969*	*2002*
Full Owner	74.5°	55.8	51.4	52.9	48.3	37.6	36.1	35.3	38.0
Part Owner		7.9	14.9	15.2	18.3	24.9	36.4	51.8	52.8
Tenant	26.5	35.3	23.2	25.8	27.7	31.0	18.3	13.0	9.2
Manager		1.0	10.4	6.1	5.7	6.4	9.2		
Total Rented			31.6	31.6	33.3	43.6	37.7	35.7	40.6§

Notes: * on number of farms ° includes part-time owners and manager § 1998
Sources: **1880–1900:** Historical Statistics 1975, series K 142–K147; **1910–69:** USDA 2000, Tables 9.6 and 9.8. 2002 USDA 2002.

The conventional wisdom about landownership in long-settled areas does not tally well with the evidence. Family farms were largely diffused, if not predominant, all over Asia.[56] In Chao's (1986) words, "for the better part of two thousand years, China remained basically an economy comprised of free peasants working independently on their own small pieces of land."[57] The surveys by Buck and by the National Land Commission show that in the 1930s family farms accounted for about two-thirds of the total land, with a huge dispersion by region, from 23 to 81 percent. Esherick (1981) argues that these figures are a bit on the high side: in his opinion, rented land accounted for some 42 percent of the total—still, substantially less than a half. It is unclear what had happened before. According to the traditional (Marxist) view, land concentration had been growing since the mid-nineteenth century, and this expropriation motivated the peasants' support for the communist revolution.[58] In contrast, Brandt and Sands (1992) surmise that the share of family farms had remained stable from the 1880s. In Japan, after the 1869 Meiji land reform, peasants owned about 70 percent of the land, but their share declined to about a half of the total on the eve of World War I—and remained stable thereafter.[59] The case of India, the other great peasant country, is more difficult to interpret. In most parts of the country, the peasants (*ryotwari* in central and southern India, and quasi-commune in the northwest) were largely independent, while the classification of those in the northeast depended on the status of the *zamindaris* (see section 8.2). If these latter, as is now believed, were indeed mainly tax collectors and not quasi-feudal landlords, then, Bengal cultivators could also be considered as independent peasants.

The case of Europe is rather complex. At the end of the nineteenth century, large estates prevailed in the United Kingdom, possibly prevailed in some Mediterranean countries (Italy and Spain), and were quite diffused in central and southeastern Europe (Austria-Hungary, Eastern Germany, Romania, etc).[60] The family farm prevailed in Scandinavia, the Netherlands, and also in France, where, however, in spite of its reputation as the quintessential country of small peasants, tenanted and managerial estates accounted for about two-fifths of total acreage as late as 1930.[61] But, above all, family farms also prevailed in Eastern Europe. The peasants owned 90 percent of the land in Bulgaria, and an increasing share of private land in Russia—about 60 percent in 1877–78 (about 120 million hectares), two-thirds (158.5) in 1906, and three-quarters (185) on the eve of the revolution.[62] Peasant ownership was on the increase in many other countries, including France (from the nineteenth century) and the United Kingdom. Most African land was held under traditional property rights, which could not, formally, be classified under the modern categories of ownership or rent. In practice, however, tribal land is undoubtedly closer to the family farm than to either managerial or tenanted farms, and thus, arguably, "ownership" dominated in most of Africa—with the exceptions of Egypt and South Africa.[63] Thus, the available evidence, albeit partial and not homogeneous, points to a clear conclusion: before 1940, family farms accounted for most of world land and their share of the total had been rising.

Unfortunately, the World Agricultural Censuses lump together "family farms" and estates managed by the landlord in the same category ("farms operated by the owner"). Thus, the data are likely to bias upward the true share of the former. To compensate for this bias, table 8.5 omits tribal land ("other tenure") and squatted land. These data show how widely the share of family land still varies among countries and continents. Even excluding African countries, where most of the land is registered as "tribal" in 1990, it ranged from a minimum of 34 percent in Panama to a maximum of 99.2 percent in India, and the coefficient of variation of a simple average for the thirty-five countries is 0.36. It also shows that the share of family land

TABLE 8.5
Share of Owner-Operated Farms, by Continent, 1930–90

	ca.1930	*ca.1950*	*ca. 1970*	*ca.1990*
All Countries				
Europe	84.7	81.9	72.6	77.4
North and Central America	53.6	50.1	60.0	67.0
South America		66.0	83.5	92.7
Asia	53.8	77.6	91.1	95.9
Africa		78.5	26.1	70.6
Oceania		47.0	91.7	
World	56.7	58.4	77.8	81.2
1930–50 sample§				
Europe	71.2	62.7		
North and Central America	50.0	55.7		
Asia	53.6	75.4		
World	52.3	56.8		
1950–70 sample*				
Europe		69.2	66.5	
North and Central America		50.0	58.6	
South America		86.2	84.1	
Asia		71.6	73.2	
Oceania		49.1	48.5	
World		61.5	67.6	
1970–90 sample°				
Europe			87.2	66.5
North and Central America			59.7	67.0
South America			83.5	92.5
Asia			91.8	95.9
Oceania				
World			72.6	80.5
1950–90 sample#		60.0	67.2	78.4

Notes: § Austria, Belgium, West Germany, the Netherlands, Sweden, Switzerland, USA, the Philippines, and Japan.

* Austria, Belgium, West Germany, Greece, Luxembourg, the Netherlands, Sweden, Switzerland, the UK, Costa-Rica, the Dominican Republic, Honduras, Mexico, Panama, USA (with Puerto Rico and the Virgin Islands), Brazil, Uruguay, Venezuela, Iraq, Israel, Jordan, Japan, the Philippines, New Zealand, and Samoa.

° Finland, Italy, Luxembourg, Portugal, UK, Canada, Honduras, Mexico, Panama, USA (with Puerto Rico and the Virgin Islands), Brazil, Colombia, Peru, Uruguay, India, Pakistan, the Philippines, and the Reunion Islands.

Mexico, Panama, USA (with Puerto Rico and the Virgin Islands), Brazil, Uruguay, and the Philippines.

Sources: **1930 and 1950:** FAO 1961, **1970:** FAO 1981. **1990:** FAO 1990.

had been rising in all the comparable samples (i.e., those with constant country coverage). The only exception is Europe from 1970 to 1990. Most of the decrease is accounted for by the halving of the share in Finland, but it reflects a generalized trend in many "advanced" countries. The heirs of farming families prefer to let the land instead of selling it.[64] On the other hand, in the "advanced countries," the largest family farms are now quite substantial operations and managerial farms are also coming back. In the United States, in 2002 "corporation not family held" farms owned less than 1 percent of the land with 6.2 percent of the sales, while Gardner (2002, 60) estimates that the large family farms employed, on average, six or seven full-time equivalent workers.[65] Even the largest corporate farms are small, however, when compared with large manufacturing and service companies.

Contracts

The quantitative evidence on the diffusion of different types of contracts is very scarce, to say the least. It suggests that sharecropping accounted for a minority of all contracts and, possibly, that its share has been declining. In India according to the National survey (Hayami and Otsuka 1993, 102), share tenancy accounted for 7.5 percent of all land in 1953–54 (i.e., a third of land under tenancy and for 5.6 percent in 1970–71; i.e., almost a half). In China, it declined from being the most common contract under the Ching dynasty, down to 15–30 percent of rented land (according to the sources) in the mid-1930s.[66] In the United States, it was widely adopted in the south, but nationwide it accounted for 8.5 percent of tenanted land in 1920 (and for 2.4% of the total), 7.4 percent in 1940 (2.2%) and then declined to 3.1 percent (0.5%) in 1960.[67] According to the 1930 World Agricultural Census, share-tenancy was unknown in the United Kingdom, Eire, Norway, Sweden, Denmark, the Netherlands, Belgium, Germany, Austria, Czechoslovakia, and Bulgaria. It accounted for a sizeable proportion of rented and total land in Spain (37.5% and 15%), France (25% and 10%), and, above all, in Italy (some 60% of rented land and 20% of the total).[68] Successive censuses seem to be much less helpful than for other "structural" features, as they report data for only few countries (table 8.6). Share-tenancy accounted for a sizeable share of tenanted land in some countries, including the United States, but only in Pakistan and in the Philippines (before the reform) did it account for a non-negligible proportion of total land.

To sum up, this section shows that (1) the average size of farms had been decreasing in LDCs, while it was constant and, after 1950s, increased in the "advanced" countries; (2) family farms were fairly widespread all over the world at the beginning of the period, and their share of total land had been increasing—with few, albeit important, exceptions; and (3) fixed rent, either in kind or in cash, seems to have been the prevailing contract, and its share seems to have risen at the expense of sharecropping. The data does not discriminate between tenanted and managerial farms, but the anecdotal evidence suggests that the latter accounted for a minor

TABLE 8.6
Share of ShareCropping Contracts, by Country, 1950–90

	% Share on Rented Land			*% Share on Total Land*		
	ca.1950	*ca. 1970*	*ca. 1990*	*ca. 1950*	*ca. 1970*	*ca. 1990*
W. Germany	0.0			0.0		
Finland			1.1			0.6
Portugal		12.8	6.1		1.8	1.6
Costa Rica		9.4			0.1	
Cuba	15.9			6.1		
Dominican Republic	48.5			1.5		
El Salvador		1.9			0.1	
Honduras	13.4			0.8		
Mexico	6.1			0.3		
USA		31.5			3.8	
Colombia		49.4			2.6	
Ecuador	11.5			1.1		
Peru		0.0			0.0	
Uruguay	5.4	4.7		2.4	0.9	
Venezuela	39.6			2.2		
Paraguay			8.2			0.1
Cyprus	38.5			6.2		
Israel	6.7			3.0		
India			26.3			0.1
Jordan		10.6			1.3	
Nepal			57.0			0.8
Pakistan		86.6	74.8		25.5	12.0
Philippines	65.1	79.3		17.7	58.9	

Sources: FAO 1950, 1970, 1990.

share of the total land, and that their importance had been decreasing in the long run (possibly with some recent reversals). Needless to repeat, these statements rely on incomplete and sometimes biased data, and thus may be called stylized facts only in an extensive interpretation of the expression.

8.5 The Development of Markets

The exchange of agricultural products and inputs, and thus the "institutions" to regulate them, long predate 1800. These institutions changed greatly in the past two centuries, but this section will not deal with this issue. In fact, a thorough analysis would entail a wider discussion of the evolution of the legal system. Suffice it to say, that markets for goods have usually been less regulated and more flexible that those for input, and that the overall degree of regulation, especially in the

TABLE 8.7
Share of Laborers on Agricultural Workers, by Country, 1881–1951

	France	*UK*	*Germany*	*Italy*	*India*	*USA*
1881	33.9	77.0		65.6	26.1*	
1911			34.6°			24.9#
1921	44.7	71.9	22.9	44.0	24.4	25.4
1951	47.4	72.8	22.5	30.1	27.9	23.4

Notes: * 1901 ° 1909 # 1910

Sources: **India:** Khrishnamurty 1983, table 6.2. **USA:** Historical Statistics 1975, series K174–K176. **Others:** Vitali 1990, table 8.

market for goods, greatly increased in the twentieth century, as we will see in detail in the next chapter. We will, instead, focus here on commercialization, as measured by the increase in market transactions. Such a rise is commonly believed to have been a key component of the modernization of agriculture and of society at large (Fei and Ranis 1997, 94, 100–102; Rosengrant and Hazell 2000, 64). "Traditional" farmers, it is said, were largely self-sufficient. They used, whenever possible, in-house labor (free or forced) and capital, hiring additional hands and borrowing only if it were strictly necessary; they produced all the input for the farm (seed, hay for animals, even tools) and most of what the household needed: food, wood for heating, and textile fibers for domestically produced clothing. In contrast, the "modern" farm resorts to outside labor and capital, purchases industrial inputs from outside, and sells almost its whole output.

Commercialization is a multifaceted process that involves all factors (labor, land, capital) and goods. In theory, one should measure it with the share of market transactions on the relevant stock or flow—that is, the share of marketed goods on total output, the share of (services by) hired workers on total labor, and so on. Unfortunately, the data on markets for factors are much less abundant than those on goods. There are almost no data on land. The creation of modern property rights (section 8.2) has undoubtedly made the transfer of land easier, as market sales are much more flexible than the traditional procedures of transfer, such as the periodical redistribution within a commune or the discretionary allocation by the ruler. But it is almost impossible to assess whether transaction under modern property rights has been growing or decreasing as a proportion of the stock of land (section 8.9).

The only data on the development of the labor market refer to the share of "laborers" on the total work force from some national censuses (table 8.7).

These data clearly, and not unexpectedly, show that the resort to hired hands was inversely proportional to the diffusion of family farms. One would infer that the worldwide increase in family farms (section 8.4) must have reduced the proportion of permanent landless laborers on the total work force of the sector and

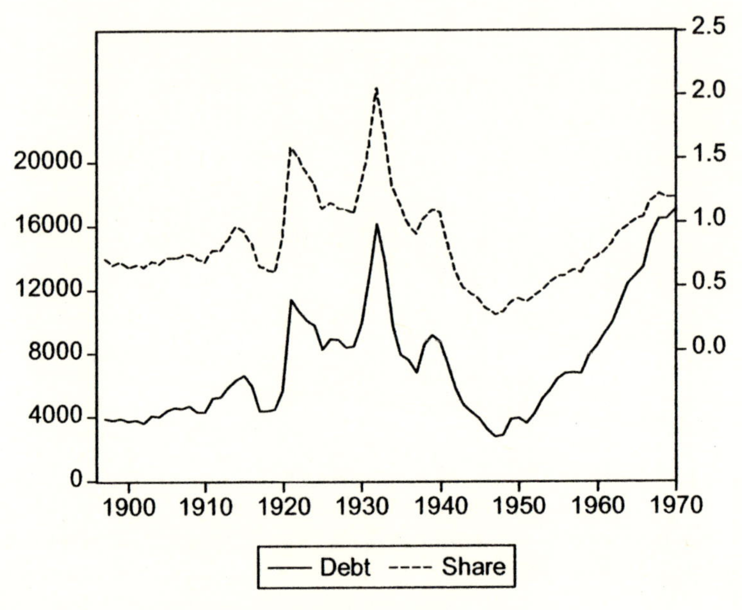

Graph 8.1 Agricultural Credit in the United States, 1910–70

thus also reduced the thickness of the labor market.[69] On the other hand, census data, referring to full-time workers, are likely to neglect a substantial part of the total transactions—the resort to non-farm workers, the custom work for the use of such specialized machinery as threshers, and the moonlighting by registered farmers, which also includes the exchanges of labor among neighboring farms. The resort to non-agricultural manpower undoubtedly declined in the long run (section 4.4), the custom work is likely to have increased while it is impossible to say anything about trends in moonlighting.[70]

The development of the market for credit ("capital deepening") is usually measured with the ratio of financial assets to total wealth (FIR), by the ratio of new issues to the GNP (ANIR), or by the ratio of private credit to GNP.[71] These measures may be biased and misleading if the numerator takes only the assets or the issues of the "formal" sector into account. In this case, they understate the amount of credit in agriculture and, therefore, overstate the extent of capital deepening in agriculture as well as in the whole economy. For example, the best available data, those for the United States from 1910 onwards, do include the mortgage-backed loans from individuals, which belong to the "informal" sector, but still exclude other loans from individuals and "others," such as the producers of equipment. Figure 8.1 shows the total amount of credit (right-hand scale, 1913 dollars) and its share of gross output.[72] The changes in total credit and in the credit-gross output ratio seem to reflect more the short- and medium-term trends from the demand side (notably during the Great Depression) than a long-term capital deepening. Such a statement is, however, subject to qualification: in theory, total credit could have risen if unsecured informal loans had been falling. The only (indirect) evidence on them is the decline in the share of "non-institutional holdings" loans on the secured market, from almost 90 percent of the total at the end of the nineteenth century, to around 45 percent after World War II to less than a fifth in 2002.[73] The initial share is revealingly high. In fact, these loans were the choice market for formal institutions: the overall share of informal credit, including short-term commercial and consumption credit, must have been higher.[74] It is thus possible that the data undervalue the true increase in the total amount of credit.

The evidence on long-term trends in other countries is very scarce, but it confirms the relative growth in the share of formal credit—and, by implication, the bias of conventional measures of capital deepening. In Japan, another (relatively) "advanced" country, formal institutions (banks, state-sponsored banks, and co-operatives) provided about 30 percent of total credit in 1912 and about a half in 1929.[75] The share of formal credit was much lower, almost negligible, in traditional agricultural societies. In China during the 1930s, commercial banks accounted for a mere 2.5 percent of total credit, and credit co-operatives for a further 2.5 percent (Feuerwerker 1983, table 18). Table 8.8 summarizes the results of an extensive survey by the World Bank in the 1970s. By then, informal credit was still very important in the LDCs, especially the Asian ones, but the data have to be framed in a decreasing trend. For instance, in India, it declined from 93 percent of the total in 1950, to 75 percent in 1970, and to 39 percent in 1981 (Hoff and Stiglitz 1993, table 2.1; World Bank 2002, 39–40).

This development of the formal institutions for credit and insurance was strongly supported by many governments, from the late nineteenth century

Table 8.8
Diffusion of Informal Credit in LDCs, around 1970

	(a)	*(b)*	*(c)*
Asia	0.72	0.31	8
Latin America	0.15*	0.62	14
Middle East	0.63	0.54	10
Africa	0.72		15
All countries		0.42	13

Note: * possibly underestimated

(a) Share of 'informal' credit on total credit to agriculture.

(b) ratio of informal credit to agricultural income.

(c) share of credit to agriculture on total credit to private sector from formal institutions.

Source: Wai 1977, 294, table 1.

onward. In Indonesia, the Dutch colonial administration exerted "gentle pressure" to set up district banks for consumer credit (paddy banks) beginning in the 1850s, which had some success in the twentieth century.[76] Many other governments, from Burma and Russia in the 1880s, to the United States during the New Deal, set up financial institutions (banks and insurance schemes) for agriculture.[77] The number has increased a lot after World War II. A database of the FAO lists eighty-three financial institutions that provided rural credit, and gives some additional information on forty-two of them: only nine of these were established before 1940, and only five are classified as private institutions (i.e., not owned by governments or co-operatives).[78] Their assets total some 71 billion dollars, and loans to the agricultural sector some 23 billion. These initiatives had mixed results, to say the least. The Russian State Peasant Bank was quite successful, as it financed a substantial part of the increase in the landownership of the peasants.[79] This success, however, owed much to the peculiar nature of these loans: they could be backed by a pledge on the land, and, until the Stolypin reform, they were often granted to the *mir* as a whole, which was less risky than a single household. Formal institutions have often succeeded in increasing their market share only thanks to the massive help by the state. For instance, in the Philippines from 1973, commercial banks were enticed to lend to peasants with the promise of free rediscount from the Central Bank.[80] Loans increased, but, as is predictable, the rate of default was extremely high. Many official lending institutions are plagued by a high rate of nonperforming loans and by a tendency to lend money to well-connected people (including the administrators) or to large farmers, who often use the money for purposes other than the original ones.[81] Thus, these schemes often end up being a form of disguised subsidization. In fact, an authoritative FAO survey signals this "quite unsatisfactory experience" and suggests giving support to credit co-operatives and quasi-informal networks.[82] The latter have developed a great deal in recent times. A recent survey of microfinance in the LDCs (undoubtedly an incomplete one) names some 1,500 institutions (700 of which are in Indonesia) with 54 million members and a total outstanding credit of $18 billion.[83]

Trends in the market for goods are relatively well documented. Purchases of non-agricultural inputs have increased a lot as a consequence of technical progress, both in absolute terms and as a share of total gross output (table 4.14). Furthermore, the figures are a lower bound of the level of commercialization from the input side, as they omit all the transactions within agriculture, from the exchange of hay and manure among households, to the long-range trade of animals to be fattened, and so on.[84] The attention of historians has focused mainly on the development of the market for agricultural output. Some of them consider the increase in marketed output as the measure of commercialization and modernity, and sometimes they pinpoint a specific figure as the threshold between subsistence-oriented "peasants" and market-oriented "farmers."[85] The underlying theory is very controversial, to say the least (section 8.9), and estimating the share of marketed output is far from easy. One can proxy it with the urbanization rate, under the assumption that city dwellers purchase all their food on the market. Indeed, the share of urban population rose about 10 percent in 1800 to 65 percent in 1980 in Western Europe, and from 6 percent to 73 percent in the United States.[86] In the past fifty years, the worldwide urban population has increased by four times, from about 750 million in 1950, to 2,860 million in the year 2000—that is, from a third to slightly less than a half of the total (UN, statistical year book). Unfortunately, the urbanization rate is a very crude proxy for the level of commercialization. It assumes that per capita consumption of food is equal between urban and rural people, and it neglects foreign trade and, above all, the transactions between farmers and non-farming rural households and among farmers. The effect of the two first factors cannot be determined a priori.[87] The omission of transactions within the countryside biases downward the estimate, and, in all likelihood, it is large enough to cause the urbanization rate to underestimate the overall level of commercialization. Indeed, an exact measure of the latter should include all transactions, irrespective of their geographical scope (within a village, national, or international trade) and the means of payment (in cash or in kind).[88] Unfortunately, such a measure needs data on production and consumption at the household level, which are hard to find in historical perspective.[89] Thus, scholars have used a wide array of proxies and indirect measures, such as the difference between output and an estimate of the consumption of peasant families (Gallman 1970), the amount of goods transported by rail or sea outside the area (Gregory 1980), the share of production from "market-oriented" farms (e.g., slave plantations or managerial estates), the amount of claims from outsiders (rents and taxes), and the share of "cash crops" (e.g., textile fibers). None of these methods is really accurate. The share of market-oriented farms on total output may overstate commercialization to the extent that slave plantations also produced food for their workers, as did the American plantations (but not the Caribbean ones).[90] The share of cash crops would bias the estimate downward if food crops were sold, and upward if some of the cash crops (e.g., cotton) were consumed at home.[91] Both the consumption-based and the transportation-based measures are bound to understate the level of commercialization. The former assume that peasants do not trade goods for their own consumption, while the latter omit all within-village transactions and, possibly, also the amount of goods shipped by traditional means of transportation.

These difficulties can explain the scarcity of actual figures on commercialization. As often happens, the research is more advanced for the United States than for any other country. Historians have long debated whether New England farmers in the eighteenth and early nineteenth centuries were self-sufficient or, at least aimed to be. Rothenberg (1992) provides strong, albeit not unchallenged, evidence that late-eighteenth-century Massachusetts was

TABLE 8.9
Commercialization of Agricultural Output, China, 1840–1936

	Rice Only	*All Goods*
1840s	10	
1894	16	
1929	22	38
1936	29	44

Source: Xu 1992, table 8.7, 8.8, p.130.

already quite a commercialized economy.[92] Atack and Bateman (1987) estimate that, in 1861, the median northern farm produced sizeable surpluses, ranging from a minimum of the consumption-equivalent of an adult male in New England, to a maximum of nine consumption-equivalent units in the Midwest. In 1910, when the official series began, the marketed output already accounted for more than 80 percent of American gross output, and today it accounts for more than 99 percent.[93] The United States may seem an exception, but markets and long-range trade were also fairly well developed in great peasant civilizations such as China, and probably had been for many centuries.[94] The share of products that they handled increased quite substantially in the nineteenth and early twentieth centuries (table 8.9). As expected, the level of commercialization was higher in "advanced" areas such as the Yangtzi Valley than in the rest of the country.[95]

There are quite a few estimates by country for the period leading up to World War II (table 8.10). These data have to be handled with care. They have been obtained with different methods, and, in many cases, the source gives a figure without specifying how it has been computed. From the 1930s, the situation of the data improved slightly, as statistical offices published "official" estimates, which were collected by Abercrombie (1965; see table 8.11).[96] As expected, the share of marketed output is roughly proportional to the level of development, but it turns out to have been quite high also in "backward" countries. There is abundant evidence of developed network of markets even in extremely "backward" countries, such as those of pre-colonial Africa.[97] Furthermore, as expected, in almost all cases, the share of marketed output exceeds the urbanization rate.

This evidence on commercialization, although not abundant, is sufficient to highlight two points. First, market transactions were quite widespread even in traditional societies, especially for goods and capital, and possibly for labor as well (the market for land may have been different, but we know very little). Second, markets have developed since then as part of the overall process of modern

TABLE 8.10
Commercialization of Agricultural Output, by Country, 1800–1940

Country	*Period*	*Share*	*Sources*
Finland	1910s	70	Hjerrpe 1989
	1930s	50	
	1980s	97	
Serbia-Bulgaria	1910s	38	Palairet 1997, 307; Berend-Ranki 1978, 68
Balkans#	1920	35–40	Lampe-Jackson 1980, 364
Canada	1926	80	Historical Statistics Canada 1965 L67–L70
	1938	77.7	
	1960	88.8	
France	1852	60–70a	Postel-Vinay-Robin 1992 table 2
	1938	75	De Cambiaire 1938, 172
Switzerland	1901–5	72	De Cambiaire 1938, 196
	1906–9	78	
	1931–46	83	
Italy	1880s	80	Federico 1986, table 4
	1930s	80	
Turkey	1840	21	Pamuk 1986, 239
	1912	45	
Russia	ca.1750	9–10b	Moon 1999, 131
	ca.1850	18–20b	
	ca.1900	25	
	1880s	30–40	Gregory 1982, tables D.2, D.3
	1910s	30–40	
Japan	1860s	50	Booth 1988, 241
	1921–24	58.3	Institute of Developing Economics 1969, table 47
	1935–38	54.3	
	1961–64	77.3	
Indonesia	1930s	15–35c	van der Eng 1994, 176
Thailand	1850s	7 c	Manarungsam 1989, 10
	ca.1910	50	Ingram 1971, 71
India	1950s	40d	Chandavarkar 1983, 764

Note: a: wheat only; b: grain; c: rice only; d: omitting transactions in kind. #Bulgaria, Romania, Greece

economic growth. This growth was particularly impressive for markets for goods, including agricultural inputs (see section 4.3), and for credit. Agriculture enjoyed a growing supply of capital at a decreasing cost, as interest rates fell (see section 4.5) and the share of the formal sector over the total supply grew. More credit at lower cost undoubtedly favored technical progress in agriculture.

TABLE 8.11
Commercialization of Agricultural Output, by Country, 1930–60

	Late 1930s	*Early 1950s*	*Late 1950s*	*Early 1960s*
Austria		80	83	86
France	78	81	83	86
W. Germany	77	78	84	87
Ireland		83	83	87
Italy		67	70	81
Norway	66	72	80	86
Switzerland	78	79	84	
UK		99	99	99
Canada	86	92	93	95
USA	87	94	95	97
Japan	74	68	72	79
Philippines		72		
Thailand		55		
Cameroon			31	
Chad			35	
Ethiopia		18		
Guinea			25	
Kenya		38	40	46
Malawi		33	35	38
Zimbabwe		71	72	81
Tanzania		39	41	43
Togo			45	
Uganda			59	56
Zambia		25	28	30

Source: Abercrombie 1965, table 1.

8.6 Self-help: The Growth of the Co-operative Movement

Informal associations of producers had existed for a long time (e.g., among the cheese producers in the Swiss Alps), but the first modern co-operatives were established in the nineteenth century.[98] Since then, their number and size have greatly increased, but it is surprisingly difficult to document this growth. In fact, the literature abounds in (usually laudatory) stories about single institutions or networks, including edifying portraits of the "apostles" who established them, but it is scarce on hard data.[99] Table 8.12 reports some data on the development in the United States in the twentieth century, when co-operatives were particularly strong in dairy products and fruit. Unfortunately, the available evidence for other countries is very thin. For Europe, however, we know the number of dairy and credit co-operatives (table 8.13).[100]

TABLE 8.12
Development of Co-operatives in the United States, 1915–2002

	Number		*Members (000)*		*Turnover**	
	Marketing	*Supply*	*Marketing*	*Supply*	*Marketing*	*Supply*
1915	5,149	275	582	59.5	11,791	172
1929	10,548	1,454	2,830	470	19,520	1,606
1939	8,051	2,649	2,300	900	17,982	3,723
1959	5,828	3,297	3,622	3,600	46,370	11,968
2002	1559	1,201	1,049	1,637	80,560	21,677

Note: * In million (1993) dollars (deflated with the consumer price index).
Sources: **1915–1959:** USDA 1998. **2002:** USDA 2004.

On other continents, the co-operative movement started later and was, on the whole, much less successful. It was a big success in Japan, but only because co-operatives were used as a tool of agricultural policy and membership was thus made compulsory (Francks 1999, 75; Hayami and Yamada 1996, 72–73; esp. Sheingate 2001, 150–60). The first co-operatives among native African peasants were set up only in the 1920s and 1930s.[101] Some Western colonial governments considered co-operatives to be the solution to the plight of the peasants, and supported their development with advice or money. The results were mostly disappointing, in spite of some cases of impressive growth, at least on paper, such as Burma and Indonesia (an 850-fold increase in the number of co-operatives from 1921 to 1966).[102]

At the beginning of the 1990s, at least one-sixth of agricultural workers (and perhaps one-third of households) all over the world belonged to a co-operative (table 8.14).[103] The survey omits many small countries (but it includes all major ones), and the data are not homogeneous. In some countries, they cover only agricultural co-operatives (marketing and purchasing), in others they also cover credit or multi-purpose ones. Yet, the difference among continents remains striking, especially if one considers sales and not membership. The turnover of the 202,000 Indian co-operatives, with 140 million members, was a mere $8 billion—i.e., roughly as much as the 111 Danish co-operatives with some 110,000 members. In fact, 90 out of the 120 billion dollars of the total turnover in the whole of Asia was produced by Japanese co-operatives. In the main, co-operatives still remain a movement of the "advanced" countries.

The membership and the turnover, although easily comparable among countries and across time, are not the best measures of the importance of co-operatives. A much more accurate measure is the market share on total production or consumption, for marketing co-operatives, and on the value of off-farm inputs for purchase co-operatives. The available data confirm the success of co-operatives in the United States and in Western Europe. In the United States,

Table 8.13
Number of Co-operatives in Europe, 1900–60

	1900	*1910*	*1937*	*1960*
Dairy				
Finland		307	1,179	527
Sweden			816	575
Denmark	942		1,330	
Netherlands	113	424	1,299	1,329
Belgium	266	286	1,165	831
Switzerland	25	108	640	1,066
France	840	2,636	5,798	
Germany		3,594	8,844	10,840
Czechoslovakia	1,347	2,201	6,080	
Austria		1,500	1,839	
Hungary	246	549	1,008	
Italy		1,526	2,372	
Spain	49	121°		1,785
Bulgaria	68	1,400		
Russia	100	3,748		
Credit				
Eire			229	195
Finland		343	670	333
Norway		51	168	342
Sweden	430	550	862	235
Denmark	1,056	1,157	1,417	1,171
Netherlands	507	801	482	
France	2,200		2,213	2,600
Germany	9,793	15,517	6,926	5,337
Hungary	716	2,425		
Italy	1,386	1,763	1,126	725
Russia	1,431	11,882		

Sources: **Denmark:** (dairy co-operatives in 1900) Jensen 1937, 315. **Bulgaria:** (1905 and 1914), Lampe and Jackson 1980, 199. **Russia:** (1905 and 1914) Kotsonis 1999, 138 and 141. **Hungary:** (1900 and 1910) Voros 1980, 96 and 107. **Netherlands:** (dairy co-operatives in 1907), Knibbe 1993, 151 and 189. **Italy:** (credit co-operatives): Fornasari and Zamagni 1997, tables 3.8 and 4.4. **Germany:** (credit co-ooperatives, 1900 and 1910) Guinnane 2002, table 3; (dairy co-operatives 1906–07) Rolfes 1976a, 497; and (dairy co-operatives 1936) Rolfes 1976b, table 7. **All others:** Dovring 1965, tables 36 and 38.

they accounted for about 10 percent of agricultural gross output in 1915 and about 35 percent– 45 percent in the 1990s, and for a somewhat lower share on purchases outside agriculture.[104] In Denmark in the 1930s, co-operatives accounted for 90 percent of sales of butter, 85 percent of bacon, 25 percent of eggs, 10–20 percent of cattle, 67 percent of purchases of seeds, and 35–40 percent

TABLE 8.14
Development of Co-operatives Throughout the World, Early 1990s

	Countries	*Number*	*Members (Thousands)*	*Turnover (Billions US $)*
Europe	23	53,315	19,288	215.6
Asia	6	243,375	148,404	121.0
Americas	8	12,249	6,001	104.5
Africa	8	22,226	6,649	8.6
Oceania	2	151	100	5.4
World	47	331,316	180,400	455.1

Source: Cote and Carre 1996, table 1.

TABLE 8.15
Share of Products Marketed by Co-operatives in Some European Countries, 1997

	Pork	*Beef*	*Poultry*	*Eggs*	*Milk*	*Sugarbeets*	*Cereals*	*All Fruit*	*Vegetables*
Belgium	20				53		30	75	85
Netherlands	34	16	9	14	83	63	65	76	73
Denmark	91	66		52	94		60	70–80	70–80
Germany	27	28			52	80	45–50	40	28
Spain	8	9	25	28	30	23	22	45	20
France	85	30	30	25	47	16	68	40	25
Italy	13	12	35	8	40	6.5	20	43	8
UK	28		25		65		24	67	26

Source: European Commission 2000, table T 152.

of fertilizers.[105] Nowadays, the market share of co-operatives in Europe is quite high for most products (table 8.15), and it has been growing also in recent times.[106]

These achievements contrast with the poor performance in the LDCs—notably in Africa. According to Eicher and Baker-Doyle (1992, 175) "despite early optimism, the record on co-operatives [in Africa] has been one of almost uniform failure under civilian, military, capitalist and socialist governments except in a few cases involving an export crop."[107] Credit co-operation, especially informal networks, fared relatively better (section 8.5). It is difficult to assess, without an in-depth analysis, to what extent the causes of failure listed in section 7.5 (sensitivity to correlated shocks, inefficiency and corruption of managers, political interference, or "cultural" factors) contributed to insufficient development. It is fair to say, however, that many of these problems are more serious in LDCs nowadays than they were in most European countries during the nineteenth century.

The disappointment about the overall poor development of co-operation in the LDCs is justified if one considers its positive effects on the welfare of Western farmers and the dire needs of most Third World peasants. On the other hand, the development of co-operatives might not be entirely beneficial for society at large.

Marketing and credit co-operatives would unambiguously increase overall welfare if they fostered technical progress (e.g., by allowing farmers to exploit scale economies) and/or if they increased the degree of competition in the economy (fighting successfully monopolists or monopsonists). In contrast, a monopolistic co-op might damage the overall welfare like any other monopolist. This possibility is remote to the extent that membership is voluntary and each member has the right to deliver the quantity he wants.[108] Thus, in principle, it might be difficult to restrict supply—but, in practice, co-operative officials have several means of enforcing reluctant growers to comply if the co-operative is a local monopsonist. As in many other cases, the issue is an empirical one.

8.7 Institutions and Agricultural Growth: The Creation of Property Rights and "Structural" Interventions

In theory one could deal with the creation of property rights and land reforms as an endogenous consequence of agricultural change. In a "Boserupian" framework (section 7.2), the creation of property rights is a key component of the population-driven process of agricultural change. The gradual evolution that Boserup posits has been disrupted by the intervention of "outsiders," but one can also model the latter as being determined by the features of area.[109] Abundant land and scarce population in temperate climates attracted Western settlers, who had to be given full ownership in order to convince them to stand the hardship of frontier life. Settled agricultural areas attracted warriors and conquistadors looking for a surplus to be extracted with violence. From this point of view, the least attractive case was sub-Saharan Africa, where the population was scarce and the environment unfamiliar and disease-ridden. In addition, land and tenancy reforms would be endogenous, to the extent that they were motivated by the quest for efficiency gains and that these gains depended on technology. This line of reasoning is intellectually appealing, but we cannot pursue it in this section, for different reasons. The concrete process of the creation of property rights has been affected so much by social and political events that it would be impossible to deal with its causes without a full discussion of the history of each country or region, and this is plainly impossible. The efficiency gains from land and tenancy reforms are controversial, as we will discuss later in this section, and anyway the whole process was driven mainly by political and social considerations. The key reason for land reform was the redistribution of income, wealth, and power. The slogan "land to the tiller" has been one of the most powerful rallying cries for political change in the twentieth century, and the American government coaxed many "friendly" countries to implement land reforms in the 1950s and 1960s as a preventive move against communist propaganda. In the words of R. King (1977), the author of the most comprehensive (and quite a sympathetic) book on the issue, "in the last resort [the political situation] is often the most decisive. It is the balance of political power in a country, which ultimately determines the extent of a reform, and political factors help to explain the frequently wide discrepancy between the provisions of a reform law and their eventual practical effects."[110] Powelson and Stock (1990, 4), who are much less sympathetic, state that (most) land reforms were "bestowed upon peasants—without their having participated in forming it—by a gracious government" and thus attribute them "by grace instead of by leverage." In the following, we will therefore assume that the creation of property rights and land reforms was exogenous, and focus on their effects mainly on agricultural growth, but also on income distribution.

As noted in section 7.2, economists and economic historians tend to assume that modern property rights are more efficient and growth-enhancing, if not necessarily more equitable, than traditional ones. The difference, however, might be small or large, and the effect of their establishment may vary accordingly. In theory, one should assess the differences on a case-by-case basis, but this is plainly impossible. It is possible, however, to put forward some rather general statements, supported by a short review of the literature on a few representative cases.

The expropriation of natives in the countries of Western settlement was clearly unfair, but very few historians would doubt that it boosted agricultural production in the long run. It caused a huge increase in the amount of labor and capital, and on top of this, the settlers, when they had adapted to the new environment, proved to be more efficient, both in static and dynamic terms, than the natives. The debate has focused on the benefits of the homestead (the free concession of farms to cultivators) versus the alternative methods of distribution, such as the sale by auction and the free concession of large estates to "powerful" individuals (the Spanish *encomienda*). Historians used to consider homestead as the only efficient and equitable method, as it transferred the land directly to the tiller. It also reduced costs of defending the settlements against natives.[111] In contrast, historians blamed *encomienda* for the concentration of landownership in Latin America, with all its negative effects (King 1977, 77–83; Solberg 1980, 60–64; Binswanger and Deininger 1997, 1968; Engermann and Sokoloff 1997, 267–70, 279–80). They criticized auction sales in the United States for offering rich businessmen (or railway companies) the opportunities to purchase large extensions of land and resell it later to farmers at a higher price (Atack and Passell 1994, 264–72; Atack and Bateman 1987, 121–26; Atack et al. 2000b, 303–10). This speculation earned them undeserved rents and, in the meantime, it withheld valuable agricultural land from production. These arguments are now being challenged. Speculators had a clear interest in selling as soon as possible: they tried to attract settlers with often overoptimistic claims, thereby accelerating the process of settlement. Furthermore, auction is theoretically more efficient than homestead, as it forces all purchasers to pay their reservation price. The poor reputation of the *encomiendas* is largely based on the assumption that the *latifundia* were backward and inefficient, but the evidence for this statement is very weak (section 7.6). Furthermore, the argument neglects the possibility of adjusting the size of farms via the markets for labor or land/credit. If small farms had really been more efficient, the landlords would have sold or rented the land. Many large Argentinian farms were rented as soon as the agricultural population became dense enough as to make the switch from extensive ranching to agriculture profitable.[112] Major inefficiencies in the markets for credit or labor could indeed prevent this adjustment, but, in this case, it is they who have to be blamed and not the initial method of distribution. In the long run, the latter did not matter that much: in the United States there was no difference whatsoever in farm size and efficiency, ceteris paribus, between land that had been given to farmers for free and that which was granted to railway companies.

In some historical cases, the creation of modern property rights has been both very inequitable and inefficient. In nineteenth-century Iraq (Jwaideh 1984, 337–43; Mahdi 2000, 77–81) and Syria (Sluglett and Farouk-Sluglett 1984), merchants and tribal chiefs seized most of the land, thereby swindling peasants, who were often unaware of the registration process and unable to file applications properly. This created conflict with traditional property rights, which allowed some freedom to sale for individual households, while cultivation did not improve.

On the other hand, economic historians used to consider the final demise of traditional property rights in Europe (i.e., the enclosures, or the Stolypin reform) as socially very unfair but necessary for the progress of agriculture. British enclosures, abolishing the *vaine pature*, deprived workers of a common right, which was essential to their livelihood, but, at the same time, they allowed the adoption of continuous rotation, which would otherwise have been impossible, as all village animals would have grazed on the innovator's grass.[113] In Imperial Russia, argued Gerschenkron in his seminal contribution to the Cambridge Economic History of Europe, the periodical redistribution of land discouraged household investment, prevented the necessary rearrangement of fields from achieving the optimal size of family farms, and created an incentive to stay in the commune (so as not to lose the right to land redistribution), thus reducing the supply of manpower for industry.[114]

Recent historical research, however, tends to downplay the advantages and disadvantages of the modernization of property rights.[115] The whole discussion on the dynamic benefits of enclosures rests on the premise that England enjoyed a great increase in productivity in the late eighteenth and early nineteenth centuries. The very existence of such an "agricultural revolution" is now put in doubt, and, consequently, as Clark (1993, 1998b) points out, this reduces the "need" for an explanation.[116] Even scholars who do believe in a fast technical progress in agriculture, such as Overton (1996) and Turner et al. (2001), barely quote enclosures as a necessary condition for it.[117] Many historians downplay the negative effects of enclosures on the welfare of agricultural workers. The demand for labor did not fall, or even rise after enclosure and the total income from common rights was smaller than assumed.[118] The Gerschenkron thesis about the perverse effects of the collective property (*obscina*) is strongly criticized by Gregory (1994). He points out that improved land was exempt from redistribution, and argues that (rational, utility-maximizing) peasants could easily circumvent the formal rules.[119] This modern debate, incidentally, echoes the well-known controversy between

Lenin and the populists. The latter, like Gerschenkron, deemed the system quite stable and able to cope with population growth, especially if peasants were not overburdened with taxes (for them, unlike Gerschenkron, stability was a great virtue). In contrast, Lenin (slavishly followed by Soviet historians) argued that the communal institutions were being eroded from within by the differentiation between a few dynamic rural capitalists (the *kulaki*) and a very large number of landless laborers.

The effects of titling on contemporary agriculture are no less controversial.[120] To be sure, the procedure is nowadays much more carefully monitored from a "social" point of view, and this makes widespread expropriation unlikely, although it does not rule out the discrimination of some holders of traditional rights (notably women). In many cases, the empirical analysis confirms the assumptions about the advantages of modern property rights. The registration of squatters had a positive effect on investments in the Amazon basin, while, in China during the 1990s, the risk of land redistribution within the village (as in Imperial Russia) reduced investments in organic fertilizers, which in turn yielded medium-term benefits.[121] The efficiency gains, however, have often proved smaller than the (somewhat overblown) expectations. Customary property rights are often more flexible, efficient, and open to technical progress than is assumed. For instance, Place and Otsuka (2001) show that, in Malawi, matrilocal settlement with no tenure rights for the husband does not prevent an efficient allocation of resources or of most, but not all, productive investments.[122] In the overwhelming majority of cases, however, modern property rights are superior, if they are properly enforced and supported by an efficient market for credit.

Land reforms allegedly aim to improve the welfare of peasants. As Griffin (1981, 10) puts it, "a land reform, in isolation, is not sufficient to remove rural poverty, but it is a *conditio sine qua non* in many countries." There is no doubt that a reform increases the workers' income by increasing their share of total production, but its effect on output is ambiguous. The supporters of reform have always argued that the effect must be positive because family farms are more efficient than *latifundia* (Lipton 1970, 288–90; Berry and Cline 1979, 7, 129; Booth and Sundrum 1985, 276–86; Binswanger et al. 1995, 2730–33; Alexandratos 1995, 30, 320). This assumption is probably true if efficiency is measured by land productivity, but not necessarily true if it is measured in terms of total factor productivity (section 7.6). An increase in production achieved by working harder and having less leisure may not increase the peasants' welfare, although one should not neglect the immaterial benefits of ownership (section 7.3). On top of this, a botched reform could even reduce output, ceteris paribus—that is, decrease the overall welfare. In really perverse cases, the reduction in output might be so great as to overcome the redistribution effect and also cause the peasants to be worse off.

It is quite common to assess the economic effects of land reforms with a simple post hoc–propter hoc macroeconomic approach—that is, by looking at trends in production or productivity before and after them. For instance, King deems the Mexican reform to be a success because productivity grew quickly in the period

1940–60, while others hold it to be a failure because productivity stagnated afterward and because it seemed to be inversely related to the diffusion of *eijdo* by area (common land) (King 1977, 109; Powelson and Stock 1990, 57–59; Thiesenhusen 1996, 40–43; Johnson 2001, 294). Needless to say, these inferences are crude and theoretically weak, as the overall performance depends on many factors other than the reform itself. However, designing and testing a counterfactual, no–land reform world seems to be an almost impossible task.

As a sensible alternative, one might adopt a microeconomic approach, by focusing on the interaction between the design and implementation of the reform, and on the performance of newly created family farms.[123] This latter would depend on (1) the characteristics of the new farms (amount of capital, size, access to markets, etc.); (2) the technical and managerial skills of the new owners; (3) the amount of financial and technical support that they receive; and (4) the compensation (if any) that new owners have to pay to the previous landlord. All these criteria imply that a reform is much more likely to succeed if it only has to transfer the ownership of existing farms from the landlord to the tenant (as in most of eastern Asia), than if it has to carve family farms out of a large estate and to transform workers with no previous management experience into independent farmers (as in Latin America). Furthermore, long-term effects can differ from short-term ones. For instance, many reform bills stipulated that farms could not exceed a maximum size, to prevent new concentrations of land ownership, or the bills even forbade new owners to sell the land (as in Mexico under the *eijdo* system). These clauses hampered the post-reform adjustment of farm size and thus hindered technical progress. The microeconomic literature on the effects of reforms is huge, although quite uneven, and, at times, heavily politicized.[124] In a strongly opinionated book, Powelson and Stock (1990) consider sixteen cases and argue that, in fourteen of them, reforms benefited the state and well-connected individuals, such as bureaucrats or wealthy farmers, but not the mass of poor peasants. Their analysis is probably too pessimistic, but reforms have not been the panacea that Griffin and other supporters hoped them to be. The eastern Asian reforms enjoy the best reputation: South Korea and Taiwan are the two exceptions in Powelson and Stock's otherwise bleak picture. Most of this reputation is based on the countries' very good agricultural performance in the 1950s and 1960s (e.g., King 1977, 199–205, 219–30; Ban 1979, 112–14; Binswanger et al. 1995, 2684; Ray 1998, 458–62). This conventional wisdom, however, is being challenged nowadays, and the positive long-term effect of land reform is being questioned even for Japan. "There is no doubt that the land reform promoted more equal assets and income distribution among farmers, thereby contributing critically to the social stability of the rural sector. However, the farm size distribution did not change, and the small-scale family farm remained the basic unit of agricultural production. Although land reform contributed to an increase in the level of living and consumption, its contribution to capital formation and productivity growth in agriculture have not been clearly visible or are not significant in terms of quantitative analysis" (Hayami and Yamada 1991, 85).[125]

Tenancy reforms differ from land reform on one very important ground: they are less invasive. Reinstating former *latifundistas* after a land reform would be politically too expensive, even if economically rational, while the parties can maintain (or return to) traditional contracts, possibly paying a lip service to new ones, if the former prove to be more efficient. Thus, reforming governments are often faced with the unpleasant choice between letting the reform fail or committing themselves to the constant and expensive job of monitoring and enforcing contracts. Many social reformers consider this flexibility a major drawback of tenancy reforms, relative to pure land redistribution (Lipton 1974, 275; Binswanger et al. 1995, 2715, 2729). The World Bank (2003), in a recent major document on land policy, deems it evidence of failure: tenancy reforms are either ineffectual or harmful.[126] Empirical analysis of the effects of tenancy reforms is complicated by the ambiguity of the category, which includes widely different measures. The same reform may score differently according to different criteria, as Besley and Burgess (2000) show in a comprehensive analysis of tenancy reforms in post-independence India. The reforms alleviated poverty and increased the real wages of landless laborers, but they had a negative impact on income and yields.[127]

The efficiency gains are of paramount, almost exclusive, importance in the adoption of consolidation policies. But the latter can improve efficiency on two conditions. First, the problems cannot be solved by the land and savings markets, either because they are imperfect or because the adjustment is made impossible by other policies (e.g., generalized subsidies to all farmers). Second, small size and fragmentation must be a problem. The farm must be too small to exploit economies of scale—possibly related to the use of indivisible input (section 7.6). Fragmentation has to be the outcome of exogenous circumstances such as partible inheritance and the haphazard process of farm-building, and not of a spontaneous choice by the peasants to spread the risks (Blarel et al. 1992, 234–40; Binswanger et al. 1995, 2728; World Bank 2003, 127–29). For instance, Blarel et al. (1992) surmise that this was the case in Ghana and Rwanda in the late 1980s, because fragmentation had very little effect on yields.[128] The authors admit that peasants may lose time moving from one field to another, but they consider this to be an acceptable price to be paid. In this case, consolidation policy would be useless if not harmful. Clearly, this example is not necessarily representative, and fragmentation may have been a serious problem in other cases. Perhaps one can conclude that consolidation policy may be sensible if it is performed on a purely voluntary basis.

8.8 Institutions and Agricultural Growth: Landownership, Farm Size, and Contracts

The interactions between structural change (section 8.4) and agricultural performance are exceedingly complex and remain, unfortunately, quite obscure. Development economics is not very helpful: it does provide some bits and pieces of a mainly static theory, notably on contracts (section 7.3) but, as far as this author knows, offers no comprehensive model of "structural" change.[129] Indeed, interpreting the change could be relatively easy if a given type of farm (as defined by size, contract, or pattern of ownership) were more efficient than others. If so, it would have increased its share (unless prevented by state intervention), and this trend would have augmented the efficiency of agriculture. However, this was not the case: as argued at length in section 7.6. Each type of farm may be more suited to given market conditions, factor endowment, state of technology, and so on. One possible alternative is the econometric test of some plausible causal relations with a simultaneous equation approach. This approach is followed by Huffman and Evenson (2001), in a recent paper on the United States from 1950 to 1982.[130] Farm-size depended on the composition of output (a negative relation with specialization in cattle-breeding), on the diffusion of off-farm work (positive) on factor prices (a strong negative relation with the capital-wage ratio in agriculture), on the allocation of R&D funding (a positive relation with expenditure in livestock), and on governmental support (mostly negative). The authors also find that size affected specialization (toward livestock and away from crops) and technical progress (positively for livestock and negatively for crops). These results show how complex interactions are, and how difficult it is to draw any inference about the cause of structural change. Furthermore, one doubts whether the state, which can include widely different agricultural areas (e.g., in California), is the appropriate unit of analysis. Anyway, the approach is so data-intensive that it would be unfeasible in other countries or periods, and, clearly, the results cannot be easily generalized for different countries. The lack of a grand theory for structural change and the dearth of empirical work strongly constrain the scope of this section. It will put forward some general hypotheses to interpret the "structural" trends as endogenously determined by factor endowment (section 4.5) and technical progress.

The farm size ultimately depends on the factor endowment—notably the land-labor ratio. Ceteris paribus, farm size will fall when population grows faster than the available land (as in most developing countries in the past). It will remain stable when work force and land remain constant (as in most countries of western Europe during the nineteenth and early twentieth centuries) or grow roughly at the same rate (as in the countries of western settlement during the nineteenth century). It will increase when the agricultural work force declines (as in the "advanced" countries after World War II). But *absolute* size can vary a lot, for any given land-labor ratio, according to the prevailing type of farm: a single managerial estate can employ all the members of several households. Therefore, as a first approximation, farm size must fall wherever the land-labor ratio decreases *and* the share of family farms does not fall.[131] In other words, an interpretation of changes in the distribution of farms by category is necessary not only for its own sake but also for the understanding of trends in farm size.

The main trend to be explained is undoubtedly the increase in the share of family farms, and the parallel decline in tenanted and managerial estates in long-settled traditional societies, such as Europe or China.[132] These processes can be interpreted as the outcome of the interaction between the changes in factor endowment and technical progress. Population growth with a given supply of land reduces the average land endowment per household. Without technical progress, labor productivity would fall and land productivity would rise: the landlord could pay lower wages, extract higher rents from his tenants, or perhaps force them to work for free to increase the capital embodied in land (e.g., by building irrigation works). This process, however, could not go on forever without technical progress: the law of diminishing returns would cause the growth in rents to tail off and eventually stop. But the total rents must be divided among members of (nonworking) landowner households, whose numbers are growing as well, possibly even faster than that of workers. Thus, in the long run, the per capita income of landowners is likely to decline even if the aggregate returns to land-cum-capital are growing. Sooner or later, the income would fall beyond the minimum deemed sufficient for an adequate lifestyle for a landowning household, which would then decide to sell and invest the proceeds elsewhere.[133] Thus, landlords as a group would own a decreasing share of total land, although each single family might experience a cycle of accumulation and decline. Summing up, in the case of "poor" countries (i.e., countries with little or no technical progress), population growth is bound to increase peasant ownership and to cause farm size to decrease.

This simple prediction does not hold true for the "rich" countries, where the evolution in the agricultural sector is determined by technical progress and the factor prices are set in the whole economy. Migration out of agriculture (section 4.4) and the increase in wages relative to capital (section 4.5) brought about labor-saving technical change and mechanization, which is likely to increase farm size.[134] The increase in wages is bound to hit managerial estates harder than family farms or tenanted estates. In fact, working on a family farm, or even being a tenant, offers advantages in terms of safety and personal satisfaction (section 7.3), while managerial estates must offer salaries competitive with market wages in industry and services and pay for supervision.[135] This competition is likely to become stronger to the extent that modern economic growth brings about a greater integration between the urban and rural labor markets and, possibly, the development of welfare state (which, ceteris paribus, reduces the expected risk of unemployment in the non-farm sector). Modern economic growth affects the decisions of agents in other ways as well. The development of the financial institutions reduces transaction costs in the land market, making the mortgaging of land easier and less expensive, and also improves the access of family farms to credit for investment. Last but not least, modernization increases the incentives to sell large estates in that (1) it reduces the social and political role of landownership, (2) it offers landowners plenty of alternative opportunities to invest their money outside agriculture, and (3) it increases the minimum income to maintain an "adequate" lifestyle.[136] Thus, pure market forces could have accounted for the rise in family farms both in poor and in rich countries, albeit with different mechanisms. But the process has been fostered and shaped by two other factors, inheritance rules and state intervention.

Inheritance rules determine the number of claimants on income from land. The previous stylized model implicitly assumes that all members of the landowning families have the same rights on the rents and on the estate upon the death of the owner (partible inheritance). If the rural population is growing, unfettered partible inheritance is bound to reduce farm-size. In fact, the equitable division of land among heirs (*fen-chia*) is widely considered the main cause of the disappearance of large estates and of the ensuing fall in the share of rented land in Imperial China.[137] In Europe, large estates survived better, because the landowning families had enacted, from the late Middle Ages, strategies to restrict the effect of partible inheritance and reduce the number of claimants to the sharing of the estate.[138] Only males could inherit the estate, while females were given money as dowries, and, in many countries, families transmitted the estate, undivided, to the first male born (*majorat*). The Napoleonic code formally outlawed these strategies, but they survived well into the nineteenth century in many European countries.[139] They have disappeared, however, and one wonders how much this change reflects the progress in jurisprudence and in society at large, or perhaps how much it reflects a declining interest, brought about by modern economic growth, in the integrity of estates.

Both land and tenancy reforms increase the share of family farm. Such an increase is the very aim of most land reforms, and it is a likely outcome of tenancy reforms. The latter reduce rents, possibly below the minimum acceptable level, and can also be considered the harbinger of land reform. Indeed, in many instances—such as Russia after the 1905–7 Revolution, or in the United Kingdom during the 1920s–1930s, or in South Korea after World War II, or in India during the 1980s—the fear of reform or of heavy taxation was sufficient to induce landowners to sell.[140] The effect of state intervention on farm size is somewhat less clear-cut, because the (size-diminishing) effect of land reforms and other pro–family farming policies might have been balanced by the (size-augmenting) consolidation policies.

The other major trend, the decline in sharecropping, is even more difficult to explain, beyond the trivial case of an outright ban as part of tenancy reform. The competing theories of contracts (section 7.3) suggest, in a nutshell, that sharecropping may be preferred to the (potentially more efficient) fixed-rent if the landlord's monitoring costs are low enough and/or the prospective tenants are strongly risk averse. Thus, the relative decline of sharecropping can be accounted for by an increase in monitoring costs (i.e., in the amount of supervision necessary or in the opportunity cost of the landlord's time) and/or by a decrease in risk. According to Chao (1982, 278–91; 1986, 172–73), sharecropping declined in China, even without any substantial productivity growth or change in risk, because the intensification of production and the growing fragmentation of farms increased supervision costs. The weakening of community ties and the growing social mobility may have increased the monitoring cost as well, by reducing the amount of available information and the economic value of reputation. It seems likely that risk and risk-aversion have been decreasing. Actually, yield variance, the simplest measure of risk, may have not decreased in spite of technical progress (section 2.2). On the other hand, the increase in the income and wealth of farmers, combined with the diffusion of formal insurance, may have reduced risk-aversion (Huffman and Just 2004, 634).

This section has to end on a disappointing note. Structural change altered the lives of hundreds of millions of people, but we still do not have a really satisfactory explanation. This section has put forward some plausible hypotheses, but any sort of econometric testing is very far away. Actually, even estimating the dependent variable, the share of different categories of farms would entail a huge amount of research work. Such research would seem to be worthwhile, however, as the results could have implications well beyond the realm of economic history.

8.9 Institutions and Agricultural Growth: The Development of Markets

Modern economic growth has undoubtedly been a major cause of commercialization. Urbanization has increased the market demand for agricultural goods, the development of the non–farm sector has created a market for salaried labor, technical progress in agriculture has required more purchased inputs, and so on. Yet, as section 8.5 shows, the diffusion of market transactions largely predated modern economic growth. How is it possible to account for commercialization in preindustrial societies? Did peasants trade goods and labor on their own accord or were they forced to do so by an unequal distribution of landholding or by the intervention of powerful outsiders? Did commercialization harm or benefit them? More broadly, did it foster or retard modern economic growth?

It seems likely that the higher the share of tenanted and managerial estates is, the more developed the markets for labor and goods are, ceteris paribus. They both need a market for labor, although they demand different types of workers. Managerial estates mainly require short-term laborers, while tenanted ones mainly need long-term tenants, that is, households with some managerial skills. And both types of estate need a market for goods to convert rents into money for the landlord's consumption.[141] Managerial estates sell their products directly, whereas in the tenanted ones, the sales can be either entrusted to tenant households (who will pay their dues in money) or centralized by the estate (with rents paid in kind). This difference may matter a great deal for tenant/landlord relations, but it hardly affects the total sales. The effect of land distribution on markets for land and capital are much less clear-cut. Unfortunately, theory does not provide any a priori insight into the relative stickiness of the market for estates and for family farms, and the data on this issue are, as said, almost nonexistent. It is likely that wealthy landlords need more credit than peasant households and undoubtedly they can have access to a different type of credit (section 7.5). The land distribution can explain some of the differences in the initial level of commercialization. For instance, the high proportion of agricultural workers and the high share of marketed output in Italy in the 1880s (tables 8.7 and 8.10) undoubtedly have to do with the prevalence of estates (Federico 1986). On the other hand, the diffusion of family farms (section 8.4) is bound to have reduced, ceteris paribus, the scope for transaction in the market for goods and labor.[142] The effect of structural change on the market for credit is uncertain: the rise of family farms would increase, ceteris paribus, the share of informal credit, but it might increase or decrease the total amount.

In the 1960s, 1970s, and early 1980s, the attitude of peasants toward trading and markets was heatedly debated.[143] Mainstream economists, such as the Nobel laureate Schultz, argued that the peasants were always eager to trade if this could increase their income. In contrast, some development economists and, above all, most anthropologists (following K. Polanyi) held that peasants preferred the traditional ways of redistribution of goods (sometimes called "moral economy") to the market exchange. Thus, commercialization had to be thrust upon them by force—for example, through an increase in taxation. This theoretical debate is now largely over. Hard-line anthropologists, faced with the evidence of market development in traditional societies, had to admit that the peasants' rejection of the market was not so adamant, or else that their attitude somewhat mysteriously changed, transforming "peasants" into market-oriented "farmers." Such a change may leave very little evidence, and indeed its existence and timing in American New England have been spiritedly and inconclusively debated.[144] Economists have admitted that rural households may show allegedly "irrational" behavior for perfectly rational motives: (1) they may prefer production for self-sufficiency because production for the market entails substantial transaction costs, which exceed the gains from specialization;[145] (2) they may prefer more leisure to more income, and thus reduce output when prices rise (if they cannot hire additional workers);[146] and, (3) they may shun full specialization for the market because they consider it to be too risky. In fact, subsistence production is subject "only" to the risk of fluctuation in the output of "food" crops, while specialization for the market adds three sources of risk: fluctuations in the output of "cash" crops and in prices of both "cash" and "food" crops.[147] In this case, peasants might prefer to continue to produce the food they need on their farm and use the rest of their resources to produce for the market ("safety first" strategy).[148] This debate on peasants' behavior is not just a theoretical curiosity. The widespread belief about their aversion to trading inspired the design of the agricultural policies in many LDCs in the 1950s and 1960s, contributing to their overall anti-agricultural and anti-market bias (section 9.5).

Many authors argue that market-averse family farmers were forced to trade, either by merchants and/or money lenders, or by the state. According to Ransom and Sutch (1977, 2001), shopkeepers in the postbellum American South forced former slaves to specialize in cotton, an easily saleable but risky and ultimately unprofitable crop.[149] According to Gerschenkron (1967), Imperial Russia extracted a growing share of a stagnant agriculture output to finance industrialization, and this policy was one of the major causes of the peasant uprising, and ultimately of the Revolution. Colonial administrators, such as the Dutch in Java (the *cultuurstelsel*, or Java system), the East India Company in Bengal, King Leopold of Belgium in the Congo (described by Iliffe as the "most brutal exploitation of the colonial period"), and also Muhammad Ali in Egypt, are accused of having mercilessly forced the native population to increase exports.[150] These statements have to be taken with a lot of caution. The whole thesis of Ransom and Sutch is based on the controversial assumption that shopkeepers enjoyed a local monopoly on credit (section 7.4), but there are alternative explanations for the specialization in cotton.[151] Simms (1977) and, above all, Gregory (1994) remarks that Russian taxation was not so heavy, and that the peasants not only could afford to pay taxes without worsening their standard of living, but also managed to save and purchase land.[152] Van der Eng (1996) raises strong doubts on the role of coercion in the Java *cultuurstelsel*, noting that exports continued to grow even after its abolition.[153] Furthermore, taxes decreased in several other cases—in India (from about 10% of output in the first half of the 19th century to 3–5% in the 1930s), in Japan after the Meiji restoration, in Thailand throughout the nineteenth century (from 15% to 4% of gross output), and also in Bulgaria after the liberation from Ottoman Empire.[154] However, the level of taxation in agrarian societies was very low by modern standards—in all likelihood lower than the share of marketed output. For instance, according to Rifkin, in China, total surplus (rents and taxes) accounted for 19 percent of agricultural NDP in the 1930s.[155]

The previous analysis suggests that, no matter how important they were in many cases, neither structural change nor claims by powerful outsiders can account for the entire process of premodern commercialization. By default, one has to conclude that, in many cases, commercialization was a choice made by the peasant—that is, they were not so consistently market-averse as is sometimes assumed. Faced with new opportunities and new challenges, peasants reacted by trading more of their crops, by changing their crop mix, and perhaps by emigrating to cities or abroad (see Hatton and Williamson, 1998). The challenges came from different quarters. Technical progress or changes in input endowment increased the potential production of food crops beyond the need of the household, or made the production of cash crops more profitable. The construction of canals in India (see section 4.3) made it possible to settle previously thinly populated semi-arid areas such as Punjab, and to intensify agriculture in many areas (Stone 1984, 105–8, 295–302; Roy 2000, 69–73; Attwood 1987, 343–52). The growth of all big cities created a new demand for perishables, which, before the development of refrigeration in the late nineteenth century, had to be satisfied by the production in the nearby surroundings.[156] In Japan, admittedly an extreme case, the prices of silk and tea relative to rice doubled after the forced opening of the country to world trade in 1859, which fostered a boom in the export of raw silk.[157] The improvement in transportation—notably the construction of railways—offered new opportunities for exports in places as far away as Piedmont in the southern United States, the Saga and Yamanashi plains in Japan, Hopei and Shantung in China, the Bombay presidency in India, and the coffee-growing areas around Saõ Paulo in Brazil.[158] It has been estimated that, in India, Kenya and the Sudan, a 10 percent reduction in access costs to market caused changes in the crop mix and an increase in use of modern inputs, which augmented income from land by between 1 and 6 percent (Von Onne et al. 1997). In fact, there is ample evidence that peasants seized opportunities to produce for foreign markets with very little or no encouragement by the state, just for the sake of a higher income. Spanish and Southern Italian wine growers in the 1880s made extensive investments in new vines to export to France, which was by then hit by the phylloxera. Chinese peasants increased their production of cotton and cocoons (the raw material for silk); and from the end of the nineteenth century, West (and, albeit less enthusiastically, East) African peasants cleared land and started to produce groundnuts, cocoa, cotton, and other cash crops.[159] The recent literature on India stresses the relevance of expected profits to explain commercialization, in opposition to the traditional view of the process being forced on them from outside (Tomlinson 1993, 51–52; Roy 2000, 92–93). Last but not least, almost all the available estimates of the price elasticity of agricultural production or of marketed surplus only are positive, and, moreover, some coefficients are quite high as well.[160] In theory, piling up evidence of market-friendly behavior does not conclusively prove the point. In fact, all these cases may be exceptions to the rule: the majority of peasants may have not seized their opportunities, and their failure to expand/grow may have gone unnoticed. This, however, seems rather implausible.

The debate on the causes of commercialization has clear implications for its effects on welfare and long-term growth. Forced commercialization is likely to have reduced the welfare of the peasants, whereas spontaneous one would have increased it. Some authors have tried to measure the effect of welfare directly by comparing the conditions of the rural population before and after commercialization. For instance, microeconomic research on eleven areas in the LDCs in the 1980s shows that, in the majority of cases, an increase in the production of cash crops augmented employment (by 10–40%) and income (by 15–25%), and did not reduce the production of food crops.[161] Such a comparison entails a very strong ceteris paribus assumption, and the approach is so data-intensive that it is hardly feasible in historical perspective. A widely used proxy for the standard of living is the stature, and its long-run increase is powerful evidence of the benefits of modern economic growth (Fogel 2004). As said (section 3.2), however, average heights have fallen in some areas/periods, such as the United States in the early nineteenth century. In an article on rural Georgia, Komlos and Coclanis (1997) attribute this decline to commercialization and urbanization. Specialization in cash crops forced peasants to have a poor and unhealthy diet, while the urban population had to pay dearly for its food because of the transportation and transaction costs (see also Komlos 1998).

The peasants' well-being is undoubtedly important, but it is not the only yardstick to assess the effects of commercialization. In fact, the latter is a precondition for "Smithian" growth via specialization according to comparative advantages. Even a forced commercialization might be beneficial to the whole economy, if it improved the allocation of resources. The gains are the greater the bigger the gap between the actual and the optimal allocation of resources is—that is, in most cases, the more isolated the household or the area is at the beginning of commercialization. Thus, it is quite common to assess gains and losses from commercialization from a macroeconomic perspective. Economists are, on average, quite sanguine. For instance, Grantham argues that specialization in production for urban markets was the most important factor of growth in French agricultural production before 1850.[162] In the nineteenth century, the spread of cotton cultivation greatly boosted income in rural Egypt (which achieved one of the highest rates of production growth).[163] Reynolds extends this statement to all the present-day LDCs before 1914: in these countries, exports were the main or the sole source of economic growth.[164] In contrast, radical historians view (forced) commercialization as a colonial plot and the main culprit for underdevelopment.[165] Huang states that ("survival-driven") commercialization in China was an outcome of population growth: peasants, faced with the prospect of dwindling per capita income, had no alternative but the risky bet of starting to grow cash crops for the market.[166] This view sparked a lively debate on commercialization and Chinese economic growth, which ultimately bears upon the causes of the Communist Revolution.[167] Huang's pessimistic appraisal has been contested by Rawski, Faure, and, above all, by Brandt. He argues that markets were quite well integrated and competitive, a fact that Huang denies, and that "increasing specialization was the source of much of the growth in agriculture during this period [1870–1937]" (and also of a fall in income inequality). A parallel debate is ongoing on India, and a revisionist, "pro-market" interpretation of the consequences of commercialization seems to be prevailing there as well.[168] These arguments are plausible and undoubtedly much more convincing that the alternative radical position, but they are difficult to prove. In fact, they praise Smithian growth on the assumption that it accounted for most or all the TFP growth—i.e., that technical progress and/or efficiency gains were very small or negligible. This assumption might be plausible for nineteenth-century China or India, or, stretching it far, for France before 1850. It is not plausible in most cases. One would need a method to estimate the effects of commercialization separately from those of other efficiency gains. So far, this is not available, but recent advances in nonparametric measurement of TFP growth (section 5.4) offer some hope that they will be available in the future.[169]

8.10 Conclusion: Did Institutions Really Matter?

By now, it must be clear that the author strongly sympathizes with the economists' view of the role of institutions in the process of agricultural change. By and large, they have, in the long run, proved to be flexible enough to adapt to technical progress and to changes in the land-labor ratio. This conclusion is, however, subject to three caveats: (1) Institutions, at each moment of time, can be suboptimal simply because the process of institutional change is, by its nature, slow. In some not so extreme cases, it can be retarded by vested interests (the history of agricultural policies offers plenty of examples). Some recent work in institutional economics argues that institutional change may not yield a socially optimal outcome.[170] For instance, grazing land in Georgia was not enclosed in the 1880s, even though it would have increased the overall income, because enclosure harmed the interests of the majority of voters (it was adopted later, in a modified version, which, in the spirit of the Coase theorem, included some compensation for the losers). To some extent, flexibility is a question of time horizon. The longer it is, the greater the scope for adjustment, and thus the more flexible institutions appear. (2) State intervention, which deeply affected the long-run change, largely escapes this rule. It has, more frequently, been inspired by political and social motivations than by the quest for efficiency gains (section 9.8). The results are mixed, but, as we will argue in the next chapter, on the whole, they are not so positive. (3) The political situation mattered a great deal in some, not necessarily exceptional, circumstances. Poor security discouraged farmers from investing in, or even, in the worst cases, from cultivating their land in the Balkans during the dissolution of Ottoman Empire in the late eighteenth and early nineteenth century, or in China in the 1920s and 1930s.[171] Sub-Saharan Africa after independence is often quoted as being the paradigm of the negative effect of a political situation, although this latter is only one among several causes of backwardness.[172]

Last but not least, it is important to stress that the discussion is limited to economic issues only, as fits the general theme of the book. The distribution of landownership may have had a limited effect on agricultural production and productivity, but undoubtedly shaped the political and social life deeply. Sokoloff and Engermann (1997) have argued that the strong initial inequality in the distribution of land has had a long-lasting impact on Latin American performance. The elites have shaped political institutions to perpetuate inequality, with negative consequences on industrialization via underinvesting in primary education and the lack of a mass market. This type of reasoning would lead us too far from the core argument, but it is worth bearing in mind.

Chapter Nine

THE STATE AND THE MARKET

9.1 Introduction: On the Design of Agricultural Policies

The nature of agricultural policies has changed dramatically in the past two centuries. Traditional states followed a policy of benign neglect, limiting themselves to extract men and money for their political pursuit, and to arrange the supply of as much food as possible to the urban population. In contrast, nowadays, as G. Libecap (1998, 181) points out at the beginning of his review of long-term changes in American policy, "Agriculture is among the most regulated sectors of the American economy. The production and sale of almost all its commodities are affected by some government policy through a complex mix of programs." This statement can be generalized to all advanced countries, but also to most LDCs, and, of course, to the now-defunct Socialist economies. This chapter will deal with this major transformation.

As a starting point, one can list six main areas of "state intervention" or "regulation".[1]

1. Structural policies affecting ownership of factors of production (creation of property rights, land reform).
2. Provision of public goods to the population (health regulation, information about products).
3. Provision of public goods to farmers (R&D, infrastructures, marketing support, well-enforced property rights and so on).
4. Transfers to farmers (subsidies, low-cost credit) or from farmers (taxation).
5. Interventions on the domestic market of agricultural products (purchases by marketing boards, e.g.) or of agricultural factors (provision of low cost credit, regulation of the agricultural labor market.
6. Interventions on international trade of agricultural products (tariffs, taxes, quotas).

Land reform, the creation of property rights, and the support for R&D have already been discussed at length in the previous chapters. The provision of other public goods has gained importance in recent times, especially in the advanced countries. Environmental legislation sets limits to the freedom to pollute (e.g., on the use of chemicals), while a growing stream of legislation has regulated the methods of production and the protection of local traditional products. These concerns will, undoubtedly, be at the heart of agricultural policy in the future. The first laws against fraud in the provision of inputs (seeds, fertilizers, etc.) and food adulteration were approved in the late nineteenth century.[2] For instance, in 1905, France introduced the *Appelation d'origine* to protect the reputation of high-quality wines, in spite of the opposition by mass producers.[3] Although pioneering and important for some categories of producers, these regulations have affected agriculture only marginally in the long-term, worldwide perspective.

The analysis in this chapter will concentrate on the three last categories: subsidies/taxation, market intervention, and trade policy. Each of them consists of several different tools, the combinations of which enable a government to pursue any aim it wants, including fine-tuning the crop mix. Just as an example, category (5) includes policies to reduce supply in order to increase prices (setting aside part of the land or reducing the size of the herds), open-market interventions in otherwise free markets (purchases to raise prices, sales to depress them), state monopolies for the distribution of inputs (e.g., fertilizers) or monopsonies for the purchase of output (or marketing boards), compulsory price-setting (valid for all transactions among private parties), and so on. Some of these instruments are substitutes (for instance, state monopoly of goods and open-market transactions), other are complements (restrictions on imports may be necessary to keep prices high). Ceteris paribus (i.e., for the same amount of transfer to farmers, and also for the same administrative costs), the choice of instrument may have quite different effects on the level of agricultural production, on the welfare of producers, and on the economy at large.[4] For instance, subsidies and minimum guaranteed prices for agricultural commodities have the same (positive) effect on farmer's income, but opposite effects on agricultural output. Subsidies reduce output because the additional income reduces the labor supply of the farmer's household. The minimum guaranteed price eliminates the risk of falling prices in the case of a bumper harvest, and thus it increases output.[5] Furthermore, prices higher than the free-market level reduce consumption, and thus aggregate welfare. The difference between the fall in total welfare and the benefits to farmers is called deadweight loss.

In spite of its wide scope, the above definition of agricultural policy is, arguably, a narrow one. A. Krueger (1992), in the introduction to the final volume of a major comparative research on "pricing policies" in the LDCs, states that the work intends to examine "the entire array of governmental policies that affect agricultural incomes relative to what they would be in the presence of a *laissez-faire* system."[6] The research deals with policies toward manufacturing (which affect agriculture via the changes in agricultural terms of trade) and macroeconomic policy (via the effect on exchange rate, interest rates, etc.), but this definition of agricultural policy could be extended to cover infrastructure policies (via their effects on the commercialization of agriculture), education policies (via the effect on the skill endowment of rural workers), and so on. Clearly, a comprehensive overview of all these policies would be impossible within the limits of this book. Thus, the chapter will only hint at them very briefly.

The chapter is arranged by area/period. As stated at the beginning, the degree of intervention has been increasing in the long run (with the Great Depression as a key watershed), and, on top of this, policies have differed according to the level of development (rich and poor) and to the political system (socialist countries being in a class of their own). The next section deals with the period up to 1914, while section 9.3 deals with the interwar years. The two next sections illustrate the policies after World War II, separately for the "advanced" countries (section 9.4) and for all the other market economies, conveniently, if inaccurately, lumped together under the label "LDCs" (section 9.5). Section 9.6 deals with the story of socialist countries, from the beginning (the seizure of power by the communist parties) to the reforms of the 1980s and 1990s. The last two sections take stock: section 9.7 discusses the effects of agricultural policies on welfare and on technical progress, and section 9.8 concludes, dealing with the causes of the policies (the political economy of agricultural protection).

9.2 Before 1914: The Era of Laissez Faire

The production growth of the nineteenth century was hardly affected by state intervention, if one excludes the creation of property rights (section 8.2). If anything, during the first half of the century, the level of intervention decreased. The Chinese government became weaker for various political reasons, while, in Western Europe, the Enlightment and Napoleon dismantled the regulations to supply cities. Imports of wheat were subject to duties in France, Spain, Prussia, and the United Kingdom, but Germany was a net exporter of wheat at that time, and France and Spain were more or less self-sufficient.[7] The British Corn Laws, which imposed a duty inversely proportional to the domestic price, remained the only real obstacle to free trade in wheat. They were abolished in 1846 after a fierce political battle: the next three decades probably marked the historical peak for freedom from state intervention.

The period of nonintervention did not last for long. The improvements in transportation and communication reduced the gap in the prices of wheat and other agricultural commodities between western Europe and its potential suppliers in the eastern part of the continent and overseas (O'Rourke and Williamson 1997, 29–55). The decrease in wheat prices triggered a protectionist movement in most European countries. Duties were imposed and subsequently raised in Germany (1879–1902), France (1885–94), Italy (1887–94), and other minor countries (but not in the UK).[8] Trade was also hampered by the first experiments in non-tariff barriers (e.g., against meat import in Germany from 1900), by trade wars (e.g., between France and Italy from 1887 to 1898, and between Austria-Hungary and Serbia from 1895 to 1911).[9] On the other side of the globe, Japan imposed duties on rice, as a war-time measure (soon to become permanent) in 1904. American farmers also complained, and rallied behind the requests of silver coinage (i.e., of a currency devalution to help exports) and of measures against speculation on land and excessive railway fares.[10]

According to the conventional wisdom, the 1880s ushered in a new era of state intervention. There is no doubt that states were more active after this period than before, but the idea of an epoch-making change seems a bit overblown in the long-term worldwide perspective. The populist movement, Atack, Bateman, Parker (2000a, 282) argue, was mainly a political one, created by a "few vocal losers [who] can win the support of a majority if the majority perceives itself just one step away from joining the losers." Actually, it did not achieve anything, and it quickly dissolved when wheat prices started to rise in the 1900s. European duties were real enough, but their impact should not be overstated. They only affected wheat and were not so high—on average about 30 percent on the eve of World War I (against 142 percent in the European Community before the 1994 Uruguay Round).[11] Wheat producers would get an equivalent increase in their income if domestic wheat demand was totally inelastic and protectionist countries were "small" ones (that is, their duties did not affect world prices). On the eve of World War I, wheat accounted for maybe one-sixth, and certainly not more than one-fifth, of gross output: thus, the total subsidy from protection (or producer-subsidy equivalent—PSE), amounted to some 5–8 percent of gross output.[12] Given the assumptions, this is an upper-bound estimate of the PSE, which, nonetheless, appears quite low if compared with the level of support after World War II (see table 9.1). The effects of protection can be computed more precisely with the CGE models, which can take all effects into account. In a path-breaking article, Williamson (1990) showed that the abolition of British Corn Laws had a modest impact on the level of agricultural production (and, a fortiori, on total GDP) but a great impact on the distribution of income and also on the allocation of manpower.[13] Without the Corn Laws, agricultural production in 1841 would have been 6 percent lower and agricultural "surplus" (rent and profits) 22 percent lower. Furthermore, agricultural manpower would have been 21 percent lower—that is, the (subsequent) abolition accelerated migration from agriculture. O'Rourke, in two recent articles, explores the effects of the return to protection in the 1880s in France, Sweden, and Italy.[14] The results tally well with those of Williamson. Protection did increase, relative to the counterfactual free-trade policy, the share of wheat on total output, and the share on rents on total Value Added, but it affected total output only marginally, by less than 5 percent in all cases. Without protection, agricultural occupation would also have diminished by about 5 percent in both France and Sweden. In other words, protection changed the crop mix and, above all, the distribution of agricultural income, but it was not essential for the survival of agriculture. Furthermore, one has to recall that the European "protectionist" countries accounted for a very small proportion of world agricultural output—possibly only about a sixth in 1910.[15] In all the other countries, the trade of agricultural products was free, and farmers were allowed to work with only the minimum interference of the state.

9.3 The Interwar Years: The Great Discontinuity

The outbreak of the war seriously reduced the quantity of inputs for European agriculture.[16] Fertilizer plants were converted to the production of ammunition, large tracts of agricultural land in France and Russia became battlefields, and men and horses were sent to the front line. According to the League of Nations, between 40 and 50 percent of the prewar work force was mobilized in all major belligerent countries. All countries resorted en masse to "marginal" manpower (women, children, etc.), and to war prisoners, but nowhere, except perhaps in the United Kingdom, did these new recruits fill the gap entirely. Thus, output fell, in as far as it is possible to tell from the incomplete data available (many countries suspended the publication of agricultural output statistics, so as not to give the enemy valuable information). The average wartime output was one-third lower than the 1913 one in Austria, France, and Germany, one-fifth lower in Russia and Hungary, and one-sixth lower in Italy, while it remained constant in the United Kingdom. Production stagnated in European neutral countries and in the countries of western settlement, and increased somewhat in Asia—so that the total output fell, perhaps, by 8 percent (Federico 2004a, table 3). This reduction was distributed very unevenly among consumers, according to their different access to world markets under wartime circumstances. Russia, cut off from its Western customers, ceased to export wheat. Its total wheat consumption did not fall, but urban consumers suffered a severe shortage of food, as distribution networks were collapsing, estates were unable to find manpower, and the peasants preferred to eat their food instead of selling it in exchange for almost worthless money. The Allied countries, in contrast, could resort to imports from overseas, in spite of the high cost of transportation. Food consumption remained roughly stable in the United Kingdom and increased by 10 percent in Italy.[17] Germany and Austria, which were subject to the Allied blockade, and had to rely on domestic production plus the requisitions from the conquered areas in the east, suffered the most. In Germany, average urban consumption during the "turnip" winter of 1917 was about half the prewar level. European governments intervened quite clumsily to prevent prices from rising, for fear of social unrest.[18] The German government set maximum prices for cereals in October 1914, and, as early as January of the next year, it imposed rationing for bread and flour. It was soon realized, however, that the combination of set prices for cereals and free (and thus rising) prices for livestock products was bound to shift resources toward cattle-breeding, a notoriously inefficient way to produce calories. Thus, the government progressively extended price controls and rationing to other goods, and made it comprehensive from 1916. As expected, this policy caused the development of a black market, which, in 1918, supplied between 15 and 50 percent of consumption at prices ten times the official maximum ones. The Allied countries, on the other hand, could afford to leave the market free for some time. In the end, however, they resorted to price controls (France in 1915, Italy in 1916) and rationing (France and the UK in 1917). However, in the United Kingdom, at least, the purchase prices were kept high to stimulate production.

In the early 1920s, while Eastern Europe was ravaged by civil war and by the beginning of the communist regime (section 9.6), the situation in the West was apparently back to normal. The farmers and horses were back in their fields, the chemical factories resumed producing fertilizers, and the market regulations were dismantled as soon as the economic conditions of the urban population had improved sufficiently (i.e., Italy in 1921, and Germany in 1922). "World" output recovered to the prewar level by the mid-1920s (section 3.2), but business was not "as usual" and farmers complained loudly. They had fared quite well during the war, exploiting the rise in (official or black market) prices, and deemed the postwar prices too low, blaming "overproduction" as the cause. Their complaints are echoed by historians: in his classic book, Arndt lists agricultural overproduction among the main causes of the Great Depression, and this claim is often repeated.[19] Actually, the evidence is not so compelling. In the late 1920s, the increase in the output of the "Atlantic economy" barely matched potential demand, and prices were almost back to the prewar peak, after having dipped in the period 1920–22 (Federico 2004b). American farmers, however, had borrowed heavily in 1920–21, and many of them could not repay their loans in the following years. About a sixth of farms were foreclosed from 1921 to 1929 (Alston 1983, table 1). Their discontent, and the fear of a future fall in prices, bred lobbying for state support. In the United States, a (bipartisan) farm bloc asked for the creation of a federal agency or a semi-monopolistic co-operative movement to control the market for main products. The proposed agency had to purchase goods until prices had reached the prewar level ("parity"), dumping any surplus in excess of domestic consumption on the world market.[20] After several failed attempts, in 1927 and 1928, both houses of Congress approved the McNary-Haugen Bill, which President Coolidge vetoed. At the end of the day, American farmers received only some measures to ease the credit crunch as well as the already-quoted Capper-Volstead Act, which partially exempted co-operatives from antitrust laws. The main importers of western Europe—France, Germany, and Italy—resorted to the well-trodden path of protection, reinstating duties, which had been suspended during the war, in 1925–26.[21] Newly established countries in eastern Europe also raised their duties (e.g., Austria in 1925), fragmenting the previously integrated market of the former Austro-Hungarian Empire, and implemented land reform as described in section 8.3 (Berend and Ranki 1974, 248). Agricultural tariffs in 1927, however, were not higher than in 1913, especially in the major countries.[22] In short, by 1929, agricultural policy had not strayed very far from traditional laissez faire policy, at least not in the "advanced" countries of the Atlantic economy. Innovative policy had been adopted elsewhere however, in Japan and in some LDCs.

The Japanese policy was a reaction to the bloody rice riots of 1918, which were triggered by a 30 percent increase in real rice prices.[23] The government reduced taxes, increased protection on foreign rice, introduced a rash of measures to stimulate production in colonies (Korea and Taiwan), and, above all, set up a marketing board to sell and purchase rice on the market in order to stabilize prices (see Francks 1999, 82–85; Anderson et al. 1986; Hayami and Yamada 1991, 78–80; Sheingate 2001, 80–84). The effect of this policy is controversial. Anderson and Tyers (1992) argue that duties were not necessary to protect domestic production. Consumers preferred the local (Japonica) variety and resorted to imports of Indica rice only because the domestic output was insufficient. Brandt (1993) disagrees: consumers would have had Indica rice if duties had been lower, and this would have released factors of production to industry, where they could have been used more efficiently.

Less developed countries faced a different problem: how to cope with the fall in price of their exports. According to Lewis, the prices of tropical crops relative to manufactures collapsed from 1920 to 1921 and, in spite of a recovery ove the next four years, it never attained the level of the early 1890s and of the immediate eve of World War I. Many governments tried to prop prices by purchasing the crop at a low price after the harvest, hoping to resell it gradually—imitating the "shadowy international cartel of planters and merchants from Ecuador, Brazil, the Dominican Republic, Trinidad and Saõ Tome" (Clarence-Smith 1995, 159), which had operated in the coffee market during 1896–97.[24] During a new crisis in 1906–7, three Brazilian states, led by Saõ Paulo, organized a massive purchase of coffee in prevision of a bumper harvest (the Taubatè Agreement).[25] Other governments followed this example: Japan in 1914, 1920–22, 1926, and 1929–30 for silk; Brazil again (this time the federal government) in 1917–18 for coffee; Egypt in 1921–23 and 1926 for cotton; Cuba in 1926 for sugar.[25] These open market operations were intended as a quick fix to stave off an actual or predicted price fall. The next step was to set up a permanent agency (marketing board), which could manage the whole crop and stabilize prices permanently with a skillful use of inventories. Such a task was clearly beyond the reach of private traders, but, it was then hoped, not of a state, which could borrow on the world capital market and force individual producers to cut their output if necessary. As early as 1924, the state of Saõ Paulo set up a board to finance coffee planters (with funds from BANESP, a state-owned bank) and to arrange sales.[27] Roughly at the same time, the British government tried to restrict further plantations of rubber trees in Malaysia and Ceylon, hoping to prevent a further price decrease (the Stevenson plan) (see Aldcroft 1977, 228; Stillson 1971, 590; Overton 1994, 77). The Brazilian example was widely imitated. Many eastern European countries (e.g. Greece in 1928, Bulgaria in 1930, Romania and Yugoslavia in 1931) reacted to the crisis by setting up marketing boards to manage (and often subsidize at the expense of domestic consumers) exports of cereals.[28] British and, less enthusiastically, French colonial administrators set up marketing boards to manage exports of primary products, especially those of white-owned estates (starting with Kenya in 1933 for coffee).[29] The number and the powers of these boards substantially increased in the early

1940s as a wartime measure. Many were given the sole right to purchase export crops. Most open market operations succeeded in raising prices of primary products for a while, but, in several cases, prices fell again when the consortium (or state agency) sold its stocks. The Brazilian BANESP managed the 1927 bumper harvest successfully but was unable to cope with the 1929 harvest, and an attempt to prevent the fall in silk prices in 1929–30 cost Japan 120 million yen (i.e., about 1 percent of its GDP). These losses highlighted the weakness of national schemes, which were exposed to fluctuations in world prices and to the risk of free-riding by other producers. Thus, some countries explored the possibility of an agreement to regulate the whole world market for a single commodity.[30] For instance, voluntary agreements among major tea producers were signed in 1930–32, but they all failed because the Dutch East Indian plantations increased their output. In 1933, a new plan, backed by a state-level agreement, did succeed because the Dutch tea was cut out from the British market, the largest in the world. But, in the long run, no scheme, domestic or international, succeeded in stabilizing prices or regulating the market. In contrast, short-run successes often bred long-term failures, as the perceived reduction in risk raised the supply curve in both the producing country and abroad. Thus, the experiments were, by and large, a failure. They left the legacy of marketing boards, which many LDCs governments were to use for "developmental" purposes in the 1960s and 1970s (section 9.5).

In the "advanced" states, the real quantum leap forward in state intervention was brought about by onset of the Great Depression. This crisis did not affect production, which remained roughly constant or even increased in most countries, with the exception of the USSR during the collectivization period (section 9.6).[31] Relative prices, however, did fall quite abruptly from 1927–29 to the minimum in 1931–33, hitting farmers hard, especially the indebted ones. Many countries felt a political and economic obligation to do something. Continental European countries first resorted to protection: according to Liepmann's estimates, on average, the duties on agricultural products were two times higher in 1931 than in 1927, but those on cereals were five times higher.[32] Furthermore, several countries imposed quotas and other quantitative restrictions such as the obligation to use a minimum proportion of domestic products. For instance, in December 1929, France stipulated that millers had to use at least 97 percent of domestic wheat (including that from its North African colonies). These policies succeeded in reducing imports, but not in stopping the slide of prices and of farmers' incomes.[33] Thus, governments intervened in the domestic market. France adopted measures to cut wine output as early as 1931, set a minimum price for wheat in 1933, and eventually (in 1936) organized a control of the whole wheat market to stabilize the price (1936).[34] Germany had pioneered market control from 1930 (for maize, milk, and other products) and the new Nazi government boosted market regulation.[35] It set up a *Reichnarstand* with wide-ranging powers to set prices and even to determine output, at least on paper. The avowed aim was to increase output and the degree of the national self-sufficiency, in order to prepare the country for war. In spite of the ostensibly pro-rural Nazi rhetoric, however, the interests of farmers were sacrificed, as purchase prices were kept fairly low to maintain the consensus of urban population. The case of the United Kingdom was somewhat different (see Rooth 1992, 89–94, 212–31; Tracy 1989, 150–56; Ilbery 1984, 144; Perren 199, 53–61). Its long free-trade tradition prevented the adoption of protection until 1931, and, even later, imports from the Dominions (and from Argentina) were free of duties under the system of "imperial preference" (adopted at the Ottawa Conference in 1932). British producers were compensated by subsidies and guaranteed minimum prices, which were managed by marketing boards (the first, for milk, was set up in 1933). The deal was not bad at all, as the PSE was about 30 percent.

European protection had quite different effects on overseas suppliers. The Dominions (and, to some extent, Argentina) depended on agricultural exports much more than the United States, but they had the lifeboat of the British market. Thus, they continued exporting, trying to extract as many concessions from the United Kingdom as possible, even with voluntary restriction of the exportation of some products, such as beef.[36] The United States had no captive export market to rely on, and, in fact, exports of agricultural products fell by one-third, to one-half of the total.[37] Farm crisis was a major component of Great Depression, even if its role is still controversial.[38] In 1929, the Hoover administration organized a Farm Bureau to finance purchases of agricultural products by co-operative associations, but, predictably, the scheme caused huge losses and had no detectable effect on prices. The crisis deepened, and Roosevelt treated it as a major priority of his administration. It had the Agricultural Adjustment Act (AAA) approved in May 1933. The law aimed to increase relative prices up to the 1910–14 level ("parity"), which by then impoverished farmers looked back on as a golden age. Production had to be cut by inducing farmers to leave idle ("set aside") a total of 35 million acres. However, farmers set aside their worst land and increased yields on the rest, using more fertilizers and more productive seeds (hybrid corn). Production of corn, wheat, and cotton fell in 1934, but recovered immediately.[39] The AAA also included other measures, such as help to refinance debts, market interventions for specific crops, and, above all, the provision of low-cost credit upon pledge of the crop via the Commodity Credit Corporation (CCC). When the loan expired, the farmer had the option of not repaying it, and leaving the crop as state property. In other words, the CCC guaranteed farmers a minimum price for their products. The CCC and, to some extent, the "set aside" policy were to remain the twin pillars of American agricultural policy for half a century. The New Deal agricultural policy, "provided more benefits to farmers than would otherwise have been possible" (up to a quarter of their income) and "established the foundation for all farm programs that came thereafter. Both farmers and politicians have been unwilling to depart from the security and stability it provided" (Hurt 1994, 295).[40] This last statement also holds true, with the necessary qualifications, for other "advanced" countries, for some of their African colonies and for Latin American countries. Factoring in the collectivization in the Soviet Union (section 9.6) and the land reform in Mexico (section 8.3), the 1930s stand out as the key decade of intervention in agriculture. Among the great agricultural countries, only India and China, which was then under invasion by the Japanese, did not join the movement.

The Second World War affected agriculture even more than the First. War operations devastated a much wider area, and war involved Asian countries, such as China and Japan, which had stayed neutral in the previous conflict. The overall pattern, however, was quite similar. Production rose in the United States and in other countries of Western settlement, while it fell in Europe.[41] Production fell by 20 to 30 percent in France and Italy, while, in Germany, it remained stable until 1944, thanks to the massive employment of war prisoners, and then it fell abruptly. The winter of 1944 was particularly harsh, with starvation in many areas of occupied Europe. Most countries tackled the supply problems through rationing and market planning, as in 1915–18. These measures, however, were not, as they had been in 1915–18, an absolute novelty. The governments improved on their previous experience, and used the market intervention boards they had set up to fight the 1930s overproduction in order to increase output. Nowhere was the difference clearer than in the United States, which had to feed all its allies, including the USSR. The government raised prices above the mythical 1910–14 "parity," so that World War II was a new golden age for American farmers.[42]

9.4 The OECD Countries after 1945: The Era of Surpluses

At the end of the war, in theory, "developed" countries could have decided to phase out the support to agriculture, as a temporary response to the exceptional circumstances of the 1930s and 1940s, and to return to the pre-1929 low intervention, if not to nineteenth-century laissez faire. This did not happen. The United States did discuss a reform of their agricultural policy, but no one questioned the principle that farmers deserved special support.[43] At the end of the day, all "advanced" countries opted for strengthening the emergency support to agriculture and making it permanent. The Treaty of Rome listed agricultural policy as one of the issues of competence of the future European Union, and stated, quite precisely, the aims of the future Common Agricultural Policy (Article 39): (1) to increase agricultural productivity; (2) to ensure a fair standard of living for the agricultural community, in particular, by increasing the individual earnings of persons engaged in agriculture; (3) to stabilize markets; (4) to assure the availability of supplies; and, (5) to ensure that supplies reach consumers at reasonable prices (Ritson 1997, 3). This list of goals is not only ambitious but also, to some extent, internally inconsistent. A big productivity increase could, indeed, deliver "reasonable" prices for urban consumers and provide farmers with a "fair standard of living," but productivity could increase only if the number of farmers fell drastically. But a massive out-migration to cities was bound to change the traditional "agricultural community" deeply.[44] The United States and Japan did not enshrine agricultural policy in their constitutions, nor did they state its aims so clearly, but they did share the goal of increasing and stabilizing the income of farmers, with or without an explicit promise to close the gap with the other sectors. In theory, as stated in section 9.1, the incomes of farmers can be raised either by leaving prices free and subsidizing farmers (the prewar British solution) or by setting a minimum price and purchasing whatever the farmer produced (the prewar American system). Almost all countries choose the second option. The "low-cost" countries, such as the United States, could set the minimum price at any level that their state coffers could afford to pay, dumping any excess production on the world market. The "high-cost" European countries set not only the minimum intervention price and but also (implicitly) a maximum one, equivalent to world price plus duty.

Although inspired by similar principles, the agricultural policies of the OECD countries show a mind-boggling range of differences in the details of implementation, by country, period, and product.[45] Here, it is possible to present only a very brief account of policies in the three main "countries," the European Community, Japan, and the United States, which do, however, account for most of the OECD output.[46]

The United States stands out for the frequent changes in its policies.[47] Major agricultural acts were approved in 1954, 1956, 1965, 1970, 1970, 1973, 1977, 1985, 1996, and 2002, and, on top of them, there were commodity-specific laws (e.g., for wool in 1954, and for rice in 1975). The first of these laws (PL 480, also known as "Food for Peace") introduced a major technical innovation, subsidies to export (initially to less developed countries), in order to get rid of surpluses. The others changed the level of support (prices, amount of loans, etc.), the system of distribution of resources, the product coverage, and so on. For instance, the "set aside" program lasted from 1933 to 1939 (after having been abolished by the Supreme Court in 1936 and revived two months later as the Soil Conservation and Domestic Allotment Act), was resumed in 1956, phased out in the 1970s, resumed in 1977, and finally abolished in 1996—each time with different coverage and characteristics. For instance, according to the 1983 Payment in Kind program, farmers who set part of their land aside were paid in commodities from the stocks. This move did reduce stocks, but the federal budget lost all the money it had anticipated to farmers. Sheingate (2001) argues that, in spite of all these gyrations, policy did move toward liberalization from the 1970s onward, as in Japan and in the European Community.[48] His main evidence is the so-called FAIR Act of 1996, which abolished intervention in some key markets and proclaimed the principle of "decoupling"—that is, that subsides have to be independent of market conditions and actual production. The historical record experience, however, would suggest the utmost caution—as confirmed by the very generous 2002 Farm Bill, which almost doubled subsidies.[49]

If compared with American wavering, the Japanese policy seems a paragon of straightforwardness.[50] Its cornerstone remained, for most of the period, the 1942 Food Control Law: it had set up a state monopoly of rice trade, which was managed by farmers' co-operatives. The country had to adjust to the loss of Korea and Taiwan, which, on the eve of the war, supplied some 15 percent of its rice consumption. The government ruled out an increase in imports, allegedly fearing that exports would not increase enough to provide the necessary foreign currency. It forbade private imports of rice and set the purchase price at a level high enough to cover the production costs of marginal producers. This included the opportunity cost of the farmer's work, computed according to urban market wages, which rose rapidly during boom years. Thus, from 1950 to 1985, the real purchase price of rice increased by almost three times. Such an increase would have been unbearable for urban consumers, and thus resale prices were kept lower than purchasing ones. In other words, the government effectively subsidized rice farmers. They responded very brilliantly: rice output boomed, and Japan became self-sufficient. Imports of other products were allowed, but were subject to very high duties, and producers received a combination of minimum price-setting measures and direct subsidies. From 1969 onward, the monopoly for rice was phased out, very gradually (its demise lasted until 1987), but co-ops still controlled supply. In the 1970s, the government introduced some incentives to set aside rice land and somewhat reduced protection on other products. The real change was brought about by the abolition of the law prohibiting the importation of rice under the GATT rules after the Uruguay Round (1994). Imports were allowed up to a maximum of domestic consumption, and the rice market was liberalized, ending the monopsony of the co-operatives. The story of Japanese policy shows a clear trend toward less state intervention, even though the initial level was extremely high.

As in the United States and Japan, the roots of postwar European agricultural policy can be traced back to the 1930s. After the war, no European government dared to liberalize its domestic market for agricultural products. Under the double stimulus of the wartime experiences of shortages and of the obsession with the deficit in the balance of payments, they set as paramount the aim of increasing total output to achieve, whenever possible, self-sufficiency and to raise farmers' income.[51] On the Continent, these aims were pursued with a combination of protection (especially on cereals) and guaranteed minimum prices, while the United Kingdom continued its prewar policy of free imports and subsidies to agriculture. By and large, these policies achieved their aim: as early as 1949–50, agricultural output in western Europe regained the prewar levels, while balance of payment constraints were becoming less and less severe. The end of the emergency, however, did not imply the end of support. In contrast, both France (with its Second and Third Plans) and Germany (with the Agricultural Laws of 1950 and 1955) extended the scope of support, adding subsidies, provision of credit, support for

R&D, and so on. Thus, agriculture policy became a sort of "extended public welfare" (Milward 1984, 229).

Agriculture did not feature prominently in the complex negotiations that led to the Treaty of Rome, even though the issue had been widely discussed in the 1950s.[52] Article 39 (already quoted) committed countries to have a Common Agricultural Policy (CAP), but without any further details. The actual policy was drafted in the following fours years, from the Conference in Stresa (1958) to the final compromise in January 1962. It was inspired by two basic principles: prices had to be the same all over Europe and all compensating payments had to be managed by the European Community via the so-called EAGGF (or FEOGA).[53] The common prices were set for the first time in 1964, and the system was fully operational from 1970, with the abolition of all the remaining intra-European duties. From 1969, prices were converted in special monetary units (or MCAs), to dampen the effects of currency floating. Each year the ministers of agriculture of the member states met and set both prices and the "green" exchange rates—usually at a level more favorable for "North-European" products (wheat, butter, meat) than Mediterranean ones (wine, fruits). The CAP was a success in terms of output: from the early 1970s, the European Union became self-sufficient for the main products (including cereals) and the second world-exporter of agricultural products.[54] The European taxpayers paid quite dearly for this success, and proposals of reform were mooted in the early 1970s.[55] Serious reform, however, had to wait some fifteen years. It aimed at curtailing expenditures by reducing production growth, either directly, with maximum production quotas, as for milk (from 1984) or indirectly with the reduction of support prices if production exceeded a given threshold—the principle of "coresponsibility" in Euro-jargon (1988).[56] In 1992, the then-Commissioner for Agriculture, Ray McSharry, succeeded in getting a wide-ranging reform proposal approved. It aimed at achieving a convergence of prices for the main products to world level by reducing protection, as stipulated in the GATT. Producers were compensated with subsidies, partially related to the extension of land set aside. In short, the European Community was to adopt the traditional British policy, which the United Kingdom had willingly abandoned upon entering the EU in 1973. The reform was successful but, as in Japan, the market is still far from liberalized.

To sum up, in all the OECD countries, state intervention was quite heavy throughout the whole postwar period, and, in spite of the liberalization in the last decade, still remains so. In a nutshell, it aimed at raising domestic prices beyond the "world" level. This aim was achieved: American wheat prices were 4 percent higher than Australian ones from 1870 to 1932, and 37 percent higher from 1933 to 1970 (Libecap 1998, table 6.A.3). According to the FAO, in 1993, domestic prices exceeded world prices by 10 percent in Australia, by 48 percent in Canada, by 29 percent in the United States, by 93 percent in the European Community and by 193 percent in Japan.[57] As stated repeatedly, these ratios are very crude measures: the potential impact of liberalization must be assessed in a CGE framework. Tyers and Anderson estimate that, in 1985, an overall liberalization of trade would have caused prices to fall in the European Union by 15 percent for wheat and by 19 percent for meat, in Japan by 68 percent for rice and by 71 percent for meat, while in the United States prices of wheat would have decreased only by 11 percent and those of meat would have increased by 13 percent.[58] The Uruguay Round of the GATT (approved in 1994) reduced the allowed level of protection and thus nowadays the gap between domestic and free-market prices is smaller. (see Summer and Tangermann 2002, 2009–10; OECD 2002)

High prices were bound to stimulate production, and indeed output growth was a key target of Japanese and European policies in the 1950s and early 1960s. The American policy was somewhat contradictory, as it contemplated both production-enhancing measures, such as minimum guaranteed prices and production-cutting ones, like the "set-aside." There is little doubt, however, that agricultural policies contributed to the spectacular growth in output (section 3.2). Unfortunately, in all OECD countries, consumption was not increasing fast enough, and thus a growing share of production ended up as inventories owned by government institutions.[59] The problem emerged earlier in the United States, where domestic production could not substitute for imports. In 1960 (the historical peak), the total value of stocks amounted to $6 billion—about half the sales of all commodities subject to the program (with shares up to 94 percent for corn and 112 percent for wheat). The situation was similar in Japan, where, in 1970, rice stocks amounted to one year's consumption. The easiest way to get rid of these stocks was to export them to the LDCs and/or to socialist countries at the world price. As said, this solution was first adopted by the United States, and later imitated by the European Union.

The costs of stockpiling and exporting at "market" prices were borne by the taxpayers. The CAP as a whole accounted for 80 percent of the European Community's budget in 1973, and, in spite of the liberalization, for about half in the year 2000; and the EAGGF accounted for 90 to 95 percent of this sum (the rest being allocated to the "structural" policies) until the mid-1990s.[60] In Japan, rice marketing alone absorbed, in 1970 (its historical peak), about half the total agricultural budget—6 percent of total state expenditures. These costs were a powerful, if often insufficient, incentive for reform. Transfers still account for a substantial part of the farmers' income, as shown by the data on PSE (table 9.1). In spite of the reforms of the early 1990s, the dependence on state handouts remains very high. In Japan, South Korea, and, above all, Switzerland, farmers are all but in name state employees, and only in New Zealand and, perhaps, Australia, do farmers get less than their predecessors in the allegedly "protectionist" countries of western Europe in the 1890s (section 9.2).

As said, the total cost of the agricultural policies exceeds the amount of transfers to farmers by the amount of the deadweight losses. Economists have produced

TABLE 9.1
PSE as Proportion of Agricultural Production, Advanced Countries, 1980–2000

	1979–81	*1986–88*	*1991–93*	*1998–2000*
USA	31	29.4	20.5	26.7
EU	35	45.0	44.0	41.6
Japan	45	66.1	59.5	64.2
Korea		70.7	75.8	67.9
Switzerland		83.8	82.0	100.0
Canada	28	40.8	34.3	19.7
Australia	7	9.3	10.1	4.8
New Zealand	17	13.0	2.0	1.1
Mexico		−1.5	29.8	18.7
OECD		42.0	39.1	36.9

Sources: **1979–81:** Fanfani 1998 table 9.7. **Others:** OECD Statistical Database.

several estimates of these losses (or, equivalently, of the gains from the abolition of agricultural policy). For instance, Gardner estimates that, in the late 1980s, American producers received (in various forms of support) a total of $16.5 billion, while consumers lost $4.8 billion and taxpayers $17.7 billion—with a net deadweight loss of $6 billion dollars (Gardner 1990, table 1.10). Blandford (1990) reviews some studies that promise total net gains from total liberalization ranging from $7.5 to 34.7 billion for the European Community, from $5.2 to 36.6 billion for Japan, and from $5 billion (i.e., a loss) to 8.6 billion for the United States in the early and mid-1980s.[61] The huge range of figures shows how much the estimates are sensitive to the methodology (partial equilibrium or CGE modeling), the scope of the analysis (i.e., including the whole world or for some countries only), the structure of the model, the assumptions about elasticities, and also the base-line year. There can be no doubt, however, that agricultural policies have reduced overall welfare in the OECD countries.

9.5 The Less Developed Countries after Independence: The Green Revolution and the "Development" Policies

Lumping together very different countries under the label LDCs is always a doubtful exercise. It is even more so when the subject is agriculture, which is so sensitive to local conditions and environment. Tropical Africa has little to do with central Asia. On the other hand, many countries shared a colonial past and all faced the same problems: feeding a fast-growing population and achieving modern economic growth. The Green Revolution (section 6.2) gave most countries the opportunity of a big leap forward in production, provided, of course, that they invested in irrigation, R&D, and extension.

Almost all countries after independence adopted aggressive development policies, trying to industrialize as soon as possible via import-substitution. Agriculture played a subordinate role—but with a wide range of alternative policies. Binswanger and Deininger (1997, 1960) distinguish four cases: (1) countries (mainly in south Asia) with a prevalence of family farms, pursuing agriculture-friendly policies (infrastructure, R&D, credit, etc.); (2) countries (mainly in Africa, with some South American examples) with a prevalence of family farms pursuing policies hostile to agriculture (such as high taxes, heavy-handed market regulations); (3) countries (like India, Mexico, the Philippines) with unequally distributed land and that pursued mixed policies (unfavorable trade and macroeconomic policies, but also land reform and investment in agriculture), and (4) countries (including Brazil and South Africa) with very unequal land distribution and a very unfair agricultural policy (overall taxation, with compensation for the elites, and timid and ill-designed land reform projects). It would be impossible to describe in detail these policies. Just for an example, one can sketch out the case of Indonesia.[62] State intervention started during the colonial period, in 1911, with a prohibition on the exportation of rice in periods of shortage. The prohibition lasted until the 1920s, and was substituted in the 1930s by a restriction on imports (plus a tax on rubber exports). In 1939, the government set up a monopoly on imports of rice. After independence, the Indonesian government nationalized the Dutch-owned estates, which, however, accounted for only a minor share of the total acreage, and it nationalized sugar production, with poor results, as Indonesia turned into a net importer. It also used its monopoly on rice imports to stabilize domestic prices (the so-called BULOG scheme). Finally, it provided some support for technical innovation, with low-cost credit and extension. The main measure, however, was the monopoly of the distribution of fertilizers, imposed for imports from 1957–58 and then extended to national products, from 1964 to the liberalization in 1976. Prices were kept so low that subsidies absorbed about half the agricultural budget. In the 1990s, the government slowly reduced its intervention in the rice market, but imports are still a state monopoly, and urban prices have remained lower than world market ones. As can be seen from this short account, the Indonesian government intervened quite heavily, but, on the whole, it supported agricultural development—and, indeed, Indonesia belongs to the first group in the taxonomy by Binswanger and Deininger.[63] Policies in Africa were much less friendly: Eicher and Baker-Doyle (1992, 45) conclude their survey of African agricultural policies stating that "all countries under all types of governments—civilian, military, capitalist and socialist—have exploited and controlled the agricultural sector."[64] They raised taxes, especially on exports, used marketing boards to set the prices of products and inputs, and also imposed different exchange rates between imports and exports.

TABLE 9.2
Policy-Determined Net Income Transfers in Some LDCs as a Percentage of Agricultural GDP, 1960–84

		(a)	(b)	(c)	(d)	(e)	(f)	(g)
Ivory Coast	1960–82	1	−49.5	−48.5	13	−35.5	−28	−63.5
Ghana	1962–76	1	−64	−63	3	−60	−84.5	−144.5
Zambia	1971–84	4	−21	−17	5	−12	−200.5	−212.5
Argentina	1960–83	4	−20	−16	0	−16	−68	−84
Colombia	1960–83	3	−8	−5	4	−1	−37.5	−38.5
Dominican Republic	1960–84	−1	−18	−19	12	−7	−29.5	−36.5
Egypt	1966–84	−5	−10	−15	7	−8	−12.5	−20.5
Morocco	1963–84	0	−9	−9	8	−1	−3	−4
Pakistan	1960–86	1	−23.5	−22.5	14	−8.5	−30	−38.5
Philippines	1960–86	−1	−8	−9	4	−5	−18.5	−23.5
Sri Lanka	1960–85	3	−16	−13	19	6	−75	−69
Thailand	1962–83	0	−24.5	−24.5	8	−16.5	−19	−35.5
Turkey	1961–83	2	9	11	4	15	−33	−18
Brazil	1960–83	20	−0.5	19.5	12	31.5	−10	21.5
Chile	1960–82	2	−1.5	0.5	4	4.5	−60	−55.5
Malaysia	1960–83	1	−14.5	−13.5	9	−4.5	−3.5	−8
Korea	1962–84	1	24.5	25.5	7	32.5	−45	−12.5
Portugal	1960–82	−2	12.5	10.5	1	11.5	−3	8.5
All Countries	1960–69	0	−8	−8	4	−4	−28	−32
	1970–79	2	−17	−15	8	−7	−40	−47
	1980–83	3	−6	−3	9	6	−44	−38
	1960–83	2	−13	−11	7	−4	−42	−46

Note: (a) price support to input purchase; (b) price support to marketing of products; (c) total price support (a + b); (d) public investments; (e) total direct policies (c + d); (f) indirect support; (g) grand total (e + f).
Source: Schiff and Valdes 1992, tables 7.1, 7.2.

In the 1980s, the World Bank published a massive research project on agricultural policies in a sample of eighteen countries, with estimates of intersectoral transfers of income (table 9.2). In all the columns, a negative sign means a net extraction of resources: agricultural GDP would have been higher without any intervention. The two columns on the left measure the effect of price policies (including taxation and marketing boards) on inputs (column a) and on output (column b) as the difference between domestic and world prices for the major commodities at the official market rates.[65] Most countries subsidized the purchase of inputs (but the sums were relevant, as a proportion of total output, only in Brazil), and almost all taxed output. Actually, taxes only hit export crops. The import-competing food crops enjoyed a positive price support in most cases, contrary to the conventional wisdom that agricultural policy protected urban consumers.[66] The sum of protection on inputs and output (column c) is similar (albeit not perfectly coincidental) to the PSE. The difference with the OECD countries (table 9.1) is striking. The total is negative in thirteen cases out of eighteen. Column d includes expenditure in irrigation, R&D, and extension, and thus is, by definition, positive or, at worst, zero. In some cases, the outlays were quite substantial, but the net effect of agricultural policies (column e) was still negative in two-thirds of the countries. However, the absolute figures are not so impressive, with such notable exceptions as Ghana and the Ivory Coast. Column f sums transfers from the so-called indirect protection—that is, the combined effect of trade policies (most notably industrial protection) and of macroeconomic policies—these latter being measured by the difference between the official and the actual (i.e., market-clearing) exchange rate.[67] This indirect protection was negative in all countries and quite high in some cases. Thus, it reinforced negative direct support in the majority of cases, while in others (notably Korea and Chile), it turned a positive direct support into a heavily negative one. As a result, agriculture was a net provider of resources (column g) in sixteen countries out of eighteen. The two exceptions, Portugal and Brazil, are hardly representative of the average LDC.

To sum up, the pro-active "development" policy of the 1960s and 1970s harmed agriculture in the LDCs, but it did so much more via macro-policies than via specific agricultural ones. As is well known, the macroeconomic policy of most LDCs drastically changed from the 1980s, with a massive process of liberalization of trade and financial markets, which led to a realignement of exchange rates to their real values. Indeed, the World Bank's research has been quite influential in shaping these policies and creating the so-called "Washington consensus."[68] Macroeconomic liberalization was bound to reduce the overall anti-agricultural bias even with unchanged agricultural policies. But the latter changed as well, albeit later and less comprehensively than the macroeconomic policies.[69] The markets for goods and inputs were liberalized, marketing boards were dismantled, import duties were slashed, and domestic prices were aligned to world market ones. The extent of the reforms and the degree of liberalization differed widely among countries. For instance, subsidies on (imported) fertilizers were gradually phased out in Mali from 1981 to 1987, while, in Nigeria, they remained at around 85 percent of the market price from 1976 (the start of a nationwide program) to 1983, when they were cut, under the pressure of international donors, to 28 percent in 1985, they jumped to 80 percent again in 1986, because the purchase price for farmers was kept fixed at a time of inflation, were progressively abolished from 1994 to 1997, and were reintroduced in 1999. In India, subsidies (free irrigation water and electricity, low-cost fertilizers, etc.) increased throughout the 1980s, with an estimated welfare loss of up to maximum of 2 percent of GDP in 1989–90, and were reduced from 1991.[70] By and large, liberalization has advanced more in southeast Asia than in the south of the continent (India and Pakistan), and more in Asia than in sub-Saharan Africa. Almost no country, however, has entirely renounced market intervention, and state-run marketing boards still survive alongside private traders, sometimes with a high share of total transactions. The partial implementation of reforms and the wide intercountry differences make it impossible to assess their effects in a conclusive way. It is, perhaps, fair to say that liberalization has improved overall welfare without being a panacea, and it has had some negative consequences (e.g., on the conditions of the more disadvantaged rural households or on the urban poor) without being a disaster.

9.6 The Socialist Countries

The "socialist" countries differ from the "market" economies considered so far because, at least in principle, the economic decisions pertain to the state (ruled by the Communist Party), whose official aim is to build a "socialist"—and later a "Communist" society. In the major book on this issue, Pryor (1992) lists thirty-three of these countries in 1980, which then accounted for 36 percent of the world population, 30 percent of the land, and 25 percent of GDP.[71] By then, most of these countries were still—or had been at the beginning of the socialist experience—poor and backward (the main exceptions being East Germany, Czechoslovakia, and, possibly, Hungary).[72] Thus, almost all Communist parties opted for an aggressive strategy of industrial development, based on an accelerated build-up of large-scale industries. As in the market-economy LDCs, the role of agriculture was clear: it had to provide the necessary raw materials, capital, and manpower. The peasants had to work for the greater glory of the country: if they did not comply, they would be forced to by the full power of the state.[73]

The first stage of the construction of socialist agriculture entailed the expropriation of all noncultivating owners (Pryor 1992, 175–79). In Russia during World War I, many noblemen sold their estates because of a shortage of manpower and the unstable political climate, and the peasants seized many of the remaining estates in two waves, in spring and autumn of 1917 (see Medvedev 1987, 25–28; Gregory 1994, 84–85; and Fitzpatrick 1994, 23–25). The new Bolshevik government legalized these seizures immediately after the October Revolution. In China, too, noncultivating landowners were also expropriated immediately after the Communist seizure of power, and their land was distributed to the peasants (Maddison 1998; 70). All these movements still remained within the limits of "traditional" land reforms, albeit on a grander scale and with more spontaneous violence than in other cases.[74] The slogan was still "the land to the tiller," and the tiller was fairly free to do what he wanted.

The path of "socialist" agriculture started to diverge from the "capitalist" one when the states began to take care of the supply of food to cities. In the Soviet Union after the revolution, peasants refused to sell products in exchange for practically worthless paper money. The authorities started to requisition agricultural goods in May 1918 and instituted a compulsory system of delivery by village quotas in January 1919. This "war communism" policy was a failure, as the peasants reacted to forced deliveries by setting aside land and slaughtering their livestock. The output collapsed to perhaps one-half of the prewar level, causing a widespread famine with at least one million deaths.[75] The government was forced to back down, and to adopt a more liberal policy, the New Economic Policy (NEP), in March 1921 (see Volin 1970, 170–76; Gregory and Stuart 1986, 47–57). Requisitions were substituted by a tax, and local markets were largely liberalized, but the state retained some control of wholesale trade and prices, especially for the most sensitive goods, such as wheat (it also kept a monopoly of foreign trade, but, by then, the USSR had no wheat to export). This interlude lasted for a few years. Worsening terms of trade and an insufficient supply of consumer goods induced peasants to reduce their sales, and their behavior gave the hard-line Communists the pretext to return to compulsory purchases in 1928.[76] The Chinese experience replicated the Soviet one thirty years later. The state started to purchase agricultural products in 1950, forbade professional trading in 1953, and abolished any residual private marketing (i.e., selling by the peasants) in 1958.[77] This socialist brand of market intervention was much more invasive than the methods employed in the LDCs. A marketing board purchases at a fixed price whatever the peasants decide to produce, while the socialist planners decided how much the peasants should produce. They set a quota to be delivered each year. In theory, peasants should have been left enough food for their consumption, but the quota had to be filled even if production fell below the target. Furthermore, the peasants had to buy all modern inputs—fertilizers, machinery, and so on—and most manufactures from state-owned enterprises. Thus, the planners could set the relative prices at any level that they deemed consistent with their development strategy, as long as it could be politically sustained, if necessary with a heavy dose of repression.

The culminating stage of development of socialist agriculture was collectivization—or the transfer of the right of use of land from the peasants to the state.[78] In some cases, the land was expropriated and transferred to state farms (or *sovchoz* in Soviet terminology), while in others, peasant households kept the formal ownership, and were "only" forced to join co-operative farms (*kolkhoz* in the Soviet Union, "communes" in China). The peasants had to relinquish most of the tools and, above all, the cattle, even though they were usually left some tiny plots to cultivate for their own personal consumption or for sale to markets (sometimes legal, sometimes illegal). The Soviet collectivization was announced in November 1929 and it started in 1930, to be completed six years later.[79] After the war, the Soviet model was exported to all of Eastern Europe with the exceptions of Yugoslavia and Poland, where an early attempt at collectivization had aborted and the land remained in the control of peasant households.[80] In China, collectivization started in 1955–56, when the peasants were urged to join co-operatives by a mixture of coercion and tax incentives. In 1958, they were deprived of all rights to land, without even a plot for household consumption (see Lardy 1984, 41–48; Becker 1996, 47–53, 83–96; Zong and Davis 1998, 7–12; Maddison 1998, 70–73). This policy (called the Great Leap Forward) caused a disastrous famine, and thus, in

1962, peasants were given a respite. The so called Sixty Articles allowed them some freedom, including that of having private plots. Only four years later, however, the start of the Cultural Revolution brought back radical policies, such as the abolition of private plots, brigade accounting (i.e., a redistribution of income according to political merits), and the forced recruitment of peasants' labor for building projects (notably irrigation). Some self-declared socialist countries did not follow this path: many tropical countries limited themselves to expropriating and managing large-scale plantations, leaving the peasant farms untouched. Thus, in 1980, collective farms accounted for more than 75 percent of the total land only in fifteen out of thirty-three socialist countries, and for 50 to 75 percent in three others.[81]

Most peasants did not like the idea of collective farming. In some cases, especially in Eastern Europe after World War II, the Communist parties tried, with some success, a relatively soft approach by luring peasants with incentives or credit write-offs and/ or threatening the reluctant ones with tax increases (Pryor 1992, 109–11). But in China and, above all, in the Soviet Union, the collectivization was imposed forcibly, and thus it was quite bloody. Peasants resisted with all the means at their disposition: just as they had done during the period of "war communism," they refused to sow, and slaughtered their cattle instead of delivering it to the *kolkhoz* or to the commune.[82] The Communist Party reacted with repression, and the deportation of "anti-social" elements (the so-called *Kulaks*). Production fell. Statistics have to be handled with care, as the chaos in the countryside made the collection of data objectively difficult, and farm managers, local party representatives, and the central bureaucracy had strong incentives to "cook" the figures. For both countries, however, the best guess suggests a fall of about 25 to 30 percent from the pre-collectivization peak.[83] The shortage of raw materials such as cotton hampered industrial growth, and, above all, the fall in food production plus the disruption of distribution networks caused widespread famine in the countryside. Hunter (1988) and Allen (2004) estimate, with quite different methods, that Soviet output on the eve of World War II would have been 15 to 20 percent higher had the NEP continued.[84] According to the latest estimates, excess deaths totaled 7 to 8 million in the USSR and around 30 million in China (i.e., about 4.5 percent of the population at the start of collectivization in both countries).[85] Thus, collectivization was a disaster in the short run. What about the long-run? How did socialist agriculture fare? How much did collectivization affect economic development?

As said, in a socialist system, the state (i.e., the party) controls the economy and makes all the essential production and marketing decisions.[86] In practice, the exact amount of control varied according to the circumstances. Some scholars stress that the state control was not as strong as one is led to believe by official statements and regulations. Davies (1998) argues that the Soviet economy of the 1950s was a "quasi-market" one, and lists, among the market features, the existence of private plots and of the right to sell the produce.[87] Zweig (1989) shows that the Chinese peasants in the 1960s and 1970s had some scope for resisting the orders from above. In many areas, the most "radical" policies were implemented only partially or not even implemented at all.[88] These works are a welcome addition to our understanding of the historical reality, but they do not alter the basic feature of socialist agriculture. The peasants had to produce what the planners said and had to sell their product to the state at fixed prices. If a collective farm met its targets, the managers were rewarded, and the whole collective farm could gain (e.g., by being allowed to sell the surplus at higher prices). Success nurtured failure, however, as meeting or exceeding the target in one year entailed the serious risk of being assigned a higher target for the following year. If, however, the farm failed to comply, the managers might be punished, and their political career terminated. In theory, ordinary peasants would also suffer, since their income was related to the farm performance. Only in the harshest times of the initial collectivization, however, was their livelihood really threatened. In the "mature" system, the state financed collective farms to guarantee all workers a minimum income (soft budget).[89] Thus, in a socialist system, workers had little incentive to work hard. Most of their income was fixed, while the variable part (a productivity-related bonus) depended on the performance of the whole team, which usually numbered hundreds if not thousands of workers. There were no incentives for peer-monitoring, and external monitoring was quite expensive in terms of time (i.e., of forgone output) and open to corruption and patronizing.[90] In other words, the system left ample scope for shirking and free-riding. On top of it, monetary income was much less appealing in a socialist system than in a capitalist one. Manufactures were relatively expensive, and/or of poor quality, and sometimes rationed. There was no chance of climbing the social ladder within the countryside (e.g., renting or buying land): any show of "wealth," however modest, entailed the risk of being singled out as a *kulak* or an anti-socialist element or being forced to redistribute to other households. Thus, rational peasants would increase leisure or, if allowed, work harder on their private plots. In fact, the latter were an outstanding success. They accounted for 3 to 5 percent of land, were allocated almost no modern input (at least officially), and yet they produced 20 to 30 percent of the livestock output.[91] Clearly, the peasants worked more intensively on them than in the collective fields. In short, socialist agriculture was plagued by a severe structural incentive problem. The problem was even worse if really, as suggested by some estimates, wages were inferior to marginal productivity of labor.[92] Such a divergence would fit well with the concept of the Soviet system as "second serfdom," popularized by Lewin.[93]

This problem of incentives was compounded by some specific planning mistakes.[94] The belief in Lysenko's theories of evolution caused a huge waste of precious resources in the Soviet Union and China, while the compulsory adoption of three-year crop rotation in China in during late 1960s and early 1970s reduced productivity in spite of the increase in the use of fertilizers (Lardy 1984, 82–85; Huang 1990, 222–36; Becker 1996, 64–79). From 1965, the Chinese government adopted a policy of regional self-sufficiency in basic foodstuffs, which unraveled the traditional division of labor between rice-producing and rice-consuming areas. Inter-regional transfers fell from 5 percent of the grain crop in the 1950s, to a minimum of 0.1 percent twenty years later.[95] Traditionally surplus areas had too much rice to consume, while the country as a whole experienced shortages of livestock products and raw materials as traditionally specialized areas (e.g. the northwest for cattle-breeding; Hopei, Shantung, and Honan, in the northeast for cotton; or the Fukien, in the south, for sugar) were forced to grow wheat or rice.

The combined effect of incentive problems and inefficient policies strongly suggests that socialist agriculture did not exploit its full potential. Some authors have tried to estimate the loss of efficiency by comparing Soviet farms with Western ones, but the results are not totally convincing.[96] In fact, the implicit counterfactual (e.g., "capitalist" USSR in the 1960s) is extremely difficult to design and manage. The official data on output, if one trusts them, show that socialist agriculture could match the increase in production in capitalist countries. From 1961 to the end of the socialist regime, gross output grew in the USSR by 1.57 percent yearly, in Eastern Europe by 1.99 percent, and in China by 2.99 percent (admittedly starting from a very low level).[97] This growth, however, was mainly an extensive one.[98] Wherever possible, the socialist planners extended the cultivated acreage—notably during the Soviet virgin land campaign in the 1950s (section 4.2). Most of this land had to be irrigated, which needed a huge amount of capital. The Chinese government started a massive campaign to build irrigation schemes in the 1960s and 1970s, even though it is unclear how many of them actually worked (see section 4.3). It also provided the communes with increasing quantities of fertilizers: consumption rose very fast (an yearly rate of 8–10 percent), albeit starting from a very low level.[99] The Soviet authorities believed in the virtues of large-scale mechanized cultivation, and thus from the 1930s invested heavily in tractors to replace the horses slaughtered during the first stages of collectivization.[100] This strategy was clearly unsuited to a "backward" and capital-scarce country, with relatively abundant agricultural labor. In fact, in both the Soviet Union and eastern Europe, agricultural manpower declined very slowly until the 1960s, whereas in China it is still growing to the present day. This delay undoubtedly reflects the backwardness of socialist countries, but also the deliberate policy of keeping the peasants out of the cities.[101] The combination of an unimpressive growth in production and an increase in inputs caused technical progress (as measured by the TFP) to be decidedly slower than in capitalist countries.[102]

The shortcomings of socialist agriculture were recognized well before the start of the political crisis and the ultimate demise of the system. Time and again, the planners tried to stimulate production and to increase efficiency by raising the purchase prices of agricultural products or by creating two-tier pricing systems (with higher prices for the above-quota deliveries). They also repeatedly tinkered with the institutional arrangements, with fairly poor results.[103] In Hungary, the most reformist country, planning was formally abolished in the late 1960s, and the state committed itself to purchasing (at different prices) whatever the collective farms decided to produce (Pryor 1992, 417–22). This was probably the most advanced reform that a 'Socialist' system could stand without changing its nature.

The process of dismantling socialist agriculture consists of three main steps: (1) the transfer of management from collective units to peasant households; (2) the liberalization of the markets for outputs and, most importantly, for inputs, which otherwise would have offered juicy monopoly profits to private entrepreneurs; and (3) the privatization of land. The process has always been complex, and arguably it is not yet over. It is impossible to describe it in any detail. It is just possible to sketch out its main features in the two most important cases, China and the Soviet Union, with some hints to the cases of other Eastern European countries (the so-called "transition economies").[104]

In China, the reform started after the return to power of Deng Xiao Ping in 1976.[105] Relative prices of agricultural products were increased by 40 percent in four years, but the real watershed was the transfer of management to households. In some areas, it started in the winter of 1977, and it became a nationwide policy (under the name HRS, or household responsibility system) in 1979–80. The share of HRS households soared from 14 percent in 1980, to 95 percent in 1984.[106] The peasant markets were reopened in 1979, and trade in all agricultural commodities (except grain) was allowed in 1982–83. The state, however, maintained some sort of control of the grain marketing and some planning rights until 1992. The reform process stopped short of full privatization. Formally, most land still belongs to the state, full property cannot exceed 15 percent of the acreage, and collective land is subject to the risk of expropriation and redistribution. Peasant households, however, can rent land with long-term contracts (15 to 50 years), and have the right to sublet and to bequest the farm.

In Eastern Europe, reforms followed the political collapse of the Soviet regime.[107] Markets for goods and inputs were liberalized quite quickly, and imports of agricultural goods soared. In contrast, property rights are not yet complete (in many countries there is no official market for land) and, above all, collective management still prevails, under different names (e.g., joint-stock, co-operatives). The process of carving up collective units into household farms has proven difficult because of the imperfections of the markets for labor and capital and of the resistance of managers and, sometimes, of workers themselves, who have showed little interest for independent farming.[108] Thus, most former Soviet countries have privatized land simply by distributing vouchers to workers and pensioners instead of returning it to owners.[109] In 2000, "individual" land accounted, on average, for about half the total acreage in the eastern European countries but for only about a fifth in former USSR.[110] This has created a dual pattern, dominated by very large (ex-collective) farms and very small plots of mixed origin (former household plots, *etc.*), with comparatively few middle-sized family farms.

As is clear from the data on output (section 3.2) and productivity (section 5.3), the reforms have been much more successful in China than in the transition economies. The growth rate of gross output almost doubled, from about 3 percent per year during the socialist period, to 5.3 percent in the 1980s and 1990s, and the increase was decidedly faster in non-grain production, which had been sacrificed during the socialist period.[111] The consumption of modern inputs, notably fertilizers and machinery, increased much less than output, and thus, according to all estimates, the TFP grew, while it had been stagnating or even declining in the previous period.[112] There is no doubt that reforms contributed enormously to the growth in the early 1980s and little to the subsequent growth, although the exact figures differ from one estimate to another.[113] The performance of the transition economies has been much worse: according to the FAO, their gross output has fallen by a third from 1989 to the present day. Actually, the quantity of inputs—notably fertilizers (table 4.15) fell more than output, so that the TFP shows a modest improvement (Lerner et al. 2003, tables 5, 6; Brooks-Gardner 2004, 581).

Such a difference has understandably attracted a great deal of attention. In a survey paper, Rozelle and Swinnen (2004) list some initial conditions for a successful process of reform. Ceteris paribus, the reforms are bound to be the more successful the more intensive the cultivation (and the smaller the scale of the economies from large-scale farming), the shorter the period under the socialist system, the less strict the economic relationships with other socialist countries (and the stronger the economic relationships with the West), and the lower the degree of state support collective farms received. They also stress that success depended a great deal on the design of the reforms, which had to be suited to local conditions. They emphatically deny the existence of an ideal path, which each country should have followed. Marcours and Swinnen (2000) try to be more precise with an econometric analysis of the agricultural performance of twenty-six former socialist countries in the five years following the reform.[114] The level of development negatively affects output but not productivity, while pre-reform distortions negatively affect both. Policy variables, such as the share of privately owned land and a dummy for the creation of effective property rights, are not significant. The outstanding performance of China could thus be explained by the relatively short time-span of the socialist regime and by the country's "backwardness". In this "structuralist" interpretation, the design and speed of the reforms did affect performance, but they cannot really be considered an ultimate cause, because they were largely determined by the initial conditions. This interpretation may be plausible, but it seems prudent to wait for further work, and perhaps allow the former Soviet countries some additional time to complete their adjustment.

9.7 On the Effects of Agricultural Policies

The defects of socialist agriculture are too evident to dwell much upon them here, but the policies in the market economies have also been extensively criticized as well. The already-quoted Krueger report concludes that in the LDCs "there is ample evidence that the heavy taxation of agriculture both depresses living standards and slows the growth of agricultural output."[115] In the introduction to his book *Agriculture in Disarray*, D. Gale Johnson (1973, 17) sketches out the prevailing conditions of agriculture in OECD countries:

> Products from the land are being produced at high cost in some parts of the world while elsewhere farm products that can be produced at low cost cannot be sold at all or only with great difficulty. The prices of farm products are manipulated by most governments and without any real knowledge of the consequences of such manipulation. In some countries, consumers are forced to pay extremely high prices for many items on their food bill, when comparable products could be made available at much lower prices. Economic relations among many friendly nations are soured by rigid adherence to economically unjustified restrictions upon trade in farm products.

Bovard (1989, 7) puts it in a somewhat more colorful way: "The main effect of [American] farm programs is to force farmers to do inefficiently what they would have done efficiently without subsidies, to force Americans to pay more for food, to drive up prices of farmland (thereby decimating American farmers' competitiveness), and to squander pointlessly tens of billions of dollars a year." But, arguably, the worst consequence of agricultural policies in the OECD countries is on international trade. In the short run, the liberalization of world trade and the likely fall in prices may harm producers in some LDCs. There is no doubt, however, that a real opening of Western markets could improve the growth prospect of the LDCs and also of many of the transition economies.[116] The exact amount of benefits is, however, very difficult to estimate: a quasi-official study of the World Bank suggested an increase of $430 billion (1992 US dollars) in 1993—i.e., roughly the size of the South Korean economy, but other authors promise "only" $128–180 billion in 2000.[117] Last but not least, agricultural protection has been and still is one of the main stumbling blocks in negotiations for the overall liberalization of trade, with potentially far-reaching consequences, which go well beyond agriculture.[118] Thus, prima facie, the price of agricultural policies seems quite high. But, have they done any good?

One can start by recalling that policies were allegedly aimed at reducing the income gap between farmers and/or at stabilizing domestic prices. These are socially desirable aims, and there is evidence of substantial progress toward both targets. The income gap between farmers and other employees has been steadily reducing in all OECD countries, and they (almost) disappeared in the 1990s.[119] In the United States, the total income of rural households increased from a mere 40 percent of non-farm ones around 1940, to gain parity at the end of the 1980s. In the eighteen countries considered by the World Bank research (section 9.4), the variance of domestic prices was substantially lower than that of comparable world ones (Schiff and Valdez 1992, 225–27). This result, however, holds true mainly for food prices, whereas stabilization policies have been much less successful for the domestic prices of exportable goods.[120]

It is not clear, however, to what extent agricultural policies contributed to these achievements. The diffusion of off-farm jobs (section 4.4), the accumulation of physical and human capital, and technical progress, which was faster in agriculture than in the rest of the economy (section 5.3), contributed to the reduction of the income gap. The evidence so far, although quite thin, shows that the liberalization of markets for food crops has not brought about the feared increase in price volatility (Kherallah et al. 2002, 75–79). Finally, one should stress that the same targets could be pursued with more cost-effective and transparent policies—such as direct subsidization.

One could defend, in a somewhat more convincing way, the state intervention stressing its dynamic gains in terms of technical progress and productivity gains, which might offset the static losses from misallocation of resources. As stated, after World War II, agricultural production and productivity increased quite rapidly in the OECD countries and in some LDCs (notably from Eastern Asia). The support for R&D has undoubtedly contributed to this success (section 6.6), even if it is difficult to state to what extent. This may also hold true for land registration, or for the creation of modern property rights, although perhaps not so much for land reforms (section 8.7). But, what about the policies sketched out in this chapter?

An economist would argue that the state support marginal and inefficient farmers, who would otherwise have been forced to change or disappear, afloat, with negative consequences on the aggregate rate of technical progress. Some years ago, Sally Clarke (1994), in a provocative book, argued otherwise: the New Deal policies fostered technical progress because the guaranteed sale of products at high prices changed farmers' expectations and stimulated investment in the latest techniques.[121] Her thesis does not tally well with the data: the increase in investment during or immediately after the New Deal was not so large (investments boomed in the late 1940s), and it accounted for a small part of total output growth (Paarlberg and Paarlberg 2000, 142–45; Gardner 2002, 262–66). A similar argument, however, could be put forward for the effects of the Common Agricultural Policy or any other price-support policy. For instance, Thirtle and Bottomley (1992) attribute the acceleration in TFP growth in the United Kingdom after 1974 to the increase in the relative prices of agricultural products brought about by the CAP, while, according to Huffman and Evenson (1993, 2001), the effect of the support policies in the United States on technical progress (and farm size) was so small as to be negligible.[122] Van der Meer and Yamada (1990, 21–24, 37–39, 63–67) try to reconcile the two views, by suggesting that the relationship between prices and labor productivity follows an inverse U-shaped curve. "High" prices do stimulate technical progress, they argue, but only up to a certain threshold: beyond it, high prices could slow down the transfer of manpower out of agriculture, and thus hamper productivity growth. The authors buttress their idea with a simple econometric test of the effect of high prices on productivity growth. They find a positive correlation, but the test is, admittedly, crude, while the dearth of comparable data on state support prevents a more sophisticated testing.

The support for the Green Revolution seems to be a prime candidate for a positive role of the state. No one reasonably doubts that the adoption of High Yielding Varieties (HYVs) enormously increased the productivity of land and labor, although the performance in terms of TFP has not been similary good.[123] In the 1960s and 1970s, the Green Revolution was highly praised, and Norman Borlaug, one of its leaders, received a Nobel peace prize. In the 1980s, the conventional wisdom changed radically.[124] Critics argued that the Green Revolution widened the disparities between areas well endowed with water, where HYVs thrived, and drier ones, where they could not be used; and that it deepened social inequality within each village. Only rich farmers could afford to buy the HYV seeds, and subsidies and technical advice accrued to the richest, most educated, or simply best connected households. The poor, who really needed help, were left to themselves. Recent research has, however, strongly downplayed these criticisms.[125] The increases in interregional or intravillage inequality have proved to be temporary: eventually, all farmers have adopted HYV if suited to their conditions, albeit with some delay. Moreover, landless laborers have also benefited from the Green Revolution, which has intensified cultivation and thus augmented the demand for labor. More recently, critics have stressed the negative effects of the excessive use of fertilizers and pesticides, the loss of biodiversity, and so on, on the environment as part of the wider discussion of the ecological consequences of modern agriculture (Gleaser 1987; Juma 1989, 100–120). These are undoubtedly serious problems, but, without the increase in yields from the Green Revolution, agricultural production would have never been able to cope with population growth, and the consequences would have been dire.[126] Thus, on the whole, the past benefits of the Green Revolution have greatly outweighed the losses. This conclusion, however, does not necessarily vindicate the present agricultural policies: subsidies for consumption of fertilizers reduce welfare in comparison to the optimal allocation of resources, which, in theory, could be achieved with perfectly free markets. Clearly, such a faith in the markets may be excessive. In the real world of imperfect markets, risk-averse peasants do need some incentives to start using modern fertilizers. But, in many cases, the subsidies have been too high and have been maintained for too long, causing consumption to exceed the optimal level, with an unnecessary waste of precious resources.[127] The (partial) liberalization of the market for fertilizers in Africa in the 1990s did not cause prices to soar and consumption to plummet, as was feared, and thus the negative effects on output have been small.[128]

To sum up, the assessment of agricultural policies must be largely negative, although not entirely so. Indeed, on paper, there was an alternative, a "no intervention," "free market" policy. In this scenario, production growth and inelastic demand would have caused worldwide agricultural prices to fall. Marginal farmers would be have been driven out of the market, and they would have either transferred to other sectors or retired. The least productive land would have been abandoned, and thus could have been converted into woodland, meadows, or other "natural" uses, while farming would have concentrated in the best areas. Labor productivity would have grown, increasing the remaining farmers' income to the skill-weighted level of other sectors. The general fall in the prices of agricultural goods would have increased the welfare of consumers—and possibly their demand for products of a higher quality, made by traditional methods. Last but not least, all countries would have gained from the specialization in the international trade of agricultural products—and clearly these gains would have been much bigger, in comparative terms, for poor countries. The state would only have had to support R&D and extension (like public goods), deal with the environmental consequence of the abandonment of land (e.g., with re-forestation), help former farmers to retrain (or retire if appropriate), and, if necessary, foster competition among buyers in agricultural markets (i.e., prevent oligopsony on the part of a handful of big traders). The outcome, on paper, seems enormously better than the current situation on all grounds. Yet, no country so far has adopted, or seems to be planning to adopt, this free market alternative policy.

9.8 Conclusion: The Political Economy of Agricultural Policies

The previous analysis raises some obvious questions: why has state intervention been so widespread and persistent, in spite of its obvious costs? Why have agricultural policies differed so much among countries, products, and periods? More precisely, why is agricultural protection—both nowadays and in historical perspective—positively related to the level of income, the "developmental pattern" in Lindert's (1991) words?[129]

The literature on these issues is huge.[130] The most theoretically consistent strand is inspired by a rational choice approach—i.e., it assumes that agricultural policies are determined by the interaction between rent-seeking farmers' lobbies and vote-seeking politicians. The most traditional strand of this literature, which also imbues most historical work, assumes that farmers take the initiative. They organize associations and lobbies, which exchange the farmers' vote for subsidies and other transfers. In contrast, a more recent strand argues that politicians actively seek the voters' support for transfers, even without any lobbying by farmers. Some game-theoretic models try to merge the two approaches, modeling policies as the outcome of the interaction between politicians and lobbies. They infer the utility function of the government (i.e., the weight it attaches to the welfare of each competing lobby) from the measures that it adopts. All these models imply that the amount of subsidies is positively related to the gap between income per capita in agriculture and other sectors. The greater the gap, the more likely farmers will organize a lobby or the more sensitive they will be to politicians' bribes. This may explain the "development pattern"—as modern economic growth, at least in its initial stages, increases the income gap. The two main classes of models diverge on the prediction of the effect of the size of the agricultural sector. In his classic book, *The Logic of Collective Action*, Olson pointed out that the larger the group, the easier it is for a potential member to avoid committing himself and, at the same time, to obtain the benefits of lobbying (free–riding).[131] Thus, the level of support should be negatively related to the number of farmers—again, consistently with the developmental pattern. This inference might not be true if politicians take the initiative: ceteris paribus, the more numerous a group (in this case, farmers), the more attractive it is for an enterprising politician. However, the number of beneficiaries cannot increase beyond a certain threshold, because the cumulative cost of protection would arouse the opposition of other voters—consumers or taxpayers.

Many authors have set out to test these models with long-term analysis of the policies of a single country or with cross-sections for groups of countries. For instance, in a well known article on the determinants of U.S. farm policy for seventeen commodities from 1912 to 1980, Gardner (1987) finds a negative relation between his estimate of net gains by product and variables such as the geographical concentration of production, the average product per farm, and the degree of import competition, which he interprets as proxies for the cost and benefits of setting up a lobby. The variable number of farmers, however, is not linear, with the maximum protection awarded to mid-sized lobbies. In a similar work, Swinnen et al. (2001) explore the determinants of protection for eleven main agricultural commodities in Belgium from 1877 to 1990.[132] It was positively related to the share of the product on GNP, negatively to the share in consumption and to world prices (which the authors, *faute de mieux*, interpret as a proxy for trends in the farm to non-farm income ratio). The comparative works do confirm the existence of a developmental pattern.[133] In fact, the level of protection is inversely related to GDP, to various measures of comparative advantage in agriculture (such as the share of agricultural products on exports), and, also to the share of agriculture on GDP or on employment (measures of the size of the sector). In many cases, however, this latter relation is nonlinear, with a peak of around 3 to 6 percent. Protection for agriculture also comes out negatively related to the elasticity of demand for agricultural products, and positively to the farm to non-farm income gap and to the geographical concentration of farmers. This literature, unfortunately, cannot discriminate among competing models. The results differ somewhat from one study to another, as the authors use different sets of explicative variables, which often proxy quite poorly the theoretical ones. The nonlinear association between the size of the agricultural sector (or of the production by crop) is compatible with both the Olson-type argument (small lobbies are easier to organize) and the "active politician" one (too small groups are not worth pandering to). Arguably, the distinction is too subtle to be of much use in the interpretation of the real world. On a more positive note, one can stress that econometric work does show that agricultural protection is a luxury good, which only countries with a relatively small agricultural sector can afford.

The historical analysis can enrich this basic insight in at least six ways.

1. The literature on political economy implicitly assumes that a vote-seeking politician could choose any combination of instruments that he may find suitable to his aims.[134] Actually, the range of feasible agricultural policies is constrained by the capabilities of the state, which depends on the level of development. Tools such as the control of domestic markets or the monitoring of a set-aside program need a huge, expensive organization, with an extensive network of local branches (especially for food crops, which are usually cultivated all over a country). The staff of the USDA grew from one employee for every 330 farms before 1930, to one for every twenty-two in the 1960s (Libecap 1998, 208).

In the "long" nineteenth century, the public opinion resented any government encroachment on its economic behavior; the civil service was relatively small, even in the most "advanced" countries and could not reach millions of farmers. Therefore, the range of feasible policy tools was limited to tariffs and possibly to some sanitation regulation.[135] On top of this, custom duties were a major source of revenue for all Western states and all over Europe agricultural goods accounted for a substantial share of imports. Thus, fiscal needs were a powerful argument for agricultural protection. The same argument holds true for LDCs after independence—with the important difference that taxable trade flows consisted mainly in exports.[136]

The outbreak of World War I did change the role of the state, but the increase in state bureaucracy and market intervention was accepted as part of wartime mobilization, to be dismantled as soon as possible. In fact, most countries did not have the means to manage the emergency support in the 1930s, and thus they resorted to farmers' associations, with different degrees of success. Sheingate (2001) argues that this corporatist approach worked much better in France and Japan than in the United States (after a promising start). He infers, perhaps slightly overstretching his argument, that this quasi-official role can explain at least part of the current power of agricultural lobbies in those countries.[137] Even nowadays, many LDCs are unable to manage their agricultural policies with a minimum of efficiency and equity—and marketing boards are a hotbed of corruption and favoritism (Bates 1981, 40–43; Krueger 1992, 74–87).

2. Agricultural policy is deeply affected by the political system. The theoretical models tend to neglect this point: one major recent survey states that "theoretical and empirical analysis on the influence of political institutions on agricultural policy outcomes is still in its infancy."[138] Indeed, most models, focusing on the behavior of the median voter, implicitly assume a democratic political system, with universal suffrage and proportional representation. This case has been quite an exception in history. In some cases, rural constituencies, and thus farm interests, were overrepresented in the House of Commons before the 1832 Reform Bill in Great Britain, and they are still overrepresented in the Japanese Diet and the U.S. Senate. The right of veto, which each government of the European Community has

on decisions about the CAP, projects the influence of its farmers well beyond their absolute number.[139] In other cases, rural interests were underrepresented. Poor education, scarce information, the power of the landlord or racism (e.g., African Americans in the South) may prevent rural workers from using their vote to defend their interests. In several cases, they were denied voting rights: suffrage was restricted in most Western democracies until well into the twentieth century. For instance, in Italy the right to vote was limited to males with a minimum estate (i.e., mostly landowners) from the Unification to 1882, when it was extended to all literate males (i.e., mainly to urban skilled workers), while the majority of (illiterate) peasants had to wait for another reform in 1911.

Democracy, however, even a limited one, has hardly been the rule in the history, especially outside the Western World. The most common system has been government by traditional rulers or by modern dictators, who base their power on the army, or on small urban elites. They need very little support from the rural masses, while they may find it convenient to keep the prices of food low (at the peasants' expense) in order to prevent unrest in the urban population (see Bates 1981, 30–44, 106–08; Kherallah et al. 2002, 75–79). Dictators also tend to consider any independent organization as a threat to their power, making the organization of peasants' interests very difficult if not dangerous. Long-standing civilian authoritarian regimes, such as the PRI in Mexico, may need more consensus among peasants than brutal military dictatorships, and, to this aim, they could avail of opportunities of patronage from the market-controlling institutions. This reasoning would suggest a positive relationship between democracy and the level of agricultural protection, ceteris paribus—but this hypothesis is not entirely supported by some recent econometric work.[140] Swinnen et al. (2001) find that domestic political variables, such as the widening of franchise, had a modest impact on the level of protection in Belgium (although the CAP did make a difference). Olper (2001) shows that the effect of democracy is positive but quite weak for a sample of thirty-five countries in the 1980s. Protection is affected much more by the quality of institutions—i.e., by the enforcement of property rights. He argues that lobbying is not worth the cost if property rights are poorly enforced.

3. The organization of interests may be easier in agriculture than in other sectors. As a rule, pressure groups can be organized either by sector (e.g., the steel industry) or by factor (e.g., capitalists, workers).[141] The two cases coincide for land, which, unlike labor or capital, cannot move to another sector. Landowners, in theory, have no choice but to organize by sector. In fact, associations of farmers and landowners were among the earliest lobbies in all of Western countries, possibly also because farmers felt more threatened by modernization at large than did others. The British Central Chambers of Agriculture (for landowners) and the French Société des agriculteurs de France were set up in the 1860s, the Farmers Alliance (among conservative tenants) and the (Republican) *Société nationale de*

encouragement à l'agriculture in 1880, the Italian *Società degli agricoltori italiani* in 1893, and the very successful *Bund für Landwirtschaft* in the same year, while, in the United States, farmers constituted the backbone of the populist movements (the Grange, the Farmers' Alliance) in the 1870s and 1880s.[142] These organizations dominated the political discourse about agricultural policies, at least according to the empirical literature on the development of farm policies. It is said that the Agricultural Adjustment Act of 1933 was inspired by the president of the American Farm Bureau Federation, while French and German farmers' associations pushed their respective governments toward a confrontation about wheat prices, which almost rocked the CAP at its beginning.[143]

This lobbying story is plausible, but it should not be overblown. As said at the beginning, lobbies are subject to free-riding, and producers may have different interests according to the factor ratios, which determine their vulnerability to foreign competition. A classic instance is the clash between cattle-breeders and farmers on the price of feedstuffs, which led to the split of the British Central Chamber of Agriculture in 1892 (Tracy 1989, 49). Sheingate (2001) argues that, in the 1950s and 1960s, these divergences caused the AFBF to lose the central role it had had during the New Deal in favor of single-product associations, and that this process greatly weakened the farmers' position versus other organized interests.[144]

4. Interventionist agrarian policies have often been justified by some ideological principle and/or by the superior interests of the nation. In the nineteenth and early twentieth centuries, protection for wheat-growing was often justified with the need for self-sufficiency in the event of war (Koning 1994, 151). Collectivization was inspired by hostility to private ownership and by the belief in collective farming as a step toward the ultimate goal of a communist system. LDCs in the 1960s and 1970s justified their Import-Substituting Industrialization policies and the sacrifice of agriculture as the only way to achieve fast development: the ruling elites and their Western advisors argued that agriculture specialization was a dead-end for development because peasants were insensitive to economic incentives and/or because exports of primary products were bound to grow more slowly than world income.[145] Nowadays, some of these pro-intervention arguments are discredited, although self-sufficiency in food is still a powerful argument in some LDCs, as shown in recent debates in the WTO. These issues have been substituted by others—such as the need to curb the power of food multi-nationals, or the need to support farming in the OECD to preserve both a traditional way of life and the environment (needless to say modern farming has little to do with traditional farming and may be quite harmful to the environment). Ideology, however, has also been important for the fostering of support for trade liberalization. The belief in the beneficial effects of free trade in wheat for urban consumers and the economy at large motivated Robert Peel's repeal of the British Corn Laws, a move that hurt the interests of his own Tory Party. In fact, the party split over the

decision and lost power, which it did not regain for almost thirty years.[146] It seems unlikely that the modern "free-market" revolutionaries of the 1980s (notably Ronald Reagan) took Peel's fate into account, but undoubtedly they refrained from pushing a really radical reform agenda. The influence of mainstream, pro-market, economic thought has been felt much more in the LDCs. Their (macro) economic conditions were much worse, and liberalization was backed by international organizations such as the IMF (the "Washington Consensus"), while their agricultural lobbies were much weaker than those in the OECD countries (Kherallah et al. 2002, 20–22).

5. History matters, probably more than a purely rational choice approach would admit.[147] Agricultural policies have very often been adopted to react to political and economic shocks, such as wars and revolutions, the fall in transportation costs in the late nineteenth century, or the Great Depression of 1929. These circumstances have often been used to justify intervention, which might have otherwise been unacceptable to the public. However, once adopted, these policies have outlasted the emergency, and their scope has been more often extended than reduced, following a well-known pattern of growth in state intervention (the "ratchet" effect à la Peacock and Wiseman [1967]).[148] The opposite case, that of external constraints helping liberalization, is less frequent, but not unheard of. For instance the reforms of the 1990's were motivated also by the need to reach an agreement for the Uruguay round of the GATT (Moyer and Josling 2002, 55–76, 118–43).

6. Last but not least, one has to take into account institutional inertia. Organizations have a vested interest in the perpetuation of state intervention. Milward (1992, 317) speculates that the power of the Brussels bureaucracy has contributed to the survival of the CAP in spite of the decline in the domestic power of agricultural lobbies.[149] Institutions have often perpetuated themselves by changing their role: for instance, the east African marketing boards, set up in the 1930s to foster exports, were widely used to tin agriculture ax them after independence.[150]

The issue is clearly very complex. The theoretical models can account for the overall pattern, but the history of each country can be understood only with a detailed empirical analysis. The issue would be even more complex if one wanted to go beyond the restrictive definition of agricultural policies (section 9.1) and take into account the political economy of macroeconomic policies.

Chapter Ten

CONCLUSIONS: AGRICULTURE AND ECONOMIC GROWTH IN THE LONG RUN

10.1 Fifteen Stylized Facts

The results of this book can be summarized in fifteen stylized facts:

1. Output has increased in the long run, enough to provide more food per capita to a population six times greater than that of 1800.
2. The relative prices of agricultural products rose until the 1850s and remained constant or declined slightly (depending on the series) from then on.
3. The quantity of all factors grew quite fast until the early twentieth century; after (about) 1950, the growth of capital continued unabated, while those of land and labor slowed down.
4. The growth in Total Factor Productivity accelerated throughout the period, achieving very high rates in the OECD countries after World War II.
5. Agricultural production grew thanks mainly to the increase in inputs ("extensive" growth) in the nineteenth century and to TFP growth ("intensive" growth) in the twentieth century.
6. Public investments in R&D and extension have played a major role in fostering technical progress.
7. Agriculture has always been a very competitive sector, because economies of scale are modest, and large farms are plagued by serious incentive problems.
8. "Traditional" property rights on land, which still prevailed throughout the world in 1800, have gradually been substituted by "modern" ownership, but the process is not yet over.
9. Most states implemented land and tenancy reforms in the twentieth century, with mixed results.
10. "Family farms" were already fairly diffused in the nineteenth century, and their share substantially increased in the twentieth century.
11. The average size of farms has fallen in the LDCs throughout the whole period, while, in the "advanced" countries, it remained constant until about 1950, and it has increased fast since then.

12. Markets for factors and goods were quite developed even in traditional agrarian societies and they developed further, well in advance of modern economic growth.
13. The 1930s marked a watershed in agricultural policies, from a period of almost perfect "benign neglect" to an era of massive intervention.
14. After 1950, agricultural policies in the "advanced" countries favored agriculture, at the expense of consumers, while, in the LDCs, they sacrificed agriculture for the mirage of fast industrial growth.
15. Collective socialist agriculture proved to be very inefficient, and the process of collectivization wrought havoc in agriculture, causing great suffering.

The analysis also suggests three general, "theoretical," statements:

i. In spite of technical and institutional progress, agriculture has retained some peculiarities, mainly relating to its unique relationship with the environment, which have constrained the intercontinental transfer of technology.
ii. The adoption, and maybe also the production, of innovations has mainly been determined by the relative prices of factors, but it has been deeply affected also by the peculiarities of agriculture (most notably the interrelatedness and/or the public nature of many key innovations).
iii. Institutions have proved to be quite flexible, although not perfectly so, and have successfully adjusted to the needs of technical progress.

Needless to say, these statements are, by their nature, controversial, and all the stylized facts need further confirmation. Thus, they entail a huge research agenda for future economic historians of agriculture, but the author will refrain from suggesting his own priorities. All issues need further work. It is likely that it will prove the need to revise, possibly drastically, some, or even all, of these results. So far, however, they point to a clear conclusion: agriculture has been a huge success story, somewhat neglected, as it lacks the glamour of other sectors. This last chapter will discuss the relevance of this success story for modern economic growth, shifting its attention from agriculture to the economy as a whole. The question is thus: how much did agricultural growth contribute to overall development?

10.2 Agriculture and Economic Growth: Some Theory

The concept of modern economic growth was put forward by Nobel prize-winner S. Kuznets in his 1966 book, *Modern Economic Growth*, which summarized a huge body of empirical research that he had undertaken in precedent years. He defines the process as jointly featured by the growth in income per capita and by a change in the sectoral shares of GDP and employment.[1] Before the start of modern economic growth, agriculture accounted for most of the economy and, thus, by definition, this structural change implies a *relative* decline in its share of GDP and employment. But, it is said, agriculture, far from being a backwater waiting to shrink, has to perform a key role in the whole process.[2] The definition of this role varies somewhat among authors, but there is a broad consensus on the three essential tasks. The agricultural sector must (1) provide goods to feed the population and to earn foreign currency ("product" role); (2) purchase manufactures, both for consumption and for investment ("market" role); and (3) supply manpower and capital to industry and services ("factor" role). These "roles" are related to each other in the framework of the "trade" balance between agriculture and the rest of the economy. Agriculture receives the proceeds from the sales of products and the returns to the use of its labor (off-farm work) and of its capital (e.g., horse and carts for transportation) in other sectors. It uses the money to purchase manufactured goods for investment and consumption, services from the rest of the economy, possibly including factors (e.g., labor) and to pay taxes (net transfers to the government). The balance between these two flows (the "surplus") appears as a net capital export (or import), which can be used for investment and/or for transfer outside agriculture. Thus, the more balanced the inter-sectoral flows of goods ("product" and "demand" roles) and labor are, the smaller the amount of agricultural capital available for long-term investments in the rest of the economy.

How can agriculture perform its three-pronged role? Clearly, the TFP growth is both a necessary and sufficient condition. Productivity growth makes it possible either to increase output with the same quantity of inputs (jointly performing the product and market roles) or to release capital and labor with the same level of output (the factor role)—or to achieve any range of combinations of higher output and less inputs (the optimal mix depending on demand). It is less trivial to stress that agriculture can perform some, but not all, its roles even without TFP growth (extensive growth). If all factors grow at the same rate (balanced extensive growth), then the total demand for manufactures increases, but agriculture cannot release factors to the rest of the economy, nor increase its supply of goods, unless taxation reduces the standard of living of the agricultural population (or the farmers' preferences shifts drastically—e.g., from leisure to consumption of manufactures).[3] The case of different rates of growth of factor (unbalanced extensive growth) may be more favorable to overall growth. Let us consider, for instance, a very simple two-sector, three-factor economy. Agriculture produces "food" with land and labor, while the rest of the economy "manufactures" with labor and capital. Workers consume food, while landowners consume food and manufactures, and save. Exogenous growth in the agricultural population—that is, work force—causes a shift along the production function (a technical change à la Boserup) and a fall in both labor productivity and in the income per capita of workers. The share of landlords (capital and land) would rise, and they could use their additional income to purchase manufactures (market role) and/or to invest in the non-farm sector (factor role). The per capita consumption of food in agriculture falls, as the decrease in workers consumption cannot be compensated by the increase in consumption by landlords, given the disproportion in numbers. Thus, marketed surplus grows (product role). In this case, agriculture would be able to perform much of its three-pronged role, with the (relevant) exception of the release of manpower. However, this pattern of growth cannot go on indefinitely, as the per capita consumption of agricultural workers cannot fall below subsistence level.

This three-pronged-role framework is simple and intuitively appealing, but it implies two strong assumptions. First, it assumes that agriculture accounts for most of the work force and GDP—that it refers to an economy at the beginning of modern economic growth. In low-to-middle income countries, agriculture must still provide food and perhaps exports, but the agricultural population can no longer be the main market for manufactures, or the main source either of labor or of capital for the rest of the economy. Further on in modern economic growth, agriculture becomes a sector like any other—without any "special" role in further development. Second, the reference to exports not withstanding, the country is assumed to be closed or semi-closed to trade and to factor movements. In an open economy, both capital and manpower can come from outside, manufactures can be sold abroad, and food can be imported.[4] In an open economy framework, the TFP growth can increase the comparative advantage in agriculture and thus push the country toward specialization in agricultural goods, thereby retarding industrialization.[5]

The three-pronged-role framework would raise serious measurement problems even if these two conditions were satisfied. It is relatively easy to measure separately its components—such as the share of agriculture on the demand for manufactures or the proportion of investments of agricultural savings out of the total ones in the non-farm sector. It is more difficult, however, to devise an encompassing measure of the contribution of agriculture to modern economic growth. The research on the issue has been going on for quite a long time. The latest product is the Morrisson and Thorbecke's "surplus," which is the net balance of all product and factor flows between agriculture and the rest of the economy.[6] Although this is useful as a first approximation, especially for comparisons among countries and across time, the "surplus" has some serious shortcomings: (1) As a net measure, it does not take the size of transfers into account. A zero surplus can correspond to no intersectoral transfers at all or to huge, but perfectly balanced, flows. (2) As a static measure, it neglects all past transfers of factors: a household, which had moved from agriculture to industry in the previous year, is treated as a "non-agricultural" one. Such a transfer, the essence of the factor role, may even reduce surplus if the household previously worked part-time in proto-industry. (3) By definition, the surplus cannot measure how important the contribution is for the overall growth of the economy—or, in counterfactual terms, how much faster the economy would have developed if the surplus had been greater. A fortiori, it cannot discriminate among the three roles. It treats all flows as if they were interchangeable, which may not be the case. For instance, a developing industrial sector may need a transfer of manpower from domestic agriculture more than the investment of agricultural capital because, nowadays, labor is internationally less mobile than capital.

To assess the role of agriculture in modern economic growth, one needs a deeper understanding of the mechanism of growth. This is the promise of growth economics, which enjoyed a strong revival in the 1990s, as part of the more general development of dynamic macroeconomics. Two classes of models seem especially relevant for the problem at hand.

The first group consists of long-run models, which aim to explain the big discontinuity in world history, and the beginning of modern economic growth.[7] They assume that there is only one good, which can be produced with different technologies. The "traditional" one (variously christened "agriculture," "primitive," or "Malthus") is featured by diminishing returns to labor, which keep the economy trapped in stagnation. However, there is also a dynamic force—such as exogenous technological progress (Hansen and Prescott), accumulation of human capital (Tamura), the production of "ideas" (Jones), population growth via its effects on market size and returns to specialization (Goodfriend and McDermott), population growth via its effects on technical progress (Galor-Weil), the "Darwinian" selection of growth-enhancing individuals (Galor-Moav), and so on. Given suitable conditions, this dynamic force makes alternative technology ("advanced," "industry," "Solow," etc.) feasible, initiating transition. Once the process has started, "traditional" (inferior) technology is bound to disappear. Thus, with some imagination, one can argue that, in these models, agriculture contributes to growth by "nurturing" the dynamic force. The models are so stylized, however, that they bear little resemblance to the real process of growth.

Other models are somewhat more realistic, or at least they aim to replicate some of the key features of modern economic growth, notably the structural change in the composition of GDP and occupation (i.e., the relative decline of agriculture).[8] By definition, agriculture, in these models, is a separate sector, producing "food." Its consumption grows less than income, or does not grow at all beyond a given (subsistence) level. Under this assumption, any increase in per capita income reduces the share of food in consumption, and thus, in the total GDP, but not necessarily the share of agriculture on the total agricultural work force. To replicate the actual transition pattern, it is necessary to hypothesize that agricultural productivity grows as well.[9] Most authors simply assume the pattern of TFP growth that best suits the problem they want to tackle. Kongsamut et al. (2001) aim to prove that structural transformation is compatible with an otherwise balanced growth path, and thus assume that productivity grows at the same rate in all sectors. In their model, different elasticities of demand are sufficient to produce transition.[10] Kogel and Prskawetz (2001) want to highlight the role of agriculture in starting the transition, and thus they simply assume that its rate of TFP growth jumps from zero to a positive value. Such a jump causes the population to begin growing, and population growth increases the rate of technical progress in the rest of the economy. The agricultural productivity continues to grow at a constant rate, causing the sector to shrink (given constant per capita consumption of food). Interestingly, all these models assume, as does the three-pronged approach, that productivity growth in agriculture is positively related to modern economic growth. The exception is a paper by Matsuyama (1992). He argues that, in an open economy, high productivity growth in agriculture would cause a process of specialization in agricultural goods, and thus slow down the structural change.

As a whole, therefore, growth models do not seem to be that interested in the actual historical evolution. However, a paper by Caselli and Coleman (2001) stands out from this point of view. In fact, the authors set out to explain the structural change in the United States from 1880 to 1990, which they estimate to have accounted for about one third of the long-term convergence in wages between the agricultural South and the industrial North of the United States. In their model, the structural change is the combined effect of inelastic food demand and of TFP growth in agriculture. Contrary to the usual assumption in these models, but realistically (section 5.3), they assume that TFP growth was faster in agriculture than in the rest of the economy. However, relative productivity growth with inelastic demand should have caused agricultural prices to fall. Prices did not fall, Caselli and Coleman argue, because manpower moved out of agriculture (reducing its productive capacity) thanks to an exogenous fall in the cost of education, which made it cheaper to acquire the necessary skills in the non-farm sector.

These dynamic models hint at a future grand theory, which could account for all stylized facts about agriculture and economic growth and also put forth testable hypotheses for historical analysis. Unfortunately, this seems quite far off. Each model tends to focus on a specific set of stylized facts, which do not always tally well with historical reality (Caselli and Coleman are a felicitous exception). The empirical side is limited to theoretical calibration in order to reproduce the selected features. In short, the future might be bright, but the present belongs to the conventional, three-pronged Johnston-Mellor framework. As we will discuss in the next section, it has, directly or indirectly, inspired much of the historical work on agriculture and economic growth.

10.3 Agriculture and Economic Growth: Debates and Historical Evidence

The role of agriculture is one of the key issues in the discourse on development, as much as other "great" topics such as the role of banks, the effect of foreign trade, and so on. It would be impossible to survey the debate on agriculture and modern economic growth in all countries. In this section, we will concentrate on three examples: the United Kingdom during the industrial revolution, France in the late nineteenth century, and the Soviet Union in the 1930s. These countries are relevant by themselves, and the debate can, to some extent, be considered representative of different positions on the issue.

The alleged "agricultural revolution" (section 6.2) is a key component of the traditional narrative on the British industrial revolution.[11] Agriculture, it is said, fed a growing population (avoiding a surge in imports), absorbed a substantial amount of manufactures, and also provided capital for infrastructures, while it released very few workers, as early factories employed mostly immigrants, or women and children. Thus agriculture fulfilled brilliantly the product and market roles, and also some of the factor role. In his well-known book on the industrial revolution, Crafts puts some quantitative flesh on these statements (1985, 43, 84, 125, 139; tables 6.4, 6.7). He estimated that (1) net investments of agricultural savings accounted for about 20 percent of all non-agricultural investments in the 1820s; (2) domestic production accounted for 85 percent of total food consumption in 1801 and for 68 percent fifty years later; (3) the agricultural workforce increased, albeit slowly, so that the sector released very few workers; and (4) agricultural demand accounted for a smallish and declining share of consumption of industrial goods, from about one-quarter in the last decades of the eighteenth century, down to one-sixth in 1851. These figures by themselves downplay the relevance of the three-pronged role, and the author hints at a "Dutch" effect—high agricultural productivity slowed down structural transformation. Crafts's estimates rely on the old Deane and Cole figures on production and productivity growth, and, thus, they are subject to the fallout of the recent debate on the growth of agricultural production (section 3.2) and productivity (section 5.3). The downward revision of earlier estimates implies a parallel downsizing of the contribution of agriculture to the industrial revolution. Robert Allen (1994), a leading revisionist, writes that "most of these functions [the three-pronged role] were not performed by the British agriculture."[12] Agriculture contributed only to the feeding of the British population—and not sufficiently, as imports substantially increased.[13] Some recent papers, however, question even this role. Stokes (2001) argues that the increase in domestic agricultural production was not necessary to feed the British population as the country enjoyed a huge comparative advantage in manufactures.[14] Harley and Crafts (2000) go even further, with the help of a CGE model. England exported textiles and other industrial goods *because* its agriculture was unable to meet the domestic need for food and agricultural raw materials. Total Factor Productivity grew, but the diminishing returns to labor and capital, given a fixed stock of land, prevented a sufficient growth in total output. In short, the recent research turns the traditional interpretation back on its head: agriculture, far from being essential, was, at best, a sideshow. One can only wait for the next round of revisions.

In the 1970s, it was fashionable to blame agriculture for the alleged sluggish growth in France in the nineteenth century.[15] Hohenberg argued that technical progress had been slow, although by no means nil, because landowners were not interested in innovation and the peasants were too fearful of risk. Thus, the demand for investment goods was insufficient. The demand for manufactured goods and the supply of agricultural products for both domestic consumption and export was insufficient because the peasants merely aimed at self-sufficiency. The supply of manpower for industry was insufficient as well, because the peasants were reluctant to leave their family farms, where they felt safe, for the perils of the city. The peasants did save a good amount, but the slow overall growth of the economy reduced the need for capital, which was, in fact, exported.[16] Agriculture was trapped in a vicious circle, as the abundance of manpower discouraged the adoption of labor-saving technology, and kept labor productivity and, thus, also peasants' incomes low. Heywood (1981; 1992, 38–42) and Ruttan (1978) contested this interpretation, shifting the blame for the poor performance from agriculture to the non-farm sector. The slow population growth reduced the demand for manufactures, industry developed slowly, and thus it did not need much manpower and capital from agriculture. Some recent work on labor markets has failed to settled the issue: labor mobility was high for temporary work, but the growing rural-urban wage gaps in mid-nineteenth century suggest that the long-term adjustment was not so smooth.[17] The whole debate, however, rests on the assumption that French agriculture was backward and stagnant. This image owes much to Arthur Young's criticism, but is largely unfair. In fact, French TFP grew as much as that of continental European countries, outperforming the United Kingdom in the second half of the nineteenth century (see Statistical Appendix table IV) and probably also in the first half, at least according to most estimates (table 5.4).

It is possible to interpret the Soviet policy of the 1930s (section 9.6) as an extreme version of the import-substitution industrialization policies pursued by the LDCs in the 1950s and 1960s (section 9.5). The Soviet regime used both its planning powers and brute force to implement the three-pronged role. It sold farmers the product of state industries (the market role); it forced them to deliver agricultural products for the urban consumption (the product role); and it ordered workers to move to cities, and it extracted capital for industrialization by using its monopoly of trade to set unfavorable relative prices for agricultural products, which acted as a tax (the factor role). This policy imposed terrible costs on the rural population, but it also achieved a fast rate of growth in industrial production. The Communist Party held that industrialization was essential to stave off the external threats to the regime, and this imperative justified any sacrifice. This claim was, albeit very cautiously, endorsed by Alec Nove (1969, 378–79) in his authoritative economic history of the Soviet Union. Conventional wisdom nowadays is, however, quite different. Recent works by Russian historians cast doubts upon the effectiveness of collectivization to extract resources for industrialization. A large number of former agricultural workers did move to cities, but it is by no means clear that agriculture provided capital as well. Taxation was not heavy, agricultural terms of trade may have not fallen as is assumed (trends depend on complex measurement issues), and the Soviet state invested massively in tractorization.[18] Gregory (1994, 119–22, 136–37) argues that collectivization was unnecessary to speed up industrialization. Had the NEP continued, without collectivization, the Soviet economy could have achieved similar or even higher aggregate growth rates and undoubtedly a higher aggregate welfare. In a very recent book, Allen (2004, tables 8.3, 8.4) tests the counterfactual of a NEP-style free market in agricultural goods with two different hypotheses on the market for industrial labor.[19] In his baseline case, industrial employment is assumed to be as high as the actual one, because it was determined by the planners. Without collectivization, in 1939 the total GDP would have been 5 percent lower, and non-farm GDP 20 percent lower. In the so-called "capitalist employment model," workers are left free to migrate to cities, and thus urban wages must be high enough to attract workers. In this case, the total GDP would have been 20 percent lower than the actual one, but non-farm output would have been half the actual one. Thus, Allen concludes, collectivization did foster industrialization, although its benefits for overall growth were not that big. In contrast, planning and forced migration to cities were indispensable for both industrialization and GDP growth. As always happens in these exercises, the results depend on the assumptions of the model. In the case at hand, the key assumption seems to be the existence of a large pool of unemployed workers in agriculture. As said in chapter 2, this kind of assumption should be taken with great caution, the more so for a system plagued by a serious incentive problem (section 9.6).

These debates are interesting and intriguing but also highly country-specific, and thus it is difficult to draw any wide-ranging inferences from them. Furthermore, with the exception of Crafts, the authors never try to measure the contribution of agriculture to modern economic growth. There is, however, some comparative work on measurement of the three-pronged role. Broadberry (1998, 390) estimates, with a sort of shift-share analysis, that, from 1870 to 1990, the transfer of manpower from (low productivity) agriculture to (high productivity) industry accounted for 20 percent of total growth in labor productivity in the United Kingdom, 45 percent in Germany, and 50 percent in the United States.[20] This, however, is purely a statistical exercise and does not imply causation. In fact, with a perfectly integrated labor market and no policy interferences, any productivity gap would reflect differences in the sectoral endowment of human or physical capital. The transfer of manpower can increase productivity only to the extent that it is accompanied by an accumulation of capital. The demand role can be estimated with the so-called "growth linkages"—i.e., the increase in local non-agricultural demand from a unit (exogenous) increase in agricultural production. In five African countries in the 1980s, the multiplier ranged between 1.30 and 2.[21] Several authors have measured financial flows, although the Morrisson-Thorbecke "surplus" has not yet been used in historical perspective. Karshenas (1993, 196–221) reviews studies on India (1950–71), Taiwan (1911–60), Iran (1963–77), Japan (1888–37), and China (1950–80).[22] The results differ quite widely, but, in most cases, contrary to the conventional wisdom, the net flows of goods (i.e., the surplus) comes out to have been modest or even negative—in other words, agriculture was financed by the rest of the economy.

Even the best static measure of intersectoral flows cannot capture the dynamic role of agriculture. In theory, one would need dynamic counterfactual analyses, but unfortunately, there are very few of them. Most growth models are backed by simulations, which, however, often show that the model can reproduce the stylized facts that the author has selected. One can, however, quote two relevant historical works. Harley and Crafts (2004) use a CGE model to estimate the effect of productivity growth in agriculture from the 1780s to the 1840s.[23] The growth augmented real wages and GDP by reducing food prices, and slowed down industrialization somewhat—but the differences were small. The second work is Levy-Leboyer and Bourguignon's (1985) macroeconomic model for nineteenth-century France.[24] They assume agricultural production to be determined in the short run by farmers' desired production (function of lagged prices) and by weather, and in the long run by the shift of production function, which is driven mostly by exogenous technical progress. They estimate this supply function separately for 1825–59 and 1887–1913 (but, curiously, not for 1860–1886), yielding a TFP growth rate of 0.9 percent and 1 percent in the two periods. The authors estimate that this growth accounted for three-quarters and one-tenth of the overall increase in French GDP during the two periods. The results are not outlandish: the TFP growth is a bit higher than Grantham's estimates (section 5.3) and a decline in the contribution of agriculture to growth is plausible. This particular model has been strongly, and rightly, criticized (Mokyr and Nye 1990), but dynamic, open-economy modeling is the approach one should follow to estimate the so-far elusive "contribution" of agriculture to modern economic growth.

Waiting for the flourishing field of growth theory to provide a general (and operational) model, one can look for a short-cut. For instance, one could argue that the faster the TFP growth in agriculture, the higher its contribution to modern economic growth. With this standard, one would state that, in the nineteenth century, agriculture contributed more in, say, France or in the Netherlands, where TFP growth accounted for more than all the growth in agricultural output, than in the United States, where TFP growth was very slow until World War I. This argument, however, does not take the stage of development and the degree of openness of the economy into account, and thus, it holds true only in the simplest version of the three-pronged-role framework. An alternative is to assess the role of agriculture with purely empirical comparative econometrics. Timmer (2002, table I) shows that, in a sample of seventy LDCs from 1960 to 1985, the growth of agriculture positively affected the rate of growth in the rest of the economy.[25] The result is suggestive, but the test is very crude, as the author himself admits. The correlation might be spurious, as both agricultural and non-agricultural growth could be caused by some third factor (e.g., macroeconomic policies, or perhaps exogenous technical progress). Temin (2002) finds a positive relation between the rate of growth by country during the postwar "golden age" of the European economy, and the percentage of the agricultural work force in 1950—that is, the potential for the release of manpower. The estimated coefficient of this variable implies that the release accounted for a substantial share of the total growth (e.g., for about half the catching up of Germany relative to the UK). Temin interprets this result as a correction of a disequilibrium in the labor market, which had been cumulating in the 1930s and 1940s as an effect of agricultural policies and of the war. Temin's argument has an interesting implication for the econometric testing of the role of agriculture in modern economic growth. It is likely that the share of the agricultural population in 1950 was correlated with TFP growth in agriculture. In fact, the higher this share, the faster the subsequent transfer of manpower and thus the faster the productivity growth. It is thus likely that substituting TFP growth to the share of agricultural population as explicative variable yield good statistical results. Most authors would interpret this result as agricultural TFP growth causing overall growth. If Temin is right, this conclusion would be wrong: the real cause of growth in both agricultural productivity and economy-wide GDP is the correction of the disequilibrium in the labor market.

This brief survey has to finish on a somewhat disappointing note. It seems likely that agricultural growth did contribute to modern economic growth, but the exact mechanism of this contribution is still uncertain. The conventional three-pronged-role framework is highly questionable, but so far no alternative theoretical framework has emerged. The work on endogenous growth theory offers some promising ideas, but still much work has to be done to have a general model. Moreover, testing such a model would be an even greater and more challenging task.

10.4 Concluding Remarks: A Look to The Future

The upbeat tone of the overall book might seem to be a revival of the optimism of the 1960s and 1970s, when the Green Revolution seemed to open the prospect of a famine-free world. As such, it is at odds with the prevailing mood about agriculture, which is largely negative, at least in the discourse of the "advanced" countries. Modern agriculture is strongly criticized for the adverse consequences of development on the environment—such as excessive use of fertilizers and pesticides, pollution, salinization, loss of biodiversity, and, the latest scare of all, the risk of genetically modified food. The buzzword is "sustainability"—that is, in the original definition of the Brundtland Report, a "development that meets the needs of the present without compromising the ability of future generations to meet their own needs."[26] Some cynical economists are voicing irreverent doubts about the practical meaning of the word.[27] The proponents strongly argue that "sustainable" agriculture is feasible, although they disagree on the acceptable level of compromise with modern technology.[28] On the other hand, there is also a widespread concern about the possibility of increasing agricultural production sufficiently to cope with the likely growth in demand, fueled by population growth, by the need to increase the nutritional standards of the still substantial number of undernourished people, and also by the change in food composition toward beef and dairy products (which are more land-intensive than cereals).[29] Official or quasi-official organizations such as the FAO and the IFPRI (part of the CGIAR network) try to strike a difficult balance between these concerns.[30]

The author, as an economic historian, has no particular competence to assess the risks of traditional, "science-based" research, the potential of alternative "sustainable" methods, or the trade-off between the environment and an increase in output. But history is useful to dampen the enthusiasm for traditional agriculture, which sways the more radical critics of modern agriculture. Traditional agriculture was undoubtedly "sustainable," as it went on for millennia in spite of the occasional environmental disaster. But it had a serious drawback, which is often neglected in the political discourse. It offered a miserable life to hard-working peasants during "normal" years, and was prone to harvest crises with disastrous consequences on the population. In 1800, world population was about one billion people: feeding the current 6.5 billion people with the traditional techniques would need more land than is available, even with a monotonous cereal-only diet, and it would need the work of 75 to 80 percent of the active population.[31] Thus, a return to the past is simply impossible. A purely "biological" and totally fertilizer-free agriculture is bound to remain a dream, or a way for rich people in "advanced" countries to quell their consciences, unless science provides a new breakthrough. The author is totally unable to assess the scientific likelihood of such a breakthrough or the probability of it being accepted by public opinion. The recent scare about genetically modified food does not bode well, but as a (optimistic) citizen, however, the author still hopes that such a breakthrough will be achieved and accepted. The future of humankind is at stake.

STATISTICAL APPENDIX

TABLE I
Production
a) Before 1938 (1913 = 100)

	GDP	*Output*									
		Total	*Livestock*	*Crops*	*Europe*	*North-western Europe*	*Southern Europe*	*Eastern Europe*	*Asia*	*South America*	*Regions of Western Settlement*
1870	53	51.5	44.8	55.3	58.4	70.3	62.9	41.7	64.9	13.4	34.1
1871	52.2	50.9	44.9	54.7	56	67.9	62.7	38.4	65.8	14.2	34.8
1872	53.6	52.1	46.3	56	57.6	70.3	66	38.3	66.8	15	36.3
1873	53.2	51.8	46.4	55.4	56.5	66.5	67.2	39.3	67	16.1	36.9
1874	56.8	55.1	48.6	58.8	62.9	77.5	67	43.3	67.1	15.6	37.4
1875	56.6	55.1	50.4	58.9	61.8	77.8	67.1	39.7	67.5	15.4	38.8
1876	55.6	54.2	50.2	57.5	58.4	70.6	66.5	39.8	67.4	16.3	41.8
1877	58.4	56.7	51.4	60.1	61.6	71.6	70.4	45.2	67.7	16.6	45.6
1878	59.7	57.9	52.6	61	63	73.6	71.4	46.3	67.4	16.9	47.6
1879	57.6	56	51.1	59.1	57.9	65.5	68.4	43.8	68.5	18.1	49.3
1880	59.8	58.2	53.4	61.3	60.4	70.3	72.2	42.9	68.7	18.8	52.3
1881	60.2	58.6	53	61.9	62.4	71.2	72.4	47	69.1	19.2	49.5
1882	62.9	61.1	54.7	64.6	64.1	73	73.1	49.1	73	22.3	53.2
1883	63.7	62.1	57.1	65.2	65.1	75.5	73.6	48.6	72.9	23.8	54.4
1884	64.9	63.1	58.5	66.1	65.9	76.2	72.6	50.2	73	24.8	57
1885	65.4	63.6	58.7	66.9	65.2	77	72.1	47.7	77.5	26	57.1
1886	65.1	63.4	59.4	66.5	65	76.7	74.2	46.6	75.5	26.8	57.7
1887	67.5	65.7	60	69.1	67.8	76.5	74.6	54	80.2	28	58
1888	68.6	66.7	61.8	70	69	77.4	75.6	55.6	80.8	30	59.1
1889	66.7	65.1	62.6	68	64.7	75.8	72.4	47.8	77.5	27.1	63.1
1890	69.8	68	63.5	71.4	68.1	79	73.4	52.5	84.3	28.5	62.7
1891	66.9	65.6	64	68.3	65.2	76.3	76	46.7	74	32.5	65.4
1892	70.3	68.9	64.4	72	69.6	79.8	79.3	52.8	82.2	36.5	63
1893	72.9	71.2	65.9	74.2	74.5	82.9	77.7	62.8	83.9	39	61.9
1894	74.7	72.9	68.1	75.9	75.7	83.9	79.1	64	86.4	45.9	64.2
1895	75.5	73.9	70.1	76.3	76.2	83.6	79.7	65.7	83.5	51.5	68.2
1896	75.6	74.2	72.9	75.4	77.8	86.8	76.5	67.5	74.6	49.1	72.3
1897	77.7	76	73.5	78.5	72.9	81.9	79.6	58.9	91.2	43.7	77.3
1898	82.5	80.7	76.4	83.3	79.3	87	82.2	68.6	94.2	44	80.3
1899	81.1	79.6	78.2	81.3	79.8	90.1	80.4	67	84.3	52.8	81.5
1900	83.3	81.8	79	83.7	82.4	94.2	82.3	68.2	87.8	48.6	81.8
1901	81.8	80.8	80.3	82.6	79.7	89.7	88.1	63.6	86.9	56.3	82.3
1902	84.7	83.6	80.6	85.6	84.4	88.8	87.1	77.8	91.9	53.6	80.2
1903	86.5	85	81.1	86.8	84.5	89.9	86.8	76.9	94.6	67.7	84.7
1904	87.4	86.1	83.3	87.4	85.8	94.2	87.9	74.6	93.9	75.9	85.9
1905	87.1	86.2	85.2	87.3	85.9	93.3	88.6	75.7	89.9	73.9	87.9
1906	89.7	88.7	87.6	90.4	86.3	91.6	92.1	77.1	96.2	74.3	91.9
1907	88.9	88.4	87.6	89.6	89.2	95.6	93.8	79.4	90.1	70.2	87.6

TABLE I (*continued*)

	GDP	*Output*									
		Total	*Livestock*	*Crops*	*Europe*	*North-western Europe*	*Southern Europe*	*Eastern Europe*	*Asia*	*South America*	*Regions of Western Settlement*
1908	91.1	91.1	90	91.8	90.9	98	95.4	80.2	93.4	88.5	90
1909	94.3	94.3	91.7	96	92.8	97.7	97.3	84.7	105.3	85.2	89.4
1910	93.5	93.7	93	94.9	90.7	92.3	90.7	88.8	104.6	80.2	90.8
1911	94.1	94.4	95	95.9	89.9	95.5	97	79.9	103.6	69.8	95.3
1912	98.1	98.6	96.8	99	95.7	97.8	92.8	94.4	103.2	101.9	99.1
1913	100	100	100	100	100	100	100	100	100	100	100
1914							92.4		105.2	90.1	95.6
1915							88.9		107.7	106	105.8
1916							94.7		112.4	93.5	104.9
1917							94.2		110.7	66.9	97
1918							95.4		94.5	108.5	103.1
1919							92.9		112.8	105.5	105.5
1920	85.3	87.8	88.2	88.6	75.5	80.4	97.9	59.3	98.9	111.3	94.3
1921	88.7	92.1	93.5	92.8	75.3	82.3	96.1	57.1	108.2	111.6	100.3
1922	93	96.9	97.4	97.6	81.4	86.4	101.8	66	111.9	112.7	101.4
1923	94.9	98.5	100.8	98.7	84.9	86.4	105.9	73.4	106.7	120.1	105.6
1924	98.6	102.5	105.2	101.8	87	90.1	102.2	76.4	109.9	144.1	112.5
1925	102.7	106.4	108.4	106.4	95.7	93	111.5	91.8	109.6	125.1	111
1926	103.3	107.1	112.2	106.4	94.6	88.8	108	95.7	110.1	146.7	114.7
1927	107.8	110.8	114.9	109.8	100.6	98.2	108.5	99.8	110.7	153.1	119.1
1928	108.8	113.1	117.3	111.6	103.3	101.6	107	103.5	113	163.2	115.5
1929	112.4	116.9	121.5	115.9	108.4	104.9	117.2	108.8	115.1	162.6	117.5
1930	109.2	113.4	119.6	112.6	104.1	102.8	104.2	105.7	117.3	141	112.2
1931	110.8	114.9	120.3	113.4	104.8	107.5	109.5	99.3	114.3	159.7	119.5
1932	109.9	115.1	117.7	114	102.6	105.6	120.2	90.9	115.8	155.5	118.8
1933	112.8	116.9	117.9	115.7	106.5	114.3	109.5	95.5	118	148.1	121
1934	111.1	116.2	117.7	114.5	106.5	114.4	111	94.8	113.3	167.1	117
1935	109.8	115	117.1	113.2	107.3	110.4	115.1	100	114.2	179.2	110.4
1936	110.6	114.8	118.8	113.4	102.7	112.5	94.2	94.5	122.4	169.7	116.2
1937	114.9	121	121.2	118.8	111.6	108.1	107.2	117.9	121.1	191.3	114.1
1938	116.6	122.9	129.3	120.4	112.6	116	106.4	111.2	114.3	178.4	123.3

Source: Federico, 2004a.

Table I (*continued*)

b) Post-1950 (1989–91 = 100)

b.1) Whole World

	Total	*Per Capita*	*Crops*	*Livestock*	*Non-food*
Prewar	30.8	75.3			
1948–52	36.7				
1950					
1951					
1952	38.7	77.5			
1953	40.3	78.3			
1954	40.8	78.3			
1955	42.4	79.9			
1956	44.0	81.5			
1957	44.0	79.9			
1958	46.9	83.1			
1959	47.7	83.8			
1960	49.4	84.6			
1961	49.8	85.0	49.6	49.2	58.6
1962	51.3	86.0	51.4	50.8	61.6
1963	52.7	86.5	52.7	52.4	64.2
1964	54.3	87.5	55.1	53.3	65.8
1965	55.2	87.1	55.7	55.1	69.3
1966	57.2	88.5	57.9	57	66.9
1967	59.4	89.9	60	59.2	67.6
1968	61.0	90.5	61.9	60.9	68.8
1969	61.1	88.9	61.8	61.9	69.4
1970	62.9	89.6	64	63.3	69.8
1971	64.9	90.6	66.4	64.7	71.7
1972	64.4	88.2	65	66.4	73.7
1973	67.8	91.0	70.5	66.6	74.2
1974	68.8	90.6	69.6	69.4	75.4
1975	70.4	91.0	70.8	71	72.7
1976	72.3	91.8	72.8	72.9	70.9
1977	73.8	92.1	74.6	74.5	76.6
1978	77.2	94.7	79	76.7	76.9
1979	78.1	94.1	79.5	78.3	77.8
1980	78.5	93.1	78.5	80.1	77.3
1981	81.4	94.9	82.4	81.2	84.1
1982	84.0	96.2	86.1	82.2	84
1983	84.2	94.8	84.5	84.9	82
1984	88.7	98.3	90.8	87.2	91.6
1985	90.7	98.8	92.4	89.5	94.1
1986	92.3	98.7	93.4	92.1	86.6
1987	93.2	98.0	94	93.8	93.7
1988	94.9	98.1	93.9	96.4	97.4
1989	98.1	99.7	98.4	97.9	95.4
1990	100.7	100.7	101	100.5	99.7
1991	101.2	99.6	100.6	101.6	104.9
1992	103.5	100.4	103.8	101.6	99
1993	104.1	99.5	103.8	103.3	94.9
1994	107.1	100.9	106.7	106.2	95.2
1995	109.3	101.6	108.3	108.9	96.6
1996	113.7	104.2	114.8	110.6	98.6
1997	116.6	105.4	117	113.2	101.2
1998	118.2	105.5	117.7	116.3	95.5
1999	121.4	106.9	120.4	119.2	96.3
2000	122.8	106.8	121.2	120.7	98.3

Table I (*continued*)

b.2) By Continent

	Western Europe	*Eastern* Europe*	*Total° Europe*	*North America*	*Oceania*	*Asia excl. China*	*Asia incl. China*	*Latin America*	*Africa*	*USSR*	*China*
Prewar	40.5			36.6	37.7	25.7		25.4	27.8		
1948–52	43.6			51.5	41.9	26.8		31.6	34.7	36.2	
1950											
1951											
1952	46.2			54.7	44.7	29.3		34.2	38.0	39.7	
1953	50.1			54.7	45.1	30.9		34.5	39.7	41.5	
1954	50.1			53.6	45.6	31.9		36.4	41.3	42.3	
1955	50.6			55.8	48.4	33.2		37.5	41.7	46.3	
1956	51.1			56.9	48.9	34.1		38.9	43.8	50.7	
1957	52.6			54.1	47.0	34.4		40.4	44.2	52.0	
1958	54.1			58.5	54.9	35.7		42.9	45.0	56.5	
1959	55.6			59.1	55.4	37.3		42.9	47.0	57.3	
1960	59.1			60.2	56.8	38.6		44.0	49.9	58.2	
1961	62.0	62.3	62.0	57.9	57.8		38.3	44.1	50.1	63.4	29.7
1962	64.2	61.5	64.1	59.5	63.0		39.3	46.1	52.8	64.1	30.8
1963	65.4	64.3	65.4	62.4	64.7		40.7	47.4	54.7	61.8	32.4
1964	66.2	66.8	66.2	63.4	68.2		42.3	47.2	55.5	68.4	34.9
1965	66.9	66.9	66.9	65.1	67.1		43.4	49.8	56.9	65.5	38.0
1966	69.3	74.6	69.4	66.0	70.5		44.8	49.9	56.6	75.4	40.7
1967	72.8	75.5	72.7	68.0	68.0		46.4	52.8	60.1	76.6	41.4
1968	73.9	77.0	73.9	69.4	74.6		47.8	53.0	61.4	80.6	41.1
1969	73.4	77.4	73.4	69.4	74.5		48.6	55.5	64.0	75.5	41.2
1970	75.1	74.4	75.1	69.0	76.4		50.6	56.8	65.9	81.0	42.9
1971	76.4	77.9	76.4	74.2	78.2		52.0	56.9	67.7	82.6	45.2
1972	77.1	81.8	77.1	74.1	79.6		51.7	57.7	67.1	77.2	44.6
1973	80.9	85.0	80.9	75.5	81.3		54.8	58.2	66.4	91.2	48.7
1974	83.6	86.8	83.6	76.4	75.0		55.2	62.3	71.5	87.3	49.0
1975	83.1	88.4	83.0	80.6	81.6		57.9	64.5	71.5	83.3	50.0
1976	83.4	92.7	83.4	83.9	86.4		58.5	67.7	71.7	90.5	49.7
1977	84.4	92.7	84.4	87.7	84.2		60.6	70.4	71.5	88.0	50.4
1978	88.7	95.3	88.6	87.9	91.8		64.2	71.5	73.5	96.2	55.3
1979	91.7	96.8	91.7	90.9	89.7		65.2	74.0	73.8	87.5	58.5
1980	92.7	93.4	92.7	88.8	84.0		66.6	76.7	75.8	85.3	59.1
1981	91.4	93.8	91.4	97.0	88.4		70.4	81.1	78.1	82.1	62.8
1982	96.2	99.7	96.2	96.4	83.8		73.7	81.5	78.1	88.4	68.6
1983	95.0	98.4	95.0	85.1	93.9		77.8	81.1	77.8	93.0	72.9
1984	100.0	104.1	100.0	95.0	91.8		82.6	84.5	78.3	92.3	80.5
1985	98.1	99.9	98.1	99.8	94.9		84.2	89.6	84.4	92.9	80.6
1986	99.8	104.4	99.8	95.9	95.5		87.1	88.5	88.2	98.6	84.0
1987	99.6	99.9	99.6	95.8	94.7		88.8	92.1	88.6	99.1	88.9
1988	98.8	104.4	98.8	90.8	98.5		93.3	96.6	93.7	99.8	91.8
1989	100.3	99.9	100.3	97.3	97.9		96.2	98.0	97.0	104.5	93.8
1990	99.8	101.2	99.8	101.3	100.9		100.7	99.5	98.0	104.5	101.1
1991	99.9	97.4	99.9	101.4	101.3		103.1	102.6	104.9	91.0	105.1
1992	97.8	84.2	97.8	107.6	104.7		108.0	105.5	102.4	87	110.9
1993	96.0	86.8	96.0	100.5	106.6		113.2	105.8	106.0	83.1	120.7
1994	93.7	80.4	93.7	113.5	101.8		117.9	112.4	108.6	71	128.0
1995	94.4	83.7	94.4	109.4	108.9		123.6	118.9	110.2	67.5	136.6

TABLE I (*continued*)

	Western Europe	*Eastern* Europe*	*Total° Europe*	*North America*	*Oceania*	*Asia excl. China*	*Asia incl. China*	*Latin America*	*Africa*	*USSR*	*China*
1996	97.9	84.1	98.0	114.0	116.0		129.0	120.8	121.8	65.4	144.5
1997	98.3	85.2	98.4	117.5	118.3		133.7	125.1	120.3	66.5	153.9
1998	98.4	84.9	98.5	119.0	121.9		137.3	127.8	124.4	59.3	159.8
1999	99.6	83.0	99.8	121.6	125.9		141.1	135.2	126.8	61.4	164.8
2000	98.2	79.3	98.9	124.2	126.2		143.7	138.0	126.5	61.6	170.7

Notes: * Excluding former URSS and East Germany. ° Excluding former USSR.
Source: FAO: before 1961, estimated from Yearbook; after 1961, FAO Statistical Database.

TABLE II
Rate of Change in Prices, 1870–1938
a) Real Prices

	1870–1913	*1913–38*	*1870–1938*
Denmark	0.00	−2.33	−0.35
France	0.05	0.23	0.21
UK	−0.08	−0.05	−0.22
USA	1.05	−0.77	−0.2
Netherlands	0.05		
Germany	0.00		
Italy	−0.1	−1.65	−0.19
Uruguay	0.18	−0.93	0.48
Japan	−0.21	−0.46	−0.76
Canada	0.11	−0.73	−0.17
Spain	−0.28	−0.87	−0.28

Sources: **USA:** Williamson 2002 and personal communication. **UK:** Mitchell 1988 prices 4 and 5 (series by Sauerbeck and the Board of Trade). **France:** Toutain 1997, series V5 and V43 (implicit deflators). **Germany:** Hoffman 1965, ii, tables 137 and 148 (ratio of implicit deflators). **Uruguay:** Bértola 1998. **Italy:** Ercolani 1969, table XII.1.1.B (ratio of implicit deflators). **Spain:** Prados de la Escosura 2004, cuadro A.4 (ratio of implicit deflators).

TABLE II (*continued*)
b) Terms of Trade

	1870–1913	*1913–38*	*1870–1938*
Argentina	0.96	−1.30	−0.73
Uruguay	1.30	−2.55	1.26
Australia	−0.71	−1.48	−0.78
Canada	1.6	−0.99	0.16
USA	0.63	−1.06	0.15
UK	−0.16	−0.93	−0.46
Denmark	0.39	−3.14	−0.41
Sweden	0.86	0.93	0.67
Finland	0.73	0.79	0.85
Ireland	0.43	−0.79	−0.36
France	0.59	1.18	0.56
Germany	1.28	0.14	0.8
Spain	−0.19	−0.17	−0.21
Italy	0.04	−0.86	−0.51
Austria-Hungary	0.17		
Netherlands	0.49		
Egypt	0.87	0.18	0.28
Thailand	0.67	1.47	0
India	0.76	−0.4	0.06
Burma	1.94		
Taiwan		1.17	
China	2.04	0.08	0.61
Korea		0.28	
Japan	0.6	1.05	0.46
Punjab	0.84	−0.28	0.36

Sources: **China:** Brandt 1989, table 4.3. **Netherlands:** Van Zanden 2000 (ratio of implicit deflators). **Denmark:** Hansen 1974, tables 3 and 4 (ratio of implicit deflators). **Austria-Hungary:** von Jancovich 1912. **Finland:** Hjerrpe 1989, table 6 (ratio of implicit deflators). **Italy:** Ercolani 1960, table XII.1.1.B (ratio of implicit deflators). **Spain:** Prados de la Escosura 2004 Cuadro A.4. **France:** Toutain 1997, series V5/V16. **All others:** Williamson 2002 and personal communication.

Table III
Rate of Change in Factor Price Ratios
a) Wage/Rent

	1870–1913	*1913–38*	*1870–1938*
Argentina	−5.62	0	−4.53
Uruguay	−5.66	2.58	−1.42
USA	−1.52	3.32	2.57
Australia	−2.31	0	0.12
Canada		2.89	
Thailand	−12	−2.26	−0.27
Egypt	−0.95	1.76	−2.7
India	−2.4	−0.84	−2.3
Burma	−3.51		
Japan	2.49	1.1	2.19
UK	1.57	1.27	2.18
Ireland	1.66	5.43	1.7
Denmark	2.78		
Spain	1.41	−1.01	−0.09
France	1.03	1.92	1.69
Germany	0.4		
Sweden	2.37	−0.78	2.15

b) Wage/Interest

	1870–1913	*1913–38*	*1870–1938*
Japan	1.13	3.58	1.64
Canada		1.63	
USA	1.76	12.5	3.1
UK	−0.5	7.23	1.07
Germany	0.95		
Sweden	1.74	0.9	1.46
France	2.07	3.69	1.02

c) Rent/Interest

	1870–1913	*1913–38*	*1870–1938*
Japan	−0.4	2.37	−0.54
United States	3.36	7.51	2.86
Canada		−1.00	
France	−3.43	0.37	−0.53
Sweden	−0.62	10.3	−0.71
Germany	0.35		
United Kingdom		0.64	−1.34

Sources: **Rents and Wages:** Williamson 2002 and personal communication (deflated with country wholesale price indexes to estimate ratios to interest rates); **Interest Rates:** Homer and Sylla 1991.

TABLE IV
Rates of Growth in Total Factor Productivity, by Country[1]
a) "Historical Estimates" (before 1945)

	Before 1870	*Van Zanden*	*1870–1910*	*1910–38*
Europe				
Austria		1.21		
Hungary		1.11		
Portugal			0.71	1.01
Belgium	0.17	0.83	0.38	0.96
France	0.56	0.46	0.75	
Denmark		1.31		
Germany		1.53		
Germany (a)			0.82	0.64
Germany (b)			0.76	1.31
Ireland		0.36		
Ireland (a)	0.61		0.71	
Ireland (b)	0.56		0.59	1.55
Italy		0.37		
Italy (a)			0.9	0.46
Italy (b)			0.7	0.42
Netherlands		0.82		
Netherlands	0.6		0.83	
Norway		0.48		
Poland		0.9		
Russia		0.34		
Russia/USSR (a)			1.18	
Russia/USSR (b)			0.8	
Russia/USSR (c)			−2.3	
Spain	0.34		0.75	1.11
Sweden		1.03		
Switzerland		0.78		
UK		0.19		
UK (a)			0.45	
UK (b)	0.9		0.4	2.1
UK (c)			0.56	
UK (d)	0.55		0.3	−0.05
UK (e)			0.3	1.17
Western Settlement				
Canada			1	0.27
USA (a)	0.4		0.52	
USA (b)			0.53	1.08
USA (c)			0.17	0.5

Table IV (*continued*)

	Before 1870	*Van Zanden*	*1870–1910*	*1910–38*
Asia				
Japan (a)			1.3	0.77
Japan (b)			0.69	0.75
Korea (a)				0.39
Korea (b)				0.33
Philippines (a)			4	−1.4
Philippines (b)			1.34	−1.94
Taiwan				1
China	0.12		0.04	0.04
India				−0.02
Other				
Argentina			−1.9	0.94
Mexico				2.2
Egypt	3.41		0.83	−0.21

Sources: Van Zanden 1991, table 1 **Argentina:** (1900–04, 1910–14, and 1935–39) Diaz Alejandro 1970, table C.3.2 (an average of the two sets of weights). **Belgium:** (1812–1846) Goosens 1992, table 56; and (1881–1913 and 1913–38) Blomme 1992, table 54. **Canada:** (1870–1910 and 1910–1920) McInnis 1986, table 14.8 (average of two different sets of input shares). **China:** (1770–1850 and 1850–1957) Perkins 1969, table S.2. **Egypt:** (1821–1875, 1875–1912, and 1912–36) computed from the data in O'Brien 1968, tables 7, 9, and 10 (assuming shares 0.50 for labor and 0.50 for land and capital combined). **France:** (1789–1870 1870–1910) Grantham 1996, tables 5 and 6 (averages of the two sets of coefficients). **Germany (a):** (1871–1911, 1911–1937) Broadberry 1998 (data supplied by the author, assuming a capital share of 40%). **Germany (b):** (1878–1913 and 1924–38) computed by the author from data in Hoffmann 1965, ii, tables 20, 29, and 46–47 (assuming shares 0.33 for land, labor, and capital). **India:** (1920–1935) computed with production data from Sivasubramonian 2000, table 6.10, and input data from Shukla 1965, table V.5 (assuming shares 0.33 for land, labor, and capital). **Ireland (a):** (1850s–70s and 1870s–1910s) Turner 1996, tables 5.3 and 5.5. **Ireland (b):** (1854–70, 1870–1913, 1913–26) O'Grada 1993, table 30. **Italy (a):** (1881–1913 and 1913–38) Orlando 1969, table VIII.9. **Italy (b)** Federico 2003. **Japan (a):** (1880–1910 and 1910–40) Hayami and Ruttan 1985, table 7.2. **Japan (b):** (1890–1915 and 1915–35) Brandt 1993, table 10, 280. **Korea (a):** (1919–37) Kang and Ramachandran 1999, table 9. **Korea (b):** (1920–39) Ban 1979, table 4.5b (gross output definition). **Netherlands:** (1810–1870, 1870–1910) computed deducting from the output growth from Van Zanden 2000 the input growth from Van Zanden 1994, tables 3.5 and 4.1. **Portugal:** (1865–1913 and 1913–1938) Lains 2003. **Philippines: (a)** (1902–18 and 1918–38) Crisostomo and Barker 1979, table 5.1. **Philippines: (b)** Hooley-Ruttan 1969, table 7.4. **Russia/USSR (a):** (1880–1914) estimate's of the author, with output from Federico 2004a, land and capital from tables 4.1 and 4.7, and work force from Gregory 1994, table 2.3, assuming shares 0.33, 0.33, 0.33. **Russia/USSR (b):** (mid 1880s to 1914) Allen 2004, 28. **Russia/USSR (c):** (1910–1940) Gregory and Stuart 1986, table 20. **Spain:** (1800–1870, 1870–1913, and 1913–1931) computed from data in Brigas Gutierrez 2000, table IV.2 (dual measure) and p. 147.[2] **United Kingdom (a):** (1862–1913) O'Grada 1994, table 6.2. **United Kingdom (b):** (1856–73, 1873-1913, and 1924–37) Matthews et al. 1982, table 8.3. **United Kingdom (c):** (1880–1910) Turner 2000, table 3.33 (simple average of the rates based on two output series, deflated with different price indices). **United Kingdom (d):** (1800–70, 1870–1910, and 1910–40) Clark 2002b, table 8. **United Kingdom (e):** (1871–1911, 1911–37) Broadberry 1998 (data supplied by the author, assuming a capital share of 40%). **United States (a):** (1840–70 and 1870–1900) Craig and Weiss 2000. **United States (b):** (1870–1910 and 1910–40) Kendrick 1961, table 34. **United States (c)** (1880–1910 and 1910–40) Hayami and Ruttan 1985, table 7.2. **Taiwan:** (1913–37) Lee and Chen 1979, table3.5 (gross output definition). **Mexico:** (1922–39) Reynolds 1970, 118.

TABLE IV (*continued*)
b) OECD Countries, after 1945

	(A)	*(B)*	*(C)*	*(D)*	*(E)*	*(F)*	*(G)*	*(H)*	*(I)*	*(L)*	*(M)*	*Others*
Austria			2.57		3.92	3.14		2.2		0.41	1.6	
Belgium	1.7	1.1	4.11	3.7			1.24	3.3	1.7		2.93	
Denmark	1.6	2.3	3.73	4.1	5.34		2.35	2.1	2.6		3.5	
France	1.6	2.3	3.05	4	6.74		2.48	2.6	2.2		1.67	
Finland			3.73	2.2	2.65			3.3			0.05	
Germany	1.4	1.6	3.63	4.3		2.55	1.79	2.8	1.2	0.86	3.59	
Greece			0.2		2.91	1.71	1.55	2.7	0.2	2.01	0.79	
Ireland	1.7	1.3	1.29			2.61	1.93	2.1	1		1.12	
Italy	2.2	2.2	0	2	4.38	1.44	2.29	2.6	0.9	1.04	0.75	
Netherlands	1.5	1.6	2.18	4.4	2.2		1.76	2.5	2		3.09	4.6
Norway			3.05	2.1				2.6			−0.21	
Portugal			−1.61					0.6		−0.29	−1.38	0.7
Spain			−0.2			3.21		3.9		1.37	1.01	
Sweden			3.15	2	4.32			2.8		0.59		
Switzerland			0.5					1.9				
UK	2.2	1.8	2.57	3.6	2.7		1.21	2.6	1.4	1.69	1.42	1.79
UK (a)												1.7
UK (b)												1.88
EC—9	1.7											
Israel			1.29		2.74			4.1			2.59	
Canada			1.29	0.9	2.52			2.0			2.02	
USA			2.08	1.5	1.8	1.76	2.27	2.7	3.1		1.75	1.83
USA (a)												1.76
USA (b)												1.93
USA (c)												2.3
USA (d)												1.57
USA (e)												1.58
Japan			−0.9	−0.2	3.46			1.3	0.3		−0.1	2.4
Japan (a)												1.6
Japan (b)												2.9
Japan (c)												2.7
South Korea					2.89	0.98		3.3		−2.76	−2.73	1.3
South Korea (a)												−0.75
South Korea (b)												3.3
Australia			0.4	1.8	0.84			2.7			2.67	
New Zealand			−0.3					1.3			1.06	

Sources: Column **(A):** (1965–85) Henrichsmeyer-Ostermeyer and Schloder (Oskam and Stefanou 1997, table 9.1). Column **(B):** (1973–89) OECD (Oskam and Stefanou 1997, table 9.1). Column **(C):** (1961–91) Trueblood and Coggins 2004. Column **(D):** (1970–87) Bernard and Jones 1996, table 1. Column **(E):** (1967–92) Martin and Mitra 2001, TL-CRS estimates. Column **(F):** (1970–87) Pryor 1992, table 8.3. Column **(G):** (1973–93) Ball et al. 2001, tables 8 and A-2. Column **(H):** (1960–80) Kawagoe and Hayami 1985, table 1. Column **(I):** (1973–93) Barkaoui et al. 1997, table 1. Column **(L):** (1965–94) Nin et al. 2003b, table 1. Column **(M):** (1962–92) Arnade 1998, table 3 (long-term estimates). Column **Others USA (a):** (1949–91) Craig and Pardey 2001, table 4.1. **USA (b):** (1938–94) Huffman and Evenson 1993, table A7.1 and USDA 1998. **USA (c):** (1949–91) Gopinath and Roe 1997, table 1. **USA (d):** (1950–82) Capalbo and Vo 1988, table 3.1 and 3.5. **USA (e):** (1947–85): Jorgenson and Gollop 1992, table 2. **USA (f):** (1960–90) Ball et al. 1999. **USA (g)**: Acquaye et al. 2003, table A.3. **Japan (b):** (1945–85) Hayami and Yamada 1991, table 1.14. **Japan (c) and the Netherlands:** (1950–85) van der Meer and Yamada 1990, tables 4.9 and 4.14 (final output definition, 1985 prices). **UK (a):** (1949–83) Zanias 1987. **UK(b):** (1967–90) Thirtle and Bottomley 1992, table 1. **UK(d):** (1950–90) Clark 2002b, table 8. **Portugal and South Korea (a):** (1961–85) Fulginiti and Perrin 1997, table 4. **Japan (d) and South Korea (b):** (1965–96) Suharlyanto, Lusigi, and Thirtle 2001, table 14.2.

Table IV (*continued*)

c) Less Developed Countries, After 1945

	(A)	*(B)*	*(C)*	*(D)*	*(E)*	*(F)*	*(G)*	*(H)*	*(I)*	*(L)*	*Others*
Asia											
Afghanistan										−1.61	
Bangladesh	0.66					−4.0	−0.42		−2.01	−0.6	0.78
India (a)	1.1	1.9				−1.6	−0.5	−0.01	−2	−1.51	
India (b)	1.19	1.9									
Indonesia		2.94					0.17		−0.91	−0.8	
Iran								−0.38	−0.76	−0.4	
Iraq								−0.98	−2.59	−2.22	
Jordan										−0.8	
Malaysia					0.4		3.55		1.51	1.69	
Myammar							−0.02		−0.83	−1.01	
Nepal			−0.61		−0.7			−0.9			
Pakistan (a)	1.11				−3.5	−2.5	−0.48		−2.12	−1.21	
Pakistan (b)	1.03	2.3									
Philippines	1.33	1.64			−0.3	1.2	1.33		−0.44	1.19	
Saudi Arabia										0.2	
Syria						−0.8				−2.12	
Sri Lanka		2.38			0.3	−1.1	0.67		−1.29	−0.6	
Taiwan	0.8	2.64				0.9			−0.39		
Thailand	3.17				−6.2		−1		−1.35	−1.11	
Turkey		3.37			2.3	−0.5		−0.76	−1.77	−1.31	
Central and South America											
Argentina					−4.8	0.6			−1.65	−2.63	
Brasil	1.71				−0.5	−0.8		−0.52	−1.68	−0.6	
Bolivia								0.93	6.11	0.4	
Chile		2.73			1.1	1.2		0.84	1.12	1.39	
Colombia		2.97			0	1.5			2.11	1.59	
Costa Rica		−2.82	0.91					1.81	4.25	2.66	
Dominican Republic	2.49			1				−1.09	−0.4		
Ecuador									−0.97	−0.6	
El Salvador		1.43						−0.19	−0.82	0.3	
Guatemala								0.26	−0.5	0.9	
Guyana			−1.46								
Haiti										−0.8	
Honduras		1.28						−0.49	−0.52	−1.31	
Jamaica		0.93								0.4	
Nicaragua									−1.83	−3.56	
Mexico	2.8					0.6		0.9	1.19	0.5	
Panama			2.03					−0.1	1.12	0.4	
Paraguay						0.8			−0.65	−1.11	
Peru		1.99				0.3		0.74	0.51	−0.1	
Suriname										1.69	
Trinidad and Tobago										−1.01	
Uruguay		1.58							−1.25	−0.1	
Venezuela		−2.61				1.6		0.64	−0.1	0.7	
Africa											
Angola				−0.82			−2.57	−1.37		−2.84	
Algeria				2.65			0.77	0.57		−0.4	
Benin				1.22			0.92			1.29	
Botswana				1.35			0.22	−0.22		1	
Burkina Faso	1.57			0.8			−0.83			−1.51	

Table IV (*continued*)

	(A)	*(B)*	*(C)*	*(D)*	*(E)*	*(F)*	*(G)*	*(H)*	*(I)*	*(L)*	*Others*
Burundi	2.68			3.39			0.54			−13.93	
Cameroon	0.98			1.82			1.72			−0.5	
Central African Rep.		1.76		2.75			1.92			0.6	
Chad				0.22			−1.42	−1.42		−1.51	
Congo	−1.4			1.23			−1.46			−0.6	
Ethiopia	0.74			−1.66			−3.85			−1.01	
Egypt		1.86		0.48	0.9	−0.3	−0.38		0.17	0.2	
Gabon				−2.32			−1.58			−16.72	
Gambia				−1.5			−2.12				
Ghana	0.58			−0.51	−4.9		−1.42			−1.61	
Guinea				1.18			0.66	−0.73		−0.9	
Guinea Basu				−2.13			−0.78				
Ivory Coast	1.3			0.9	−6.6		0.63			−0.1	
Lesotho				−1.74			−1.99			−1.71	
Liberia				0.03			−0.74			−2.53	
Lybia				5.6		2.5	4.88	0.81			
Kenya	0.59	2.36		1.85			1.87		−0.87	1.19	0.34
Magadascar		−0.06		−0.11			0.19			0.2	
Malawi	0.57	0.68		0.28			0.2	0.03		−0.1	
Mauritania			0.57	−0.3			−4.93			−10.31	
Mauritius				1.79		−0.9	0.51			0.7	
Morocco		1.02		2.43	−0.1		0.52	1.05		1.39	
Mozambique				0.28			−0.16	−0.74		0.5	
Mali	2.2			0.83			0.84			0.3	
Namibia				0.98			0.13				
Nigeria	1.01			−0.35			−7.98		−2.57	−2.63	
Niger				1.49			−0.69			−4.4	
Reunion				1.16			1.45				
Rwanda	3.16			6.15			1.58			−10.87	
Senegal	−0.04		0.39	1.52			0.93			−1.31	
Sierra Leone	0.32			0.54			0.62	0.08		1	
Somalia			−0.45	1.82			0.51			0	
South Africa		2.76		1.34	2.4		1.48	0.28	0.62	1	1.2
Sudan	−0.31			0.11			−0.08	0.39	2.65	−1.21	
Swaziland			2.11	3.32			3.25				
Tanzania	2.04	5.67		0.24			0.5	0.05		1.19	
Togo	0.08			−1.34			−0.92			−5.55	
Tunisia		2.93		3.78			1.93	0.07		1	
Uganda				7.83			0.79			1.78	
Zambia	−0.98			1.53	−0.1		0.82	−0.01	−2.28	0.3	
Zaire	1.54			8.05			0.61		−2.98	−1.31	
Zimbabwe	0.3	−0.36		2			1.1	0.59	−1.36	0.9	
Sub-Saharan Africa											0.54

Sources: Column **(A):** Pingali and Heisey (2001), table 5.2 (dates India (a) 1950–80, India (b) 1956–86, Pakistan (a) 1950–83, Pakistan (b) 1956–85, the Philippines 1950–80, Thailand 1951–81, Bangladesh 1960–89, Taiwan 1951–81, Brazil 1968–87, and Mexico 1960–91; other countries 1971–86). Column **(B):** (1967–92) Martin and Mitra 2001, table 2 (Column TL-CRS), TL-CRS Column **(C):** (1970–87) Pryor 1992, table 8.3. Column **(D):** (1961–91) Lusigi and Thirtle 1997. Column **(E):** (1961-85) Fulginiti and Perrin 1997, table 4 (Malmquist estimates). Column **(F):** (1960–80) Kawagoe and Hayami 1985, table 1. Column **(G):** (Africa 1961–91 and Asia 1965–96) Suharlyanto, Lusigi, and Thirtle 2001, table 14.2. Column **(H):** (1965–1994) Nin et al. 2003b, table 1. Column **(I):** (1962–92) Arnade 1998, table 3 (long-term estimates). Column **(L):** (1961–91) Trueblood and Coggins 2004. Column **Others:** South Africa (commercial agriculture, 1961–91) Schimmelpfenning et al. 2000, 2; Bangladesh (1948–81) Pray and Zafar 1991, table 4.1; Kenya (1964–96) Gerdin 2002, table 4; and sub-Saharan Africa (39 countries, 1963–1988) Block 1995, table 4.

Table IV (*continued*)

d) Socialist Countries, After 1945

	(A)	*(B)*	*(C)*	*(D)*	*(E)*	*(F)*	*Others*
USSR	0.25	−2.23		−0.1	−0.74	−0.1	1.21
Albania	0.55					−0.5	
Bulgaria	1.39	−1.18	0.6	2.48	0.29	1.98	
Czechoslovakia	1.93	−0.47	0.1		4.29	1.59	0.84
E. Germany	2.09	−0.26	0.3				
Hungary	1.76	−0.53	0.6		2.89	3.83	
Romania		−2.23		0.4	−1.42	1.19	
Poland	0.02	−3.47	−0.9	−0.08		−1.71	1.3
Yugoslavia	2.77	−1.18	3.4			−0.5	1.3
Cuba	0.35					−1.21	
Korea, North	1.13					0.9	−0.3
Mongolia	0.13						0.51
Sao Tome	−3.41						
Laos							1.75
Cambodia					−1.68		−1.86
Vietnam					−1.14		−0.18

Sources: Column **(A):** (1970–87) Pryor 1992, table 8.3. Column **(B):** (1970–1986) Lazarcik (Pryor 1992, table SF1). Column **(C):** (1960–80) Wong and Ruttan 1990, table 2.5. Column **(D):** (1965–94) Nin et al. 2003b, table 1. Column **(E):** (1962–92) Arnade 1998, table 3 (long-term estimates). Column **(F):** (1961–91) Trueblood and Coggins 2004. Column **Others:** USSR: (1950–77), Gregory and Stuart 1986, table 20; Czechoslovakia and Poland: (1967–92) Martin and Mitra 2001, table 1 (column TL-CRS).

e) China, After 1945

	Before Reforms	*After Reforms*	
Wiens 1982	0.8		
Feng 1987		1981–85	0.9
Tian 1987		1981–84	6.4
McMillan et al. 1989		1978–84	5.9
Liu 1990		1978–86	2.8
Zhu 1991		1979–86	2.8
Stavis 1991		1980–89	0.5
Zhu 1993		1986–90	1.3
Wu 1992		1985–91	4.3
Wong-Ruttan 1990	−3.2		
Kalirajan et al. 1996	−5.6	1978–87	6.1
Lambert-Parker 1998	−1.4	1979–95	3.3
Maddison 1998	0.6	1978–94	3.7
Wen	−1.2	1978–87	6
Nguyen-Wu 1999	1.7		
Fan-Zhang 2002	0.12	1989–97	3.28

Sources: Wiens 1982; Feng 1987; Tian 1987; McMillan et al. 1989; Zhu 1991; Stavis 1991; Zhu 1993; Wu 1992; and Kalirajan et al. 1996 from Wu and Yang 1999, table 3.3. Maddison 1998 and Wen from Maddison 1998, table 3.14, Wong and Ruttan 1990, table 2.5, Lambert and Parker 1998, table 2, and Nguyen and Wu 1999, table 4.2.

TABLE V
Contribution of Total Factor Productivity Growth to Production Growth, by Country
a) "Historical Estimates" (Before 1945)

	Pre-1870	*Van Zanden*	*1870–1910*	*1910–1938*
Europe				
Austria		0.86		
Hungary		0.69		
Portugal			0.61	0.82
Belgium	0.19	0.86	0.52	0.59
France	0.7	1.24	1.28	
Denmark		0.74		
Germany		0.91		
Germany (a)			0.5	2.2
Germany (b)			0.48	0.47
Ireland		2.4		
Ireland (a)	*		*	
Ireland (b)	*		2.01	2.62
Italy		0.43		
Italy (a)			0.63	0.79
Italy (b)			0.57	0.53
Netherlands		0.64		
Netherlands	0.58		1.28	
Norway		0.92		
Poland		0.47		
Russia		0.32		
Russia (a)			0.45	
Russia (c)				*
Sweden		0.8		
Switzerland		0.98		
United Kingdom		1.26		
UK (a)			*	
UK (b)				1.12
UK (c)			2.46	
UK (e)			*	2.89
Western Settlement				
Canada			0.39	0.1
USA (a)	0.15		0.21	
USA (b)			0.34	1.83
USA (c)			0.1	0.85
Asia				
Japan (a)			0.74	0.77
Japan (b)			0.39	0.81
Korea (a)				0.16
Korea (b)				0.21

TABLE V (*continued*)

	Pre-1870	*Van Zanden*	*1870–1910*	*1910–1938*
Philippines (a)				0.25
Philippines b)				*
Taiwan				0.28
China	0.2		0.09	0.09
India				−0.03
Other				
Argentina			−0.56	0.32
Mexico				
Egypt	0.74		0.38	*

b) OECD Countries, After 1945

	(A)	*(B)*	*(F)*	*(G)*	*Others*
Austria			2.61		
Belgium	0.77	1		0.84	
Denmark	1.07	1.05		1.11	
France	0.89	1.28		1.56	
Germany	0.82	1.33	1.74	1.89	
Greece			0.81	1.06	
Ireland	0.59	0.62	1.09	0.89	
Italy	1.38	1.38	1.04	1.67	
Netherlands	0.36	0.55		0.61	1.21
Spain			1.18		
UK	0.76	1.29		1.58	1.1
UK (a)					
UK (b)					1.1
EC—9	0.89				
USA			1.08	1.12	0.97
USA (a)					1.11
USA (c)					1.04
USA (d)					0.89
USA (e)					0.83
Japan					1.27
Japan (a)					0.76
Japan (b)					1.78
South Korea			0.41		

Table V (*continued*)
c) Less Developed Countries, After 1945

	(A)	*(C)*	*Others*
Asia			
India (a)			0.33
India (b)	0.38		
Pakistan (a)	0.51		
Pakistan (b)	0.36		
Philippines		*	
Sri Lanka	0.28		
Thailand	0.3		
Turkey	0.55		
Central and South America			
Brasil	0.68		
Chile	0.43		
Dominican Republic		0.38	
Peru		0.7	
Africa			
Burundi	0.42		
Cameroon	2.16		
Central African Rep.	0.59		
Chad	0.82		
Congo	0.33		
Ethiopia	*		
Egypt	0.57		
Magadascar	0.18		0.13
Mauritania	0.24		
Mauritius		0.43	
Namibia	0.67		
Niger	0.72		
Senegal	0.79		
Sierra Leone		0.21	
Somalia	0.2		
South Africa		*	
Sudan			0.43
Swaziland	*		
Tanzania		0.53	
Togo	0.64		
Tunisia	0.05		
Zaire	*		
Zimbabwe	0.68		
Sub-Saharian Africa	0.28		

Table V (*continued*)
d) Socialist Countries, After 1945

	(A)	*Others*
USSR	0.22	0.27
Albania	0.19	
Bulgaria	0.99	
Czechoslovakia	0.83	
E. Germany	0.82	
Hungary	0.79	
Romania	0.02	
Yugoslavia	1.43	
Cuba	0.17	
North Korea	0.25	
Mongolia	0.09	

e) China, After 1945

	Before Reforms	*After Reforms*	
Feng 1987		1981–85	0.08
Liu 1990		1978–86	0.37
Zhu 1993		1986–90	0.28
Lambert-Parker 1998		1979–95	0.53
Maddison 1998	0.29	1978–94	0.79
Fan-Zhang 2002	0.05	1989–97	0.72

Note: * negative growth
Sources: Same as in Statistical Appendix, table IV

NOTES

Chapter One: Introduction

1. Cf. Johnson 2000, p.1.

2. The world population around 1800 was 890 million, according to Clark (1977, table 3.1), 954 million according to Biraben (1979 Table 1), and rose to 1,041 million in 1820 according to Maddison (2001 Table A-c). The figure on the population in 2000 is from the United Nations.

3. FAO Statistical Database.

4. According to the FAO estimate (2001) in 1997–99, there were 815 million undernourished people—777 in developing countries (17% of the population), 27 in 'transition' economies (6% of the population), and 11 in advanced countries. For a critique and an alternative, substantially lower, estimate, see Svedberg (2003, table 6). In 1969–72, undernourished people totaled almost one billion (Runge et al. 2003, table 2.2). All the improvement is accounted for by the spectacular performance of Asian agriculture, which halved the number of undernourished while the population increased by two-thirds.

5. Sen 1981, pp. 1–8, 45–51. See also, on the structural inferiority of tropical agriculture, Gallup and Sachs 2000.

6. Sum of "agricultural workers" in Western Europe, the USA, Canada, Australia, New Zealand, and Japan in 2000 (FAO Statistical Database).

7. Martin 1992.

8. The data are updated to the latest available version on the relevant website, but the period covered is limited to the year 2000 even if the source reports data for later years.

Chapter Two: Why Is Agriculture Different?

1. This truly epoch-making change can be appraised by comparing the description of "traditional" agriculture by Grigg (1982, 51–68), Timmer (1988, 291–300), and Binswanger (1995, 115–17) with the survey of modern environmental economics by Lichtenberg (2002).

2. Cf. FAO Statistical Database. These data are obtained by adding up national statistics on land use. An independent estimate (Wood et al. 2000, table 2), based on the analysis of satellite maps, is about 25% lower (3.6 billion in 1992–93 instead of 4.9). However, it omits fallow.

3. Cf. the survey of early estimates by Cohen (1995, 161–212) and the recent ones by Parikh (1994, 37), Alexandratos (1995, 16), Rosengrant and Hazell (2000, 290–94), Runge et al. (2003, 45–46), and FAO (2003).

4. Cf. Janick et al. 1974, 219–27; National Research Council 1993, 191; Wood et al. 2000, 49; Lindert 2000, 10–13, and Runge et al. 2003, 56–58. On the importance of soil to determine agricultural production, see Gill 1991, 87–88.

5. Cf. Parson and Palmer 1977, 5; Pingali 1989, 255; Pearson 1992, 244; Eicher and Baker-Doyle 1992, 35; and (on a somewhat more sanguine tone) Kates et al. 1993, 15.

6. Cf., Atack and Bateman 1987, 175–82 (yields in American Midwest); Grantham 1989, 56–57 (yields in France); Eschelbach Gregson 1993; and Gregson 1996 (crop mix in Missouri).

7. In a very recent paper on late 18th-century England, Brunt (2004, 209) finds a significant relation between wheat yields and soil quality only for the poorest soil.

8. Anthony et al. 1979, 27. Cf. also, Gill 1991, 85.

9. Wood et al. 2000, 55.

10. Gill 1991, 31–36.

11. Duggan 1986, 69.

12. The foregoing discussion assumes that these constraints are roughly symmetrical and that no type of climate is intrinsically superior to the other. Masters and McMillan (2001) disagree on the basis of an empirical growth model. The rate of growth in GDP in 1960–90 turns out to have been positively related, inter alia, to the number of days of frost—i.e., a temperate climate is more conducive to development than a tropical one.

13. Grigg 1982, 83.

14. Cf. Shiel 1991, 61–63. On diffusion in nineteenth-century England, see Turner et al. 2001, 81–83, and Brunt 2004, 212–14; for nineteenth-century France, see Grantham 1978, 321. The farm consumption of lime increased tenfold in the United States from 1910 to World War II (Historical Statistics 1975, series K194).

15. Cf. the discussion of the concept of "distance" by Sunding and Zilberman 2001, 236; for an application, see Englander 1991.

16. Cf., Russell 1973, 23–59; Shiel 1991; Smil 2001, 21–35. Nitrogen and phosphorus constitute the plant fabric, while potassium regulates the photosynthetic process and the synthesis of aminoacids and proteins. The effect of nutrients depends on a variety of factors—notably the supply of water, the amount already available (beyond a certain level additional quantities would be useless), and the amount of other nutrients. Liebig (Section 6.2) stated the so-called law of minimum: the plant growth is constrained by the minimum amount of each of the three basic nutrients.

17. Boserup 1951, 15–16; Allan 1965, 30–35; Grigg 1982, 37–43; and the useful table 1 of Pingali 1989.

18. Cf. the descriptions by Morgan 1969, 248–59; Hopkins 1973, 31–34; Anthony et al. 1979, 116–35, and Cleary and Eaton 1996, 52–55 on Asia; cf. also the references in note 2 of chapter 7.

19. The whole cycle (crops and fallow) is called rotation. The growth cycle for all major plants lasts less than one calendar year, and this leaves some time to grow a minor crop (intercropping). In traditional European agriculture, this practice was known but rather unusual.

20. Cf. Bray 1986, 16–18; for case studies, Huang 1985, 57–58, Francks 1984, p.106–107 (and Marks 1998, p.283 on Westerners' reaction to eighteenth-century China). Taiwanese farmers since the 1950s have succeeded in growing rice four times per year (Ishikawa 1981, 51), but with the help of fertilizers and modern, early-ripening varieties of plants.

21. Cf., Gleave and White 1969, 285; Miracle 1967, 40. The supply of animal manure was structurally scarce. In many areas, cattle-raising was impossible because of the diffusion of the the tse-tse fly or for lack of suitable food (especially in forests), while in others it was the task of specialized tribes (cf. the literature quoted in note 3 of chapter 7).

22. Chorley 1981, 80–81. Constant and abundant manuring is sufficient to maintain soil fertility: the same fields of Rothamstead, a famous English experimental farm, have been cultivated with wheat since 1843 (Russell 1973, 33).

23. Van Zanden 1994, 92, and also Knibbe 2000, 43, de Vries and van der Woude 1997, 203–5. Cf., Turner 2001 et al., 83, and Thompson 1968, 77–79 for Great Britain; and Goosens 1992, 287–92 for Belgium.

24. Cf., Bray 1986, 48, Kostrowicki 1980, 289, Perkins 1969, 70, and Pomeranz 2000, appendix B, 226–27 and 302–6 on China; and Francks 1984, 109 on Japan. According to Hayami and Yamada (1991 Table A.15), these nonchemical "commercial" fertilizers accounted for about 10% of the total supply of nitrogen in Japan in 1883–87, and for about 18% in 1933–37 (when artificial fertilizers accounted for 27% of the total).

25. Sivasubramonian (2000, 125) estimates that only 40% of the available manure was used in Indian agriculture. Furthermore, night soil was considered impure, and only a special caste could touch it. Blyn (1966, 194) and Richards et al. (1985, 727) argue that the available quantity has been falling in the 19th and early 20th centuries, as population was growing and forests were being cleared.

26. Cf., Eicher and Baker-Doyle 1992, 83.

27. The seasonal fluctuations would be even greater in an integrated definition of "cattle breeding," which also includes the production of feedstuff.

28. Cf. ISTAT 1934, I, 7, and Gill 1991, 100–102.

29. Cf. the thorough discussion of seasonality by Gill 1991, 104–27.

30. The exact shape of the transformation curve is controversial (cf., Earle 1992 and Temin 1982).

31. For further data, see section 4.4.

32. Cf. Nurkse 1953 and Lewis 1954, and, for further discussion and surveys of the literature on overpopulation and productivity in agriculture, see Reynolds 1977, 11–14; Grigg 1982, 21–27; Ghatak and Ingersent 1984, 50–62; Booth and Sundrum 1985, 226–35; Eicher and Baker-Doyle 1992, 83–85; Fei and Ranis 1997, 120–21.

33. The opportunity cost of peasants' labor remains positive even if they have no productive work on their farms, as they have to forfeit leisure. The seasonal unemployment in winter does not mean that labor productivity over the whole year is zero, and thus that workers can be permanently employed in industry. In this case, the total agricultural work force would fall short of demand at peaks, and thus agricultural production would fall.

34. Solomou and Wu (1999, 2003) estimate that between 1850 and 1913, weather accounted for about one-third of fluctuations in agricultural output in Germany and France and for two-thirds in the United Kingdom, with substantial effects on total GDP and prices. See also Stead 2004, 335–39.

35. Francks (1984, 110) quotes an even higher figure—on average 20–30% losses—for the rice crop in the Saga plain (Japan). Cf., also, the estimates of potential losses without pesticides by Wood et al. 2000, 36.

36. If there were no imports from other areas, the price increase would perfectly offset the (exogenous) fall in output, and peasants would neither gain nor lose if the demand elasticity were unitary. They would gain if the demand elasticity were less than one—a case that is far from implausible for food crops.

37. From 1815 to 1818, real prices of agricultural products rose by 15% in the United Kingdom, by 20% in the United States, and by 30% in France and in the Netherlands (see table 3.4).

38. The monsoon arrives in southern India around May 25 and then proceeds northward to the eastern border around July 15—retreating four to five months later (from September

1 to January 1, according to location). Cf. McAlpine 1983 (22–26, 205–10 and tables 3.13, 6.2) and the list of famines in Visaria and Visaria 1983, appendix 5.2 (the estimates on death from Roy 2000, 294). Blyn (1966, 183) argued that the 1910s and 1920s, as a whole, were unusually dry, causing a reduction in rice yields. The delay of the monsoon negatively affects production even nowadays (Rosenzweig and Binswanger 1993, 64–65; Lamb 2003, table 3).

39. For China, see Ashton et al. 1984; for Rwanda and Burundi, see Martin 1987, 113. Needless to say, all these figures are highly imprecise.

40. Cf. Mitchell 1998a, table C5.

41. The Irish population dropped from 8.2 million in 1841 to 6.8 million in 1851 and has never recovered.

42. For diseases affecting vines, see Agulhon et al. 1976, 389; Moulin 1991, 93–95; Loubère 1978, 79–81 and 154–80; Unwin 1991, 283–296; Voros 1980, 103–5; Olmstead and Rhode 2000b. For the boll weevil, see Osband 1985; Hurt 1994, 226–28. For *pebrine*, see Federico 1997, 36–39. For the rinderpest, see Duggan 1986, 117 (South Africa); Parson and Palmer 1977, 17 (East Africa); Iliffe 1979, 123 (Tanganyika); McCann 1995, 92 (Ethiopia). For the swine fever, see Gunst 1996, 45 (Hungary). For the banana disaster, see Bulmer-Thomas 1995, 175.

43. This statement refers to increasing variability from poor cultivation practices. An altogether different issue is the degree of vulnerability to diseases of different types of seeds. It is often argued that "modern" varieties (section 6.2) are more vulnerable because they all share the same genes. This claim is probably unfounded (Smale 1997, 1263–65).

44. Cf. Hazell 1989 (all cereals, 1960–82) and Singh and Byerlee 1990 (wheat only, 1951–86), and, for a general discussion, Anderson and Hazell 1989. Yield variance is clearly higher in dryland areas.

45. The American coefficient was computed from Historical Statistics 1975 series K 506 and K 507 and FAO Statistical Database, the British one from Austen and Arnold 1989, 102. See Hazell 1989 for an analysis of worldwide trends in the 1960s and 1970s.

46. Cf. the vivid description of the American Midwest in Bogue 1963, 105–106.

47. Cf., Loubère 1990, 40.

48. Cf. the description in Biagioli et al. 2000.

49. Cf., Duckham and Masefield 1970, 100–107; Pearson 1992, and Wood et al. 2000, 24–30. See for further examples, Grigg 1974, 4, and Kostrowicki 1980, chap. 3.

50. In contrast, the correlation is weakly positive (0.33), based on the data on labor and land productivity in 1995 dollars for the period 1990–94 for 135 countries, available from world development indicators (www.worldbank.org). The sample, however, includes tropical and temperate countries, advanced and traditional, so that the assumption of the same production function is questionable.

Chapter Three: Trends in the Long Run

1. Cf. Federico 2004a for additional references and information.

2. The countries are Australia (1828–70), Austria (1830–70), Belgium (1812–70), Denmark (1818–70), France (two alternative series, from 1803–12 to 1870), Germany (three series, for different periods), the Netherlands (two series, 1808–70), Greece (1848–70), Poland (1809–70), Portugal (1848–70), Spain (two series, one since 1800 and another,

1850–70), Sweden (1800–70), and the USA (1800–70). There are also several benchmark estimates for England.

3. Trends in British production are actually quite uncertain, but the recent estimates have revised downward the rates of growth—below the population growth. See Allen 1999; Clark 1993, 2002b. For a recent survey of the discussion, see Turner et al. 2001, 9–26.

4. Bairoch 1999, table 1.2; population data (on Europe, North America, Oceania), is from Biraben 1979 and Maddison 2001, Appendix A.

5. Heights depend on net nutrition during the period of biological growth (until 17–18 years of age)—that is, the difference between caloric consumption (determined by agricultural production) and claims from workload and disease.

6. According to Maddison (2001, table B-17), from 1820 to 1870, the Japanese population grew by 10% and production by 20% (cf. Hanley and Yamamura 1977, 70–74). The Chinese population rose from about 340 million in 1800 to 410 million in 1840, and then fell, after the Tai'ping Revolt, to less than 360 million in 1870, to exceed 410 million again only forty years later (Maddison 1998, table D-1). Production may have increased a bit less than population (Richardson 1999, 20).

7. From 1800 to 1850, the combined total of Asia, Africa, and South America rose from 750 to 925 million people, according to Biraben (1979), or from 700 to 880 million people, according to McEvedy and Jones (1978)—corresponding to growth rates of 0.42% and 0.46% respectively. According to Maddison (2001, table B-10), from 1820 to 1870 the population of the overseas LDCs increased from 805 to 895 million—i.e., at 0.21% yearly only (due to the consequences of the Chinese disaster). In the same years, the population of Eastern Europe increased from 95 to 145 million (0.85% yearly). Needless to say, all these figures are highly tentative and have to be used as a rough order of magnitude.

8. Cf., Federico 2004a. The gross output is computed as the total market value of all products, net of re-uses within agriculture itself for seed and feed but inclusive of the farmers' domestic consumption (Rao 1993, 12–14). It differs from the Value Added (used to compute the economy-wide GDP) because it includes the cost of input purchased out of the sector (fertilizers, fuel, etc.). In short, it measures the capability of agriculture to provide food, clothing, and heating, while the Value Added its capability to create income. Both indices are computed as weighted averages of country series—the weights being the country share on total value of world output in 1913 (converting national currencies into British pounds at the 1913 exchange rate).

9. The figures refer to gross output to be comparable with the data published by the FAO. Value Added has been growing a bit more slowly for the increase in expenditures (discussed in chap. 4). For more details, including an analysis by country, see Federico 2004a.

10. Cf. respectively Chao 1986, 216 (1880–1950) and Wang 1992, table 4.1 (1887–1933). For further details, see Federico 2004a and the survey by Richardson 1999, 31–39.

11. The share of world production in 1970 (for successor states) is from Rao 1993, table 5.4. This alternative weighting would be more accurate if production per caput grew parallel in the twenty-five countries and in the rest of the world from 1913 to 1970.

12. Computed by the author from data in Maddison 2001, tables B-10 and B-18. Reynolds (1985) also argues that intensive growth had already started by 1870 or was about to start in all countries. In east Africa, production per capita of grains in interwar years remained constant and exports per capita boomed (Mosley 1983, 71).

13. Data are from the WTO website (www.wto.org) and UN Demographic yearbook.

14. According to the latest estimates (Fan and Zhang 2002, table 1), the production in 1961–62 was 30% lower than the 1957–58 one, which, according to Perkins (1969, table ii.8) was 25% greater than in 1933. If both estimates are right, output in 1961–62 was about 6% higher than in 1933, with a population 30% greater. The causes of this different performance will be dealt with in section 9.6.

15. Wiggins (2002, 101–3) cautiously raises some doubts about the FAO data and consequently on the very pessimistic conventional wisdom.

16. Cf. Evenson and Pray 1994, Minde et al. 1999, 29–31.

17. Cf., Federico 2004a, table 6; Rao 1993, table 5.4.

18. The use of urban series adds a further bias, which is unfortunately impossible to assess. Farm-gate prices differ from urban ones by the amount of transportation and marketing costs—which can decline or increase relative to commodity prices. Furthermore, not necessarily an urban index, weighted with the consumption basket of urban consumers, reflects preferences of rural consumers and the actual crop mix. Last but not least, the national market might not be fully integrated—so that the prices in a couple of big cities would not represent those in all rural areas. These differences can account for some divergences in price indices.

19. This hypothesis relies mainly on the greater share of administered prices in services and on the higher labor content in services. If productivity growth in services lagged behind manufacturing and agriculture à la Baumol, service prices would rise faster (or fall slower) than manufacturing prices.

20. The series for Germany and Spain, which start in 1850, are therefore hardly representative. For further data and a discussion of the political consequences of the price shock, see Berger and Spoerer 2001, 296–306.

21. This estimate belongs to a long-standing discussion about long-term trends in external terms of trade (the ratio of export to import prices) of LDCs, which started in the 1950s. Prebish and Singer argued that they had been declining since the late nineteenth century, and inferred that specialization for exports was harmful in the long run. The latest contributions to the debate (Bleaney and Grenaway 1993; Hadass and Williamson 2003) find little evidence of a long-run downward trend and stress the negative impact of some short periods of fall—such as the interwar years or the 1980s. Foreign terms of trade differ from the agricultural ones considered here, as they also include other goods both on the export side (e.g., minerals) and on the import side (e.g., agricultural products).

22. As expected, terms of trade varied more than real prices. A direct comparison is possible in 21 cases (11 for 1870–1913 and 10 for 1913–38): in all but one (Italy in 1913–38), the two series have the same sign or (one or both) are not significantly different from zero, and in all but three (Italy in both periods and the United States in 1870–1913), the absolute rate of change is greater for terms of trade than for real prices (see Federico 2004b).

23. Terms of trade were lower in 1920–22 than in 1894–96 in 9 out of 20 countries but the difference was substantial in 4 cases only, Australia (a 22% fall), Ireland (23%), Italy (30%), and Thailand. This latter series features an unusual (and somewhat suspicious) steady decline in terms of trade from a peak during the late 1890s onward.

24. There was no full recovery in 8 countries—Argentina, Australia, Denmark, Germany, Italy, UK, Ireland, and Taiwan—and only in these two latter cases did the difference exceed 10%. Real prices confirm this trend: at their peak, they were lower than in 1911–13 in 3 countries out of 10 (Germany, Italy, and Japan) and higher in the other six—Canada,

France, Denmark, USA, UK, and Uruguay. In the last two cases, real prices exceeded the prewar peak throughout the 1920s. The Lewis index of terms of trade for cereals in 1925–27 exceeded the 1911–13 level by 5%, with a 30% recovery from 1921–23.

25. Real prices decreased by 23%—i.e., as much as terms of trade for the same group of countries.

26. Cf., Williamson 2002. The reasoning assumes that price convergence affected more agricultural products than manufactures. It also assumes that the underlying world terms of trade were rising.

27. See Grilli and Yang 1988 (table 3.5), World Bank 2003 (ratio of implicit deflators of agriculture and industry), and WTO (www.wto.org—prices of agricultural products and manufactures in international trade).

28. The implicit real prices of agricultural products (i.e., the ratio between the implicit deflators of industrial VA and GDP) fell even faster than the terms of trade—at 2.82%. This fact does not necessarily contradict the hypotheses about the ranking of rates of change among different indices. In fact, after 1972, the weight of agriculture on total VA in the "advanced" countries (which account for most of world output) has been minimal. The real prices are thus closer to the "true" index.

29. Mundlak and Larson (1992) show that, from 1968 to 1978, world commodity prices transmitted almost entirely to domestic prices, and this accounted for a substantial share of changes in domestic prices.

30. In theory, demand can shift for the same aggregate income, even if the consumers' preferences and/or income distribution change. The demand-shift could exceed the income-increase (times income elasticity) if distribution of income-shifts becomes more equitable or if consumers shift their preferences toward food (quite an implausible change). For a discussion of the determinants of income elasticities, see Booth and Sundrum 1985, 40–53.

31. The "ceteris paribus" terms of trade would worsen the more the less elastic demand and supply are in agriculture (relative to the same parameters for manufactures and services). The price demand for agricultural goods is relatively inelastic, as food consumption is clearly constrained for biological reasons. Agricultural supply may be inelastic as well, as it is constrained by the stock of land, which is fixed in the short run.

32. The countries are Australia, USA, France, Italy, Russia, Japan, Spain and the UK. See Federico 2004a.

33. See FAO, Statistical Database. The source does not report data on the value of output, and thus it is not possible to compute the share of raw materials on output.

34. The share of livestock products on gross output in purchasing power parities can be computed in 1913 for a total of 50 countries, the 25 in the sample and for many others, including China (22%), Mexico (26%), and Turkey (35%). It comes out to have been 45.8% in the first 25, and 25.2% in these additional 25 countries. See Federico 2004a, Appendix A. In 1970, livestock products accounted for 46.4% of gross output in the first 25 countries (or in their successor states, if appropriate), for 25.8% in the additional 25, and for 29.7% in the rest of the world (Rao 1993, table 5.4).

35. The computation assumes that livestock products always accounted for 25% of output in the rest of the world (see note 34, this chapter) and the latter accounted for 45% of world gross output in 1913 (note 11 of this chapter). Needless to say, these assumptions are questionable. The share of the rest of the world on the total output might have been lower, or that of livestock products on the output of the rest of the world might have risen. However, this case seems implausible, if, as argued in the text, the increase in the livestock products

was demand-driven. Per capita income surely grew much less in the rest of the world than in the 25 countries.

36. In 1970, livestock products accounted for 39.4% of the world gross output if the USSR is included and for 34.1% if it is excluded (Rao 1993, table 5.4). The share without the Soviet Union grew to 38.8% in 1990. Also the total share must have grown in the same years, because according to the FAO indices, which include the USSR, livestock production rose slightly faster (2.32% per annum vs. 2.28% for crops).

37. The ratio between input (the calories in feedstuffs) and output (calories from dairy and meat products) may vary greatly according to the animals' diets. Estimates for cattle range from a minimum of 4.5 to 16 calories (Cohen 1995, 180). On the likely effects of changes in composition of food demand and the methods to cope with them (such as encouraging the consumption of poultry), see Pinstrup-Andersen et al. 1999, 13–14; Runge et al. 2003, 53–56; Rosegrant and Hazell 2000, 178–90.

38. This long-term reconstruction uses the Lewis series for 1850–1913, the Vidal series for 1925–38 (spliced to 1913 with data from Vidal 1990), and the WTO data for 1950–2000 (linked to 1938 with data on exports of market economies from the UN Statistical yearbook).

39. The value of agricultural trade in 1870, 1913, and 1938 is estimated by assuming that agricultural products accounted for 75% of Lewis's primary products in 1870 and 1913, and that Vidal's foodstuffs accounted for 60% of agricultural trade in 1938 (these figures are from Yates 1959, table A 17). The value of world gross output is obtained by dividing the estimate of output of the 25 countries from Federico 2004 by their share of world production. Total agricultural exports amounted to 46% of agricultural GDP in 1996–2000 (World Bank 2003)—i.e., to 37% of gross output if worldwide expenditures accounted for 20% of the latter, or to 28% if they accounted for 40%. Peterson et al. (2000, 371) quote an official USDA estimate of 25% for 1990.

40. See Yates 1959, table A.17 for 1913, 1929, and 1937; and UN data as elaborated by the World Trade Analyzer for 1997–99.

41. In a recent article Peterson et al. (2000) show that the pattern of net trade for some key agricultural commodities in 1992 was determined by factor endowment, while policy variables are not significant. However, policies might have still affected the level, if not the composition, of trade.

Chapter Four: Patterns of Growth

1. These data take boundary changes into account as much as possible: the figures for "India" are the sum of present-day India, Bangladesh, and Pakistan; those for "Russia" include all the post-Soviet states; and those for "Austria" include present-day Austria, Hungary, (former) Yugoslavia, the Czech Republic, and Slovakia. It is impossible to take all minor boundary changes into account, and thus the data are not exactly comparable.

2. To be precise, it would remain constant unless and until land was taken away from native tribes and transferred to the state. This act moved land out of the productive stock and into the public domain just to return it to the stock when it was assigned to new settlers.

3. See Hurt 1994, 99, 178–80; Grigg 1974, 267–76; and the maps of location of agricultural production by Paullin reproduced by Atack and Passel (1994, 284, 404–6).

4. Historical Series 1975, series J 51. The figures probably understate the true amount of "used" land, as they are likely to omit some land occupied by squatters without a legal title, and also the public domain used for grazing.

5. For a general overview of the process, see Grigg 1974, 276–283. For more information and data for Argentina, see Cortes-Conde 1997, 47–50; for Australia, Shaw 1990, 4–12; for Canada, Buckley 1955, table VI, and Adelman 1994, 176–77; and for New Zealand, Bloomfield 1984, table I.4.

6. Pavlovski 1968, 161–79; Grigg 1974, 265–66; Treadgold 1957 (the data from pp. 31–35); Gattrell 1986, 64–66; Moon 1999, 49–59. In the 1950s, the Soviet regime engineered another wave of colonization, toward Central Asia, the so-called "virgin land campaign" (see note 28 of this chapter).

7. See Gottschalk 1987; Eckstein et al. 1974, tables 1, 2.

8. A series of large-scale resettlement projects moved some 2.3 million people from Java to other islands in Indonesia from 1950 to 1984. After 1984, migration continued on a spontaneous basis. Up until 1993, some 4 million people migrated (van der Eng 1996, 152–54; Cleary and Eaton 1996, 94–99).

9. The total acreage under the control of the Buenos Aires government rose from 2.1 million hectares in 1779 to 85.8 million in 1890 (Cortes-Conde 1997, 50).

10. See Buckley 1955, table VI, Historical Statistics Canada 1982, table M34; Green 2000, 202–4 and Norrie 1975. Norrie argues that settlement was delayed by the lack of suitable varieties of wheat (section 6.2).

11. See Harley 1980 and the survey of the debate in Atack and Passel 1994, 417–19.

12. Historical Statistics 1975, series J54, J55, J59, J51. In Canada, the ratio rose from 48% in 1871 to 56% in 1951 (Historical Statistics Canada 1965, series L7, L12).

13. In theory, cropland includes artificial meadows and fallow, which are both indispensable stages of rotation. On the other hand, it should not include the land left idle in slash-and-burn cultivation, which do not entail any cultivation practice. The distinction would be quite difficult to implement if there were statistics for Africa, but these are not available anyway.

14. This statement does not rule out an excessive use of natural pastures (overgrazing) if the number of pasturing animals exceeds the carrying capacity of the land.

15. The acreage under tree crops in 1800 is estimated assuming the same growth rate as in 1860–90; cf., also, Simpson 1995, table 1.4 (and pp. 62–70) and Bringas Gutierrez 2000, table IV.1.

16. Data on "cultivated area 1890–1910" are extrapolated backward to 1858 according to the total of area under major crops (Bloomfield 1984, table V.9) and root and green fodder crops (Bloomfield 1984, table V.11). The procedure might yield a downward bias in the earlier years, as the share of ploughland seems to have declined in the long run.

17. Richards (1982, 38) reports different figures for 1800 (0.8 million hectares) and 1882 (2 million hectares), which imply an even faster rate of growth.

18. On the abundance of land in South America, see Bulmer-Thomas 1995, 92; Holloway 1980, 26, 112; Merrick and Graham 1979, table IX-2 (on Brasil), Bauer 1975, 127 (on Chile); Dye 1998, 184–86 (on Cuba). On the Balkans, see Lampe and Jackson 1982, 16; Palairet 1997, 22, 91–94.

19. In the whole period 1600–1868, farmers created 1,789 "shinden" (new fields), increasing the total arable acreage from some 2 million to 4.2 million *cho*, roughly as many hectares (Hanley and Yamamura 1977, 74). Almost a half of these shinden (788) were created since

1800. However, the figure may overstate the proportion in the increase of acreage, as the later "shinden" were usually smaller or less productive. The areas under irrigated rice (paddies) grew from 1.3 million hectares in 1700 to 1.6 million in 1836, and to 2.6 in 1889 (Ishikawa 1967, 98). Nakamura (1966, 22–23) argues that the growth in acreage before 1890 was spurious, as official figures are underestimated.

20. Perkins 1969, table II.1 and appendix B (with a margin of error plus or minus 3 million hectares). According to Chao (1986, table 5.1), the acreage increased from 59 million hectares in 1784 to 77 million in 1887; according to Maddison (1998, table 1.5) it rose from 73.7 million hectares in 1820 to 109.7 millions in 1952. On the expansion of cultivated land in the South (a 64% increase in Guanxi from 1793 to 1838), see Marks 1998, table 9.1; Lin 1997, 29. These are official figures, which may undervalue the acreage (Feuerwerker 1980, 15). The current acreage is about 145–50 million (see note 27 of this chapter).

21. According to Richards et al. (1985, 700), the total cultivated acreage increased from 85.4 million hectares in 1860 to 134.5 million in 1920. Roy (2000, table 3.6) reports slightly lower figures for the area labeled "net of states and Burma": 75 million hectares in 1885, 102 in 1921, and 104 in 1938 (see also Blyn 1966, table 6.1 and Sivasubramonin 2000, table 7.5 for the twentieth century). The current total acreage of India, Bangladesh, and Pakistan exceeds 200 million hectares. On the case of Bombay Presidency, see Charlesworth 1985, 23–40.

22. See, in general, Booth and Sundrum 1985, table 1.6. For Malaysia, see Booth 1991, 26–27; Grigg 1974, 93–104; Hill 1997, 177–78; Overton 1996, 55–57. For Anatolia, see Quataert 1981, 76; Aricanli 1986, 35–40.

23. Clarke and Matko 1984, table 56 (see also Volin 1970, 484; McCauley 1976, tables 2.4, 2.5). This long-run increase is the outcome of a steep fall to a minimum of 77.7 million hectares in 1922 and of the subsequent recovery. The losses from World War I are crudely estimated, adding up the arable land in Poland and the Baltic states in 1938.

24. For China, see Perkins 1969, table II.1 (a further 10%, up to 98 million hectares) and Feuerwerker 1983, 63 (but see note 20 of this chapter). For Burma, see Saito and Kiong 1999, table II-1 (from 6.9 million hectares in 1910 to 8.6 million hectares in 1939). For Thailand, see Manurungsam 1989, table 2.7 (from about 2 million hectares in the 1910s to 3.8 million hectares in 1936–37).

25. Historical Statistics 1975, series J 52 and J 62.

26. For an analysis of the same data by country, see Mundlak et al. 1997.

27. According to Perkins (1969, table II.1), who relies on early official statistics, total acreage rose from 98 million hectares in 1933 to 112 million in 1957. The original FAO Yearbook reported a much lower figure (90 million hectares) for the late 1950s, while the current FAO Statistical Database suggests a figure of 105 million hectares for 1961. The acreage declined to a minimum of 100 millions hectares in 1980, to rise steadily to 148 million in 2000. Such a massive increase is somewhat suspicious, and many scholars believe that official Chinese data were undervalued (for the debate, see Nguyen and Wu 1999, 54–73; Wu and Wang 2000; Lindert 2000, 145–52, esp. table 6.1.B).

28. The African population increased from 170 million in 1940 to 277 million in 1960 (UN Statistical yearbook). The total acreage in Soviet Union increased by 31.4 million hectares—from 188.6 million in 1953 to 220 million in 1960, plus 10 million hectares that proved unsuitable for agricultural use and was left idle (McCauley 1976, table 4.3; cf. Medvedev 1987, 167–170).

29. For the rise and fall of ranching in general, see Grigg 1974, 243–50. On South America, see Diaz-Alejandro 1970, 148; Cortes-Conde 1997, 60; Chonchol 1995, 119–24, 128–32. On New Zealand, see Evans 1969, 24–30. On Canada, see Solberg 1987, 40–50. On the similar case of Serbia, see Palairet 1997, 91–100, 298–305. The change from cropland to arable can be seen as part of a general process of intensification (section 6.3).

30. The sum of pasture in farms (Historical Statistics 1975, series J55) and "grazing land not in farms," the open grassland (series J 62) rose from 1880 (406 million hectares, 12% of which is in farms) to 1900 (422 million, 26% in farms) then declined very slowly until 1940 (390, but 48% in farms) and fell during the next thirty years (to 335 and 65%). The FAO Statistical Database reports different data, which seem to include all pasture in farms plus some of the grazing land. From 1960 to 2000, the acreage declined by about 12%, as much as cropland did during the same years.

31. For the fall in yields, see the discussion in section 5.2. The effect of poor soil quality is quite controversial: see Blyn 1966, 179; Islam 1978, 80.

32. See Parker and Klein 1966, 532–35 (and the critique of this approach by Olmstead and Rhode 2002, 934–36). See also the similar analysis by Grantham 1991 for France, where regional differences in soil quality and climate were, however, much smaller than in the United States.

33. For example, see Alexandratos 1995, 17, 355; Pingali et al. 1997, 79–90; Lomborg 1998, 104–6; Pretty 1995, 69–78; Greenland et al. 1998, 47–49; Gardner 2002, 121–22. The tricky issue of measurement is dealt with by Mortimore (1998, 5–7, 25–37), Rosengrant and Hazell (2000, 299–310), and Wood et al. (2000, 45–53).

34. On Egypt, see Richards 1982, 77–80, 120–28; Owen and Pamuk 1998, 142–56. On China, see Huang 1985, 62 and 1990, 25; Pomeranz 1993, 121–25. On India, see Stone 1984, 134–44; Whitcombe 1983, 708–10. On precocious (and excessive) alarms about the desertification of African drylands, see Mortimore 1998, 17–22.

35. Mortimore (1998, 21–25) criticizes the conventional wisdom about desertification in Africa: he argues that boundaries of desert move back and forth without any clear trend, and the movements depend on rainfall and not on human action. A well-known historical case of successful restoration is the American Dust Bowl, the areas of the Great Plains that had been hit by very serious drought in the 1930s (Hurt 1994, 300, 311; Hansen and Libecap 2004a).

36. Following the so-called principle of the single farm, expenditures do not include the value of items (e.g., manure or feed) produced within agriculture, even if they entail a monetary outlay. Seeds are included only if purchased from specialized firms. In this case, they appear both among sales (at the farm-gate prices) and purchases—including the value added in processing (cleansing, sorting, packaging, etc.) and the transaction costs (transportation, sales, etc.). In theory, capital should also include wages, or the equivalent for the consumption of the working members of rural households, but no estimate includes them.

37. Grantham's data (1996, 77) on capital include only "vineyards and land improvements" (after 1789)—in other words, new reclamation. In fact, the share of land on total capital is a mere 10%.

38. The official series may underestimate the value of capital in American agriculture because they use too high a depreciation rate, which assumes still-functioning machinery as worthless (Gardner 2002, 45).

39. These investments raise a nice "theoretical" problem for the estimate of national accounts (the work performed by hired workers or building companies unquestionably belongs to the Value Added of construction). They can be dealt with in two different ways. The American accounts (Towne and Rasmussen 1960, 275–76) have a specific item for "improvement to land," while the European ones (e.g., Grantham 1996) subsume the work by farmers in the total value of their final product (as the value of, say, legal services of the in-house lawyer of a car manufacturing company is included in the cost of the car). The two methods yield quite different estimates of total capital if computed with the method of perpetual inventory (i.e., as the cumulated sum of past investments, suitably discounted). The "American" method correctly allocates new improvements to investments, whereby they transfer to capital stock. In contrast, the "European" method underestimates current investments, as agricultural products by definition are consumption goods. The larger the net investments are in comparison with repair and maintenance, the greater the bias is.

40. For the historical data, see Gallman 1986, table 4.A.1 (the share at constant prices declined from 61% in 1840 to 54% in 1900). For the 1992–94 estimate, see the USDA 2004. Land and buildings accounted for 75% of capital in Canada (Historical Statistics Canada 1965, series L15–L38).

41. According to Primack (1966), clearing an acre of woodland required 32 days of work in 1860 and 25 days in 1910, while plowing an acre of prairies (a task entrusted to specialized teams of up to eight oxen) needed 1.5 days in 1860, and 0.5 days fifty years later.

42. Craig and Weiss 2000 estimate that from 1840 to 1900 land clearing, other improvements, and maintenance absorbed about 40% of the total work effort.

43. On the cost of settlement, see Atack and Passel 1994, 277–78; Bogue 1963, 67–85, 169–70 and Adelman 1994, 177 on Argentina (a cost of $500–2000 in the 1880s). On income, see Atack and Bateman 1987, 129, 243; Historical Statistics 1975, series D705. Clearly, the cost was substantially smaller for pastoral estates.

44. Holderness 1988, tables 1.6, 1.7. The figure excludes the cost for the construction of farm buildings. The gross output is from Ojala 1952, 208–9.

45. The French tree-crops capital increased by a half from 1789 to 1870 (Grantham 1996, table 3. See also, Agulhon et al. 1976, 237; Loubère 1978, 140–43). In Italy, yearly investments increased fivefold from the early 1870s to the peak in 1877–81 (Ercolani 1969, table XII.4.22). The "intensive" crops (wine and fruit) rose from 4% of Californian agricultural output in 1879 to almost 80% forty years later (Rhode 1995, 777).

46. Holderness 1988, table 1.9; 2000, table 13.4. For the post–1945 investment, see also Brassley 2000b, 519; Turner et al. 2001, 89–90; Brassley 2000h, 67. Drainage was also used in France (Price 1983, 379) and Denmark (Jensen 1937, 159–60). The diffusion of underground drainage was greatly helped by the invention of a machine to produce tiles in the 1830s–1840s.

47. See Bevilacqua and Rossi-Doria 1984, 67 and, for further information and detailed data, the very recent book by Novello (2003, 90–103, 165–71, 261–68). Reclamations in the Netherlands added some 0.25 million hectares from 1500 to 1815 (de Vries and van der Woude 1997, 31) and all the increase in the nineteenth and twentieth centuries—a further 0.3 million hectares (table 4.1). According to Holderness (1988, table 1.8), from 1760 to 1860, 1.2 million hectares were reclaimed in England. The figure seems high: such a massive movement, involving about a sixth of the land in 1850, seems to have gone unnoticed in reference works.

48. McCorvie and Lant 1993 (the data from table 4); Bogue 1963, 84–86.

49. Actually, rice can also be cultivated on nonirrigated land. This system (known as rain-fed cultivation) is widespread in Asia. However, unlike paddy cultivation, rice from nonirrigated land cannot be cultivated more than once a year, and the yields are lower and more variable (Bray 1986, 6–11; Ishikawa 1981, 90–92; Pearson 1992, table 8.1).

50. On the "hydraulic cycle," see Perdue 1987, 201; Chao 1986, 204–8 (esp. table 9 on the number of floods). For the stagnation in acreage, see Perkins 1969, table IV.3; Feuerwerker 1980, 7; Maddison 1998, table 1.5 b.

51. In India, the irrigated acreage rose from 9.4 million hectares in 1885–86 (12% of the acreage) to 21.6 million in 1938–39 (22.1%) (Roy 2000, table 3.6). In Indonesia (Java), irrigated land grew from 2.4 million hectares in 1880 to 3.4 million in 1938 (van der Eng 1996, table A.4), but the ratio to total land declined from 55% in the 1850s (Booth 1988, 74) to 50% in 1880 and 38% in 1938. In Japan, the acreage rose from 1.8 million hectares in 1880 (38% of the acreage) to 3.4 million (55%) in 1940 (Hayami and Yamada 1991, table 4.1); in Taiwan, from 0.3 million (41% of the total acreage) in 1921 to 0.5 million (60%) in 1939 (Lee and Chen 1979, 61 and 86); in Korea, from 0.75 million in 1925 to 1.15 million in 1935 (Ban 1979, table k.4); in the Philippines, from 0.3 million hectares in 1902–18 to 0.52 million in 1938 (Hooley and Ruttan 1969, table 7.4). For Indochina, see Grigg 1974, 100.

52. See Grigg 1974, 92–93, 105–8 for information on Vietnam; Booth 1988, 77–78; van der Eng 1996, 42–66, for Indonesia; Stone 1984, 13–35, Whitcombe 1983, 677–729, and Headrick 1988, 170–96, for British India; Headrick 1988, 196–205, for Egypt; and Hayami and Ruttan 1985, 289–93, for the Japanese colonies.

53. Feeny refers to a large-scale project to irrigate all of the Central Plains, which was rejected (see also the information in Ingram 1971, 82–86). For a different view, see Manarungsan 1989, 83–84.

54. The irrigated acreage roughly doubled in Italy, from 1.4 million hectares in 1905 to 2.2 million in 1948 (Svimez 1956, table 131; Bevilacqua and Rossi-Doria 1984, 68); in Spain, from 0.85 million hectares in 1858 to 1.3–1.4 million in 1916 and 1940 (Grigg 1974, 147; Simpson 1995, table 6.1 and pp. 126–88); and in Chile, from 0.44 million hectares in 1875 to 1.1 million in 1930 (Bauer 1975, 107). In Mexico, it jumped from a mere 17,000 hectares in 1930 to 257,000 in 1940, and to 1.9 million in 1958 (Reynolds 1970, tables 4.7, 4.8). Finally, almost all the expansion of acreage in Egypt (table 4.1) was irrigated.

55. For the data, see Historical Statistics 1975, Series J86; Gardner 2002, table 6.1. See also Olmstead and Rhode 2000b, table 3; Rhode 1995, 775.

56. Estimates are from Wood et al. 2000, 53–58.

57. See Palanisami 1997, tables 1, 2. For further data and information, see Clark 1970, 59; Bray 1986, 71–87.

58. It is possible that official sources omit the areas irrigated by wells and other minor sources. Thus, the data of table 4.10 may underestimate the diffusion of irrigation and, by implication, also underestimate the rate of growth in irrigated acreage.

59. In Indonesia, technical irrigation grew from 8% of total acreage in 1880 to 65% in 1938 (van der Eng 1996, table A.4), and in India, from 34% in 1885–86 to 52% in 1938–39 (Roy 2000, table 3.6). In Japan, irrigation from major projects (subsidized under the various Arable Land Readjustment Acts) increased from 2.5% of irrigated land in 1905 to about a third in 1940 (Hayami and Yamada 1991, table A14). In China, the area

irrigated by motor-driven pumps grew from 1.6% of total acreage in 1952 to more than half in 1978 (Xu and Peel 1991, table 5.1).

60. In India, private canals accounted for about 12% of total irrigated acreage in 1885–86 and 1938–39 (Roy 2000, table 3.6). For the debates about the funding of irrigation projects, see Whitcombe 1982, 693–705; Stone 1985, 19–23. Private capital also financed the major irrigation project in Thailand during the nineteenth century in the Rangsit area close to Bangkok (Ingram 1971, 81; Feeny 1982, 60).

61. The subsidies to drainage were suspended in 1956, and, since 1990, the policy has totally changed: now the federal government subsidizes the restoration of wetlands.

62. See Ishikawa 1967, 94–98; Hayami and Yamada 1991, 136–40.

63. Data on expenditures (which omit some local outlays) are from van der Eng 1994, table A.8.2. Data on gross output and implicit deflator is from van der Eng, personal communication.

64. The first silos were built in 1873 in Illinois, and, in the 1920s, there were about 100,000 of them all over the United States (Olmstead and Rhode 2000b); sileage spread in Europe after World War II (Grigg 1974, 199–201; Brassley 2000f, 571, and 2000h, 70).

65. In France, the average stock of buildings grew from 12 square meters per hectare in 1840 to 19 square meters in 1862 and 20 square meters in 1892, and remained constant until 1913 (Grantham 1996).

66. On China, see Perkins 1969, 56–60; Feuerwerker 1980, 6; Chao 1986, 196; and the local studies by Myers (1970, 177) for Hebei and Shandung in the north, and Lin (1997, 42) for Guangdong in the south. On Thailand, see Feeny 1982, 38–40.

67. For South Africa, see Duggan 1986, 105–6. For the other areas, see Miracle 1967, 242; Wrigely 1970, 57; Mosley 1983, 80; Iliffe 1979, 286–88; Anthony 1979, 140; Eicher and Baker-Doyle 1992, 120. The plow was traditional in the area north of Sahara and Ethiopia (McCann 1995). Hoeing was quite widespread in early nineteenth-century Australia (Raby 1996, 78).

68. On India, see Roy 2000, 59. In Hungary, wooden plows accounted for about half the total stock around 1870, and disappeared 25 years later (Voros 1980, 67, 103). The Russian stock in 1910 consisted of 3 million traditional (all-wooden) plows, 7.9 so-called sokha plows (i.e., wooden plows with an iron blade), and 6 million modern iron plows (Lyaschenko 1970, 734—on modernization, see also Volin 1980, 111; Gattrell 1986, 202–3; Moon 1999, 128–29; Spulber 2003, 77). The share of households using iron plows also increased in Bulgaria and Serbia (Lampe and Jackson 1982, 183–85).

69. For data on the United States, see Historical Statistics 1975 Series K189. For Europe, see Bairoch 1999, 73. For Denmark (the pioneering country), see Jensen 1937, 173. For the United Kingdom, see Brassley 2000g, 73.

70. These figures are confimed by country sources. On the eve of World War II, there were about 50 to 60 thousand tractors in the UK, where they substituted steam plows (Collins 1983, 73–74, 92; Brassley 2000h, 74), some 70,000 in Germany (Corni 1990, 227), 40,000 in Italy (UMA 1968, table 2), 35,000 in France (Moulin 1991, 161), 7,000 in Hungary (Wald 1980, 210), and 25,000 in all of Eastern Europe, excluding Russia (Berend 1985, 166).

71. On the regional distribution of tractors, see Olmstead and Rhode 2001, table 3, and for the Soviet priority on motorization, see section 9.6.

72. On the prices of implementation in the UK, see Afton and Turner 2000, table 44.34.

73. White 2001, 495; Olmstead and Rhode, personal communication. Hayami and Ruttan (1985, 478) publish a quality-adjusted series of machinery prices, which essentially takes into account only the increase in the unit power of tractors.

74. Apart from all the usual problems about data reliability, one should draw attention to the date of the census. As stated, the number of animals, especially small ones (pigs, sheep), varies, all else being equal, according to a well-defined seasonal cycle. Thus, the outcome of a census also depended on its timing: for instance, when Argentina, in 1913–14, moved its census from summer to winter, the number of sheep halved and that of pigs fell by about 10%. Finally, the tables do not report data on horses and mules as it is impossible to distinguish the rural animals from the total stock.

75. Poultry breeding also boomed in China and Southeast Asia (Rosengrant and Hazell 2000, 178–85).

76. Manure accounted for about half of the total fertilizers in Europe in 1995 (EUROSTAT 1998/99, 223–24), while, according to Smil (2001, appendix Q), all "natural" sources (including fixation from air) supply between 50 and 60% of world total nitrogen.

77. Data from Brassley 2000g, 566, and 2000h, 73. For data on the USA, see Olmstead and Rhode 2000b, 30; Bateman 1969, 227–29. English breeds of pig, such as the Yorkshire and Berkshire, spread throughout Europe (cf., Federico 1992, 57 for Italy; Gall and Gunst 1977, 45–50 for Hungary).

78. Figures computed from Historical Statistics 1975, series K565 (cattle), K567 (hogs), and K569 (sheep) deflated with the overall index of prices of farm products (series E23, E25, E52 and E53). The change thus includes all the effects of change in composition.

79. Cf. Feinstein 1988a and b, tables 10.2, 10.5, 10.8; USDA 1999. The comparison is only tentative, as the definitions differ.

80. Cf., respectively, Historical Statistics 1975, series K247, K240 (2.5%); and Anthony et al. 1979, 68–69 (2% from a survey of 24 countries). The latter figure overstates the actual ratio slightly, as it includes the purchase of tools, which belong to the capital investment.

81. Thompson (1968, table 5) shows expenditures at current prices deflated with the price index for agricultural products. The expenditure for other inputs also increased in the USA 8 times from 1800 to 1850 (Historical Statistics 1975, series K247, "off-farm purchases"), and only by a half in the Netherlands, where the initial level was quite high, from 1807–9 to 1849–51 (Van Zanden 2000, "other inputs," series at current prices deflated with agricultural prices).

82. See Historical Statistics 1975, series K247 for 1870–1910; series K276-278 for 1910–70 (both deflated with prices of agricultural products); and USDA 1999 for 1970–96. The data are not exactly homogenous. The earlier series exclude expenditure on seed and feed and are thus likely to understate the increase in off-farm purchases. The USDA data include these and other expenditures within the sector, such as the purchase of livestock.

83. For the intensification of Chinese agriculture, see Li Bozhong 1998, 46–47, 83–86.

84. Cf. Thomson 1968 and the sceptical view by Turner et al. 2001, 140. On imports of these "natural" fertilizers, see Grantham 1984, 198–200; Smil 2001, 40–48, appendices F, G; Cariola and Sunkel 1985, tables 3, 8; Hunt 1985, 258–60; Wines 1985, 33–55. Exports of nitrates went on growing until the 1910s, while those of guano peaked in the 1870s and later declined because of the exhaustion of the deposits.

85. For further country data and information, see Historical Statistics 1975, series K193 (USA), Brassley 2000d, table 7.3 (Britain), Agulhon et al. 1976, 207–12 (France); Knibbe 2000, table 5.2; Van Zanden 1994, 207–12 (Netherlands); Hayami and Yamada 1991, table A.15 (Japan); ISTAT 1958, 131 (Italy); Goosens 1992, 287–92, and Blomme 1992, 244–51 (Belgium).

86. On the low consumption of modern fertilizers in the Thirld World, see van der Eng 1996, tables 28, A.7 (Indonesia); Blyn 1966, 195 (India), Diaz-Alejandro 1970, 163 (Argentina); Richards 1982, 126–27 (Egypt); Ban 1979, table 4.4 (Korea); Cheng 1968, 29 (Burma); Lee and Chen 1979, 68 (Taiwan); Faure 1989, 37 (China).

87. For the ecological consequences in Asian countries, see Pingali et al. 1997, 79–80, 104–10.

88. See also Bairoch 1999, 97; Brassley 2000f, table 7.11; Wood 2000, 34–39.

89. Herbicides accounted for 45% of the total outlay, insecticides for 29%, and fungicides for 16%. The data refer to the total cost, and thus the quantity is likely to have grown faster.

90. In Japan, from the 1880s to the 1930s, fertilizer prices relative to prices of agricultural products fell by two-thirds (Hayami and Yamada 1991, tables A 10, A 11). In Italy, from the 1900s to the late 1930s, the real price of soda nitrate fell by 35% and that of sulphate of ammonium by 60% (ISTAT 1958, table 96, deflated with the wholesale price index, table 91), and Indonesia by about 50 percent (van der Eng 1996, table 3.12). In the UK, the price of soda nitrate fell by a third from the 1850s to the 1870s, but then remained stable (prices from Afton and Turner 2000, table 44.35, deflated with the Sauerbeck-Statist price index). In the Netherlands, prices relative to rye fell by 40% from the 1870s to World War I (van Zanden 1994, table 4.7). In Spain, prices of superphosphate roughly halved from 1901 to 1928 (Simpson 1995, figure 5.1).

91. The table sums up men and women. The common practice to count only men (Hayami and Ruttan 1985, 450, and Bairoch 1999, 29) is bound to bias trends and, above all, intercountry comparisons as the gender ratio differed quite substantially among countries (cf. table 4.18). Furthermore, the rationale for the omission is somewhat dubious: the censuses are likely to report the number of full-time women as accurately as full-time men, and to omit part-time men and women.

92. Heston (1982, table 4.1) quotes higher figures for 1875 (85.2 million) and 1895 (94.1 million), as does Mitchell (1998a, table B1) for 1911 (103 million) and 1931 (100 million).

93. Helling (1966, 140) challenges the reliability of the census: her alternative estimates are 37% higher in 1852, 10% in 1881, and 28% in 1907. Fremdling (1988, table 6) quotes an estimate by Holhs and Kaelble: a total of 8.3 million workers in 1895—or 15% less than the Hoffman figure for the same year.

94. Higgs (1996) argues that these data are undervalued, as female workers are undercounted. See also the discussion (on England and Wales only) by Allen 1994, 107–8; Afton and Turner 2000, 941–55.

95. The timing of the decline differed markedly within the United States: the agricultural work force started to decline in the Northeast in 1900, while in other areas (Northern Plains, West, and South) it peaked in 1935 (Historical Statistics 1975, series K 17-K81).

96. See Spulber 2003, table 1.2 for Imperial Russia, excluding Poland and Finland. Clarke and Matko (1984, table 1) report different figures for 1870 (76.8 million for "pre—September 1939 territory") and 1914 (114.5 for "probably present territory"). According to

Clarke and Matko, the population within the 1939 borders was some 15% smaller than the postwar one. Adding 15% to the 1914 figure yields a total of some 132 million within the borders of Imperial Russia. The figure is not far from Spulber's, but the growth rate in the previous 40 years is decidedly inferior (1.23% instead of 1.53%). Both Gregory (1994, table 2.4) and Bairoch (1999, table 2.1) hypothesize a growth rate close to the upper bound of the range—1.4% from 1883–87 to 1909–13, and 1.48% from 1850–1910 respectively.

97. See also Wheatcroft and Davies 1994a, 129. The latter fall, from 61.6 million to 48.2, does, however, contrast with the alleged increase in the total number of hours worked in the same years, by 6% according to Nimitz (quoted on p. 129), and by 12% according to Gregory and Stuart (1986, table 20). The implicit 30–35% increase in the number of hours of work may seem excessive, given the incentive problems of socialist agriculture (section 9.5).

98. The agricultural workers/total population ratio (0.382) is computed from the data for 1950 (FAO Statistical Databases), while population is from Maddison 2001, table A-c.

99. Workers were often recruited from other areas within the same country, exploiting either the different crop mix (e.g., from mountains and plains) or small differences in the timing of operations for the same crop. The colder the climate is, the later wheat ripens, and thus laborers could reap wheat in several areas. In principle, these migrant workers should be included in the total agricultural workers of table 4.18.

100. Adelman 1994, 163–64, 112–18. The *golondrinas* left Italy in the autumn after the harvest, arriving in Argentina for the winter harvest, stayed in Argentina for a full year until the next harvest, and then returned in Italy just in time for the Summer harvest. In this way, they succeeded in harvesting four times in two years exploiting the different timing of seasons in the northern and southern hemispheres.

101. For the effect of intersectoral mobility on industry, see Magnac and Postel-Vinay 1997. On the role of migrant workers in American agriculture, see Wright 1988.

102. The American statistics are a conspicuous exception: they count as agricultural workers all household members who had worked for at least 15 hours a week, and all hired hands who had been paid at least for one hour in the week of the survey (Historical Statistics 1975, 453). In fact, Lebergott's adjusted estimates of the core labor force (1984, table 7.3) are decidedly lower than the official figures.

103. The ratio fluctuated around 0.5 in Italy from 1905 to 1933 (Arcari 1936) and was stable in Japan at around 0.75 from the 1880s to the 1930s (implicit ratio computed from Hayami and Yamada 1991, table A5). The difference reflects both the different mixture of tasks and a different wage for the same task.

104. Some of these changes, such as the rise in Germany, are somewhat dubious, and may conceal changes in definition (Helling 1966, 132–34).

105. The share of women in the world without USSR rose from 38.2 to 43.7% of the total.

106. See Gardner 2002, table 4.11; and for the 2002 figure, see USDA 2002. In 1986, the average age of agricultural workers in 18 OECD countries was 43 years, higher than the average of 36 years for the total work force (OECD 1994, table 1.3).

107. Craig and Pardey measure the skill as the difference between the average wage and the wage of the unskilled laborer. Their work is particularly detailed as they consider 30 different "types" of farmers, differentiated by age and education. See also Capalbo and Vo 1988, figure 3.1; Gardner 2002, 117–18; Acquaye et al. 2003.

108. From 1971 to 1987, workers with post-secondary education (with or without a degree) in 8 OECD countries (Austria, Canada, Japan, Norway, Spain, Sweden, the UK, and the USA) rose from 3.2% to 10.3% of the agricultural work force, but from 21.7% to 30.2% of total manpower (OECD 1994, table 1.4).

109. See also Clark and van der Werf 1998; Voth 2001, table 7. Clark and van der Werf infer the (lack of) changes in the number of hours worked from a comparison between day and piece-rate wages, and in the number of hours worked from the stability of the cost of food (in terms of days of work) from the Middle Ages to 1850 in England. For a similar result in the Netherlands, see Van Zanden 2000.

110. The decline to the 1980s is confirmed by data for the USA (20% from the 1930s to the 1970s, according to Historical Statistics 1975, series K173, K184) and Japan (a 10% fall from 1940 to 1985, according to Hayami and Yamada 1991, table A 5).

111. In Germany in the 1980s, the gap was about 25% for self-employed workers and 15% for wage workers (Planck 1987, table 8.4, pp. 170–71).

112. Part-time farmers are defined as those who work off-farm for more than 200 days a year.

113. In the European Union (European Commission 2001, table 126) in 1997, 55% of the agricultural manpower worked less than half their time in agriculture. In Japan, the number of part-time households increased from about a third in the 1930s to 84% in 1984 (Historical Statistics Japan 1987–88, vol. 2, table 4.2) and the share of agricultural income on total income of rural households declined from 72% in 1955 to a mere 14% in 1990 (Francks 1999, table 4.1). For a case study, see Jussaume 1991.

114. Before 1910, Lindert computes the implicit rents from land prices given the prevailing interest rate.

115. Before 1910, the rent-interest ratio rose quickly in the USA (consistently with the growing relative scarcity of land), rose moderately in Germany, while it fell in Japan and Sweden, and remained unchanged in the UK. In France, it rose until the 1890s, and then it fell.

116. The average rate for LDCs (table 4.23) is heavily affected by the veritable collapse in the wage-rent ratio in Thailand, which was, according to the data collected by Williamson (2002), 40 times higher in 1870 than in 1913. However, wages decreased, relative to rents, also in Egypt, India, and Burma.

117. In Williamson's econometric analysis, the wage-rental ratio comes out as negatively related to terms of trade and to land-labor ratio. The first result tallies well with expectations (agriculture by definition is more land intensive than manufacturing), while the latter does not, as one would expect that wages were higher in land-abundant countries.

118. The dearth of data may be an indirect consequence of the "structural" change in landownership toward a growing share of "family farms" (section 8.4). The wage-rent ratio is an important measure of income distribution when landlords are a separate group. It is much less relevant—and thus it draws much less attention from data gatherers—when rents and wages join in the income of landowning households.

Chapter Five: The Causes of Growth

1. For general, nontechnical overviews of measurement issues, see Overton and Campbell 1991; Gardner 2002, 30–45.

2. Japan was to reach that level of productivity only in the early 1950s. In the meantime, population growth had reduced labor productivity in Thailand by about a third relative to the beginning of the century.

3. For the comparison between China (the Yang-tzi Delta) and England, see Huang 2002, 509–11; Pomeranz 2002, 544–48. These works are part of an on-going debate on the levels of income in the most "advanced" areas of Europe and China, and, as often happens in debates, they yield quite different results.

4. From 1866, the source is Historical Statistics 1975, series K506, K507TJ.

5. Their conclusion is based a wide collection of farm accounts, which may not be representative.

6. See also Allen and O'Grada 1988, tables 3 and 4 for France and Ireland.

7. For the original data, see Blyn 1966, 150–56, table 7.1. They are endorsed by Mishra (1983) and by Sivasubramonian (2000), the author of the most recent estimate of Indian national accounts, and deemed even too optimistic by Islam (1978, 47–49). These statistics imply a 40% fall in per-capita consumption in Bengal, which Pray (1984) and Heston (1983, 387–91) deem implausible. In his estimate of Indian national accounts, Heston assumes that yields remained constant. Maddison (1985, 201) and McAlpine (1983, 29–33, appendix D) admit that the official statistics may be wrong, but do not fully endorse Heston's alternative hypothesis. For a short survey of the debate, see Roy 2000, 52–55.

8. Rice yields fell by 10% in Indonesia from 1880 to 1920, but they recovered in the next twenty years (van der Eng 1996, table 4.1). They fell by 30–40% in Thailand from 1906–9 to 1948–50 (Ingram 1971, 49–50; Feeny 1982, 24; Manarungsan 1989, 83–84, table 2.19), while they rose by two-thirds in Japan from the 1870s to the 1930s (Ishikawa 1981, table 1.5). Feeny (1982, 191, n.8) reports data for other Asian countries in the period 1910–39, with quite a mixed record: Burma –0.24, French Indochina –1.23, Korea 1.39, Taiwan 1.56, the Philippines 1.46, Malaya 0.96 (1920–40), and Ceylon 1.20 (1920–40).

9. A rise in yields is implicit in the assumption of a constant per-capita output, which underpins Perkins's preferred estimate of production (1969, 29–35). His view is shared by Pomeranz (2000, 226), but not by Chao (1986, 208–16), Feuerwerker (1980, 14), and, for Guangdong, Lin (1997, 42–45), who believe yields have stagnated.

10. For some data on yields in 19th and early 20th centuries, see Simpson 2000, table 3.

11. For instance, the production of pork per pig fell in Denmark by about 40% from 1861 to 1923–24 (Jensen 1937, tables N and Q). The fall compounds the effects of changes in the reproduction rates (number of living offspring per sow) and in the weight at the slaughterhouse. On the carcass weight of animals in the United Kingdom, see Turner et al. 2001, 174–207.

12. The fall from 1910 to 1930 reflects a (somewhat mysterious) decrease in production from 1995 liters per cow (and heifer) in 1924 to 1534 liters in 1925. Until 1924, the data include southern Ireland, while they subsequently refer to the UK at its present borders. This change in boundaries could account for the fall if Irish cows were more productive than British ones. However, this was not the case: in 1926, they were about 10% less productive.

13. For the role of better feed (jointly with better breeds) in increasing the milk yield per cow, see Grigg 1994, 194. The increase in land-use depends on the productivity of feed (grass, corn, etc.) per unit of land and thus on technical progress in arable.

14. For Indonesia, see van der Eng 1996, table 4.6; for Japan, see Ishikawa 1981, table 1.5; and, for the USA, see Craig and Weiss 2000, table 1 for 1840–1900, and Historical Statistics 1975, series K449 for 1900–40.

15. For the stagnation to 1900, see Craig and Weiss 2000; Bateman 1969, tables 5 and 6. For growth in the 20th century, see Historical Statistics 1975, series K475–K477 for the products (milk, beef, and hogs) and K411 and K415 for the number of animals.

16. Bairoch 1999, 124–26 for "advanced" countries (the rates are 0.45% for 1800–70, 1.4% for 1870–1910, 1.1% for 1910–50, and 4.7% 1950–95), and p.151 for LDCs (by 0.1% yearly 1900–50).

17. The same reasoning holds true for other measures based on physical coefficients, such as the "grain equivalents" suggested by the German Statistical Office (cf., Nutzenadel 2001, 308–9).

18. See Clark 2002b, table 5 (0.33%); Allen 1994, tables 5.1, 5.3 (0.77%); Turner et al. 2001, table 7.2 (1.02%); Deane and Cole 1962, tables 37, 31 (1.56%).

19. For Russia, see Gregory 1994, table 2.3 (1.4% from 1883–87 to 1909–13); for Indonesia, van der Eng 1996, table 2.10 (Java 0.43% and other islands 0.92% from 1880 to 1937); and for the Philippines, see Crisostomo-David and Barker 1979, table 5.1 (1.4% from 1902 to 1938).

20. On the Balkans, see Palairet 1997, 362–64, table 12.1; on Spain, Simpson 1995, 24, tables 1.4, 1.5 (according to his estimates, labor productivity stagnated until the turn of the century, increased at a 1.5% yearly rate until the Civil War, and then fell again, so that, by 1949–51, it was merely 22% higher than in 1900).

21. See Perkins 1969, table S1 for 1850–1957 (the stagnation of output per capita is his "preferred" estimate in a range from a 20% increase in the best case and a 21% fall in the worst); Pomeranz 2000, 98 (who dates the beginning of stagnation back to the 1600s); and Rawski 1989, 325–26. On the growth of labor productivity in the Yang-Tzi area in the early 19th century, see Li Bozhong (1998, 110–18), and for the period 1893–1933, Brandt (1989, 130–33).

22. The production is measured by three-year moving averages centered in the year indicated (but 1960 is 1961–63 and 2000 is 1998–2000). The use of gross output instead of the Value Added in the denominator overvalues productivity increase in comparison with the rest of the economy. Land includes only arable and pasture.

23. The production can be measured either with gross output (and inputs should include purchases from other sectors) or with Value Added (and inputs should include only capital, labor, and land). On growth accounting, see Solow 1957 and Barro 1999 and, on production functions, Mundlak 2001, 34–36.

24. Recently, scholars have pointed out that all conventional measures may overestimate the benefits of TFP growth as they do not take the negative effects of technical change on environment into account (Mubarik and Byerlee 2002, 839–44). They advocate the use of a wider measure, the Total Sustainable Factor Productivity (TSFP), which includes an estimate of these losses. They estimate that, in 16 districts of the Punjab, from 1966 to 1994, the environment-adjusted TFP growth rate was a third of the conventionally measured one (0.4% instead of 1.3%). The result is, indeed, impressive, although a skeptical reader might object to some inferences. The authors interpret a negative coefficient of a time trend in a cost function as reflecting unmeasured environmental damage, while it could reflect many other influences, such as a poor agricultural policy (see Sections 5.4 and 6.6). However, measuring degradation is extremely difficult (section 4.2).

25. For an example, see Schaefer 1983, table 7.

26. The Tornqvist-Theil index is an approximation of the Divisia one, which measures productivity in two benchmark years, weighting the rate of change in inputs with a geometric average of factor shares in the two years. A production (profit) function is a regression with output (or its price) as dependent variable, explained by quantity (prices) of inputs plus a time trend. The coefficient of the latter measures the technical progress. If applied to a cross-section of farms, the method can measure economies of scale as the coefficient of a size variable, such as acreage.

27. See Fulginiti and Perrin 1997; Nin et al. 2003b; and the survey by Ruttan (2002).

28. From this point of view, the Malmquist "distance" seems more suitable to microeconomic studies, where one can assume the same production function (see, e.g., Tauer 1998; Coelli et al. 2002).

29. The Total Factor Productivity was 76% of the British TFP in the Netherlands, 73% in Belgium, 67% in Ireland, 66% in France, 56% in Germany, 50% in Austria, 49% in Sweden, 41% in Hungary, and 34% in Russia (Clark 1993, table 4.1).

30. Growth accounting is used in 225 estimates, including almost all of the "historical" ones, flexible production forms in 72, and the Malmquist distance in 335. For more details, see Federico 2004c.

31. When necessary, shorter time intervals have been reworked to match the three periods (before 1870, 1870–1913, and 1913–40) as closely as possible to compute the averages. However, each estimate is considered as an observation of the total count.

32. For instance, both Bernard and Jones (1996) and Martin and Mishra (2001) omit land, while Van Zanden (1991 and 1988) does include it, alongside labor and capital, but not purchased inputs, as he should have done given that he uses gross output (and not Value Added) as a measure of production. The effect of the omission of an input is not predictable *a priori*, as the input share sum to 1 all the same. Thus, the omission would underestimate (over-estimate) the growth in TFP if its use grew less (more) than the use of the other inputs.

33. See Perkins 1969, table S.2. Estimates range from –0.22% to 0.35% in 1770–1850, and from –0.18% to 0.21% in 1850–1957: the preferred ones (0.12% per annum in 1770–1850 and 0.04% in 1850–1957) are based on the assumption that output grew as much as population. See fn. 21 of this chapter.

34. For tests of the effect of different methods of estimation, see Mokyr 1987, table 2 (on Britain before 1860), van der Meer and Yamada 1990, table 4.14 (on Japan and the Netherlands from 1960 to 1985), Barkaoui et al. 1997 (on 11 OECD countries, 1973–93), Fulginiti and Perrin 1999 (on 18 countries, 1961–91), Martin and Mitra 2001 (on 42 countries, 1967–92) and Federico 2004c. Some differences still persist among estimates computed with the same technique: the coefficient of correlation of country estimates from the two major comparative works by Arnade (1998) and Trueblood-Coggins (2004) is a modest 0.66, although both authors use the Malmquist distance for the same period (1962–92). The difference may be accounted for by the different set of inputs: Arnade includes irrigated land.

35. The coefficient of variation of the average is 0.31. The dispersion was similar in China after 1979, according to Fan and Zhang (2002, table 6), and much greater (a coefficient of variation 1.73), according to Lambert and Parker (1998, table 2). Huffman and Evenson (2001) report that the coefficient of correlation between state-level growth rates for crops and livestock (in 1950–82) was a mere 0.25. Also on the difference between technical progress in crops and livestock, see Nin et al. 2003a.

36. The total factor productivity grew at a rate of 1.58% per annum in the United States from 1989 to 1999 (USDA 2004) and at 1% in the United Kingdom from 1990 to 2002 (data available online at www.statistics.gov.uk/stabase/Expodata).

37. The comparison can be rather accurate when based on estimates by the same author(s) for both periods. The rate of growth in TFP in the USA increased from 0.17% per annum in 1880–1910 to 0.5% in 1910–40, according to Hayami and Ruttan (1985), and from 0.53% in 1870–1910 to 1.08% in 1910–37, according to Kendrick (1961). In the UK it rose from 0.4% to 2.1% from 1873–1913 to 1924–1937, according to Matthews et al. (1982, table 8.3). The acceleration is confirmed also by the results of an econometric estimation (Federico 2004c).

38. The rates are respectively 0.54% and 0.77% in 1870-1913; 1.74% and 1.71% in 1913–50; and 2.17% and 0.88% after 1950 (non-farm TFP growth from Gordon 2000, table 10.1; farm TFP growth from Kendrick 1961, table B-I).

39. See Matthews et al. 1982, table 8.3. The growth in agricultural TFP exceeded the overall one in 1856–73 (0.9% vs. 0.4%) and in 1924–37 (2.1% vs. 0.7%) and almost matched it in 1873–1913 (0.4% vs. 0.45%). In 1856–73 and 1924–37, agriculture outperformed manufacturing.

40. The size of the gap depends on the method of computation of TFP, but the result is robust to difference in specification. Cf. similar results on "advanced" countries in Bernard and Jones 1996, table 1.

41. Van Zanden 1991 p. 228. Cf., Statistical Appendix, Table IV.

42. For further discussion, see Allen 1991b; Coelli et al. 2001; and Thiam et al. 2001. One should take into account also environmental losses as a third component of an "ideal" decomposition, if data were available.

43. Hayami and Yamada tackle the problem in a work about Japan before 1940 (1991 pp.111–129). They proxy technological prowess with rice yields, and thus assume that difference in yields measure the width of technological gap between "top" prefectures and the rest of the country. They show that catching up was quite important in the period 1900-1930, when it accounted for half the economy-wide growth in yields. This method is ingenious, but the underlying assumption is highly questionable (Section 5.3).

44. Fulginiti and Perrin (1997, table 3; 1999) consider 18 countries in 1961–85: allocative efficiency grew in 14 of them, and technical efficiency fell in all countries but 4, because, the authors speculate, of anti–market price policies. Suharlyanto et al. (2001) deal with a total of 65 countries, mainly African and Asian ones, in 1961–91. The total TFP is negative in 27 cases, technical efficiency is negative in 19, and allocative efficiency is negative in 35 countries, including 6 socialist ones. Trueblood and Coggins (2004, appendix table 1) find technical regress in 44 out of 113 countries and decline in allocative efficiency in 55 (including five socialist ones). Both parameters decline in 27 countries.

Chapter Six: Technical Progress in Agriculture

1. Biggs and Clay 1988, 22–26; Hayami and Ruttan 1985, 56–58. On the very harsh opinion of Russian "experts" about the practices of peasants, see Kotsonis 1999, 100–104.

2. See Allen and O'Grada 1988, 113–16, and in general on Young's work, the recent reappraisal by Brunt 2003a. In the nineteenth century, the best British practice (section 6.2) became a yardstick of agricultural progress all over Europe.

3. On Italy, see Daneo 1980, 86–88; on Russia, Moon 1999, 126–28, and Spulber 2003, 80; on Australia, Raby 1996, 42–60. For a general discussion about conservatism and technical change, see Mokyr 2002, 218–84.

4. "An ancient myth pervades our agricultural writings: that whatever the stage of social development, there is one valid farming system only—as though every system that is more simple, every enterprise that adopts extensive methods to economize on labor, were proof of the practising farmer's ignorance" (von Thunen quoted in Clark 1977, 258).

5. See also the convenient summary of the different stages of intensification reproduced by Booth and Sundrum 1985, table 4.1. Boserup implicitly infers factor prices from factor endowment. This inference is reasonable as a first approximation, but is not necessarily true (section 4.5). The concept was later rediscovered by the anthropologist Clifford Geertz (and rechristened "involution") in his book on Indonesia (1963), who later inspired much work on Asia, notably the influential books by Huang (1985, 1990) on China. For a critique of this concept, see Booth and Sundrum 1985, 209–12.

6. The relation between population growth and technical progress is at the core of some recent growth models, e.g., Galore and Weil 2000. However, unlike Boserup, they assume population growth to be endogenous and the result of rational decisions by households.

7. See the book edited by Koppel (1995), which includes a rejoinder by Ruttan and Hayami and also Iftikhar and Ruttan 1988.

8. For further discussion of "casual" discovery and farmers' selection of suitable varieties, see Fowler 1994, 76–79; Brassley 2000c, 522–25; Tripp 2001, 46–53.

9. A key innovation was the use of sealed bottles to transport live plants, beginning in 1829 (Juma 1989, 46).

10. For Canada, see Norrie 1975; Adelman 1994, 28–31. For the USA, see Olmstead and Rhode 2000b, 2002, 938–43.

11. The Merinos arrived in Australia in 1797 and in South Africa in 1804. See Grigg 1974, 250; Raby 1996, 30; Duggan 1986, 35; McClelland 1997, 235.

12. For sweet potatoes, see Biggs and Clay 1981, 325; Huang 1985, 116 (for China). For maize and cassava, see Cock 1985, 14–17; McCann 1985, 102 (Ethiopia); Miracle 1966, 231–35 (Congo). For the planetary diffusion of maize, see Warman 2004, 37–41, 60–63, 66–70, 98–109.

13. Cocoa was introduced in 1870s–80s, while other cash crops in the 1900s. See Grigg 1974, 40–42; Allan 1965, 346–47; Anthony et al. 1979, 135ff, 249; Kates et al. 1993, 403; Wrigley 1970, 13–15 (Uganda); Martin 1987 (Rwanda and Burundi); Iliffe 1975, 274–90 (Tanganyika); Hopkins 1973, 137–40, 216–19 (West Africa). The groundnuts were brought from Africa to India (Blyn 1966, 109).

14. On the diffusion of viticulture in California and in other non-European areas, see Unwin 1991, 242–52 (for the early modern period) and 296–306 (for the 19th century).

15. This achievement was not matched by a similar success of "scientific" hybrid vines, available since the 1930s (Loubère 1990, 63–68, 74–75; Simpson 2000, 103–4). These yield more than traditional varieties, but their grapes are allegedly not good enough for premium wines. Thus, their use is limited to the mass production of cheap wine, while high-quality wines are still produced with traditional grapes (sometimes mandatorily).

16. See Kloppenburg 1994, 68–69, 93–110; Agulhon, Desert, and Specklin 1976, 426–28; Brassley 2000a, 603–5; Hayami and Ruttan 1985, 218; Darlymple 1988, 33–35; Fowler 1994, 48–53; Perkins 1997, 42–74. Hybridization is, at the same time, easier and more necessary for cross-pollinating plants (which need the pollen from another plant for

reproduction) than for self-pollinating ones (which can reproduce from their own seed). The latter maintain their original characteristics after replanting, while open-air pollination can add undesirable traits to second-generation seed. Wheat and rice are self-pollinating, maize cross-pollinating.

17. There is some uncertainty in dating the production and commercialization of hybrid maize seeds, but undoubtedly their use boomed after 1935. See Grilliches 1957, 517; Kloppenburg 1994, 91–92; Clarke 1994, 168; Tripp 2001, 69.

18. See, for further information, Kloppenburg 1994, 157–61; Evenson and Wesphal 1995, 2252–54; Pearse 1980, 35–40; Gill 1991, 198–213; Byerlee 1996; Bray 1986, 23–25; Darlymple 1979, 704–7; Perkins 1997, 106–12, 210–35; Smale 1997, 1258.

19. See Pingali et al. 1997, 209–10; Pingali 1998, 476; Byerlee 1996, 704–8 (who states that varieties need to be replaced every seven years); and (on a more cautious note) David and Otsuka 1994, 430-31; Ruttan 1994, 11. Hopes are especially pinned on hybrid varieties of rice (Bray 1986, 25)—a line of research pioneered independently by the Chinese in the 1960s and 1970s (Stone 1988, 790–98; Lin 1994, 377–85; Xu and Peel 1991, 85).

20. On sugar beets, see Blomme 1992, table 14; Bairoch 1999, 113, fn. 2. On cane sugar, see van der Eng 1996, 77–80; 216–17; Bill and Graves 1988, 15; Hayami and Ruttan 1985, 262–63; Headrick 1988, 237–43. For Cuba, see Dye 1998, 246.

21. See Judd, Boyce, and Evenson 1986, 91–92; Iftikhar and Ruttan 1988, 13; Alexandratos 1995, 186; Cock 1985, 34; and, on recent investments, Fowler 1994, table10. On the rates of adoption by crop, see table 6.2.

22. For instance, out of 13 international research facilities of the CGIAR network, only 2 deal with livestock (both in east Africa), 8 with field crops, and 3 with general issues such as policy. See Fowler 1994, table 10; Gill 1991, 205–8.

23. See McClelland 1997, 230–35; Huffman and Evenson 1993, 13–14; Olmstead and Rhode 2000a, 710–17; Turner et al. 2001, 96; and, for case studies on sheep-rearing, Bowie 1987; Raby 1996, 27–35.

24. The first major success of scientific hybridization was the production of the F_1 variety of silk cocoons in Japan in 1914, which had an impact on silk production as big as that of hybrid corn in the USA (cf. Kyokawa 1984; Federico 1997, 89–91).

25.The great campaign to eradicate bovine tuberculosis in the USA (Olmstead and Rhode 2004) from 1917 to 1962 cost about 260 million of 1918 dollars (equivalent to 1.6% of gross output of American agriculture in 1918), yielding benefits of some 3.2 billion dollars to agriculture and saving perhaps 25,000 lives.

26. Cf., on the history of biotechnologies and the controversies it raised, Pringle 2003 (the story of tomatoes on pp. 67–77).

27. See Norrie 1975; Solberg 1987, 117–125; Adelman 1994, 49–53, 226; and, above all, the recent article Libecap and Hansen 2002.

28. The key innovation was a technique to preserve silkworm eggs during the summer. Double cropping was widely adopted because there was enough land to grow additional mulberries to feed silkworms and enough unemployed labor to breed them during that period of the year. These conditions were not met in Italy, where the second crop never diffused. See Nghiep and Hayami 1979; Federico 1997, 136–39.

29. The traditional rotation produced 5 crops in 3 years (cotton and fodder the first year, beans the second, wheat and maize in the third), the new one produced 4 in 2 years (cotton and wheat in the first, maize and fodder in the second). Cf., on intensification in Asia, table 6.11.

30. This traditional interpretation can be traced back to Lord Ernle, and, in its modern version, has been put forward by Chambers and Mingay 1967; Deane 1967; Timmer 1969; and, more recently, by Overton 1996 and by Brenner and Isett (2002, 624–28). On the relation between agricultural growth and the industrial revolution, see section 10.4.

31. For a intermediate position, see Turner et al. 2001, 40–115. See also the survey by Hudson 1992, 64–75. In theory, the debate could be settled by good data on Total Factor Productivity. Unfortunately, the range of available estimates (table 5.4) is so wide as to support any statement. For the impact of some specific innovation on wheat yields, see Brunt 2004, 210–15.

32. Shiel 1991, 54–58; de Vries and van der Woude 1997, 203–5. Ambrosoli (1992) tells, in fascinating detail, the long story of the spreading of the botanical knowledge of grasses throughout Europe. He also argues that Great Britain in the eighteenth and early nineteenth centuries imported massive quantities of seeds.

33. The wheat was cultivated by short-term tenants (2–3 years), who, at the end of their period, sowed alfalfa. The pasture fed the landlord's cattle for some 10 to 15 years, and then the land was rented to wheat-growers again. See Grigg 1974, 248–50; Adelman 1994, 73; Solberg 1987, 64; and, for a similar development in New Zealand (in the 1900s), Hawke 1985, 31–33; and, in the Midwest, Bogue 1963, 140.

34. For examples from the Po Valley, see Cuppari 1870, 306–17; Porisini 1971, 131–32.

35. On potatoes, see the overall survey by Salaman 1949; Mokyr 1981 and O'Grada 1999, 13–18 (on Ireland); Simpson 1995, 74–78 (on Spain). On maize, see Warman 2004, 112–31; Coppola 1981 and Federico and Malanima 2004, 453–54 (for Italy); Grantham 1978, 321–24, and Clout 1980, 88–95 (for France). France is an interesting example, as the peasants chose the solution that best suited the environment and factor endowment: grass in the north and around Paris, maize in the southwest and Rhone, and potatoes in the northeast and Pyrenées.

36. In 1998–2000, the average yield was 38% higher for maize than for wheat in Europe and double that in North and Central America (FAO Statistical Database). The potato blight caused the Great Irish Famine of 1845–49 (section 2.2), while a purely maize-based nutrition caused a severe disease, pellagra, which was quite widespread among Italian peasants in the nineteenth century (De Bernardi 1984). Native Americans, and also, to some extent, slaves on plantations, escaped this fate thanks to a more varied diet, which included pork.

37. See Warman 2004, 151–71 (on maize as human food in the southern USA), 174–91 (on maize as feedstuff in the northern USA), and 197–202 (on the Soviet decision, allegedly taken personally by Khruschev, to imitate the American system). The USSR maize acreage jumped from 4.3 million hectares in 1954 to 37 million in 1962 (Gregory and Stuart 1986, 233).

38. FAO Statistical Database; Morris 1998, 23–26. About half of the maize used as animal feed is processed.

39. The book was commissioned by the British Association for the Advancement of Science. See., Moulton 1942, 1–5; Grantham 1984, 198–200; Brassley 2000a, 594–99. For a short historical overview, see Russell 1973, 3–22.

40. Information from Moulton 1942; Aftalion 1991, 26–27, 84–89; Brassley 2000a, 611; Smil 2001, 48–57, 109–30.

41. The Bordeaux mixture was successfully used also to fight the potato blight after 1882 (Brassley 2000e, 553; Mokyr 1990, 139).

42. Actually, DDT had been synthesized in 1874, but its insecticide power was discovered only in 1939, by the Swiss scientist P. H. Muller, and it was released for private use in 1945 (Russell 1999).

43. See also Hurt 1982 and 1994, 101–3, 134–40; Bogue 1953, 148–53; Brunt 2003b, 446–53 (for England); Raby 1996, 78–86 (for Australia).

44. The Whitney gin was not the first machine to strip lint (there had been antecedents in India), but it was the first to be suited to short-stapled American cotton (Lawkete 2003).

45. See Bairoch 1999, 72 (for the United States); Hayter 1939; Bogue 1963, 79–80; Primack 1969, table 1 (the estimate of the labor input); McMichael 1984, 216–18 (for Australia); Hawke 1985, 30–33 (for New Zealand). Barbed wire was not the only possible alternative to lumber, but it was by far the best one: live fences (e.g., of the Osage tree) were slow to grow, and ditches or sod fences required much work.

46. Distinguishing between agriculture and processing is easy neither in theory nor in practice, as, in the long run, the physical location of processing has been changing. Traditionally, peasants processed most of their products (e.g., grapes into wine, milk into butter and cheese), while nowadays in "advanced" countries they sell them to food industries. National account definitions set different (and somewhat arbitrary) criteria to divide the processes, which do not take the location (on- or off-farm) into account, but only the nature of the product to be processed. For instance, threshing is part of agriculture, while butter-making is not. In any case, technical progress in processing reduces the gap between consumer and farm-gate prices and thus, *ceteris paribus*, augments the farmers' income. On the other hand, the industrialization of processing reduced the demand for on-farm labor during the slack season.

47. For sugar-processing, see Dye 1998, 30–31 and Grigg 1974 p.218; for the butter separator, Jensen 1937, 174–76; for the palm-oil presses, Martin 1987, 118; for wine presses, Loubère 1979, 97–110 and 197–200, and Loubère 1990, 89–104; for oil presses, Simpson 1995, 168; for coffee processing, Holloway 1980, 30; for cotton gins, Hurt 1982, 101–4; for rice milling in Thailand, Manarungsam 1989, 68–70. All these breakthroughs were preceded by many failed attempts. See the list of major innovations, according to the date of patents in Schmookler 1966, appendix D.

48. See Hurt 1982, 40–56; McClelland 1997, 151–62. On the debate about their adoption, see section 7.6. Australian farmers in the late 1840s used a different machine, the so-called stripper (Grigg 1974, 264; Raby 1996, 86; Shaw 1990, 7). It extirpated the wheat instead of cutting it. Thus, it wasted a lot of grain, and was suitable only for very land-intensive cultivation.

49. See Hurt 1982, 77–83; 1994, 146 and 194. On the diffusion in California, see Olmstead and Rhode 1993a, 105.

50. For milking machines, see Grigg 1974, 200; Brassley 2000h, 74; Bairoch 1999, 73; Bateman 1969, 208. For corn pickers, see Burnell 2001; Clarke 1994, 170–75; and also (on the early work) Bogue 1963, 159–63. On cotton harversters, see Holley 2000, 35–57, 93-113; Whatley 1987; Heinecke 1994, table 1; Grove and Heinecke 2003, 742–45. Whatley (1987) argues that the International Harvester Company (which had bought the patent from the inventor, A. Campbell) could have built and marketed the cotton harvester much earlier, but it chose not to since institutional problems reduced the demand for a cotton-harvesting machine (section 8.8). Even if he were right, the lag with the wheat reaper would still have been substantial.

51. Cf., Burrows and Shlomowitz 1992, and Bill and Graves 1988 p.15.

52. On early machines, see Francks 1996, 790–93; Bray 1986, 57–60; Ishikawa 1981, 19; Hayami and Yamada 1991, 88. For recent developments, see Pingali et al. 1997, 273–80.

53. Holloway 1980, 28. These machines could pick only perfectly dry fruits, so that the harvest lasted too long and interfered with the next crop.

54. On the tomato picker, see Rasmussen 1968, 533–35, and, on its adoption, Gardner 2002, 18. On grapes, see Loubère 1990, 53–60; Unwin 1991, 345–46.

55. Mokyr (1990, 139) suggests that reproducing the movement of the hand (e.g., for picking fruits) is inherently more difficult than reproducing that of the whole arm (as in the reaper).

56. Mechanization was easier for processing than for fieldwork also because the innovations could use technologies developed in other sectors for similar operations, such as crushing.

57. On the early history of the motorization of American agriculture, see Wik 1953, 203; Sargent 1979, 3; Olmstead and Rhode 2000a, 704–06; Gardner 2002, 11–12. Bairoch (1999, 65) reports a figure of 90,000–110,000 "tractors" worldwide in 1910, which seems too high, even including steam-powered ones. Moreover, the figure of 3,000 tractors in the German empire in 1907 (Corni 1989, 227), probably includes the steam-powered machines.

58. The pumps were useful especially in rice-producing countries (Francks 1984, 219–23; 1996, 783–84; and Bray 1986, 105–6), while trucks long remained an exclusive of American farmers (Olmstead and Rhode 2000a, 711–13). Electricity arrived even later: in 1935, only 11% of American farms had it (Gardner 2002, 15).

59. The acreage estimate comes from Historical Statistics 1975, series J52, J 53. The reduction in land use is one of the key components of the total social savings from motorization, which White (2001, 495) estimates to have exceeded 8% of GNP in 1954.

60. Even under perfect competition among seed producers, the purchased seed must be more expensive than its own product, because it includes the cost of selection, packaging, and so on (Tripp 2001, 41–42). The price gap is, of course, much bigger for hybrids, as producers have to cover the expenditure in research, development, and production.

61. All econometric works use (ex post) market prices for factors, which may diverge from expected farm-gate prices for a series of reasons (for land, see section 7.3). This approximation introduces a potential bias in statistical analysis.

62. Hayami and Ruttan 1985, 187-197. They compute the counterfactual factor shares in the *i*th year given the elasticity of substitution and the change in factor prices between 1880 and the *i*th year. For instance, without labor-saving technical progress, in 1980 the labor share in American agriculture would have been 63% higher than the actual one (and the shares of land, capital, and fertilizers would have been correspondently lower).

63. Olmstead and Rhode 1993b. The Williamson (2001) series confirms the fall, while Hayami and Ruttan's own series of the wage-land price ratios (1985 Table c.2) show a sudden decline in relative wages at the beginning of the twentieth century, and then a substantial rise, which brought the 1940 level well above the 1880 one.

64. Olmstead and Rhode 1998. The authors claim that lagged factor price represents the induced innovation effect because they also include current factor prices in the set of variables, which should pick all short-term substitution effects. As expected, the coefficients are negative. The overall fit of the regression is much improved by adding a set of area dummies. The results are somewhat more supportive of the induced innovation hypothesis for the period 1940–80 only.

65. Thirtle et al. 2002. The design of the test differs from the Olmstead and Rhode for several reasons. They use nation-wide data instead of state-level data, they distinguish short- and long-term adjustment with a dynamic ECM approach instead of with different price data, and they add R&D expenditure and farm size. In an induced innovation framework, the latter variables are endogenously determined by (exogenous) changes in factor prices.

66. See the surveys by Capalbo and Vo 1988, 109–15, Mundlak et al. 1997, 20–24; Mundlak 2001, 33–34. See also the historical work by Hayami and Ruttan 1985, 192 (who show that, in Japan until about 1915, technical progress was labor intensive), and Wade 1981, tables 5.1, 5.2 (who finds, for the period 1870–1965, labor-saving bias in Denmark, land-saving bias in the UK, and no bias in France).

67. Not all data are very accurate, especially for Western countries. In fact, statistics omit abandoned land, crop failures, and sometimes double cropping. In traditional Western agriculture, double cropping was quite rare, but it has been growing in recent times with the massive use of fertilizers (Gardner 2002, 29).

68. See Chao 1986, 196–97; Xu and Peel 1991, 68–70; Ash 1996, table 12.8.

69. On yields, see section 5.2. The small reduction in the Chinese ratio in 1938 is to be attributed to the extension of cultivation in Manchuria, where double cropping is impossible: however, the ratio increased again, to 1.56 in 1992 (Zong and Davis 1998 p.24). For a case-study of intensification in the Philippines, see Hayami and Kikuchi 2000, 27–31.

70. On the intensification of African agriculture, Eicher and Baker-Doyle 1992, 93–97; Gleave and White 1969, 280–88; Grigg 1974, 57–69; Mosley 1983, 76–82; Anthony et al. 1979, 128–35; Pingali 1989, 244–47; Kates et al. 1993, 14–21; McCann 1995, 56–61; Gray and Kevane 2001. On the institutional consequences on property rights, see section 8.2.

71. FAO 2003, table A7. See also Alexandratos 1995, 15.

72. Cf., Tripp 2001, 55–71. He estimates (table 4.2) that noncommercial seeds (i.e., not produced by seed companies) still account for a substantial proportion of total use in some "advanced" countries. Local seed (and local suppliers) enjoyed a clear advantage in an uncertain world. For the history of long-range trade in seeds in the USA, see Fowler 1994, 36–39.

73. See Hayami and Yamada 1991, 128; and, for the USA, Olmstead and Rhode 2002, 938–46 (wheat); Grilliches 1957, 517 (hybrid corn). For European countries, see Darlymple 1988; Smale 1997.

74. The figures of the table refer to the quantity of fertilizers per hectare of arable land and tree-crops. This measure would overvalue the actual consumption of fertilizers per unit of land if fertilizers are spread also on pastures and on double-cropped acreage (Smil 2001 p.141). Differences in the denominator may account for the differences between the figures of Table 6.3 and those from other sources (e.g., Knibbe 2000 Table 6).

75. The consumption in Russia rose from a mere 1 kg/ha around 1900 to 6.9 kg in 1912 (Lyaschenko 1970, 734; see, also Berend and Ranki 1974, 71). Clearly, nationwide American figures conceal huge regional differences: most fertilizers were used on the East Coast and in California (Wines 1985, 148–61).

76. The use of pesticides in 1998 was more than double in Europe ($102 per hectare) than in the United States ($40). See Wood et al. 2000, table 9.

77. For Europe, see Collins 1973. For USA, see Turner et al. 2001, 92; McClelland 1997, 129–50. In Indonesia, the scythe substituted the traditional sickle in the 1970s only

because the new dwarf varieties of rice had to be cut close to the soil (van der Eng 1996, 182–83; Booth 1983, 180).

78. See section 4.3. For information about the factor intensity, see Boserup 1951, 23–26; Gill 1991, 127; Duggan 1986, 105–8; McCann 1996, 48–50; Eicher and Baker-Doyle 1992, 113, 123–27; Hopkins 1971, 37; Mosley 1983, 78.

79. For the diffusion of the reaper in the USA, see David 1973 (and the discussion of section 7.6). For the UK, see Brigden 2000, 511. Bairoch (1999, 55) states that in the whole of Europe there were about 100,000–120,000 reapers by 1880, one million by 1900, and 45 million by 1920.

80. Olmstead and Rhode 2001, tables 1, 3; Sargent 1979, table 1.6; Hayami and Ruttan 1985, table C2. Estimates of the share before the war are a bit higher, but the trend is similar.

81. For Japan, see Hayami and Ruttan 1985, table C3. The figures for the UK and Germany are computed assuming an average of 10 HP per tractor and 1 HP per horse (Hayami and Ruttan 1985, 474). The number of tractors in Germany is from Corni (1989, 227), while the number of horses from Mitchell (1998c, table D5) has been reduced by one quarter to take the non-rural horses into account. The number of tractors and rural horses for the UK comes from Collins (1983, 73–76), who strongly downplays the relevance of tractors. A similar computation would be almost meaningless for Asia, Africa, and also the Mediterranean countries (including Italy and France), where most of the draft power was provided by oxen.

82. For Russia, see Lyaschenko 1970, 784; Wheatcroft and Davies 1994, 112. For Hungary, see Voros 1980, 68; Eddie 1971, 307. For Anatolia, see Quataert 1981, 78. For Mexico, see Millar 1995, 65–66. Clearly, threshing and other machines were widely adopted in rich labor-scarce countries such as Australia, Argentina, and Canada (Raby 1996, 88–91; Solberg 1987, table 6.1; Adelman 1994, 235–43; Diaz-Alejandro 1970, table 3.11). For a case study on the adoption of steam threshing, see Federico 2003.

83. Bateman 1969, 208–10. The low opportunity cost of labor is only one of several possible explanations of the slow diffusion of milking machinery. It could have been slowed by the need of a minimum size of the herd (see the discussion of scale economies in section 7.6) or by the lack of suitable source of power before electricity (Bairoch 1999, 73).

84. For tractors in the USA, see Olmstead and Rhode 2000a, 705; Brassley 2000h, 74. For cane harvesters, see Burrows and Shlomowitz 1992; for power tillers, Francks 1996, 788; and for cotton pickers, Holley 2000, 94.

85. This happens if the new technique is (1) nonrival by its nature (i.e., can be used by anyone without limiting someone else's use), and (2) not excludable (i.e., its free use cannot be prevented by institutional arrangements, such as a patent). If partially excludable, it is often called an "impure" public good (or, as in the text, only partially appropriable).

86. See, among many others, Alston et al. 1998. They argue that (1) agricultural R&D is too risky and expensive to be undertaken by a single firm, (2) it can entail negative externalities (e.g., pollution), which a private firm would not take into account, and (3) it may pay off only after a long delay. These features, however, are common to other sectors, where public support to R&D is much less widespread than agriculture. One may add that the small size of most agricultural firms (section 7.6) hinders investment in R&D if the latter entails some economies of scale. Clearly, farmers can pool their resources, but this increases transaction costs and possibly the scope for free riding.

87. See Brassley 2000d, 546 (UK); Simpson 1995, 109 (Spain); Hayami and Yamada 1991, 74 (Japan); van der Eng 1996, 112 (Indonesia); Wines 1985, 126–35 (for the USA).

88. Francks 1996, 788. For the similar case of the cotton harvester in USSR, see Pomfret 2002.

89. See Fowler 1994, 77–78; Federico 1997, 88–90; and, for seeds, Tripp 2001, 70–71.

90. The story of this legislation is narrated in detail, with an unsympathetic view, by Fowler 1994, 103–20, Kloppenburg 1994, 132–50; Juma 1989, 152–69.

91. See Darlymple 1988; Byerlee 1996, 712; Kloppenburg 1994 (esp. 278–90); Frisvold and Condon 1998; Pray et al. 1998; Huffman and Just 1999, 14–18; Tripp 2001, 82–86; the World Bank 2002, 46–50; Runge et al. 2003, 91–96. Binembaum et al. (2003) discuss the effects of patenting on R&D, and Sunding and Zilberman (2001, 224–26) the optimal incentives to innovation.

92. In a quite detailed analysis of the adoption of hybrid maize seeds in Latin America in 1996, Kosarek et al. (2001) show that its market share is negatively correlated with its price, but positively correlated with the degree of legislative protection.

93. David 1974. For a review of the debate, see O'Grada 1994, 152–56.

94. On steam plows, see Brigden 2000, 510; on the straddle tractor for vineyards, Loubère 1990, 44–49; and, on the power tiller, Francks 1996. The presences of scattered trees in the fields, quite common in traditional Mediterranean agriculture, also makes the use of machinery more difficult.

95. Rasmussen 1968; Hurt 1994, 359–61. For tomatoes, see Hurt 1994, 360–61; for cotton, Holley 2000, 127–29.

96. Pingali et al. 1997, 236-237. Cf., for further evidence, Rao 1994, 13–15, David and Otsuka 1994, 410–412, Hayami and Yamada 1991, 165, Foster and Rosenzweig 1996, 942, and, for Africa, Anthony et al 1979, 256–257.

97. Cf., for some evidence about the unbalanced use of fertilizers, Johnston 1998 (especially, 45–49), and, for a historical case-study (about the American South in the 1860s), Earle 1992.

98. Evenson and Westphal 1995, 2248. On this point, see Englander 1991; Byerlee 1996, 710.

99. Evenson and Westphal (1995, 2254–56), in an analysis on India, measure this environmental distance with the difference in yields of the *same* variety of rice in experimental fields in two districts. They find a quite large effect: a 36% increase in distance reduces the spill-ins by 80%. McCunn and Huffman (2000), in their analysis of the determinants of TFP growth in the USA from 1950 to 1982, group a priori states according to geographical closeness and environmental similarity. R&D expenditure in similar states did affect TFP growth positively in crop production, but negatively in the livestock sector. They also find that TFP growth was not statistically related to local R&D expenditure. This result is undoubtedly unexpected: it may reflect some free riding on R&D by states.

100. A case in point is dry farming in the North American prairies. Its adoption in the 1900s and 1910s coincided with a succession of unusually wet years. The abnormally high yields were falsely attributed to the new technique, which enjoyed an undeserved reputation. Libecap and Hansen (2002) speculate to what extent better information on "normal" weather conditions in those areas could have helped farmers to avoid the initial enthusiasm and the subsequent losses.

101. An individual is defined as risk averse if he or she prefers a secure payment to placing a bet with the same average outcome. For some empirical evidence on peasants' risk-aversion, see Kanwar 1998, 21–40.

102. On the early sociological literature about innovators, see Grigg 1982, 154–55, and, for a general overview of the issue of risk and adoption of innovations (including the role of risk-reducing institutional innovations), see Sunding and Zilberman 2001, 235–46.

103. For additional references on the positive effect of education, see Jamison and Lau (1982, esp. 21–45); Huffman 2001, 354–64. For a discussion of the determinants of TFP growth, see section 6.6.

104. See Russell 1973, 9–11; Grantham 1984, 196–98; Brassley 2000a, 600–602. On Mockern, see Finlay 1988. Juma (1989, 54) lists some forerunners in the colonial USA, but they were too short-lived to have much of an impact.

105. For the data, see Brassley 2000a, 617, and, for a general survey of the literature, Ifthikar and Ruttan 1994, 49–60. For Italy, see Orlando 1984, 49–51; for France, Wade 1981, 237; for Canada, Estey 1988 and Solberg 1987, 109–13; for Australia, Raby 1996, 144–49, and Shaw 1990, 7; for Russia, Antsiferov 1930, 79. An exception to this flourishing of initiatives was Argentina (Solberg 1987, 111).

106. Grantham 1984, 192. In 1900, there were about 500 experimental stations all over the world.

107. See the accounts (sometimes with different details) by Huffman and Evenson 1993, 13–20; Marcus 1988, 12–26; Hayami and Ruttan 1985, 212–14; Olmstead and Rhode 2000a, 714–17; Hurt 1994, 189–94, 256–60; Kloppenburg 1994, 57–65; Clarke 1994, 28–32; Alston and Pardey 1996, 10–15; Sheingate 2001, 40–43, 47–54.

108. Hayami and Yamada 1975, 237–46; Hayami and Ruttan 1985, 231–40; Francks 1992, 123–28; Bray 1986, 23. For different, somewhat dismissive opinions, see Nakamura 1966, 83–87; Sheingate 2001, 56–58.

109. On seed production, see Eicher and Baker-Doyle 1992, 111–15, 134–38; Tripp 2001, 115–29. See also the case studies in van Veldhuizen 1997.

110. On Africa, see Headrick 1988, 215–18; Anthony et al. 1979, 249–53; Eicher and Baker-Doyle 1992, 97ff. On Indonesia, see van der Eng 1996, 77–83. The Indonesian Sugar research facilities remained private until 1928, when a looming financial crisis forced the government to make contributions by planters compulsory.

111. Van der Eng 1996, 69–74, 80–86. Other exceptions were two stations for rice in British Burma (Cheng 1968, 39–48) and the substantial network of stations that Belgium set up in the Congo in 1933 (Miracle 1965, 242–44). The Japanese also invested heavily in R&D and extension services to increase production of rice in Korea and Taiwan (Ban 1979, 110–11; Lee and Chen 1979, 85–86; Juma 1989, 76–79; Hayami and Ruttan 1985, 280–94), but their motivations were far from altruistic. They aimed at using their colonies to supply the domestic market, but Japanese consumers preferred the Japonica variety of rice, which was not traditionally cultivated outside Japan.

112. See Brandt 1997, 304–7; Feuerwerker 1983, 67; Myers 1970, 180; Pomeranz 1993, 72–88 on cotton; Federico 1997, 184–85 on silk. For Thailand, see Feeny 1982, 52–56.

113. Fowler 1994, table 10. The Soviet Union and other socialist countries decided not to join and chose to develop their own national systems.

114. See the very interesting analysis by Brunt (2003b, 466–70) on the patterns of diffusion of improved plows in England.

115. For data on the number of published books and journals in the UK, see Sullivan 1984, tables 1, 2, and for the information on print runs, Goddard 2000, 682–86. On the USA, see McClelland 1997, 203–10; Farrell 1977; Bogue 1963, 193–201; Ilbery 1985, 77; Scott 1980, 18–22; Fowler 1994, 39–40. On Belgium, see Blomme 1992, 262–63. On the parallel development of agricultural literature (*nong-shu*) in China, see Gang Deng 1993.

116. Clearly, he would run the risk of losing readers and thus ultimately ruin his business, but, in the short run, he might go unnoticed, especially if the suggestion to buy was not clearly outlandish. In fact, farmers were seldom privy to the whole range of possible alternatives. For the case-study on silkworms races in nineteenth-century Italy, see Federico 1997, 88–89.

117. See Hurt 1994, p.149; Fowler 1994, 39. By 1910, according to a survey, almost all U.S. farmers in the northeast and midwest read the agricultural press (Scott 1970, 20).

118. On American seed companies, see Fowler 1994, 36–39; Tripp 2001, 36–37. The story of Montecatini is a personal communication from F. Amatori.

119. For details and further information, see Grantham 1984, 200–203; Brassley 2000a, 606–10; Goddard 2000, 651–60; Raby 1996, 114–37 (for the UK); van Zanden 1994, 182–84 (for the Netherlands); Jensen 1937, 180 (for Denmark); Agulhon et al. 1976, 117–118; Clout 1980, 62–63 (for France).

120. A good example is the Comizio Agrario of Piacenza (Banti 1989, 139–48). After some years of decadence, it was on the verge of disbanding when it started to test and produce fertilizers with great success. For other case-studies of "dormant" societies, see Raby 1996, 121 for Australia.

121. For a discussion, see Hanson and Just 2001, 778–79. They recall the existence of a wide range of mixed systems—such as paid public extension (where farmers paid a fee, which was lower than the full cost) or public-funded private extension (where specific tasks were contracted out to specialized firms) and subsidized private extension (with tax breaks or subsidies). The privatization of the extension services, however, is only beginning.

122. For example, see van der Eng 1996, 121–25 on Indonesia; Quataert 1981, 83–84 on the Ottoman Empire; Voros 1980, 61 for Hungary; and Francks 1984, 153–54 for Japan.

123. See respectively Henriksen 1992, 162; Barral 1979a, 374; van der Eng 1996, 123. See also Kotsonis 1999, 97–101.

124. See Brassley 2000a, 625–40 (UK); Voros 1980, 61 (Hungary); Reichrat 2004 (Germany), D'Antone 1991, 404–5 (Italy).

125. See Hayami and Yamada 1991, 72–73; Francks 1999, 75; Sheingate 2001, 67–70; and, for a grass-root view, Francks 1984, 150–56.

126. On the USA, see Huffman and Evenson 1993, 22–24, and Scott 1970, 288–313; on Italy, Orlando 1969, 26–28, and Rossini and Vanzetti 1986, 567; and, on Indonesia, van der Eng 1996, 94–107. Also, on the United Kingdom, see Brassley 2000a, 640–49; and, on Bengal, the quite critical view by Islam 1978, 10–11.

127. This system aimed at establishing a permanent relationship between the extension worker and a fairly small group of peasants, with a regular schedule of home visits. See Feder and Slate 1993, 53–54; Antholt 1998, 354–56.

128. For the USA, see Alston and Pardey 1996, table 2-A.3; for Japan, see Hayami and Yamada 1991, table A.13. The difficulties of measurement are discussed by Judd et al. 1986, 79–81, and Alston et al. 1998, 1062–64.

129. Grantham 1984, 192. The sum was about a third higher than the American outlay at the same time, while European output (excluding Russia) was 50% higher.

130. The differences must reflect some difference in coverage, definition, or conversion into dollars (Alston et al. 1998 use PPP, while Judd et al. 1986 may have used exchange rates). According to Judd et al. the "transition" economies (omitted from the table for consistency) spent on R&D 992 million (1993) dollars in 1959; 2,238 million in 1970; and 2,606 million in 1980; and in extension 639, 980 and 1,305 million respectively.

131. On spillovers from non-agricultural research, see Evenson 2001, 607–8.

132. See Iftikhar and Ruttan 1994, 12–14; Alston and Pardey 1996, 38; Alston et al. 1998, 1066; Pray and Umali-Deininger 1998, 31–32, and, for the flow of information, see Wolf 1998.

133. See Judd et al. 1986, table 1; Pardey and Beintema 2001, tables 1, 3. The distribution of expenditures in public R&D is different if the world total includes the "transition economies": in 1959, they accounted for 27% of the total (reducing the share of advanced countries to 49%) and, in 1980, for 20% (with the advanced countries at 43%). China alone accounted for 2.5% of expenditure in 1959 and 9% in 1980.

134. In 1980, the cost per man/year in R&D varied from $130,000 in North America to $25,000 in Southeast Asia (Judd et al. 1986, table 5). The range was even wider for extension workers, who need less training ($52,000 vs. $1,000).

135. Judd et al. 1986, tables 1, 2. Both figures exclude transition economies: adding them would yield marginally lower rates for expenditures in extension (4.2%) and R&D (6.1%).

136. In 1989, there was one extension worker for every 400 farmers in the USA, and only one for every 2,500 in the LDCs (Alexandratos 1995, 347).

137. See Grilliches 1957 and 1958. On the methodology, see Evenson and Westphal 1995, 2272–73.

138. Schimmelpfenning and Thirtle (1999, 465–66) compute the internal rate of return to total public RD in advanced countries in the 1970s and 1990s with and without spill-ins from domestic patents and foreign public RD. The difference is huge: the addition of spill-ins lowers the rate from from 64% to 5.7%. See also the estimates by Traxler and Byerlee (2001) on spill-overs in the Indian R&D system.

139. See Alston and Pardey 1996, 207–18; Schimmelpfenning et al. 2000; Alston et al. 2001, 14–19, 27-32; Sunding and Zilberman 2001, 226–28; Evenson 2001, 612–16. The list of these errors includes misspecification of the counterfactual (arguably, it should be other R&D, not no R&D), the choice of wrong lags to compute the stock of knowledge, the underestimation of the true opportunity costs of resources for R&D (e.g., from taxation), the overestimation of benefits of the increase in production for neglect of social costs (e.g., from pollution), and so on. The net effect of these biases is difficult to assess. On balance, the authors suggest that the rates of return are likely to be overvalued but still high.

140. On the "urban" bias, see Iftikhar and Ruttan 1994, 10–13; Pray and Evenson 1991, 363.

141. Schimmelpfenning et al. 2000. This bias for both extension and R&D suggests a consistent policy, probably motivated by the desire of the apartheid regime to minimize the use of black manpower.

142. Marcus 1988, Alston and Pardey 1993, 17–23; and, for the literature on the political economy of agricultural R&D spending, Sunding and Zilberman 2001, 218–24. Actually,

the habit of granting special congressional funds was much older: in 1935, the Bankhead Act had divided the total funds proportionally to the population (Kloppenburg 1994, 86). Marcus argues, however, that political influence had been sidelined in the 1950s by the diffusion of a peer review model for the selection of projects.

143. See Becker 1996, 65–79; also, section 9.6. Other sources suggest that the Chinese research system fared better, at least after 1956 (Stone 1988, 790–94).

144. Birkhaeuser et al. 1991; Evenson 2002. Interestingly, this general conclusion is somewhat at odds with the results of two recent studies. Huffman and Evenson (2001, table 2) find a consistent pattern for the macroeconomic effect of extension on TFP growth in the USA (1950–82): all the variables, including interaction with public and private R&D and schooling, are positive for livestock and negative for crops. Owens et al. (2003) find, with microeconomic data, that extension increased land productivity in Zimbabwe in the 1990s.

145. Evenson 1991, 320–25. He also shows that national and international research are sometimes complementary and sometimes substitutes.

146. Schimmelpfenning et al. (2000) suggest that extension workers in South Africa tended to focus on maximizing output, instead of profits.

147. This conclusion is supported by empirical results by Birkhaeuser et al. 1991 (a negative interaction between extension services and schooling) and Schimmelpfenning et al. 2000 (a negative shadow price of extension in the long run and especially in the 1980s and 1990s). See also Huffman 2001, 365–66.

148. Mundlak (2001, 26) points out that the two-stage decomposition implies that variables of the first stage (the explanation of TFP growth) are independent from those of the second stage (the explanation of production growth). He argues that the condition is unlikely to be met, and thus implicitly favors an integrated approach.

149. Huffman and Evenson 1993, 203–11; see also Gopinath and Roe 1997. In a later article (2001, table 2), Huffman and Evenson confirm the different impact of public and private research on aggregate TFP growth in crop and livestock production and deal with the impact of R&D on several "structural" variables—such as specialization in crops and livestock, size of the farms, and off-farm work. McCunn and Huffman (2000) show that R&D affected the convergence in TFP levels among states—especially via the spill-ins. Thirtle et al. (2002) show that R&D also determined long-run factor ratios: a 1% increase in R&D stock increased the machinery-labor ratio by 0.4%. Cf. the microeconomic analysis by Jin et al. (2002) on China.

150. See Pingali and Heisey 2001, table 5.3; and also Schimmelpfennig and Thirtle 1994 and Arnade 1998.

151. Dummies for the method of estimation (especially for the Malmquist "distance") and for the time interval (especially those for interwar years and for post-1950) are highly significant. Unfortunately, the lack of a measure of state intervention (including support to R&D and extension) for many countries/periods prevents the direct testing of the role of agricultural policies.

152. Although, as said, this agenda of research is not directly concerned with testing Hayami and Ruttan's model, its results have some interesting implications for it. The factor endowment does not determine productivity growth because it is possible to develop and adopt (different) innovations suited to any factor ratio. Arguably, in an extended Hayami-Ruttan framework, any differences in TFP growth rates should reflect only

advances in the underlying scientific and technical knowledge, which determines the rate of shift and the shape of the "meta-production function." Thus the significance of institutional variables casts some doubt on their "induced institutional innovation hypothesis."

Chapter Seven: The Microeconomics of Agricultural Institutions

1. See Binswanger and Deininger 1997, 1964; and, in the same vein, Ray 1998.

2. See Boserup 1955, 77–88; and also Deininger and Feder 2001, 290–93.

3. Overton 1996, 61. The quotation is particularly significant, as it refers to the UK in the eighteenth century, then one of the most "modern" countries in the world. The rights were even more complex in "backward" countries like India, as shown by the (sometimes hopelessly confused) descriptions by Stokes et al. 1983; Chaudhuri 1984; McAlpine 1983, 84–98; Stein 1992.

4. For Africa, the following account is based on descriptions by Allan 1965, 360–67; Miracle 1967, 31–33; Morgan 1969; King 1977, 356–67; Gleave and White 1969, 278–80; Eicher and Baker-Doyle 1992, 82–83; Bruce 1988, 24–30. For Asia, see Feeny 1982, 37–43; Cleary and Eaton 1996, 23–24, 47–57. Pooling of products among households was practiced only in very poor and backward societies, such as Bechuanaland, where it only disappeared in the 20th century (Duggan 1986, 80–85, 109–10).

5. See Miracle 1967, 17; Morgan 1969, 256–61; Wrigley 1970, 9; Eicher and Baker-Doyle 1992, 14–16; and, for the (somewhat atypical) case of Bechuanaland, see Duggan 1986, 74–80.

6. For Mexico, see King 1977, 94–95; Conchol 1995, 154–59; Walsh-Sanderson 1980, 16; for Brazil, see Holloway 1980, 71; for the debate on India, see Goswami and Shrivastava 1991, 248–50.

7. On the economics of slavery, see Barzel 1997, 105–13.

8. See Brandao and Feder 1995; Binswanger et al. 1995, 2719–21; Deininger and Feder 2001, 294–96; Johnson 2001, 295–98; World Bank 2003, 23–35, 113–16. Specifically on the environmental impact of poorly definined property rights, see also Lichtenberg 2001, 1257. The idea has been popularized by De Soto (2000, esp. 46–54), with the slogan of "dead capital" waiting to be mobilized.

9. See Atwood 1990; Bruce 1993; Hyden et al. 1993, 418–19; and, for some empirical evidence, see section 8.7.

10. Some countries (such as Thailand, with its 1954 Land Code) adopted a mixed system, with different categories of rights, from full ownership to simple traditional usufruct. Holders of "inferior" rights had the option to upgrade them. This system was meant to give some official sanction to customary rights (and extract taxes from holders), without a full titling. In 1981, squatters in state-owned forests were given a title to land, with limited acreage per household. See Cleary and Eaton 1996, 69–71.

11. The twin assumption of exogenous technology and crop mix is adopted to keep the discussion manageable, but it is clearly unrealistic. As repeatedly pointed out in chapter 6, technology and crop mix could be and are changed to match land and labor.

12. The following discussion refers to modern property rights only. The problem was different under traditional rights. Land was not really scarce in hunting and gathering and in swidden agriculture systems, while in European feudal systems, serfs were considered as

part of the estate. Slaves could be bought, sold, and hired on the market when necessary for cultivation (Fogel and Engerman 1974, 52–58). Serfs and slaves were compelled to move with the whole plantation, if their masters wished. In the southern U.S. whole plantations with all their slaves migrated and resettled.

13. Ray 1998, 410–13. In theory, under standard conditions, one well-functioning market (labor or land-cum-credit) only is a sufficient condition to have a perfect allocation of resources.

14. Binswanger et al. 1995, 2707–9; Deininger and Feder 2001, 307–9. Unfortunately, the data on the agricultural land market (e.g., on number of transactions and prices of land) are extremely scarce, even for the USA (Lindert 1988).

15. Rawal 2001, 616–17. The turnover is higher in some Latin American countries—up to a maximum of 5% in Colombia, corresponding to a full rotation in 20 years (World Bank 2003, 113).

16. Other authors suggest more detailed classifications, which include "traditional" estates. For instance, Grantham (1989b, 1–10) considers "slave plantations", "feudal estates", "capitalistic farms" and "peasant farms". Binswanger et al. (1995, 2661–62) list "*hacienda*" (i.e., the feudal farm), the "landlord estate" (the tenanted estate), the "Junker estate" (a labour-intensive managerial farm), the "large commercial farm" (a highly mechanized, capital intensive managerial farm) and the "wage plantation" (a "Junker estate," which specialized in one cash crop). De Janvry (1981, 109–13) considers three types of large farms only (the "pre-capitalistic estate", the "large commercial farm" and the "capitalistic estate"), but four categories of small farms (the "family farm", the "subsistence farm internal to the pre-capitalistic estate", the "external sub-family farm" and the "subsistence farms in corporate communities").

17. A popular, but somewhat arbitrary, way to tackle the issue is to set criteria a priori. For example, one could define family farms as those that resort to less than, say, 50% of hired work and yield the owner a per capita net income of not greater than, say, the double of the national average, and assume that all others farms are managerial ones.

18. See Giorgetti 1974, 292–302; Biagioli 2002, esp. 84–85. Innovative landowners, such as Ricasoli, the "inventor" of Chianti (as a high-quality wine) oversaw their *mezzadri* much more closely than more traditional landowners. See Biagioli 2000, esp. 335–57.

19. Brandt 1989, 140; Huang 1990, 106; Deng 2003, 498. Sometimes, this contract was stipulated as part of a credit transaction: For example, household A, owner of land, borrowed money from B, and went on cultivating as a tenant, paying a rent. At the end of a very long period, he could redeem full ownership by repaying the loan. If he did not succeed, he would lose the ownership, and remain as a tenant.

20. For Ireland, see Turner 1996, 8–9; for the Balkans, Lampe and Jackson 1982, 189, Berend and Ranki 1974, 60–62; and, for Guangdong, Lin 1997, 127–29.

21. Allen and Lueck (2002, 69–84) argue that there can be two efficient solutions only for the division of expenditures in share–cropping contracts—either the same shares as output or all borne out by the tenant. They find some evidence for this dichotomy in a datasets of American contracts in the 1980s, but they would not find much historical support. If any, expenditures were more often borne by the landlord than by the tenant. Indeed, Allen and Lueck implicitly assume that the two parties face the same cost of capital, a case that does not seem so common in history (section 7.4).

22. For Tamil-Nadu, see Hayami-Otsuka 1993, 89, Tomlinson 1993, 81; for Argentina, Adelman 1994, 13. The two cases are indeed exceptional. In the former, "the

landlords" share included the interests on debts, while, in the latter, the tenant supplied all the capital.

23. A nice example is the *hunusan* contract for harvesting rice in the Philippines. Traditionally, harvesters got one-sixth of the crop, but the rise in yields after the Green Revolution would have raised the implicit wage above the (stagnant) market wage. Therefore the contract was reserved to workers willing to weed for free. See Hayami and Kikuchi 2000, 34–37; and, for a parallel change in Java, see Booth and Sundrum 1985, 214, 219.

24. These costs are often fixed per transaction, and thus they are proportionally greater for small plots or farms, which partially accounts for their higher unit price. For some evidence on very high costs in transition economies, see World Bank 2003, 97.

25. Ransom and Sutch 2001. They attributed to racism the difference between the actual share of African American farmers owning the land they were cultivating (a mere 20%) and the 55% they should have owned given the household features (age and education of the head, household composition, etc.).

26. See, for Nazi Germany, Corni 1989, 183; and for some examples of post–land reform restrictions see section 8.3. Clearly, all these restrictions can be circumvented, but at a price.

27. On this premium for small farms, see Clark 2002a, 286–89 (for 19th-century England), and Wade 1981, 194 (for France).

28. It is important to stress that both factors can also affect land prices negatively. Prices can fall *below* their true value if interest rates are deemed too high or if potential purchasers have pessimistic expectations about future returns.

29. This motivation, however, holds true only for farmers with some capital. In fact the double mortgage on the same land (first for purchase and then for productive purposes) is usually forbidden (World Bank 2003, 96). Deininger and Feder (2001, 307) argue that the "premium" makes full mortgaging of land impossible (thus further hampering sales). This statement implies that banks are ready to finance a purchaser only up to the true value of land. This behavior would be justified if banks had different information. Otherwise, they would lend according to the market price, inclusive of premium.

30. The literature quotes two other reasons. It is said that land is safer than other investments as it cannot be destroyed and that peasants prefer to own land in order to be self-sufficient. Neither is very convincing. Land without capital embodied is almost worthless, and capital in land is as vulnerable as any other capital to physical damage. Self-sufficiency is not necessarily such a paramount aim for peasants (see section 8.7), and production for domestic consumption is possible also under tenancy.

31. A free worker would supply labor until the return equates with his reservation cost (i.e., the wage he can earn in another job or the disutility of his work). Under a sharecropping arrangement, he would get only half the product of his additional effort, as the rest would accrue to the landlord. Thus, he would provide less than optimal effort, unless the landlord could force him by monitoring. In principle, the reasoning holds true also for the landlord: however, most of the factors he provides, land and capital, are fixed, and thus he cannot vary the supply. See Marshall 1920, 535–36. Cheung (1966) objected that, in a perfect market with zero monitoring costs, the landlord could bid his own share of product up in order to reduce the return to labor down to the tenant's reservation wage. Cheung's argument has been dismissed with two different counterarguments—that the shares are fixed and that monitoring costs are never zero. The former statement is clearly false. For

further references, see Chao 1982; Quibria-Rashid 1984; Singh 1991; Hayami and Otsuka 1993, 35–46; Ray 1998, 424–29; Deininger and Feder 2001, 312–13; Huffman and Just 2004, 620–22.

32. The reasoning owes much to Hayami and Otsuka 1993; Binswanger 1995; Binswanger and Rosenzweig 1995; Barzel 1997; Ray 1998; Deininger and Feder 2001; and World Bank 2003.

33. Any breach of contract is possible only if ex ante information is imperfect. Otherwise, the damaged party would spot the malignant intentions beforehand and not sign the contract.

34. See Berry and Cline 1979, 1–26; Schmitt 1991; Allen and Lueck 1998; Binswanger and Deininger 1997, 1964; Barzel 1997, 51–54; Binswanger 1995; Binswanger et al. 1995, 2694–700; Ray 1998, 448–50; Allen and Lueck 2002, 167–71.

35. Frisvold (1994) estimates that, in the ICRISAT sample, the productivity difference between hired laborers and family members on the same farm, all things being equal, exceeded 10% of labor product in about two-fifths of the cases. According to one estimate (Schmitt 1991, 452–53), the management of hired labor needed twice the time of family members. For a discussion of these other determinants of allocation of household labor, see Booth and Sundrum 1985, 114–22.

36. The imperfection of the agricultural labor market is widely regarded by economists (see, e.g., Berry and Cline 1979, 8–12; World Bank 2003, 81–82) and by historians (see, e.g., Koning 1994, 25–27) as a matter of fact, but it should, of course, be carefully investigated. One should not only take the local opportunities into account but also the long-term alternatives, such as emigration to cities. The design of tests depends greatly on the available data. For example, see Swamy 1998 (the simplest one) and Lamb 2003, 86–88 (a quite sophisticated one).

37. Chayanov 1966. This terminology may seem too abstract if applied to very poor peasants. However, even the poorest peasant must decide when the disutility of work exceeded the utility of additional goods—otherwise, he might end up working himself to death.

38. Hazell et al. 1986; Walker and Jodha 1986; Besley 1995, 2156–69; Ray 1998, 591–615; World Bank 2002, 43. On the origins of agricultural insurance in the UK in the first half of the 19th century, see Stead 2004, 339–46. The range of insurable events also depends on the local conditions, which determine the likelihood of their occurrence. The insurance premium would be impossibly high if an (observable) disaster happened every year.

39. This model is particularly popular for the USA (see Atack and Passel 1994, 388–91, 407–10; Atack et al. 2000b, 319–21). See also, on Argentina, Adelman 1994, 140.

40. It is usually assumed that tree crops are more subject to mistreatment and need more monitoring than wheat or other cereals (see, e.g., Ackerberg and Botticini 2002); also row crops are deemed relatively monitoring-intensive (Allen and Lueck 2002, 58–64).

41. For additional references and discussion, see Currie 1981, 102–14; Ellis 1988, 140–57; Otsuka et al. 1992; Hayami and Otsuka 1993; Ray 1998, 420–41; Deininger and Feder 2001, 309–12; World Bank 2003, 87–92.

42. Otsuka, Chuma, and Hayami (1992) extend the choice to permanent wage labor, but deal separately with uncertainty and enforceability.

43. Huffman and Just (2004) also argue that risk can account for other features of sharecropping contracts, such as the division by half of the output. The example is not so

felicitous, as the range of shares is quite wide. The model explicitly neglects monitoring and enforcement costs. Also Ghatak and Priyanka (2000) relate the diffusion of sharecropping to risk, but with a a different and somewhat unconventional approach. They assume that both landlord and tenant are risk-neutral and that, in the event of output failure, the fixed-rent tenant is not obliged to pay his dues. He thus has an incentive to adopt risky techniques, and the landowner opts for share-tenancy to prevent him from doing so.

44. See Allen and Lueck 2002, 50–55. The authors do not consider permanent wage labor as alternative because in the USA it is used only for casual unskilled labor. See also the similar model by Barzel 1997, 35–54. Banerjee et al. (2002) use a simplified version of the same framework, with risk-neutral individuals, assuming that only the tenant's effort is nonmoniterable. In this case, fixed-rent tenancy is more efficient, but the landlord would opt for sharecropping if the tenant is too poor to pay the rent in case of crop failure. Note that this model predicts a negative correlation between wealth and the diffusion of sharecropping—i.e., exactly the same prediction of the risk-based model.

45. The contract was supposed to expire at the death of the vine, but, in practice, it could be extended almost indefinitely by replanting the offspring of the original vine.

46. Hayami and Otsuka (1993, 97–98) resort to an indirect argument based on a comparison of relative efficiency and tenants' incomes between sharecropping and fixed-rent tenancy. The empirical studies that they review, mostly for Asia, show that productivity is similar, but sharecroppers earn less. According to the authors, sharecroppers forfeit the potential gains for the sake of less risk.

47. The diffusion of fixed-rate tenancy was positively related to the proportion of animals owned by the tenant and negatively to the number of supervisors. The interpretation of these results in terms of competing models is somewhat ambiguous, as one can interpret the first variable as a proxy for wealth.

48. In contrast, race comes out as not significant.

49. See Allen and Lueck 2002, 55–63, 95–118. While testing the transaction cost model, they proxy the potential damage to land in the regressions with the crop mix (with a priori assumptions about the potential for land mismanagement), the percentage of irrigated land, and the share of institutional or absentee landlords. In the risk-based model, they proxy risk-aversion with wealth and the share of farm revenues on total household income. On the characterization of owners, see Nerlove 1996, 22–24.

50. Ackerberg and Botticini's 2002 article deals with choice of contracts in 15th-century Tuscany, using an extensive database drawn from the 1427 cadastre. The authors find that both risk (proxied by the wealth of the tenant) and monitoring costs (proxied by the share of supervision-intensive vines) were important.

51. The theoretical literature on agricultural capital markets, indebtedness, and insurance, especially in the LDCs, is huge. The reasoning is inspired by Hoff and Stiglitz 1993; Binswanger 1995; Besley 1995; Ray 1998, 530–80; and, for credit in advanced countries, by Barry and Robinson 2001.

52. According to the World Bank survey (Wai 1977, table 3), in fifteen LDCs in the 1970s, about two-thirds of agricultural credit was used for "productive" purposes and the rest was used for consumption.

53. An interesting exception was the financing of the Malaysian rubber boom at the beginning of the twentieth century, with stocks issued on the London Stock Exchange (Stillson 1974, 590–97). Rubber plantations, however, were fairly big enterprises and were credited of meteoric prospects for growth.

54. For a short discussion, see Besley 1995, 2175–82, and for a tentative taxonomy, see World Bank 2002, 40. The credit co-operatives (section 7.5) are a border case. They belong to the "formal" sector as they are subject to state regulation and control, but rely on peer monitoring like the "informal" institutions (Besley 1995, 2186–85; Guinnane 2001).

55. For Maharashtra, see Charlesworth 1978, 109; for Bombay, see Charlesworth 1985, 95–116, and Kranton and Swamy 1999, 20–21; for Burma, Cheng 1968, 183–93. See also the general discussions by Amin 1982, 84–95; Chandavarkar 1983, 796–803; Bose 1994, 4–21; Tomlinson 1993, 64–65; Roy 2000, 74, 93. Moneylenders accounted for two-thirds of the total credit in Egypt in 1913 (Richards 1982, table 3.15).

56. Biagioli 2002, 84–86; and, foı similar long-term relations, see Francks 1984, 36–38 (for Japan); Amin 1982, 127–30 (for the Uttar Pradesh in India); Brandt and Sands 1992, 184 (for China). Some landlords helped their (wealthy) tenants even in the UK during the 19th century (Holderness 2000, 916–17).

57. See Unwin 1991, 325 (brand labels started to spread only at the end of the century, even for high-quality wines such as champagne); Clarke 1994, 65; Adelman 1994, 205–8.

58. From 1930 to 1933, some 10,000 banks failed (Alston et al. 1994, 411): two-thirds of them were rural ones, with assets consisting mainly in mortgages on land. The role of agricultural depression as a cause of this failure is, however, controversial (see White 1984; Thies and Gerlowski 1992; Hamilton 1985; Wicker 1996, 35, 164; and, for a survey of the discussion, see Federico 2004b).

59. For instance, Irish borrowers had to pay the co-signer (Guinnane 1994), and German ones had to pay substantial fees on all renewals of short-term loans (Guinnane 2001).

60. Cf. Charlesworth 1985, 94–116 (who does not seem particularly convinced about this conventional wisdom). This reasoning led the colonial administration to enact one of the earlier measures of tenancy reforms (section 8.3).

61. All the data refer to nominal interest rates (unfortunately authors do not provide information on price growth). They are thus potentially biased and potentially not comparable among countries/periods.

62. See Nagaraj 1984, 251; Islam 1978, 157.

63. For India, see Amin 1982, 116 (30–40% for sugarcane growers of eastern Uttar Pradesh); Stone 1984, 322 (12–25% in western Uttar Pradesh, in prosperous times after the construction of canals). For Burma (25%) see Cheung 1968, 171; for Thailand (35%) see Ingram; for China see Pomeranz 1993, 36 (6%–25% in different areas in 1910–11) and Faure 1989, 80 (30% in Guangdong in the late nineteenth century). Rates were also high in Africa—up to 150% according to Eicher and Baker-Doyle (1992, 175). In Europe, before the establishment of credit co-operatives, they were 30% in early 19th-century Germany (Guinnane 2001, 368), and 40% in the late 19th-century southern Italy (Galassi 2001, 46).

64. See Feuerwerker 1983, 87; Pomeranz 1993, 36. For local cases, see Huang 1985, 188; Lin 1997, 138. These figures refer to loans in money. Rates on loans in kind were higher.

65. For Bengal, see Islam 1984, 157; for Japan, Kato 1970, table 1; for the USA, Clarke 1994, table 3.1; and for early-19th-century France, see Price 1983, 119 (Limousin); Moulin 1991, 53 (Bourgougne).

66. The rates were to decline to 6–7% in the 1890s, almost closing the gap with gilt-edged securities. See Bogue 1963, 169–73, and, for the parallel decline in California,

Rhode 1995, 792. On the capital market in general, see Atack and Passel 1994, 412–414 (the rates on bonds are from Homer and Sylla 1991, table 41).

67. See Ingram 1971, 67. Lending from relatives was also cheaper in Africa (Eicher and Baker-Doyle 1992, 171) and in the Yang-Tzi (Huang 1990, 108), but, apparently, not in North China (Huang 1985, 190).

68. Bauer 1975, 90–100. For other examples of borrowing from city banks by landlords, see Bodenhorn 2000, 509–11 (planters in Virginia before the Civil War); Adelman 1994, 204–7 (great landlords in Argentina); Mosley 1983, 178–79 (white settlers in eastern Africa in the 1930s, at subsidized rates); Hansen 1983, 871–78 (foreign owners of large estates in 19th-century Egypt). See also, on the theory of interlinked contracts between labor and credit, Hayami and Otsuka 1993, 70–83.

69. Solberg 1987, 141–42. Adelman (1994, 204) argues that the spread was smaller, about 3%. These statements are not necessarily contradictory: they may simply refer to different periods or areas. However, they show how difficult it is to be precise on this issue. For other examples. See Federico 1997, 164–65 (China), Bose 1982, 19–21 (Tamilnadu, India), and Kherallah et al. 2002, 135 (Africa).

70. See Ransom and Sutch 1977, 2001; Goldin 1979; Atack and Passel 1994, 392–93. For similar cases of lock-in in Argentina, see Adelman 1994, 181; and, in India, Tomlinson 1993, 64.

71. In a recent paper, Kranton and Swamy (1999) argue that monopoly could not be so bad for welfare, as a monopolist money lender might be more ready than a competitive one to forgive loans to farmers hit by a production shock instead of repossessing and reselling the land to another farmer. The whole argument, however, rests on the assumption that the "old" farmer is more productive than his replacement. This assumption may or may not be true in practice, but it contrasts with the hypothesis that well-functioning market institutions screen the best entrepreneurs.

72. This latter possibility has been pointed out by Stiglitz in a series of papers, conveniently summarized in Hoff and Stiglitz 1993, 38–39 (see also Besley 1995, 2196; and Ray 1998, 553–61 for further work). By definition, the higher the interest rate is, the higher the expected returns must be, and thus the more likely that risky projects are financed. Lenders, however, would not like to scare away steady clients nor would they like to have a very risky portfolio. Thus, they may prefer to charge a lower interest rate and select randomly among the applicants for credit. The Stiglitz model has an interesting and paradoxical implication: the share of credit to innovation (in all likelihood more risky than credit to consumption) should be ceteris paribus positively related to the interest rate.

73. Rhode (1995, 790–94) shows how the fall in interest rates made fruit production more profitable relative to the traditional extensive wheat-growing.

74. A modern co-operative differs from a commune (section 7.2) on three grounds: it performs some specific, well-defined tasks, it is voluntary (with an option to withdraw), and it is regulated by written rules (statutes). Some American marketing co-operatives, such as the California Associated Raisin Company, adopted a slightly different legal framework. They were legally organized as corporations—with membership open to nonfarmers, but without voting rights. They stipulated contracts with individual farmers for the delivery of the crop (Saker-Woeste 1998, 113–16, 198–200).

75. Clearly, forced collectivization, Soviet-style, is altogether different (see section 9.6).

76. Henriksen (1999) points out that good behavior was not necessarily dictated by civic virtues. Danish peasants might have been ready to cheat their fellow members as much as

anyone else (including commercial firms), but they refrained from doing so because they risked being expelled and losing, as a consequence, the access to industrial processing.

77. See Saker-Woeste 1998, 118–20 for the case of California Associated Raisin Company. Branding was only a part of a more general strategy, which aimed at controlling the whole crop, also by heavy-handed means (126–35, 169–90).

78. See Knapp 1969, 72–77; Hurt 1994, 223; Sheingate 2001, 62–63 for the failed struggle of southern farmers against American Tobacco in the early 1900s. On competition in American agricultural markets in general (mainly after the 1930s), see Lauck 2000, 38–61; see also Solberg 1987, 131–40 for Canada; and Simpson 1995, 229–30 for Spain.

79. On the Netherlands, see Knibbe 1993, 153, and Van Zanden and van Riel 2004, 284–88; on Italy, see Pezzati 1995.

80. See Ransom et al. 1998; Sunding and Zilberman 2001, 226–29; and, for some information about the development of contracting in the United States, see Gardner 2002, 68–70. In 2002, these deliveries accounted for about 15% of the total sales (USDA 2002).

81. The most extreme case are the so-called Grameen banks, which lend to very poor households who wish to start a new activity. See Murdoch 1999.

82. This pro–co-operative legislation was further strengthened afterward by state laws in the 1920s and by the Agricultural Marketing Act in 1937 (against local free-riding). See Benedict 1953, 184–85; Constance et al. 1990, 24; Libecap 1998, 187–89; Gardner 2002, 203–5; Lauck 2000, 111–35; and (especially on the legal battles about co-operative trusts), Saker-Woeste 1998, 147–63, 196–215.

83. As late as 1937, there were a mere 106 marketing co-operatives with 32,000 members (Solberg 1987, 67, 146).

84. For the case of the network of Dutch credit co-operatives, see Van Zanden and van Riel 2004, 292–93. Italy is an interesting case, as featured by the competition of two main associations, a mostly urban socialist one (established as early as 1886) and a mostly rural Catholic one (established in 1896); see Zangheri 1987, 74–86; Fornasari and Zamagni 1997, 54–71.

85. See Islam 1980, 180–82 for Bengal; Cheng 1968, 195 for Burma; and also Guinnane 1994, 53 for Ireland.

86. See Guinnane 2002, 94–96. Auditing was initially a voluntary choice, but it was made compulsory in 1889.

87. Kotsonis 1999, 15–25. The movement had been promoted from above, by local authorities (*zemstvo*) as an alternative to the Stolypin reform (section 8.2). This case would tally perfectly with the credit argument for modern property rights (section 7.2).

88. Galassi (1998, 2001) also hints at the possibility that the lack of trust reflected an unusually high level of production risk, because of the Mediterranean environment.

89. See Lauck's (2000, 21–38) fascinating account of the "corporate farming debate" in the USA during the 1960s and early 1970s.

90. In fact, demand for harvesting skills, for example, concentrates in a short period of the year, and thus specialized workers would remain idle the rest of the year. Specialization is much easier in livestock-raising, an all-year activity that can be divided into specific tasks (distributing the food, collecting the products, checking the animal health, and so on). A related but different argument refers to entrepreneurial talent (Banerjee 2000, 255). The latter is a scarce resource in agriculture as in the rest of the economy, and it is thus likely that the number of farms exceeds that of talented individuals. In this case, consolidation

might in theory augment productivity. In practice, however, the net contribution of pure talent (as opposed to other management skills, which can be acquired with on-the-job experience) does not seem so great in farming.

91. Binswanger 1995, 128. This statement is purely speculative, and one could think of objections. For instance, tenants might obtain low-cost capital and misuse it, or a monopsonist landlord might want to squeeze extra rent from his tenants. Both types of behavior would be myopic, however.

92. See Binswanger et al. 1997, 2703–4; Ray 1998, 450–56; Deininger and Feder 2001, 303; World Bank 2003, 83–84. See also the massive collection of data about farm size and productivity in India, in Dyer 1997, 14–21, and statistical appendix.

93. See Myers 1970, 170–77; Huang 1985, 140–45; Brandt 1989, 155. For some evidence of the relation for the Balkans, see Lampe and Jackson 1982, 442; for Russia, see Gattrell 1986, table 4.6.

94. These superior incentives extend to capital as well, to the extent that it consists in cumulated work (section 4.2). On the other hand, poor households may have limited access to credit for the purchase of modern inputs (section 7.4) and/or they may renounce the purchase of indivisible inputs. Thus, the relative intensity of capital per unit of land may ultimately depend on the composition of capital—i.e., the share of traditional (land and buildings) versus modern (machinery, outlays for fertilizers, and so on) items. The greater is the share of traditional items, proportionally the greater the capital endowment per unit of land in family farms. The data on capital intensity per hectare in Asia in the 1970s confirm this hypothesis (Booth and Sundrum 1985, 195–99).

95. See Hopkins 1973, 137, 210–12; Parson and Palmer 1977, 7–8; Kaniki 1995, 393; Clarence and Smith 1995, 163; Epale 1985; Iliffe 1979, 156; Wrigley 1970, 31; Eicher and Baker-Doyle 1992, 42–44. See also, the (predictably) not conclusive comparison of partial productivity measures by Tiffen and Mortimore 1990, 51–74. Mosley (1983, 172–202) argues that productivity in the "core" large white-owned farms was very high, but the average productivity was reduced by the large number of inefficient smallish ones.

96. See Fogel and Engerman 1977, 297; Fogel 1989, 75–80, 409–11; Ransom and Sutch 1977, (esp. 44–47); the survey by Atack and Passel 1994, 310–16; and Smith 1998, 21–23, 61–67, 81–86.

97. The abolition of slavery caused a similar fall in output in all sugar-producing areas except Cuba, possibly because the abolition process in Cuba lasted about 15 years (Engerman 2003, table 1; Dye 1998, 45–48).

98. Irwin (1994) tests these competing theories, broadly supporting Fogel and Engerman. Brinkley 1997 reminds us that American Southern agriculture was also hit by the diffusion of hookworm, a debilitating disease, with severely negative consequences on work intensity.

99. Berry and Cline 1979, table 3.1. Therefore, the average size of small farms range from 1.28 hectares in Ceylon to 91 in Brazil, and that of large farms from 9.25 to 1,000 hectares.

100. See Lamb 2003. See also, Rosenzweig and Binswanger 1993, 67–70 (who use the same ICRISAT sample with a slightly different specification); Feder et al. 1993 (for China); Benjamin 1995 (for Java). Newell et al. 1997 show that, in India, the inverse relationship between size and land productivity holds among villages but not within each village.

101. See Hayami and Ruttan 1985, 146–47 (25 countries in 1960 and 1980), Kislev and Peterson 1996, 165 (USA in the 1970s and 1980s), Mundlak et al. 1999 (37 countries from 1970 to 1990), Gutierrez and Gutierrez 2003 (47 countries in 1970–92). For further evidence and references, see Hayami and Otsuka 1993, 14–15; Hallam 1993, 205–15;

Johnson and Ruttan 1994, 692; Deininger and Feder 2001, 304; Chavas 2001, 267–71; and, for a general discussion of the estimation of production functions in agriculture, see Mundlak 2001, 3–40.

102. For some anecdotal evidence on the failure of very large-scale farming in LDCs, see Johnston and Ruttan 1994, 695–99. See also Lauck 2000, 37.

103. Allen and Lueck 2002, 190–92. The process of industrialization is very advanced in broiler and hog production, and also in feedlot cattle, but not in the cow-calf industry, which supplies feeder cattle for fattening. The authors attribute the difference to the seasonality of births (section 2.2).

104. See Feder 1971; Edelman 1992, 3–23; and, for similar arguments about the Iberian peninsula, see Simpson 1995, 234–35; Reis 1992, 87–90.

105. On additional evidence about *latifundia* and innovation, see, for Egypt, Owen 1986, 75–79, and Owen and Pamuk 1998, 42; for Ireland, O'Grada 1995, 29–32; for Hungary, Eddie 1968, 217; for Spain, Simpson 1995, 236–37; for Southern Italy, Petrusewicz 1996, esp. 109–37.

106. See Millar 1995, 11–56, 72–87; and also in the same vein, Chowning 1997.

107. Bulmer and Thomas (1995, 129) and Adelman (1994, 260–69) share, albeit in a somewhat more reserved mode, Taylor's (1997) criticism of the conventional wisdom.

108. See Koning 1994, 15–20; and, for some recent German debate, see Zimmerman 1999, 97–99.

109. On the threshold model of innovation, see Sunding and Zilberman 2001, 231–33; and, for further information on actual thresholds, see Federico 2003 (threshers); Bateman 1969, and Olmstead and Rhode 2000a, 705–9 (milking machines); Loubère 1978, 199–204 (wine-making); Dye 1998, 67–102 (modern steam-powered equipment for sugarcane crushing); Amin 1982, 106–10 (traditional equipment for sugarcane crushing).

110. See the evidence by Fuijiki (1999, 390) on the excessive endowment of machines in Japan after 1950. Allen (2004, 76) argues that, in Imperial Russia on the eve of the revolution, there were too many horses, as peasants preferred to purchase an extra animal instead of sharing it.

111. See David 1966, and, for the debate, Atack and Passell 1994, 282–91. Interestingly, David put forward a different explanation for the apparently similar delay in the adoption of the reaper in the UK (see section 6.4).

112. See Sargen 1979, Whatley 1987; Clarke 1994, 90–95, 117–34 (and 170–75 for corn pickers); Lew 2000; and White 2001. See also the nontechnical discussion by Adelman (1994, 221–25).

113. According to their computations (Olmstead and Rhode 2001, 679–81), a 1% increase in the share of farms with tractors caused a 0.24% increase in the average size, while farm size accounted only for 8% of the total increase in the share of farms with tractors. Indeed, as late as 1940, about a quarter of American farms owned neither horses nor tractors (Olmstead and Rhode 2001, 668). This fact suggests a wide use of custom work and tractor sharing, contrary to a statement by Clarke (1994, 90) that the latter was "rare." Allen and Lueck (2002, 156–59) show that in the 1980s North America, leasing equipment was more common among large farms than among small ones, once controlled for total capital. This result would seem counterintuitive (one would expect small farms to share indivisible inputs). It is consistent, however, with the very high level of mechanization of the farms in the sample. Most of them owned at least one piece of each type of equipment, and resorted to renting if additional machines were necessary. In fact, the regression shows

also that crop-specific equipment was less likely to be rented than general-purpose machines, such as tractors.

114. For the theoretical argument, see Binswanger 1995, 125; Ray 1998, 413–19, 446. For examples of the sharing of animals, see Tan 2002 (bulls in 18th-century England); Francks 1984, 121 (horses in the Saga plain in Japan); and Simpson 2000, 107 (horses in southern France in the 1950s).

115. See Hayami and Otsuka 1993, table 6.1. Land productivity (from a subset of studies) comes out slightly lower in sharecropped farms than in leased ones, but according to Hayami and Otsuka, the difference is accounted for by the higher share of subsistence crops, which is not directly related to the choice of contract.

116. Cf., Shaban 1987. Cf, for further evidence, Hayami and Otsuka, 94–95 and Deininger and Feder 2001, 312–313.

117. Sereni 1968, 178–86; Giorgetti 1974, 309–15; Daneo 1980, 15–19 (Galassi 1986 offers an alternative, and much more convincing, environment-based explanation of the "failure"). For a similar argument about India, see Hoebner 1984, 12–15; Tomlinson 1993, 30–35; and, esp. Chaudhuri 1984, 146–59.

118. For further evidence about the adoption of innovations by sharecroppers, see Hayami and Otsuka 1993, 96–97.

119. Whatley's (1987) article belongs to a long-standing debate on the causes of the timing of out-migration from the South (Heinecke 1994; Holley 2000, 129–85; Grove and Heinecke 2003).

120. However, compensation might not be possible for investments of which the effects are monitorable only in the long run (such as manuring the fields) or are not monitorable at all (such as experimenting with new techniques). In these cases, the tenant will invest only if he believes that his contract will last long enough to reap the returns. In other words, the level of investment depends on the tenant's expectations about the behavior of the landlord (Mokyr 1983, 80–88; Banerjee et al. 2002, 253–54; World Bank 2003, 43–45, 93).

Chapter Eight: Agricultural Institutions and Growth

1. See Fogel 1989, 21 (with population data in 1800 from Biraben 1979 and Clark 1977).

2. On the same line, see also Miller 1995, 89–91, 111–15. He argued that "debt peonage" existed in Southern Mexico, but not in the north, where workers could flee to the USA.

3. For China, see Pomeranz 2000, 70–86; Deng 2003, 487–97.

4. See Hawke 1985, 22–24. The Maori were often swindled, or paid a pittance for their land, but at least they were given some money. Natives need not be necessarily poor tribes: many "wealthy" Mexican landlords were expropriated in California.

5. In 1970, squatters officially occupied about 7–8% of total land in Brazil (FAO 1981). For Australia, see McMichael 1984, 89–125; Shaw 1990; Butlin 1994, 129–30. For Brazil, see Holloway 1980, 110–20; Alston et al. 1999, 33–40, who present a simple model of the decision to squat.

6. The following account is based on the following sources. For USA: Benedict 1953, 3–22; Hurt 1994, 86–92, 131–34, 187–89, 240–41; Atack et al. 2000b, 287–302; Atack and Passel 1994, 249–73. For New Zealand: Evans 1969, 24–39; Hawke 1985, 24–30. For Canada: Buckley 1955, 10–15; Solberg 1980, 54–60; Adelman 1994, 23–40; McInnis

2000, 67–70; Green 2000, 209–14. For Argentina: Adelman 1994, 62–70; Solberg 1980, 60–64; Cortes-Conde 1997, 53–54. For South America: Chonchol 1995, 112–40; Bulmer and Thomas 1995, 92–95. For Australia: Alston et al. 1999, 33–40; McMichael 1984; Raby 1996, 40–45; Butlin 1994, 127–35; Shaw 1990.

7. For Siberia, see Pavlovski 1968, 172–74; for Thailand, see Ingram 1971, 12–18, and Feeny 1982, 93–94; for Burma, see Cheng 1968, 137–42; for Serbia, see Lampe and Jackson 1980, 118–19, and Palairet 1997, 81–85.

8. Cf., Parker 1972, table 4.2 for the USA (homestead data include the distribution of land to Civil war veterans), and Buckley 1955, 10–11 for Canada.

9. The figure refers to the cumulated acreage of Argentina, Australia, Canada, and the USA. It is assumed that all land in Argentina was given to landlords, and that homestead accounted for the same proportion of arable and pasture farming in Canada and the USA. Australia accounted for about 2% of arable farming in 1938 but for almost 60% of pastureland. Therefore, transferring Australia to the "non-homestead" category would only marginally affect the estimate for arable farming, while it would cause the worldwide share of homestead to fall to about 10% of pasture farming and to about 15% of all land.

10. Although the discussion refers to sub-Saharan Africa, most trends were common to southeast Asia. On the evolution of the latter, see Cleary and Eaton 1996, 29–45. As in Africa, colonial powers intervened in customary property rights to the extent that land was needed for plantations.

11. See King 1977, 334–38; Kaniki 1985, 383–89; Warman 2004, 75–79. For the case of the Boers in South Africa, probably the most extreme one, see Duggan 1986, 35, 132–34. Land grabbing by colonial powers was not a European specialty: the Japanese also expropriated large estates in Korea (Ban 1979, 109–10).

12. See Eicher and Baker-Doyle 1992, 18; Hopkins 1973, 211–13; and, for case studies of plantations, see Epale 1985, 45–56 (western Cameroon) and Wrigley 1970, 16, 31–32 (Uganda).

13. In 1958–60, European settlers owned 89% of the land in South Africa, 49% in Southern Rhodesia and Swaziland, 7% in Kenya, and 6% in Bechuanaland, and negligible areas in the rest of sub-Saharan Africa (King 1977, table 22).

14. Bassett 1993, 6–9. However, the establishment of formal boundaries among colonies belonging to different powers hampered shifting cultivation (Allan 1965, 335–37). In some cases, as among the Ganda (Uganda), the colonial policy reinforced the position of the chieftains, whom the administrators were dealing with, versus the ordinary tribesmen (see Wrigley 1970, 16–17).

15. In some countries of southeast Asia (e.g., Malaysia), the British administration experimented a double standard of registration for native land, with less formal requirements and incomplete property rights (e.g., forbidding mortgaging and sales outside the community). See Cleary and Eaton 1996, 43.

16. FAO 1981, table 5.6, and FAO 1990. Unfortunately, these two censuses report the data on four countries each (hardly a representative sample) and the countries differ—so that the fall might well be spurious. Cf., on the land registration policy in Asia, Cleary and Eaton 1996, 55–74.

17. For the data, see Holderness 1988, tables 1.6, 1.7; Allen 1994, 98. It is uncertain to what extent land was privately enclosed in the same years. Holderness estimates that private enclosures accounted for 20% of parliamentary enclosures, while Clark and Clark (2001) argue that they were almost negligible.

18. For France, see Grantham 1980 (with a useful summary of European legislation in table 1); Agulhon et al. 1976, 130; Clout 1980, 40; O'Brien et al. 1977, 385. For the Netherlands, see de Vries and van der Woude 1997, 224; van Zanden 1994, 32–35. For Belgium, see Goosens 1992, 221–29. For Germany, see Tilly 1978, 385; Koning 1994, 52; Zimmerman 1999, 94–96. For Italy, see Rossini and Vanzetti 1986, 573–79; Corona 2004, 375–83.

19. On Hungary, see Voros 1980, 100–101; on Russia, see Moon 1999, 212–14 (the data from Treadgold 1957, table 4). See also, on the dissolution of the Serbian *zadruga* (associations of cattle-breeders based on extended families), Lampe and Jackson 1980, 113–17; Berend and Ranki 1974, 66; Palairet 1997, 100–102.

20. The reform provided for reunification of scattered fields into a unitary farm. On the reform, see Gattrell 1986, 124–27, Lyaschenko 1939, 748, and Spulber 2003, 76. For its implementation, see the somewhat sceptical view by Pallot 1999, 94–126, 190–218. A clause for opt-out from a similar reform was approved in Hungary in 1894 (Voros 1980, p. 101).

21. See King 1977, 94–97; Reynolds 1970, 136–38; Walsh-Sanderson 1980, 12–16; Katz 1974; and, for a similar process of expropriation in Bolivia, see Chonchol 1995, 134–40.

22. See Aricanli 1986, 28–32; Schilcher 1991, 180–82; Owen and Pamuk 1998, 15–16; 106–8; Mahdi 2000, 74–81. Egypt was a peculiar case, as it was only partially subject to Ottoman legislation until the British conquest in 1882: see the (somewhat diverging) accounts by Richards 1982, 19–29; Cuno 1992, 198–207.

23. See Booth 1988, 67–73, 134–137; van der Eng 1994, 142–44; Cleary and Eaton 1996, 65–66. See also, on India, Roy 2000, 102–6; and, on Korea, Ban 1979, 108–9.

24. In Russia, the serfs' obligations (especially the amount of work in the demesne) had been rising in the long run, with some periods of respite, and grew until the abolition of serfdom in 1861 (Moon 1999, 72–77).

25. See Lampe and Jackson 1982, 33–38; Palairet 1997, 34–43, 129–40 for the Balkans. See Keyder 1991, 1–13; Schleicher 1991, 186–88; Aricanli 1986, 26–27 for Anatolia and the Middle East.

26. See King 1977, 279–81, 286–88; Stokes et al. 1983; Chaudhuri 1984, 102–16; the surveys by Tomlinson 1993, 42–44, 63–65; Roy 2001, 22–25. Many of the original holders of *zamindari* rights went bankrupt in the crises of the 1790s and 1800s.

27. On Europe in general, see Blum 1978, 377–400. On Russia, see Lyaschenko 1939, 377–87; Moon 1999, 109–12; Spulber 2003, 14–18. On Hungary, see Voros 1980, 23; Scott 1971, tables 1, 2; Berend and Ranki 1974, 46–47. On Serbia, see Palairet 1997, 85–88.

28. See Hayami and Yamada 1991, 63; King 1977, 194–97; and Smethurst 1986, 47, who argues that the peasants enjoyed almost total ownership even before 1859.

29. See Lampe and Jackson 1982, 134–38; Palairet 1997, 42–49, 174–78.

30. For the evolution in the 19th century, see Stokes et al. 1983; Chaudhuri 1984, 112; and, for the reform, see King 1977, 286–88.

31. This account is mainly based on Fogel 1989.

32. The number of slaves rose from 0.9 million in 1800 to almost 4 million in 1860 (Historical Statistics 1975, p.14 and series A 119).

33. The process of abolition in Thailand lasted from 1874 to 1915 (Ingram 1971, 58–65; Feeny 1982, 94–95; Manarungsam 1989, 64–67). When they conquered African

slave-based societies, the Western powers declared slavery illegal but let slaves go on living with their masters (Hopkins 1973, 226–30; Iliffe 1995, 206–7).

34. The distinction is inspired mainly by King 1977, 14–24 and Lipton 1974, 270 (but see also the different, and quite cumbersome, taxonomy by De Janvry 1981, 207–13). Land reform differs from the abolition of the feudal system, as discussed in the previous section, as it affects estates held under modern property rights. In practice, however, the difference may not be so clear. The compensation could vary from zero (pure expropriation) to the whole market value of land: any sum equal or superior to the market value entails a subsidy to the former landlord. Tenancy reform covers a wide range of measures, which may affect all the clauses of existing or future contracts, from the length of the contract (e.g., setting a minimum time or making renewal automatic but for special cases, or allowing the tenant to transfer his right), to the landlord's right to evict the tenant during tenancy (prohibiting eviction altogether or subjecting it to some outside court decision), to the division of product (e.g., setting a maximum rent or outlawing sharecropping).

35. On Bulgaria, see King 1977, 35. In Germany, uncultivated Junker estates were distributed to landless peasants, but the total amount of land was small—one million hectares from 1886 to 1932, out of 31 million hectares of total land (Tracy 1989, 103; Corni 1990, 155–56; Koning 1994, 87, 103–11, 131–35). In Ireland, tenants were given interest-free loans to purchase land (Turner 1996, 214–15; Guinnane and Miller 1997, 595–96). In New Zealand, since 1892, the state purchased large pastoral estates at market prices and resold it to farmers (Evans 1969, 29–34; Hawke 1985, 92–95).

36. The account is based on Walsh-Sanderson 1980 (the estimate of land transfer from p.6); De Janvry 1981, 123–33; King 1977, 87–97; Reynolds 1970, 136–43; Powelson and Stock 1990, 35–59.

37. For the Balkans, see the League of Nations, 50–51 and table XIX; Berend and Ranki 1974, 224–231; Lampe and Jackson 1982, 351–54; King 1977, 37–45. For some (very modest) measures in Italy, see Rossini and Vanzetti 1986, 660–61; and, in the UK, Perren 1995, 66. The massive transfers of land in the USSR and China after the Communists' seizure of power, by far the largest ones, will be dealt with in section 9.6.

38. He deals (more or less at length) with Mexico, Bolivia, Cuba, Venezuela, Colombia, Chile, Peru, Japan, Taiwan, North and South Korea, India, China, Pakistan, Sri Lanka, Indonesia, the Philippines, Egypt, Iran, Iraq, Tunisia, Algeria, and Kenya.

39. King 1977, 194–99; Hayami and Yamada 1991, 86; Bray 1994, 215–16; Sheingate 2001, 150–53.

40. On the Indian policy, see Islam 1978, 161–62; Charlesworth 1978, 100; Bose 1994, 4–5; Chandavarkar 1983, 799; Tomlinson 1993, 64. On the Irish acts, see Turner 1996, 204–10; O'Grada 1995, 255–58; Koning 1994, 87; Guinnane and Miller 1997, 596–98.

41. See the list in Besley and Burgess 2000, 396–402; and also King 1977, 283–92 (the quotation is from p.291).

42. On Italy, see Giorgetti 1974, 528–35; Rossini and Vanzetti 1986, 745–46. On the Philippines, see Hayami and Kikucho 2000, 37–39; Cleary and Eaton 1996, 76–77.

43. The standard measure of fragmentation is the number of fields per farm. The (very scanty) sources put forward quite high figures—about 20 in France in 1882 (Wade 1981, 193) and 11.6 in China in the 1930s (Chao 1986, 94). They do not specify if fields are contiguous or not—nor do they give the distance between them.

44. For the Netherlands, see King 1977, 17; and, for the Nazi policy, Corni 1990, 183–90, 311–13.

45. Fennell 1997, 36–40 (for the antecedents), 91–97 (for the debate about the minimum size of farms), and 207–39; Fearnie 1997, 33–35; Tracy 1989, 267–69, 325–30; Fanfani 1998, 254–71. On the similar policy in Japan (after 1992), see Francks 1999, 100–101.

46. See King 1977, 295–96 (the policies affected about a quarter of total acreage); and, for some African examples, Anthony et al. 1979, 147–49.

47. FAO 1961a and 1961b (Census 1930 and 1950), FAO 1971 (1960), FAO 1981 (1970), and FAO 1990. The dates are approximate, as the surveys were held in different years.

48. The whole of Africa is represented by Egypt and Mozambique in the 1950 census, by Cameroon, Central African Republic, the Reunion Islands, Liberia, and Swaziland in the 1970 census, and by Egypt, Guinea, Guinea-Bissau, Morocco, the Reunion Islands, and Uganda in the 1990 census.

49. FAO 1971, 1 Mixed farms—i.e., those partially owned and partially rented, but managed unitarily—are registered as a separate category ("mixed") in some countries/censuses, while in others, they are allocated to one category, either according to the "holding" principle (i.e., as a unit, under the prevailing tenure) or to the "area" principle (i.e., they are divided in two or more units with the same tenure). In the following tables, the acreage of "mixed" farms is divided between owned and rented tenure according to the shares of these two categories on the total acreage (net of "mixed" farms).

50. For Argentina, see Diaz Alejandro 1970, table 3.4; for China, fn. 51; for Japan (and Java), see table 8.1.

51. See Feuerwerker 1980, table 3, and 1983, 76; Chao 1986, table 5.1; Richardson 1999, 74. The average size also declined in Java (table 8.1) and, according to Stokes et al. (1983, 63), in India.

52. See Stanton 1993, table 4.5. In England, the share of large farms (over 300 acres, or 121 hectares) remained constant around a third of total acreage from 1851 to 1966, and rose to over a half in 1983 (Grigg 1987, table 2). For the increase in size in France, see Moulin 1991, 183.

53. The 2002 data are from USDA 2002 table 3. The figures for 1870, 1940, and 1970 are computed by the author with data from Historical Statistics 1975. The total gross output is obtained by deflating the current-price data (series K265 and K270, omitting government payment and non-"farm-related" income) with the index of prices of agricultural products (series E25 and E43) in 1870 and with the index of prices received by farmers (series K344) in 1940 and 1970. The output per farm is computed by dividing gross total output by the number of farms (series K16).

54. For the data on Mexico, see King 1977, 94; and, on the other countries, De Janvry 1981, table 3.1. On Argentina, see the general assessment by Bulmer and Thomas 1995, 92–95; Barsky 1991, 66–68; Adelman 1994, 131; Cortes-Conde 1997, 63; and Solberg 1987, table 4.1. On Chile, see Bauer 1975, 117, 127, table 39. On Brazil, see Alston et al. 1999, 40.

55. On Canada, see Historical Statistics Canada 1965, series L3-L8 (the data, however, include "mixed" farms—those partially rented).

56. The share of family farms in Burma declined from 85% in the 1900s to two-thirds in the late 1930s (Cheng 1968, table VI.2). In Java in 1905, more than half of agricultural workers owned some land (Booth 1988, table 2.13). In Korea in 1937–39, (after massive Japanese expropriations) family farms still accounted for 40% of the land (Ban 1979, 111; Yoong-Deok and Young-Yong 2000, table A1). Family farms are said to prevail (but without

data) in Thailand (Ingram 1971, 13; Feeny 1982, 91–95; Falkus 1991, 59), and in Anatolia (Quataert 1980, 49; Aricanli 1986, 23, 43, 49; Owen and Pamuk 1998, 4).

57. See Chao 1986, 132; and also Esherick 1981, esp. 399. The original data are reported, with some variants, by several authors: Feuerwerker 1983, 77; Brandt and Sands 1992, 189; Brandt 1989, 141–47; Chao 1986, tables 8.1, 8.2; Deng 2003, 493–95. About 5% of the land was owned and rented out by institutions, and the balance (about 25%) belonged to landlords. Some tenants with long-term fixed rent contracts (see section 7.3) were, in all but name, owners.

58. Brandt and Sands 1992, 183. On the Marxist interpretation, see Chao 1986, 102; and also Wright 1992; Huang 1985, 9–13; Richardson 1999, 70.

59. See the Institute of Developing Economies 1969, table 6 (from 1903); Francks 1984, 148–50, 200–203, 259–69; Smethurst 1984, 12–15, 58, 73, 88.

60. In the UK, managerial estates (including the so-called farmers-in-hand—i.e., landowners who personally managed their estates) accounted for 85% of total land in the 1880s and for 65% in the late 1930s (Sturmey 1968, 285, 299, table 3; Perren 1995, 45; Lindert 1987, table 1; for Ireland, Turner 1996, table 7.1). According to the data collected by Eddie (1971, table 7) farms over 20 hectares (quite a low threshold for "large" estates) accounted for 67% of acreage in Austria, 56% in Germany, 54% in Hungary (see also Held 1980, 200, 221), and 51% in Romania. In Italy during the 1930s, family farms accounted for one-third of acreage (INEA 1951, 61–62). The share was probably lower thirty years before, as Italian peasants had increased their holdings during the war (Lorenzoni 1938).

61. On France, see Clout 1980, table 2.1; Grantham 1975, 294; Heywood 1981, 362; Hohenberg 1972, 222; Moulin 1991, 140–41; and Caron 1979, 131. The figure is from Dovring 1965, table 30. On Denmark, see Jensen 1937, 128–35; Henriksen 2003. On the Netherlands, see van Zanden 1994, table 5.6. Koning (1994, 17–20) argues that the share of large farms increased in the first half of the 19th century in Europe, but he provides very little quantitative evidence.

62. See Berend and Ranki 1974, table 2.4, for Bulgaria (in 1897); and Antsiferov 1930, 20, for Russia. His figures for 1877–78 and 1905 include land owned outside the *mir* (4.5 and 22 millions hectares respectively).

63. In Egypt, property rights were better developed, and peasant land accounted for about half of the total acreage (Richards 1982, 39, tables 2.2, 2.3; Owen 1986, table 3; Owen and Pamuk 1998, 32).

64. See Planck 1987, 175–76. Interestingly, the number of rented farms in the USA decreased from 1950 onward, while their share on land remained constant (table 8.4) This divergence suggests that, when retiring, owners preferred to rent the fields and keep the house instead of renting or selling the farm with all the buildings.

65. See US Census 2002, table 58 (with an average sales per farm around $1.8 million versus less than $100,000 for all farms). Pure family farms ("individual and family ownership") accounted for 52.5% of the sales, and "partnerships and family-held corporations" for 40%. The data refer to farms with more than $250,000 in sales.

66. Chao 1982, 286–91; and Feuerwerker 1983, 86 (both reporting figures from Buck and National Land Commission). Sharecropping was less frequent in the south, where cultivation was more intense and double-cropping widespread, than in the north. See also Lin 1997, 115–31; Huang 1985, 202; Richardson 1999, 70.

67. Historical Statistics 1975, series K142–147. The share is higher in some areas (Allen and Lueck 2002, table 2.4). On the different categories of sharecroppers in the American South, see Alston and Kauffman 1997 and 1998.

68. Dovring 1965, table 30. Sharecropping in France in 1892 accounted for 11% total land and 23% of the rented one. It was the standard contract in central Italy and was also quite common in some areas of the north and, with different features (very short-term contracts), in the southern *latifundia*.

69. This general inference does not rule out an increase in the number and share of landless laborers in some very densely populated areas, such as southeast Asia (Booth and Sundrum 1985, 221–38; Cleary and Easton 1996, 65–66; but cf. Roy 2000, 77–79 on India). In India the share of laborers declined during the period of British rule (Roy 2000, 77–79) and increased after independence because many landlords, fearing the land reform, evicted their tenants and hired them back as laborers (Hayami and Otsuka 1993, 87). Managerial estates have not disappeared and some big family farms also bring in workers. Such an increase, however, seems to have been more an exception than the rule. In most cases, landless farm workers have eventually found employment in the non-farm sector.

70. The potential bias from the increase in custom work is difficult to assess. The census data should include specialized workers among laborers (no bias), but they would miss the increase in labor market transactions if the custom work were performed by farmers with their own machinery.

71. On the FIR and ANIR, see Goldsmith 1969, esp. 86–91, 178–212; and, on other available measures, Levine 1997, table 1.

72. Data on real estate debts are from Goldsmith 1955, table M 27; Historical Statistics 1975, series K361; those non–real estate loans, from Goldsmith 1955, table A-63; Historical Statistics 1975, series K376–380. They are deflated with the price index, from the Bureau of Labor (series E25 and E42). Data on the gross output are from Federico 2004a (1910–38); Historical Statistics 1975, series K266–270 and K273–275 (1939–70). The measure differs slightly from Goldsmith's FIR, which has GDP as its denominator.

73. See Fei and Ranis 1997, 94, 100–102; Rosengrant and Hazell 2000, 64; and USDA 2004, table 10.5. The informal credit (secured and unsecured) was tentatively estimated at about 8–10% of total credit in the late 1960s (Historical Statistics 1975, 478), while it accounted for 20% in 2000–2002 (USDA 2004, table 10.10). "Non-farm individuals" accounted for most of the non-institutional mortgages—at least until the 1950s (Goldsmith 1955, table M-27).

74. For some evidence on informal credit networks, see Atack et al. 2000a, 274–76; Bogue 1963, 173–82 (on the Midwest); Rothenberg 1992, 113–47, and Clark 1990, 264–270 (both on Massachusetts).

75. Ishikawa 1967, table 2.20. Total mortgages in real terms almost tripled from 1900 to 1930 (Smethurst 1986, table 1.11).

76. Van der Eng 1994, 127–38. These institutions initially accepted deposits and provided credit in kind (rice), and they started to handle cash only at a later stage. After independence, they were transformed into co-operatives and were funded by a specialized agricultural bank. They financed mainly the purchase of fertilizers.

77. See Cheng 1968, 190–93, for Burma (from 1883); Pavlovski 1968, 146–60, for the Russian State Peasant Bank (from 1882); Hurt 1994, 294–98, Clarke 1994, 181–198, and Gardner 2002, 196–201, for the federal loans to American farmers (since 1933). Similar

policies were also adopted, or planned, in other countries—such as France in the 1880s and again in the 1920s (Sheingate 2001, 66, 97); Bulgaria in 1894 (Lampe and Jackson 1980, 200); Anatolia in the 1890s (Quataert 1981, 82); Japan in 1912 (Kato 1970, 330–35); and Egypt in 1898 and 1931 (Richards 1982, 152–56; Owen 1986, 79–81). On subsidized insurance schemes, see Hazell et al. 1986, 9–11; Hazell 1992; Gardner 2002, 228–29 (for USA).

78. FAO Agristatbank. The data exclude the huge number of Chinese banks. The figure on loans refers to 33 banks.

79. See Pavlovsky 1968, 146–160, and, in a somewhat less enthusiastic vein, Kotsonis 1999, 60–70. The bank accounted for 18% of all land-backed mortgages in 1904, and 37% in 1914 (Antsiferov 1930, 74). From 1882 to 1914, it financed the purchase of some 16 million hectares—i.e., about half the total increase in peasants' land (see note 58). After the Stolypin reform, the bank loans to peasants fell to only 40% of the total assets, to climb back to 90% on the eve of the war. The banks did not finance the purchase of land if the pledge was land deemed essential for the livelihood of the borrowing household. Mortgaging already-owned land to finance investment, however, was allowed. The local authorities (*zemstvos*) heavily subsidized credit co-operatives (section 8.7).

80. Hayami and Kikucho 2000, 34; Ray 1998, 574. In Thailand, banks were forced to lend a sizeable part of their assets to peasants, and yet, in the late 1980s, formal credit accounted for only about a half of the total (Ray 1998, 534–36).

81. See Krueger 1992, 89; Eicher and Baker-Doyle 1992, 170–72; Hoff and Stiglitz 1993, 47–49; Powelson and Stock 1990, 42–45 (on Mexico). The two features are obviously related, given that well-connected individuals have greater chances of being released from their obligations. The rate of default often follows a political cycle, as borrowers hoped their loans would be written off before elections (Besley 1995, 2173; Hazell 1992, 576). The experience with public insurance schemes is even worse: the seven cases discussed by Hazell (1992, 572–73) paid out between two and half (USA, Japan) and almost six times (India) the amount of the premiums that they had received. For further American data, see Goodwin 2001, table 1.

82. Alexandratos 1995, 327; Rosengrant and Hazell 2000, 69–73. The policy of charging positive real interest rates and the reliance on social networks and peer monitoring (and the low administrative costs) account for the good performance of four successful Asian banking institutions, including the Grameen Bank (Yaron 1994).

83. Lapenu and Zeller 2001. Their database includes all the institutions with more than 500 members, that receive some support from donors, and whose average loan does not exceed $1000. About four-fifths of the members belong to 19 major institutions. The database also includes urban institutions and mixed (rural-urban ones), which account for 1.5% and for 60.5% of total assets respectively. For a symphatetic but hard-nosed discussion, see Murdoch 1999.

84. These exchanges are excluded from gross output as well, at least as long as they do not involve firms outside agriculture. According to the (slightly crazy) rules of national accounting, an exchange of, say, corn between two neighbors is excluded from gross output if direct, while it is included if farmer A sells the corn to a local feedstuff company, which then packages it and resells to farmer B.

85. For instance, Thorner (1975, 205–6) suggests 50%; Danhof (1979, 131), 40%.

86. European data are from de Vries 1984, table 3.8; American data are from Historical Statistics 1975, series A57-72. These sources use different definitions of urban population, and so figures are not comparable.

87. The omission of foreign trade biases downward the measure, ceteris paribus, if the country is a net exporter of agricultural products and upward if it is a net importer. The difference in per caput food consumption is subject to conflicting income and price effect: urban dwellers have a higher income (i.e., consume more), but food is more expensive in the cities than in the countryside.

88. This definition may seem too wide, as it pools together apparently very diverse transactions, such as an exchange of food between neighbors and the sale to a merchant for exporting overseas. However, any distinction between different kinds of transactions entails quite strong, and possibly unwarranted, assumptions on the peasants' attitude and behavior. See von Braun, Bouis, and Kenney 1994, 11–13; and also the discussion in section 8.9.

89. The household-level commercialization can be measured either from the production side (as a ratio to gross output) or from the consumption side (as ratio to consumption). The two figures would differ to the extent that consumption and income diverge—e.g., for the amount of claims on farm product (taxes, rents). At a macroeconomic level, rents are included in the numerator, which, however, can differ for the presence of foreign trade in agricultural products. See von Braun et al. 1996; Blades 1977.

90. This caveat does not apply to managerial farms since food for workers is part of their total wage bill, although paid in kind.

91. For instance, the domestic consumption of cotton fell from about a half of production in the Yangzi area during the 1890s to a sixth in the 1930s (Brandt 1989, 56; Huang 1990, 96–99).

92. Rothenberg 1992, 79–111. On the commercialisation of New England, see Hobbs-Pruit 1984; Shammas 1982; and the two case studies by Clark (1990, esp. 146–55) and Barron (1984, esp. 51–68). The estimate is from Atack and Bateman 1987, 212–19.

93. Data from Historical Statistics 1975, and USDA 2000, tables 2.39, 2.40. The figure is the ratio of home consumption (series K270) to the sum of the latter consumption and cash receipts (K267, K268).

94. See Chao 1986, 2; Pomeranz 1999, 70–80; Deng 2003, 480–82. Perkins (1969, 114) suggests that, in the 1930s, about 15–20% of output was marketed locally (within a few miles) and roughly the same amount was marketed nationally (on long-range trade, see Lardy 1983, 7–9, 30–33; Cartier 1986, who sums up the work by the Chinese scholar Wu). Perkins tentatively adds that prior to 1900 only 7–8% of output entered long-range trade.

95. Perkins 1969, 113–15. Market sales already accounted for about a third of output in the 1890s and for 45–50% in the 1930s (Brandt 1989, 75–76). In Hopei and Shantung, another fairly "advanced" area in the 1930s, the share varied, according to the crop, between 30 and 55% (Myers 1970, 199).

96. According to Anthony (1979, 58), in 24 African countries, about 40–45% of output was marketed in the late 1960s.

97. According to Ade-Ajawi (1985, 12), "There was no region of Africa so remote that the rural communities in the early nineteenth century could have been described as completely self-sufficient and self-contained." See also Jones 1980, 10–18; Anthony 1979, 80–84; Hopkins 1973, 51–55; and Eicher and Baker-Doyle 1992, 155–59.

98. The first American sale co-operative was allegedly established in upstate New York in 1810 (Knapp 1969, 11) and the first European one in Norway in 1856 (Smith 1961, 6), while the first purchase co-operative was organized in Switzerland in 1865, and the first rural credit co-operative in Germany was set up in the 1850s (Guinnane 2002, 89). As always happens, dating the birth of institutions precisely is difficult and somewhat pointless. The

discussion will, for obvious reasons, deal with agricultural and rural credit co-ops, neglecting the consumer co-ops, which were mainly urban associations.

99. On the history of co-operation, see Smith (1961) and Knapp (1969) for the USA, and Dovring (1965, 206–21) for Europe.

100. See also Simpson 2000, 106–10, for the development of wine production co-operatives.

101. The first marketing co-operative was set up by coffee-growers in Tanganyika in 1925 and collapsed during the great crisis, to be revived as a compulsory government marketing agency—see Iliffe 1979, 274–76. For some additional information, see Eicher and Baker-Doyle 1992, 175.

102. The number of co-operatives grew in India from 847 in 1909, to 12,100 in 1939 (Chandavarkar 1983, table 9.4, which covers credit only); in Burma from 71 in 1906–09, to 3,836 in the early 1920s, and then plunged to 1,573 after the Great Crisis (Cheng 1968, table VII.3); in Indonesia, from 86 in 1921, to 574 in 1939, to 14,146 in 1958, and to 73,406 in 1966 (van der Eng 1996, 130–35). For the sluggish and late development of co-operatives in Bengal, see Islam 1978, 162–74; in Guangdong, see Lin 1997, 153 (and, in China as a whole, Feuerwerker 1983, 88); and in Thailand, see Feeny 1982, 68.

103. Number of agricultural workers is from FAO Statistical Database.

104. The figure is the ratio of sales (USDA 1998) to gross output and purchases on total expenditure (Historical Statistics 1975 Series K266–270 and K276–278 for 1910 to 1970 and USDA 2004). Thus, they overstate somewhat the actual market share, as the turn-over includes also packaging and distribution. Of course, the market share was higher for some products such as fruits. The California Fruit Growers' Exchange accounted for about two thirds of the market as early as the 1920s (Hurt 1994, 235; Knapp 1969, 250–267).

105. Jensen 1937, 315–16, 338–46; and also Henriksen 1992, 171–74. On the Netherland, see van Zanden 1994, 130–36; on New Zealand, Hawke 1985, 90. The co-operative share on the wine market was decidedly lower: 1–2% in Italy during the 1930s, 13% in Spain, and 25% in France during the 1950s (Simpson 2000, 110).

106. Out of 42 figures for product/country from table 8.15, 31 were higher in 1997 than in 1975 (Ilbery 1985, table 7.2).

107. See also, for other LDCs, Alexandratos 1995, 323.

108. See Sexton and Lavoie 2002, 876–79. This theoretical suggestion is supported by the case of the apparently unstoppable "Sun-Maid" co-operative among Californian raisin producers, which failed in 1927 because of oversupply (Saker-Woeste 1998, 188).

109. Cf., for some insights on this line, Binswanger et al., 1995 pp. 2664–2672, and Engerman and Sokoloff 1997 p. 264.

110. King 1977, 12, 92; along the same line, see World Bank 2003, 146–47; on Latin America, Bulmer and Thomas 1995, 320.

111. On the cost of protection, see Allen 1991. He implicitly compares homestead, which concentrated new settlers in a specific area, with free squatting, which let them scatter all around the frontier.

112. Adelman 1994, 85–95; Cortes-Conde 1997, 55–70. They were not sold, Adelman argues (110–18, 155–60) because the Italian immigrants to Argentina (unlike British immigrants to Canada during the same period) did not want to settle permanently. This reluctance could have also been a case of self-selection. Argentina may have attracted short-term immigrants instead of prospective long-term settlers because they knew that the land market did not offer sufficient opportunities for purchasing.

113. On the whole debate regarding enclosures, see Hudson 1992, 73–75, and Allen 1994, 115–16. The traditional view of technical progress is powerfully advocated by Deane 1967, 40–45, and Chambers and Mingay 1966, 94–98. On the social consequences of enclosures, see also Clark 1993, 264–65, and 1998b, 99; and Shaw-Taylor 2001, 643–47.

114. Gerschenkron 1967. These communal obligations arguably developed in the eighteenth century as a reaction to the modernizing tax reform by Peter the Great, which imposed a per capita uniform tax on peasants. They were strengthened by landlords, who wanted to extract as much labor as they could from all their serfs (Moon 1999, 214–16).

115. For instance, the abolition of serfdom in Hungary was traditionally considered an essential precondition for productivity growth (see Berend and Ranki 1976, 15; Gunst 1996, 19) and was considered almost irrelevant by Komlos (1986, 89) and Voros (1980, 24–26).

116. See Clark 1993, 249–53; 1998b. His position is shared, albeit in a more moderate tone, by Allen (1991b, 236–37; 1992, 1–21; 1994, 115–17; and 1999, 217–20).

117. See Overton 1996 and Turner et al. 2001, 88–89. Grantham (1980) finds no negative effect of vaine pature in 19th-century France, but his results are based on a very small number of cases.

118. For general references, see those listed in note 113. See also Clark and Clark 2001 (common land was much less diffused by 1800 than previously assumed), and Shaw-Taylor 2001 (common rights were less widespread and possibly worth less than claimed).

119. The improved land was often excluded from redistribution. On the earlier debate, see Gattrell 1986, 1–20. The revisionist position has been forcefully argued for by Gregory (1994, 50–52) and, more cautiously, by Pallot (1999, 68–69) and Moon (1999, 220–26); while Spulber (2003, 82–84) endorses Gerschenkron's thesis.

120. One can contrast the cautiously positive assessment by Binswanger et al. (1995, 2719–23) and Pagiola (1999), who puts forward some rough comparisons between costs and benefits of titling—with the critical surveys on Africa by Bassett (1993, 11–18), Reyna and Downs (1988, esp. 10–15), and (on gender issues) Lastarria-Cornjiel (1997). Interestingly, the World Bank (2003), in its otherwise detailed report on land policy, does not provide a comprehensive assessment of the policies it had sponsored.

121. See respectively Alston et al. 1999, 138; Jacoby et al. 2002. They also show that expropriation/redistribution risk does not affect other decisions, such as the crop mix. For additional references on the positive effect of secure property rights on long-term investment, see Pingali 1989, 252–54; Cleary and Eaton 1996, 73–74.

122. See Place and Otsuka 2001 and also the similar results about intensification in Burkina-Faso by Gray and Kevane (2001, 581–82). In the same vein, Johnson (2001, esp. 304) finds that the applications for registration of former *eijido* (common land) after the 1992 Mexican reform were independent of the value of assets to be gained, and infers that traditional rights were deemed to be secure enough to grant credit. The efficiency and flexibility of customary rights is stressed by most empirical work from other research traditions (sociology, anthropology, etc.). See Breusers 2001, 50–51 (on Burkina-Faso); Dujon 1997 (on the Santa Lucia Islands); Kibreab 2002 (on the Sudan).

123. On the design of an optimal land reform, see Banerjee 2000, 263–68; World Bank 2003, 151–57.

124. For more general discussions, see, besides King 1977, Hayami and Ruttan 1985, 388–90; Alexandratos 1995, 320–23; Deininger and Binswanger 1999, 266–67; Deininger and Feder 2001, 318–19; World Bank 2003, 120–23, 143–50.

125. In the same vein, see van der Meer and Yamada 1990, 130, 143; Francks 1999, 77.

126. See World Bank 2003, 116–20. See the general discussions by Hayami and Ruttan 1985, 395–99; Brandao and Feder 1995; Ray 1998, 441–45; and the model by Banerjee et al. 2002. The possibility of eluding inefficient clauses is not costless. In fact, unlawful but efficient contracts are not officially enforceable, and thus they are less secure than they could be without the reform.

127. The abolition of intermediaries (i.e., the *zamindaris*—see section 8.2) had a clearly positive effect on both equity and performance. Banerjee et al. (2002) deal in-depth with one of these reforms, the so-called operation Barga in West Bengal, which protected sharecroppers from eviction and lowered rents. According to their estimates, the reform increased rice yields for sharecroppers by about 50% and also had a positive effect on investment. For a skeptical view of the same measure, see Hayami and Otsuka 1993, 101–3.

128. The estimates of effects of fragmentation vary widely according to cases. According to Wan and Chen (2000, 144–45), it causes severe output losses in post-reform China, possibly up to 10% of the potential production of foodgrains. In contrast, Besley and Burgess (2000) find positive effects on production in India. For a quite negative view of the social consequences of consolidation, see Bruce 1988, 42–44.

129. This feature contrasts with the traditional interest in the issue during the 19th and early 20th centuries among economists, from Marx onward. See the survey by Booth and Sundrum 1985, 131–38. Chapter 6 of their book, dealing with population growth and farm size, is an exception in the dearth of theoretical discussion on the issue.

130. They estimate the system with a cross-section panel dataset with the state as unit of observation. Size is measured with the total services from land and capital, and the equations also include year dummies to take short-term fluctuations into account.

131. Actually, the size of farm can increase, in these conditions, if several households join together. This case, however, seems so rare as not to have attracted attention.

132. The recent increase in rented land in some "advanced" countries is quite easy to explain. Some heirs of family farms who want to try their luck outside agriculture might be unwilling to sell because either they want to keep open the option to come back or, simply, they find the conditions in the market for land not appealing enough. In this case, letting the family farm could be a reasonable compromise.

133. It is implicitly assumed that returns to alternative investments exceed the rents from the estate. This assumption is plausible, given the relatively low level of risk of landowning and the possible existence of a land premium (section 7.3).

134. For a simple model, see Kislev and Peterson 1996, and also the analysis of optimal farm organization by Allen and Lueck (2002, 167–98), as determined by technology and crop choice via the demand for specialization, the variance of output, etc. Their model is static, but in principle one could use its insights also to interpret long-term structural change. Technology and crop choice, however, depend also on the type of farms, and assessing a priori causes and effects is exceedingly difficult.

135. Nominal agricultural wages might be slightly lower, as prices in the countryside are lower, but country-life may not be so appealing. On the difficulties of managerial estates in the late 19th century, see Koning 1994, 23–39; Grantham 1975, 308–9.

136. Also, farming households are likely to have a minimum target income and they may be willing to sell their farm if it does not meet their goal. They can, however, supplement the income from a suboptimally sized farm with off-farm work: indeed, the great growth in part-time farming (section 4.4) shows that this strategy has a strong appeal for farmers.

137. See Feuerwerker 1980, 14; Chao 1986, 111–27; Myers 1970, 159–65; Lin 1997, 36–38; Huang 1985, 78; Brenner and Isett 2002, 616. Partible inheritance is also quoted as a cause of the reduction in farm size in France, after the Napoleonic Code of 1804 (Moulin 1991, 40), and in Spain (Simpson 1995, 199), Burma (Cheng 1968, 146–50), Egypt (Richards 1982, 33), Malay (Overton 1994, 65), and the Balkans (Lampe and Jackson 1980, 354).

138. In addition, family farmers could resort to similar strategies to keep the unity of the farm. Planck (1987, 160–63) argues that the custom of undivided inheritance (with compensation to non-farming heirs) accounts for the larger size of farms in Northern Germany. It is important to point out that these strategies are much easier to implement in an "advanced" country, where there are plenty of potential occupations outside farming than in LDCs.

139. Malatesta 1999, 15–25; on the Indian *zamindaris*' similar strategies, see Stokes et al. 1983, 67.

140. On Russia, see Pavlosky 1968, 157; on Korea, Yoong-Deok and Young-Yong 2000; on India, Rawal (2001) and Banerjee et al. 2002, 257. Sturmey (1968, 290) speaks of a "campaign against landowners" in the UK, which "moved from mere vilification of landlords as a class to threats of nationalisation of land or the imposition of heavy taxes thereon."

141. The landlord can consume his rents as food, provided he has a sufficiently large retinue. The share of consumption in kind is said to have been very high during the high Middle Ages, when great landlords and kings moved their courts from one fief to another to consume otherwise useless agricultural products. This behavior, however, was dictated by the lack of markets. Even the most traditionalist landlord (and his employees) would prefer to settle in one location and enjoy a wider consumption basket, including manufactures and other luxuries, which have to be purchased with cash.

142. In theory, commercialization could have increased if rents had increased as a proportion of total product (see section 8.8). This hypothesis cannot be ruled out, as data on factor shares are scarce. However, at least in Europe, the rent-wages ratio fell (see section 4.5), and thus, the share could only have risen if technical progress had been very land-intensive. This was not the case, as discussed in section 6.3.

143. The two seminal works are Schultz 1964 and Polanyi 1944 (see also Shanin 1973). For reviews of this early literature, see Federico 1984; Eicher and Baker-Doyle 1992, 24–30. The use of the words "moral economy" (Scott 1976) is confusing in that it introduces another dimension (Booth and Sundrum 1985, 216–20). In theory, an output-maximizing choice of crop mix and allocation of factors (based on full specialization for the world market) is perfectly compatible with a "moral" redistribution of income. This is the essence of the modern welfare state.

144. See for the mid 19th century, Atack and Bateman 1987, 103–5 and for pre-revolution New England, see Henretta 1978; Clark 1990, esp. 13–15; Rothenberg 1992, 38–40, who focuses on the period before 1750).

145. These costs include marketing, transportation, and collecting information on market opportunities, plus any monopolistic or monopsonistic profits on the sale of "cash" crops and the purchase of "food" crops. See Key et al. 2000.

146. For these models, see Singh et al. 1986, esp. 3–42; Ellis 1988, 102–40; von Braun et al. 1994. Chayanov (1966) can be considered as a pioneer of modern microeconomic analysis of households, suitably disguised for political expediency. Interestingly, his work was extremely popular among Western historians in the 1970s and 1980s as an alternative to standard economic theory (see, e.g., Aymard 1982; Huang 1985, 9–15).

147. It is important to stress that an increase in risk is likely, but by no means sure (see, e.g., Finkelshtain and Chalfant 1991; von Braun 1986, 52). The actual level of risk depends on the volatility of production of the two crop mixes (the "subsistence" and the "market" ones) and on the covariance of prices. Specialization might even annul risk if prices of crops vary inversely to output, as can happen in a closed market with (unitary) elastic demand. This case, however, seems implausible.

148. See the model by Fafchamps 1992, and, for some evidence, von Braun et al. 1996; and also, for Bengal, see McAlpine 1983, 144–48.

149. See Ransom and Sutch 1977 and 2001 (published in a special issue of *Explorations in Economic History* entitled "*One Kind of Freedom* Reconsidered"). Wright (1986, 110–13) extends this argument to poor white farmers. The implicit level of commercialization in the two cases differed—as African American sharecroppers had to deliver half their crops to the landlord.

150. On Java, see Booth 1988, 196–99; Elson 1990, 25–33; Cleary and Eaton 1996, 37–40; van Zanden and van Riel 2004, 115–17, 181–82. On India, see Mishra 1982; Stokes et al. 1983, 49–50; Mukherjee 1984, 54–59; Amin 1994; and (less gloomy) Charlesworth 1978. On Egypt, see Richards 1982, 21–24. On the Congo, see Miracle 1967, 235–40; Acemoglu et al. 2001 (who state that taxation reached 60% of the output). On Tangakyika, see Iliffe 1979, 134; and, for the whole of Africa, Iliffe 1995, 203–05 (the source of the quotation). On the other hand, the Western administrators of some African colonies hindered commercialization of natives' production for fear that it could outcompete white-owned farms.

151. Farmers cultivated cotton because they had to pay increasing rents with a smaller acreage (Wright and Kunreuther 1975), or because it yielded higher returns (McGuire 1980), or because the crop fit well in rotations (Temin 1983; Harris 1994).

152. See Simms 1977; and Gregory 1994, 42–48, 52–54 (and section 8.4 for the data on peasant ownership). For some supporting anthropometric evidence on the improvement in the welfare of Russian peasants since the 1880s, see Mironov 1999 and Wheatcroft 1999. In an earlier paper, Wheatcroft (1991) puts forward an intermediate interpretation. He argues (153–61) that the total tax burden rose until 1904–05 and decreased afterward, but that indirect taxation was still growing. He adds (166–70) that the 1905 uprising was triggered by the decision to cancel all arrears on payment for the 1861 redemption in 1903–04.The peasants hoped to be relieved of further payments—and, indeed, these were abolished in 1906.

153. See van der Eng 1996, 211–12; and, for the evidence about the improvement in conditions for the peasants, Elson 1990, 33–45. For a similar debate on India, see Stokes et al. 1983, 89–90; McAlpine 1983, 120–30, 190–94; Tomlinson 1993, 51–55; and esp. Roy 2000, 92–93. Cuno (1992, 198) argues that commercialization in Egypt predated Muhammad Ali.

154. On India, see Roy 2000, 250 (the figure for the 19th century); Kumar 1983; Sivasubramonian 2000, table 6.10 (tax revenue and gross output in the 1930s). On the see Balkans, Palairet 1997, 176–78, 281–311 (the peasants reduced their workload and sales,

thereby earning a reputation for laziness); On Japan, see Karshenas 1993, 210; Smethurst 1986, 47–53 (the increase in taxation was more than compensated by the abolition of feudal levies). On Thailand, see Manarungsam 1989, table 2.11; Ingram 1971, 78.

155. See Feuerwerker 1983, 74. Deng (2003, 500–502) and Faure (1989, 82) share this opinion, while Lin (1993, 113) endorses the traditional (Marxist) view of increasing and oppressive taxation contributing to peasant unrest and ultimately to the revolution.

156. The idea can be traced back to von Thunen (Grigg 1982, 135–50). On specific cases, see, for Paris, Grantham 1989, 49–52; Price 1981, 291–97. For London, see Taylor 1987, 56; Perren 1995, 15. For Budapest and Vienna, see Voros 1980, 92, 105; for the Netherlands, see van Zanden 1994, 153–55; for New York, see Lindert-Zacharias 1999, 23–44; for other northern U.S. cities, see Danhof 1979, 140–50; Atack and Bateman 1987, 150–53.

157. Huber 1970, table1; and, for the export boom, see Federico 1997, 40.

158. For the Piedmont (the area of small white farms), see Wright 1986, 39–43 (and, for the north, Atack and Bateman 1987, 228–29); for Japan, see Francks 1984, 145–148, 180–84, and Smethurst 1986, 157; for Hopei and Shantung, see Myers 1970, 191–99; for Bombay, McAlpine 1983, 144–48, and Charlesworth 1985, 134–42; for Saõ Paulo, see Holloway 1980, 7–13.

159. See, on Spain, Simpson 1995, 70–74; on Italy, de Felice 1971, 146–69; on China, Federico 1997, 83–85. For cocoons and silk, see Brandt 1989, 89ff; Myers 1970, 184–94; see Huang 1985, 125–28 on cotton (and a different opinion by Pomeranz 1993, 78–82). On Malay, see Overton 1994, 75; and, on the Balkans, Lampe and Jackson 1980, 362. On the responsiveness of African peasants to economic incentives, see the general works by Bates 1981, 82–87; Kanicki 1985, 393; Jones 1980, 20–23; Anthony et al. 1979, 135–38; Iliffe 1995, 202–8, 212–17; and the area studies by Hopkins 1973, 172–82, 216–224 Morgan 1969, 245–46; Ingham 1979 (West Africa); Iliffe 1979, 154–56, 274–89 (Tanganyika); Wrigley 1970, 16 (Uganda); and Duggan 1986, 38 (South Africa).

160. On the earlier (and technically somewhat obsolete) literature, see the surveys by Askari and Cummings 1976; Ghatak and Ingersent 1984, 172–214; Eicher and Baker-Doyle 1992, 75–76. For further, more modern estimates, see Singh et al. 1986, table 1.1 (for market sales); Goswami and Shrivastava (1991); Abler and Sukhatme 1991, table 20.5; Federico 1996; Kherallah et al. 2002, 140; and the survey by Mundlak 2001, 47.

161. See von Braun 1994 (7 out of 11 areas were in Africa). The authors do not quote technical progress, and thus one has to infer that the increase in the production of cash and food crops was made possible by the use of underutilized input. See also the survey of microeconomic studies by Wiggins 2002, 105–106. They show that "good access to market" made successful intensification much easier.

162. See Grantham 1989. All areas grew wheat, but the north specialized in livestock the south in wine and fruits—and the areas around cities in vegetables and dairy products. See also, on France, Caron 1979, 125–27; Hohenberg 1972, 284; Price 1983, 304–6 and, in a more critical vein, Wade 1981, 177–80; and, for the USA, see Rothenberg 1992, 214–40 (on late 18th-century Massachusetts) and Danhof 1979, 179 (on the 19th century).

163. O'Brien 1968, 183–85 (and, for his estimate of TFP growth, see Statistical Appendix, table IV). Hansen and Whattleworth (1978, 458) question the reliability of his estimates of production, and thus, of overall growth. Richards (1982, 38–39, 80–95), while acknowledging the growth of income from commercialization, reminds us that its distribution worsened and that increase in irrigation affected the environment negatively.

164. Reynolds 1985; O'Brien 1985. A poor development record is not sufficient evidence to disprove this statement: a strong increase in exports might well have been insufficient to start development because its positive effect was balanced by other factors (e.g., a poor endowment of human capital, bad institutions, etc.).

165. In the words of the introductory chapter of the "official" UNESCO *History of Africa* (Bohaen 1987, 35), "The essential aim of colonial administrators was the exploitation of African resources, animal, vegetal or mineral ones, for the exclusive benefit of colonial powers, and especially of the trading, financial and mineral companies of the mother-country."

166. Huang 1985, 1990. He thus implicitly assumes that cash crops were more profitable, but that peasants shunned them as being too risky. He also assumes that there were no alternatives, such as seeking off-farm work or obtaining more food from the same land (i.e., a technological change à la Boserup). Huang admits that a minority of peasants (which he labels "entrepreneurial" farmers) switched to cash crops out of desire for profits. His work has been criticized because his bold generalization is based on a tiny sample of households in 33 villages of north China.

167. Rawski 1989; Faure 1989; Brandt 1989 (the statement on commercialization is on p.179); Brandt 1997; and the special issue of *Republican China* (Little 1992), notably the article by Hartford, who points out how it is possible to interpret the same historical case in two widely different ways. See also the reviews by Richardson (1999, 3–10) and Deng (2000).

168. Roy 2000, 12–15; and, somewhat more cautiously, Tomlison 1993, 58–63.

169. Kurosaki (2003) estimates that change in specialization accounted for about one-sixth of total production growth in Punjab. The computation, however, is based on yields (i.e., land productivity), not on the TFP growth.

170. The possibility of nonsocial optimal outcome in the presence of asymmetric information is strongly emphasized by Hoff et al. 1993 (esp. 13–15). The example is from Kantor 1998.

171. See Palairet 1997, 35–39 for the Balkans; and Feuerwerker 1983, 28–29, and Brandt 1990, 179–80, for China.

172. See Binswanger and Townsend 2000. Artadi and Sala-I-Martin (2003, table 1) estimate that "ethno-linguistic fractionalization" (the main cause for wars) reduced the growth rate of sub-Saharan Africa as much as low openness, high price of investment goods, and excessinve public spending in consumption, but much less than unfavorable environment and poor human capital.

Chapter Nine: The State and The Market

1. The taxonomy is freely inspired by Libecap (1998, 200).

2. The first regulations against food adulteration were approved in the USA (in 1856 for milk, followed by several other commodity-specific acts and, in 1906, by the Food and Drug Act) and in the UK (the Food and Drug Act in 1875); see Benedict 1953, 132; Grigg 1974, 194; Law 2003. Laws against fraud in fertilizers were approved in Britain in 1894 (Brassley 2000d, 550) and in Japan in 1899 (Hayami and Yamada 1991, 74). Britain approved an Adulteration of Seeds Act in 1869, while in the USA seed certification was

mainly voluntary, with the support of some state-level legislation beginning the 1920s (Wines 1985, 130–31; Fowler 1994, 82–83, 100–103; Tripp 2001, 107–8).

3. See Loubère 1990, 104–9, 113–25; Unwin 1991, 276–80, 311–25; Sheingate 2001, 92. This protection became really effective in France in the 1930s, and it was upheld for the first time by a foreign (British) court in 1958–59.

4. For descriptions of tools and their effects (in increasing order of complexity), see Andreosso-O'Callaghan 2003, 69–76; Tyers and Anderson 1992, 124–55; Tokarik 2003, 7–19 and Alston and James 2002.

5. The minimum guaranteed price effectively rules out the lower tail of the distribution of expected prices (the so-called Sandmo effect). Thus, it increases the average expected price and the desired output even if the support price does not exceed the average free-market one.

6. See Krueger 1992, 3; Schiff and Valdez 2002.

7. The UK raised its duty in 1818, France in 1819, Spain in 1820, while Prussia (and later the Zollverein) maintained its duty. See Tracy 1989, 39–43, 57–65, and 83–89; Clout 1980, 126–27; Agulhon et al. 1976, 115, 233; Simpson 1995, 222; Koning 1994, 50–70.

8. For factual information, see, Bairoch 1989, 52–67; Tracy 1989, 48–49, 65–69, 87–89, 110–11; Agulhon et al. 1976, 411–12 on France; See Federico 1984a on Italy; Reis 1992 on Portugal; Simpson 1997 on Spain; Webb 1982 on Germany; Perren 1995, 24–26, and Green 2000 on the UK Hayami and Yamada 1991, 77–78; Brandt 1993, 274 on Japan. The reaction to the crisis is interpreted with a very simple factor-endowment model by Rogowski (1989, 21–59). See also Verdier 1994, 106–47. Koning (1994, 29–34, 79–81, 93–103, 137–44) puts forward a totally different interpretation of the crisis, as a consequence of rising wages that hit large farmers.

9. On barriers to imports of meat in Germany, see Tracy 1989, 99–101; Webb 1982, 317–19; on trade wars, see Conybeare 1987, 183–97; Lampe and Jackson 1980, 175–76; Gunst 1996, 58–59; Palairet 1997, 300–307.

10. See Hurt 1994, 204–13; Atack et al. 2000a, 279–283; and Sheingate 2001, 59–65. The economic case against the populists' statement is developed in Atack and Passel 1994, 419–25.

11. For the historical rates of nominal protection, see Bairoch 1989, table 9; Tracy 1989, tables 3.2, 4.2 and 4.3; Federico 1984, table VIII (for Italy). For the current rates, see Sumner and Tangerman 2002, table 2. Duties on rice in Japan were about 15–20% (Brandt 1993, 274).

12. The figure is obtained using the data on nominal protection (note 11). On the share of wheat on total output in Italy, see Federico 2001, table 1A; and in France, Toutain 1961, table 76. For the definition of PSE and other measures, see OECD 2000, table 1, and OECD Statistical Databases.

13. See Williamson 1990, 135–38. All the figures in the text refer to the baseline case of the UK as a small country (i.e., that its trade policy could not affect world prices because world trade elasticities were infinite). Williamson estimates that under the alternative, large-country assumption, the total effect would be smaller. The abolition increased agricultural output by 1%. The latter set of results tallies well with the earlier estimates of the impact of trade liberalization by McCloskey (1980) and Irwin (1988).

14. See O'Rourke 1997 and Federico and O'Rourke 2000, table 10.5. The change in total agricultural output can be computed by weighting the sectoral changes of tables 7 (the

UK), 8 (France), and 9 (Sweden) with the sectoral shares from table 6. These results are hardly affected at all by different assumptions about factor mobility.

15. The figures refer to the combined gross output of France, Germany, Italy, Sweden, Spain, Portugal, and Austria-Hungary (although Hungary was a net exporter of wheat). They accounted for about one-third of the total output of the 25 countries, which represented about half of the total output of the world (see Federico 2004a).

16. See the overall synthesis by League of Nations 1943; Hardach 1977, 134–63; Koning 1994, 154–66; and the country studies by Barnett 1985 (UK), Offner 1989, 61–68 (Germany); Moulin 1991, 133–34 (France); Bachi 1926, 272–84, 315–30 (Italy); and Malle 1985, 322–25 (Russia).

17. For the UK, see Koning 1994, table 6.1 (the estimate refers to caloric content of food) and for Italy, Rossi et al. 1993, table 3.A. Italians could consume more, thanks to imports funded by Allied loans.

18. For Germany, see Offner 1989, 27–29; Holtfrerich 1986, 81–90; Tracy 1989, 147–49, 163, 181–82. For the UK, see Perren 1995, 32–36; and, for France, Moulin 1991, 134.

19. Aldcroft 1977, 223–31. For additional references, see Federico 2004b. He estimates the potential demand in the "Atlantic economy" with income data from Maddison (2003), price data from Statistical Appendix Table III and a range of plausible elasticities.

20. For details, see Benedict 1953, 173–238; Genung 1954, 5–59; Constance et al. 1990, 26–29; Hurt 1994, 266–73; Libecap 1998, 186–88; Olmstead and Rhode 2000a, 730–32; Sheingate 2001, 100–105.

21. See Tracy 1989, 149, 163–164. The fascist government adopted some additional incentives for wheat-growing, renaming as "la battaglia del grano" ("the battle for wheat"), an early example of the fascist regime penchant for warlike metaphors; see Tattara 1973, 374–86; Zamagni 1993, 259–60.

22. Lewis (1949, 214) estimates that nominal protection on agricultural products in Europe was about 25% both in 1913 and in 1927. He relies on the work by Liepmann (1938, table A.1), who computes the "potential tariff" as a simple average of the ratio of duties by product in importing countries to export prices, for a set of 144 commodities. The measure is homogeneous across countries and is thus easily comparable, but it does not take the different patterns of consumption and thus of imports into account.

23. Hayami and Yamada 1991, 78. For prices of rice, see Okhawa 1967, table 10 (prices deflated with the implicit deflator of gross national expenditure from Okhawa and Shinohara, table A. 50).

24. See also Federico 1997, 160–61 and 183 for similar operations on the market for raw silk in the nineteenth century.

25. Aldcroft 1977, 229–30; Topik 1987, 67–77; Cardoso da Mello et al. 1985, 90–95; Greenhill 1995, 192–94; Abreu and Bevilaqua 2000, 41–42. The agreement also included a tax on new plantations to curb the growth of output.

26. For silk, see Federico 1997, 161; for Cuban sugar, Tonizzi 2001, 47; and, for Egyptian cotton, Owen 1986, 71–72.

27. Topik 1987, 78–83; Cardoso da Mello et al. 1985, 108–11; League of Nations 1931, 53–55; and, for some information on other commodities, see Aldcroft 1977, 223–31. A marketing board for sugar had been set up in Indonesia during World War I, but mainly to allocate scarce shipping (van der Eng 1996, 216–17).

28. See Berend 1985, 176–80; Lampe and Jackson 1980, 344–47, 448–54. See also, on Egypt (for cotton), Owen and Pamuk 1998, 37; and, on Australia (for butter and other exports), Mauldon 1990, 312–14.

29. See Jones 1980; Hopkins 1973, 264–66; Fortt and Hougham 1973, 29–33; Mosley 1983, 176–80; Eicher and Baker-Doyle 1992, 45; Mosley 2002, table 10.3; Kherallah et al. 2002, 10–13.

30. On tea, see Gupta 2001; on sugar, Bill and Graves 1988, 10–15; and van der Eng 1996 pp. 224–226. On the International Rubber Regulation Scheme (IRRS, 1934), see Manarungsam 1989, 113; van der Eng 1996, 237–40.

31. The "world" output index declined only in 1931, and only by 1% (Federico 2004a).

32. France and Germany raised duties in 1929 (Tracy 1989, 164, 183), and Japan in 1930 (Anderson and Tyers 1992, 106). According to Lewis (1949, 214), who relies on estimates by Liepmann (1938, table A1.), the average 'European' nominal protection rose from 26% in 1927 to 65% (plus non-tariff barriers) in 1931.

33. Trade in agricultural products fell, from 1929 to 1932–33, by about a fifth—i.e., slightly less than total trade (Vidal 1990, 268–69).

34. See Tracy 1989, 169–82, and, on the politics of reducing wine output, Loubère 1990, 120–30; Sheingate 2001, 96–99. Similar policies of generalized market control were adopted in other countries, such as Italy, Czechoslovakia, Norway, Switzerland, and even erstwhile stalwarts of free trade such as the Netherlands and Denmark (in order to manage the fall in exports without a disastrous price war). See Tracy 1989, 125–29; Berend 1985, 180–82; Andreosso-O'Callaghan 2003, 78–89.

35. See Corni 1990, 101–50; Tracy 1989, 182–87; Hendriks 1991, 31–33 (the accompanying structural policies have been described in section 8.3).

36. Cf., Hawke 1985, 244–46 (New Zealand), and Shaw 1990, 12–16 (Australia).

37. Historical Statistics 1975, series U214 ("crude materials") and U215 ("crude food"), deflated with the farm product price index (series E53). Exports of wheat almost disappeared, while those of cotton, a raw material without domestic substitutes in Europe, held much better (series U275 and U279). On the American policy, see Benedict 1953, 230–401; Libecap 1998, 189–96; Constance et al. 1990, 28–30; Olmstead and Rhode 2000, 732–35; Clarke 1994, 140–61; Hurt 1994, 287–95; Paarlberg and Paarlberg 2000, 138–42. For cotton, see Holley 2000, 57–66.

38. On agriculture in the Great Depression, see Federico 2004b; and, for the role of the fall in relative agricultural prices in the international transmission of the Depression, see Madsen 2001.

39. Historical Statistics 1975, series K503, K507, K554.

40. The estimate is by Libecap 1998, 194.

41. Production rose by some 20% in the USA (Historical Statistics 1975, series K246, deflated with relative prices series—series E23, E42). On Italy, see the (apparently) precise data in Ercolani 1969, table XII.1.1; on France, Moulin 1991, 154–55; on Germany, Tracy 1989, 199; on USSR, Medvedev 1987, 126–30; and, on the Balkans, Lampe and Jackson 1980, 528–48.

42. Paarlberg and Paarlberg 2000, 141; Hurt 1994, 320; Benedict 1953, 402–59.

43. On the different plans (Brannan, Aiken, etc.), see Benedict 1953, 469–90; Constance et al. 1990, 30; Sheingate 2001, 130–39.

44. In spite of the massive migration to cities (section 4.4), a non-negligible number of people stayed in the countryside, possibly out of fear that their human capital might be

worthless in other jobs or simply out of a preference for the country life. This behavior paid off in the long run.

45. The inventiveness of lawmakers to find different ways to subsidize farmers seems boundless: American farmers in the 1980s received price support under the CCC loans, deficiency payments (if market price fell below a "target" price), and subsidies to their storage of commodities (which the farmer was compelled to sell if prices exceeded a given threshold). See Gardner 1990, 32–37.

46. The agricultural policies of other countries closely resembled those of the three major areas. "Minor" European countries (Switzerland and Norway) imitated the EU, the former Dominions (Australia, Canada) imitated the USA, and the Asian "tigers" (notably South Korea and Taiwan) imitated Japan.

47. See Hurt 1994, 320, 352–57; Constance et al. 1990, 32–52; Gardner 1990; Clarke 1994, 211–18; Gardner 1996; Olmstead and Rhode 2000a, 735–39; Paarlberg and Paarlberg 2001; Sheingate 2001, 128–50, 196–211; and Gardner 2002, 216–18, which lists all major legislation in three pages.

48. See Sheingate 2001, 201–208; and also Gardner 2002, 133, 205, 217. On the FAIR Act, see Moyer and Josling 2002, 144–73.

49. See "Bush the anti-globaliser," *Economist*, May 9, 2002.

50. This narrative is based on Francks 1999, 85–101; Hayami and Yamada 1991, 82–95, 221–29; Sheingate 2001, 150–61, 222–33; and Mulgan 2000, table 1.1, 32–33. The purchase price is from Francks 1999, table 4.4 (deflated with the consumer price index from Maddison 1991, table E-4).

51. For the policies of the 1940s and early 1950s, see Johnson 1973, 31–37; Milward 1984, 435–40, and 1992, 228–27; Tracy 1989, 216–41. See also Barral 1982, 1439–48 on France; and, on Germany, Sheingate 2001, 161–64, and Hindricks 1999, 36–40.

52. For details on the pre-Treaty plans and on the negotiations, see Milward 1984, 440–61, 1992, 284–317; Fearne 1997, 11–31; Tracy 1989, 251–55; Fanfani 1998, 75–96; Fennell 1997, 1–11.

53. For descriptions of the procedures, see Fennell 1997, 133–54; Fanfani 1998, 116–40; Andreosso-O'Callaghan 2003, 93–99, esp. tables 4.3 and 4.4 on support by product and country in 2000 (see also Tracy 1989, table 14.4).

54. For the data, see Tracy 1989, table 12.3; Ackrill 2000, table A.2.1; Andreosso-O'Callaghan 2003, tables 4.5, 7.1. In 1997–98, the European Community runs a small deficit in its trade of agricultural products (OECD 2000, table T 178).

55. See Tracy 1989, 303–6; Fennell 1997, 154–65; Andreosso-O'Callaghan 2003, 112–14. The defects of the CAP had already been pinpointed by Mansholt in the 1960s (Fearne 1997, 33).

56. See Fennell 1997, 169–72; Fanfani 1998, 175–88; Fearne 1997, 50–53; Sheingate 2001, 214–21; Moyer and Josling 2002, 96–117; Andreosso-O'Callaghan 2003, 116–26.

57. Alexandratos 1995, table 8.1. For other estimates, see Ritson 1997, tables 1.1, 1.2; Milward 1992, 161; Anderson et al. 1986, table 2.5; Blandford 1990, table 9.1; Tokarik 2003, table 1.

58. Blandford (1990, table 9.5) quotes a 1986 paper by Tyers and Anderson. These authors published later a book on the issue (Tyers and Anderson 1992, 241) but reported only simulations of partial liberalization—stating clearly that full liberalization was not likely enough to be considered.

59. For the United States, see Historical Statistics 1975, tables K303–K325, K339–K342; for Japan, Francks 1999, 81; and for the European Community, Ackrill 2000, table A.3.2.1.

60. For the European Union, see Andreosso-O'Callaghan 2003, table 4.1; and for Japan, Francks 1999, 89.

61. See Blandford 1990, table 9.8. The estimates of gains from the Uruguay Round of the GATT (which, as said, entailed only a partial liberalization) vary from $3.5 billion to 58.3 billion (Summer and Tangermann 2002, table 5).

62. The account is based on Booth 1988, 133–37, 146–52; van der Eng 1996, 114–17, 186–88; Timmer 2002, 1508–10. For other countries, see the first three volumes of Krueger 1992.

63. Morrison and Thorbecke (1990, 1086) estimate that in 1980 the total transfer to agriculture was positive by 74 billion rupiahs—about 0.6% of agricultural GDP (for more information on their methodology, see section 10.2).

64. For details, see Bates 1981; Mosley 2002; Kherallah et al. 2002, 13–16.

65. The price support on output (column a) is computed as an average of assumptions 1 and 2—i.e., assuming that the products omitted from the database received half the support of the covered ones.

66. On this point, see also Herrmann 1997, table 3; Mosley 2002, 177–79. All these results are quite sensitive to the choice of country/period.

67. See Schiff and Valdez 1992, 12–13 and appendix 2.1 for the computation; and Schiff and Valdez 2002, 1428–37. See also the estimates collected by Bautista and Valdez 1993.

68. Krueger 1992, 92–109. For critical views of the liberal policies, see Rao and Caballero 1990; Karshenas 2001.

69. On Africa, see the short syntheses by Binswanger and Townsend 2000; Mosley 2002, 180–86; and the careful survey by Kherallah et al. 2002 (the data on fertilizer reforms, 36–47). On reforms in Asia, see Rosengrant and Hazell 2000, 196–218.

70. See Abler and Sukhatme 1996, 366; Rosengrant and Hazell 2000, 202–3 for the estimate of losses. See also the recent estimate by Kalirajan and Bhide (2003, 140–43) on the effect of an increase in the price of fertilizers.

71. Pryor 1992, 25, 361–66. The "socialist" credentials of many Third World countries were somewhat shaky. Many of them defined themselves as socialist or Marxist mainly to get aid from the Soviet Union.

72. The GDP per head in Russia was $1,488 (1990) dollars in 1913, and $1,370 in 1928; in China $597 in 1936 (the pre-communist maximum) and $439 in 1950; in Czechoslovakia $3,259 in 1949 (Maddison 2003, tables 3-c, 5c). The agricultural population accounted for 85% of the total in Russia around 1900 (Gregory 1994, 42) and of China in 1950 (Lardy 1983, 2), while it accounted for only 38% in East Germany in 1950 (Wilson and Wilson 2001, 121).

73. The following account is heavily based, as the references make clear, on the cases of the USSR and China. For further information on other countries and references, see Binswanger et al. 1995, 2, 688–90, and, above all, the book by Pryor (1992).

74. According to Becker (1996, 31), the expropriation caused 3–5 million deaths throughout China, while Huang (1990, 166–71) narrates an almost bloodless process, at least in the Yang-Tzi delta.

75. In 1921, the worst year, the sown area fell by 15% from the prewar level, and the number of animals (horses and cattle) by 33%. See Malle 1985, 338–446; and also Volin

1970, 149–50; Gregory and Stuart 1986, 37–47; Wheatcroft and Davies 1994b, 63. On the famine, see Adamets 1997; Wheatcroft and Davies 1994a, 53; and Wheatcroft 1997.

76. See Gregory 1994, 92–94, 107–13; Allen 2004, 84. Given the positive elasticity of peasant supply (Gregory and Moukchtiar 1993), the Soviet policy was bound to reduce deliveries: the crisis was largely a self-inflicted wound.

77. See Lardy 1984, 30–33; Zong and Davis 1998, 7–11; Huang 1990, 171–74. The Chinese government, unlike the Soviet one, offered relatively high prices for agricultural products.

78. For a general (and somewhat sympathetic) outline of the process of collectivization, see Pryor 1992, 7–15, 97–130. Clearly, it was a political decision, supported by very weak economic arguments (Gregory 1994, 104–22; Becker 1996, 47–70). Both the alleged fall in marketed output and the alleged superiority of large-scale farms do not stand up to even superficial scrutiny. Stalin and Mao made the ultimate decisions, but it is still controversial whether they acted according to their inner beliefs in the superiority of large-scale production (as in industry) or out of the fear that social differentiation in the countryside might develop capitalism again. This issue would, however, lead us too far away from agriculture into politics, if not into the psycho-pathology of dictators.

79. The share of households belonging to collective farms rose from 4% in 1929, to 31% in 1930, then to 53% in 1931, and to 90% in 1936; see Volin 1970, 203–20 (the data are from table 22). The actual enrollment is likely to have been lower, especially at an early date, as local party bosses might have wanted to boast of their achievements. See Gregory and Stuart 1986, 99–108; Medvedev 1987, 80–85; Gregory 1994, 104–12; Fitzpatrick 1994, 49–72.

80. See Pryor, 109–11. The process of collectivisation started early and proceeded rapidly only in Bulgaria, while, in other eastern European countries, the process started only in the late 1950s and was relatively slow. For the case of East Germany, see Wilson and Wilson 2001, 112–14.

81. Pryor 1992, table 4.1 (sum of the columns state and collective farms); the figures may overstate the share in that some countries' collective farms existed only on paper.

82. In the USSR, the number of animals roughly halved (Gregory and Stuart 1986, table 11; Wheatcroft and Davies 1994b, 119; Fitzpatrick 1994, 53–54).

83. For the Soviet Union, see Clarke and Matko 1984; Wheatcroft and Davies 1994b; Allen 2004, appendix C and personal communication (and Federico 2004a, graph 4); for China, Lardy 1984, table 4.3; Ashton et al. 1984, table 5; Lin and Yang 2000, 145; Fan and Zhang 2002, table 1. The fall might overstate the actual effect of collectivization, because part of it might be accounted for by bad weather.

84. See Hunter 1988, 210–11 (grain only); Allen 2004, 100–101.

85. On the USSR, see Livi Bacci 1993; Wheatcroft and Davies 1994a, 71–74; the latest figure is from Poliakov 2000 (I thank A. Graziosi for the reference); on China, see Lardy 1984, 150; Ashton et al. 1984; Lin and Yang 2000; Becker 1996, 266–72; and also (in a less critical vein) Huang 1990, 269–86. The population figures are from Maddison 2003, tables 3a and 5a.

86. For a general description of the socialist system, see Pryor 1992, 163–90; on the USSR and Eastern Europe, Fitzpatrick 1994, 103–52; Gregory and Stuart 1986, 223–30; Medvedev 1987, 95–110; Swinnen 1988, 589–90. On China, see Becker 1996, 105–12, 175–91.

87. Davies 1998, 74–75. Fitzpatrick (1994, 174–96) argues that members of the Communist Party in the countryside were very few, and thus unable to control all the farms, especially during the Great Purge.

88. The degree of implementation depended on structural features such as the distance form the cities or the wealth of the commune, and on the outcomes of political struggles at the local level. For a case study, see Huang 1990, 269–74.

89. This trend was formally sanctioned by the transformation of many *kolkhoz* (co-operative farms) into *sovchoz* (state farms) after the war: *sovchozi* land grew from 7% to 60% of acreage from 1940 to 1985 (Medvedev 1985, 318).

90. Dong and Dow (1993) estimate that direct supervision costs in a Chinese commune absorbed about 10–20% of the output.

91. See, for the 1930s, Medvedev 1987, 361–70; Pryor 1992, 242; Spulber 2003, 214–15; and Wheatcroft and Davies 1994b, 126. These plots were legally only a temporary concession.

92. McMillan et al. (1989, 793) estimate that wage was only a third of marginal productivity. Clearly, these exercises are very difficult as they cannot be based on market prices.

93. See Lewin 1985, 183–84. He points out that the socialist system restricted mobility, as serfdom had, and assimilates collective fields to the landlord's demesne and private plots to the household manors.

94. For a more extensive discussion of socialist policies, see Pryor 1992, 197–231.

95. The following is based on Lardy 1984, 48–54. A similar policy was adopted in the German Democratic Republic (Wilson and Wilson 2001, 117–22).

96. For some additional references and discussion, see Pryor 1992, 235–46; Brooks and Nash 2002, 1554–57.

97. The computation is based on official statistics (FAO Statistical Databases) and thus the figures are subject to all the relevant caveats. The end-dates are 1980 for China, and 1990 for the USSR and transition economies. See also the revealing comparison between actual production and planning targets, in Spulber 2003, 218–19.

98. According to Gregory and Stuart (1986, table 20): from 1928 to 1938 (i.e., including all the period of collectivization) capital fell by 3%, while all other factors increase (land by 25%, labor by 12%, and purchases by 41%). From 1950 to 1977, labor fell by 19%, but all other inputs increased (land by 37%, purchases by 77%, and capital by 1116%).

99. Nguyen and Wu 1999, 69–73; Lambert and Parker 1998, table 2; Maddison 1998, table A3 (see also the data in chapter 4).

100. Wheatcroft and Davies 1994b, 119–22; Hunter 1988; Davies 1998, 64–71. Originally, tractors and other machinery were owned by specialized agencies (the so-called MTS) and rented to *kolkhoz*. The system entailed extremely high transaction costs, and was abolished in 1958 (Medvedev 1987, 177). For the case study of the (failed) mechanization of cotton harvesting, see Pomfret 2002.

101. For instance, the 1932 Soviet passport law subjected emigration to cities to an official authorization (Fitzpatrick 1994, 80–103).

102. See Statistical Appendix, table IV; Federico 2004c. In Poland, where "social" and "private" agriculture coexisted, the former decidedly underperformed in terms of TFP growth, in spite of a preferential access to inputs and technology (Boyd 1988).

103. For the Soviet case, see Pryor 1992, 210, 428–35; Gregory and Stuart 1986, 234–42; Medvedev 1987, 162–68, 177–80, 183–88, 345–47; Brooks and Gardner 2004; and Lerman et al. 2003 (who stress the great increase in subsidies during the 1970s and 1980s). See the estimates of PSE in the USSR in 1986 (and 1997, after the reforms) in Kwiecinski and Pescatore 2000. China differed, as agricultural policies in the 1960s and 1970s changed mainly as the result of a power struggle between different factions

within the Communist Party, while efficiency considerations played a minor role (Zweig 1989).

104. For some information on other Asian countries, see Rosengrant and Hazell 2000, 219–22; and Rozelle and Swinnen 2004.

105. The account of Chinese reforms is based on Lardy 1984, 88–97, 190–221; Zweig 1989, 50–75, 169–88; Huang 1990, 194–98; Lin 1992; Nguyen and Wu 1996, 54–60; and Zong and Davis 1998, 12–22.

106. McMillan et al. 1989, table 5. See also the somewhat more sceptical views by Huang 1990, 241–42; and Nguyen and Wu 1999, 59–60.

107. See Spulber 2003, 317–19, 329–337; Swinnen 1998, 590–92; Goetz et al. 2001; O'Brien et al. 2000, 16–21; Wilson and Wilson 2001, 126–34; Karlova et al. 2003; World Bank 2003, 136–40.

108. See World Bank 2003, 100–104, 136–40. On the case of Russia, see Tillack and Schulze 2000, 457–60. In a very interesting microeconomic analysis on Romania, Rizov et al. (2001) show that the decision as to whether to farm individually or to remain in the collective farm is positively affected by the development of the markets for credit and inputs, by the diffusion of household farming during the socialist regime, and also by features of the household such as education and age (albeit in a nonlinear way).

109. See Pryor 1996, 297–333; Lerman 2000 (esp. the handy table 1); Brook and Nash 2002, 1557–60, 1563–68; Brooks and Gardner 2004, 582–84; World Bank 2003, table 4.1; Karlova et al. 2003, 546–556. Returning the land to former owners is not an easy task for technical reasons as well. Tracing the heirs of families who had emigrated to cities decades before is not simple, and restituting estates to former landlords, possibly foreigners (e.g., Germans in Poland), may be politically tricky.

110. See World Bank 2003, table 4.2; see also Lerman 2001, table 3; Andreosso-O'Callaghan 2003, table 6.2; Spulber 2003, 329–30. Private farms account for most or all of the land only in the Baltic states.

111. Rates of change (from 1961 to 1980 and from 1980 to 2000) are computed with a linear regression with output data (FAO Statistical Database). See also Lambert and Parker 1998, table 2, for rates by product; and Ash 1996. As repeatedly stated, socialist production statistics might overestimate output growth, while the postreform data might undervalue it (Nguyen and Wu 1999, 73). In this case, the break in performance trends would be even greater.

112. For inputs, see the tables in Maddison 1998; Lambert and Parker 1998, table 2; and, for the estimates of TFP, see Statistical Appendix table IV. A word of caution is appropriate. In fact, if socialist agriculture was really plagued by the incentive problem, any quantity measure of labor (e.g., the number of workers), unadjusted for intensity of work, would overestimate the amount of labor. However, the effect of the estimate of TFP is undetermined.

113. Between 1979 and 1984, the reforms accounted for 80% of the growth in TFP according to McMillan et al. (1989, 790), for half of the growth in output and 90 percent of the growth in TFP, according to Lin (1992, table 6); and for half of the growth in TFP (Huang and Rozelle, quoted by Wu and Yang 1999, 40). This contribution, as expected, was lower if one considers the second half of the 1980s (0.30 of TFP according to Wu, quoted by Wu and Yang 1999, 40) or a longer time period. Reforms accounted for a quarter of output growth and two-thirds of TFP growth from 1965 to 1985, according to Fan (1991, table 4), and for three-quarters of total growth from 1980 to 1992, according to Nguyen and Wu (1999, 78). Lambert and Parker (1998) find, for the period 1979–95, that

the HRS affected technical progress positively, but efficiency negatively—i.e., they favored the "best" provinces, while the "backward" ones fell behind relative to the frontier. On a related vein, Ghatak and Seale (2001) show that the effect was positive in the north, and negative in the south.

114. The "distorsion" includes the number of years of the socialist regime, the belonging to the CMEA, and the share of collectively owned land. The "policy" variables for each country are computed as the residual of a regression with the development and distorsions as explicative ones.

115. Kruger (1992, 2), who also relies on the (somewhat crude) test of the negative relation between GDP growth and levels of agricultural taxation by Schiff and Valdes 1992, 203–7. See also the classic book by Bates 1981.

116. The word "real" is intended to stress the difference between the envisaged opening and the existing trade agreements between the EU and African countries (the Lome Convention), which excluded products that competed with European ones (Andreosso-O'Callaghan 2003, 200).

117. See Andreosso-O'Callaghan 2003, 249 (South Korean GDP is from Maddison 2003, table 5b), Tokarik 2003, 37 and Anderson et al. quoted in FAO 2003. The difference may reflect the partial liberalization in the 1990s, but also the different methods of computation. Tellingly, most official studies (cf., for example, OECD 2002) deal with the effect of a *partial* liberalization of agricultural trade.

118. Agricultural products were excluded from the original GATT and from the Kennedy Round, so that liberalization started with the Uruguay Round (Summer and Tangermann 2002; Andreosso-O'Callaghan 2003; 229–37 and 244–48). Agricultural protection is a key issue of the on-going "Doha" round of talks, which are not progressing swiftly and smoothly, to say the least.

119. See for all "advanced" countries, OECD 2000, 46; for the USA, Gardner 1992, 76, 82; Gardner 2000, fig. 1, and, for Japan, see Francks 1999, table 4.1.

120. On this point, see also Herrmann 1997, table 1; and Knudsen and Nash 1993, 269—who also discuss the stabilization policies critically. One can contrast this with the enthusiastic endorsement of Indonesian policy by Timmer (2002, 1510).

121. See Clarke 1994, 170–87. Specifically for the South American, see Holley 2000, 73–74; and the criticism by Paarlberg and Paarlberg 2000, 142–45; and Gardner 2002, 257–62.

122. See Thirtle and Bottomley 1992, table 1; and Huffman and Evenson 1993, 211–12, and 2001, tables 2, 3.

123. Pingali and Heisey 2001, 69. Arguably, the Green Revolution accounts for most of the production boom in Asia (section 3.2), which in turn accounts for the great reduction in undernourishment on that continent (note 4 of chapter 1). On trends in TFP, see section 5.3 and Statistical Appendix, table IV.

124. See Pearse 1980, 159–65; Shiva 1991; Pretty 1995, 59–92; Griffin 1974, esp. 48–85. See also the more moderate position by Grabowski 1995, 82, and by Alexandratos 1979, 366–67.

125. See the general analyses by Hayami and Ruttan 1985, 330–45, 358–62; David and Otsuka 1994; Pray and Evenson 1991, 359–61; Evenson and Westphal 1995, 2279–81; Rosengrant and Hazell 2000, 73–74; Tripp 2001, 95–101 (who strongly stresses the farmers's autonomy in choosing whether or not to use HYVs), and the review of the debate by Biggs and Clay 1988, 33–40. The impact of the Green Revolution is the subject of dozens,

if not hundreds, of case studies. See for example, Rao 1994 (India); Sharma and Poleman 1993 (Uttar Pradesh, northwestern India), Hazell and Ramsay 1991 (south India); and Booth 1988, 158–65 (Indonesia).

126. Shiva (1991) seems to miss this obvious point. She treats the Green Revolution as a Western plot to sell seeds and environment-destroying fertilizers to Indian peasants who could manage extremely well with their traditional techniques and varieties. She seems to ignore the country's previous history of famines and yield stagnation and the five-fold increase in population from 1800 to the present.

127. See Rosengrant and Hazell 2000, 144–46 on Asia; Bates 1981, 50–63; and, above all, the careful analysis by Kherallah 2002, 26–74 on Africa. The consumption of fertilizers fell in Nigeria, where the policy changed several times (section 9.5).

128. The case was different in the transition economies, where fertilizers had previously been distributed almost for free. According to Brooks and Gardner (2004, 577), in 1991 the PSE amounted to about 60% of total output—i.e., it was higher than in the European Community (table 9.1). These subsidies were abolished in 1992, and agriculture became a net contributor to the state budget. The consumption of fertilizers (section 4.3) collapsed, and production plunged (section 3.2).

129. A very nice example of this pattern in history is the recent increase of protection to Mexican farmers (table 9.1).

130. See the surveys by Anderson et al. 1986; Gardner 1991, 94–95; Binswanger and Deininger 1997; de Gorter and Swinnen 2002; and Moyer and Josling 2002, 10–28, who take into account other approaches. As examples, see also the models by Balisacan and Roumasset 1987; de Gorter and Tsur 1991; and de Gorter and Swinnen 1994.

131. See Olson 1971. One could remark that the commitment to build a lobby could still be worthwhile for "political entrepreneurs"—i.e., for individuals who plan to use the lobby for their own political career, or for institutions such as the Catholic Church in the Netherlands (van Zanden and van Riel, 288–89).

132. The distinction between shares of a product on GNP and in consumption makes sense in a small open country such as Belgium.

133. See Honma and Hayami 1986 (14 industrialized countries in 1955–87); Lindert 1991 (43 countries in 1980); Swinnen et al. 2000 (37 countries in 1972–85); Olper 2001 (35 countries in 1982–92); and, for further references, de Gorter and Swinnen 2002, 1898–99.

134. On the choice of instruments, see de Gorter and Swinnen 2002, 1918–23.

135. Just as a historical curiosity, it should be remembered that, in 1894–97, some representatives of great Prussian landowners broached a plan for a monopoly of grain imports, which anticipated the policies of the 1930s (Tracy 1989, 96–97).

136. Lindert (1991, 70) finds a positive relation between independence and negative protection to exportables, which he interprets as evidence of the importance of fiscal needs. On this point, see also Krueger 1992, 35, and, for a historical case study, Federico 2004d.

137. See Sheingate 2001, 81–99, 151–60; and, on corporatism in Germany, Hendricks 1991, 39–45.

138. De Gorter and Swinnen 2002, 1901. This statement is true in a very narrow meaning of the word "influence." For instance, it neglects the seminal work by Rogowski (1989) and Verdier (1994) about the influence of trade policy interests on political institutions.

139. A representative case is the French opposition to the reform of the CAP in the 1990s; see Sheingate 2001, 211–22.

140. See Swinnen et al. 2001; Olper 2001. According to Lindert (1991, 63), democracy is positively related to protection but only if interacted with sector size—i.e., ceteris paribus, lobbying is more effective in democratic countries.

141. See the classic books by Verdier (1994, esp. 26–35) and Rogowski (1989, 3–30).

142. See Tracy 1989, 49, 71–72, 94–96, 174–176; Koning 1994, 89–103, 144–51; Hurt 1994, 204–13; Sheingate 2001, 59–67; Puhle 1982; and Rogari 1999 (esp. 49–51, 65–92). Not all these organizations were lobbies in the modern meaning and not all survived.

143. On the USA, see Libecap 1998, 190; Sheingate 2001, 107–113; on Europe, see Tracy 1989, 233–36; Milward 1992, 237–50, 286–90; Hendricks 1991, 44; Fearne 1997, 22–31.

144. See Sheingate 2001, 130–46, 181, 196–211. The results of empirical studies on the effectiveness of lobbying are, as often happens, mixed (de Gorter and Swinnen 2002, 1916–18).

145. The negative impact of "developmental" ideology is strongly stressed by Krueger (1992, 40, 121), and also by Binswanger and Townsend 2000.

146. O'Rourke and Williamson 1999, 77–78. For a detailed narrative, see Gash 1972, 531–617.

147. The limits of a purely econometric work are exposed by the inability of Gardner's (1987) model to account for the gearing up of agricultural support in 1933.

148. See also Krueger 1992, 36; Libecap 1998, 216; Olmstead and Rhode 2000, 738–39.

149. On decision-making in the European Community and the power of the Directorate-General for Agriculture, see Grant 1997, 146–82.

150. For some examples, see Krueger 1992, 46–48.

Chapter Ten: Conclusions

1. See Chenery and Syrquin 1975; Prados et al. 1993.

2. This role was theorized for the first time by Johnston and Mellor 1961. See Kuznets 1966; and, for a further discussion and references, Myint 1975; Ghatak and Ingersent 1984, 26–70; Timmer 1998 and 2002; Rao and Caballero 1990, who also cover, quite sympathetically, the radical ("structuralist") literature.

3. Dualistic models (Fei and Ranis 1997, 120–22) assume that agriculture can release disguised unemployed workers without affecting its output, but, as argued in section 2.2, the very existence of a pool of underutilized labor is doubtful.

4. Fei and Ranis (1997, 306–8) in the latest version of their dualistic model, argue that exports of goods and imports of capital can substitute agriculture as a market and source of capital. They briefly consider the flows of manpower, arguing that restriction on immigration by the "advanced" countries prevents emigration from the labor-abundant countries that they consider.

5. This case differs from the so-called Dutch disease, which is caused by the exploitation of some non-renewable natural resources.

6. See Morrison and Thorbecke, 1990; and also Karshenas 1993; Winters et al. 1998. The measure needs an extended Social Account Matrix (SAM). The latter must consider

the *actual* use of factors by sector and not the main occupation (i.e., it should report the number of hours of off-farm work by agricultural workers and of on-farm work of non-agricultural laborers). It should also disaggregate the Value Added according to the main occupation of factors.

7. Goodfriend and McDermott 1995; Galor and Weil 2000; Hansen and Prescott 2002; Tamura 2002; and Galor and Moav (2002). See also the valiant attempt by Clark (2003) to find this last model useful for the interpretation of the British industrial revolution.

8. Matsuyama 1992; Echevarria 1997; Kongsamut et al. 2001; Kogel and Prskawetz 2001; and Gollin et al. 2002.

9. Productivity in agriculture plays a twin role in this framework, as it prevents an excessive rise in the agricultural terms of trade. For example, see Echevarria (1997).

10. One should quote another model, by Laitner 2000. It differs from the others because it does not focus on structural change per se. It aims to explain the increase in the saving-GNP ratio during modern economic growth. He argues that this increase is largely apparent, as it reflects a change in the main assets, from land to financial assets. The preindustrial accumulation from capital gains on land escapes the statistical measurement and thus the saving ratio is structurally undervalued.

11. See, above all, Deane 1967, 36–50; and, for the role of the agricultural market on engineering and steel, see Bairoch 1963. For a general overview of the debate and many additional references, see Hudson 1992, 78–96.

12. Allen 1994, 119; see also Allen 1999, 217.

13. Clark et al. (1995) argue that the consumption of foodstuffs increased less than expected, given trends in prices and wages (but the gap seems smaller with the new wage index by Feinstein 1998).

14. Stokes assumes exogenous technical progress at 0.48 percent yearly, just enough to offset the diminishing returns to land, and constant per capita consumption of agricultural goods. Agriculture could have fed the British population, but it did not: imports of agricultural products grew from 13% to 28% of food supply from 1780 to 1850. This increase was matched by a corresponding increase in the export of manufactures and thus in industrial production. The model implies that agriculture used less capital and labor than the counterfactual world without import of agricultural goods and exports and manufactures.

15. cf., the surveys by Grantham 1997, and, very briefly, Crouzet 2003.

16. See Hohemberg 1972; Agulhon et al. 1976, 252, 461–65; O'Brien and Keyder 1978, 128–29 (and, more recently, O'Brien 1996). Wade (1981, 166–247) puts forward a somewhat more sophisticated version of the same argument. He argues that risk-averse farmers preferred a diversified, labor-intensive, crop mix and labor on the family farms to specialization and migration. These preferences hampered the transfer of manpower to the non-farm sector. However, as Sicsic (1992, table 1) points out, there were many landless laborers who did not have this option.

17. Magnac and Postel-Vinay 1997; and Sicsic 1992, who also reviews the debate.

18. Gregory and Stuart 1986, 242–24; Gregory 1994, 113–17; Davies 1998, 57–60; Allen 2004, 101–2. On tractorization, see table 4.11. This revision strongly downplays the break between the Stalinist period and the 1950s and 1960s, when agriculture was heavily, and unsuccessfully, subsidized.

19. The author hypothesizes different levels of (state-planned) accumulation and savings. The results quoted refer to the so-called "high investment strategy" case.

20. According to Gollin et al. (2002, 163), structural change accounted for 29% of growth in labor productivity in the LDCs during 1960–1980.

21. Delgado et al. 1998, (which also surveys the previous literature). The exact figure depends on the share of locally produced goods on additional consumption, which depends on the elasticity of supply. The data quoted are lower bounds.

22. These studies use different versions of surplus measures, none of which is as complete as Morrison and Thorbecke's version.

23. The model is an improved version of Harley and Crafts 2000, with a 0.6% yearly growth in agricultural TFP from 1770 to 1840 as the baseline case (p. 830). The counterfactual is no growth in agricultural productivity, with different assumptions about the trade in agricultural products. The results quoted in the text refer to the hypothesis of free trade under the assumption that the UK was a large country (i.e., that its demand for agricultural products caused a rise in world prices).

24. See Levy-Leboyer and Bourguignon 1985, esp. 126, 216–223. They also argue that the growth in agricultural production was not fast enough to match non-agricultural demand, causing prices and imports to increase. Finally, they estimate that a very serious agricultural crisis (a 15% fall in output for three consecutive years) would have reduced overall GDP by 12% in 1834–36, and by only 2.5% in 1900–02.

25. See also Hwa 1988; Echevarria 1997; and, specifically for Pakistan, see Rastegari-Henneberry et al. 2000 and, for a general overview of the empirical literature on growth, see Temple 1999. These results can be contrasted with those of the microeconomic study by Foster and Rosenzweig (2004) for India. They show that non-agricultural income are negatively related to yields, which the authors use as a proxy of the level of agricultural development. This result implies "competition rather than complementarity" (p. 538) between farm and non-farm activities at least at the micro-level. Such an outcome is surely not unexpected, but it raises the wider issue of the relevant area of analysis: agricultural productivity growth might affect positively the non-farm sector in the same province, or in the same country or perhaps worldwide. The choice of the country as a unit of analysis is conventient but clearly aribitrary.

26. Quoted in Beckerman 2001, 88. For a brief historical overview, see also Ruttan 1994, 4–8; and, for the economists' approach, Lichtenberg 2002.

27. Beckerman (2001, 85) states that "it is either meaningless or morally repugnant, and in any case non-operational." Even a sympathetic author such as Lélé (1991) has to admit that the concept is extremely difficult to use in practice. For the definition of "sustainable" productivity growth, see Mubarik and Byerlee 2002, 839–44.

28. One could, for instance, contrast the "pro-science" approach by Conway 2001, Gardner 2002, 349–52; and Ruttan 2002; with the deep skepticism of Pretty 1995, 13–18 (who, however, is not the most radical of writers on the issue).

29. For some long-term predictions, see Parikh 1994; Dasgupta 1998; Runge et al. 2003. Interestingly, some of the gloomiest forecasts about food prospects come from the same people who protest the loudest about the environmental dangers of modern agriculture (Lomborg 1998, 92–109).

30. See Pinstrup-Andersen et al. 1999; Runge 2003; and FAO 2003a.

31. The estimates of the needed acreage depends on the assumptions about achievable yields. It would be necessary to cultivate about 2–2.5 billion hectares (30%–60% more than the current one) if yields were at the highest pre-fertilizers level—those of Britain around 1850 for wheat, and those of Japan in 1880s for rice. But these yields were achieved

in very "advanced" countries, blessed by a favorable environment and after huge investment, notably, in irrigation. Assuming, quite optimistically, that feasible average yields were 30% lower, the minimum necessary acreage would rise to 3.2 billion hectares—i.e., close to the maximum total cultivable area, according to the FAO, which includes all the forests. The demand for land would exceed the maximum available land if average yields were lower than two-thirds of the highest preindustrial ones.

Statistical Appendix

1. Whenever possible, the dates are reworked to have comparable periods of time. The percent columns measure the contribution of growth in Total Factor Productivity to overall growth. The figure exceeds 1 if the quantity of input has diminished during the period, while it is negative if productivity has declined.

2. The figures for 1870 are estimated by extrapolating the 1857 data with the rate of change 1857–90. Results from the primal (output) measure (from table IV.1) are quite different and apparently less plausible.

BIBLIOGRAPHY

Abel, Wilhelm (1980) *Agricultural Fluctuations in Europe from the Thirteenth to the Twentieth Centuries.* New York: St. Martin's Press. Originally published in German in 1935.

Abercrombie, K. C. (1965) "Subsistence Production and Economic Development." *Monthly Bulletin of Agricultural Economics and Statistics* 14, no. 5 (May): 1–8.

Abler, David, and Vasant Sukhatme (1996) "Indian Agricultural Price Policy Revisited." Pp. 359–85 in *The Economics of Agriculture: Papers in Honor of D. Gale Johnson*, vol. 2, edited by J. M. Antle and D. A. Sumner. Chicago and London: Chicago University Press.

Abreu, Marcelo, and Alfonso Bevilaqua (2000) "Brazil as an Export Economy 1880–1930." Pp. 32–54 in *An economic history of twentieth-century Latin America*, vol. 1, edited by Enrique Cardenas, Jose Antonio Ocambo, and Rosemary Thorp. London and Basingstoke: Palgrave.

Acemoglu, Daron, Simon Johnson, and James A. Robinson (2001) "The Colonial Origin of Comparative Development: An Empirical Investigation." *American Economic Review* 91: 1369–401.

Achilles, Walter (1993) *Deutsche Agrargeschichte in Zeitalter der Reformen und der industrialisierung.* Stuggart: Ulmer.

Ackerberg, Daniel A., and Maristella Botticini (2002) "Endogenous matching and the empirical determinants of contract forms." *Journal of Political Economy* 110: 564–91.

Ackrill, Robert (2000) *The Common Agricultural Policy.* Sheffield: Sheffield Academic Press.

Acquaye, Albert, Julian Alston, and Philip Pardey (2003) "Post-war Productivity Patterns in U.S. Agriculture: Influences of Aggregation Procedures in a State-level Analysis." *American Journal of Agricultural Economics* 85: 59–80.

Adamets, Serge (1997) "À l'origine de la diversité des measures de la famine sovietique: la statistique des prix, des récoltes et de la consommation." *Cahiers du monde russe* 38: 559–86.

Ade Ajawi, J. F. (1985) "Africa at the Beginning of the Nineteenth Century: Issues and Prospects." Pp. 1–22 in *General history of Africa. Vol. VI: Africa in the Nineteenth Century until the 1880s*, edited by J. F. Ade Ajawi. Paris: UNESCO, 1985.

Adelman, Jeremy (1994) *Frontier Development. Land, Labour and Capital on the Wheatlands of Argentina and Canada, 1890–1914.* Oxford: Clarendon Press.

Aftalion, Fred (1991) *A History of the International Chemical Industry.* Philadelphia: University of Pennsylvania Press.

Afton, Bethanie, and Michael Turner (2000) "The Statistical Base of Agricultural Performance in England and Wales, 1850–1914." Pp. 1757–2140 in *The Agrarian History of England and Wales, Vol. VII: 1850–1914*, t.2, edited by E.J.T. Collins. Cambridge: Cambridge University Press.

Agulhon Maurice, Gabriel Desert, and Robert Specklin (1976) *Histoire de la France rurale. Vol. 3: Apogée et crise de la civilisation paysanne 1789–1914.* Paris: Seuil.

Aldcroft, Derek H. (1977) *From Versailles to Wall Street 1919–1929.* London: Allen Lane.

Alexandratos, Nikos (1995) *World Agriculture: Towards 2010. A FAO Study.* Chichester and New York: FAO and J. Wiley.

Allan, William (1965) *The African Husbandman.* Edinburgh and London: Oliver and Boyd.

Allen, Douglas (1991) "Homesteading and Property Rights: Or 'How the West Was Really Won.'" *Journal of Law and Economics* 34: 1–23.

Allen, Douglas, and Dean Lueck (1992) "Contract Choice in Modern Agriculture: Cash Rent versus Cropshare." *Journal of Law and Economics* 35: 397–426.

——— (1998) "The Nature of the Farm." *Journal of Law and Economics* 41: 343–86.

——— (1999) "The Role of Risk in Contract Choice." *Journal of Law, Economics and Organization* 15: 704–36.

Allen, Douglas, and Dean Lueck (2002) *The nature of the farm. Contracts, risk and organization in agriculture.* Cambridge: MIT Press.

Allen, Robert (1988) "The Growth of Labor Productivity in Early Modern English Agriculture." *Explorations in Economic History* 25: 117–46.

——— (1991a) "Labor Productivity and Farm Size in English Agriculture: A Reply to Clark." *Explorations in Economic History* 28: 478–99.

——— (1991b) "Entrepreneurship, Total Factor Productivity and Economic Efficiency: Landes, Solow and Farrel Thirty Years Later." Pp. 203–20 in *Favorites of Fortune*, edited by Patricia Higonnet, David Landes, and Henry Rosovsky. Cambridge: Harvard University Press, 1991.

——— (1992) *Enclosure and the Yeomen.* Oxford: Clarendon Press.

——— (1994) "Agriculture during the Industrial Revolution." Pp. 96–122 in *The Economic History of Britain since 1700*, vol. 1, edited by D. McCloskey and R. Floud. 2nd edition Cambridge: Cambridge University Press.

——— (1999) "Tracking the Agricultural Revolution in England." *Economic History Review* 52: 209–35.

——— (2004) *Farm to Factory: A Reinterpretation of the Soviet Industrial Revolution.* Princeton: Princeton University Press.

Allen, Robert C. (1991) "The Two English Agricultural Revolutions, 1450–1850." Pp. 236–54 in *Land, Labour and Livestock: Historical Studies in European Agricultural Productivity*, edited by Bruce Campbell and Mark Overton. Manchester: Manchester University Press, 1991.

Allen, Robert C., and Cormac O'Grada (1988) "On the Road Again with Arthur Young: English, Irish and French Agriculture during the Industrial Revolution." *Journal of Economic History* 48: 93–116.

Alston, Julian M., Connie Chang-Kang, Michele G. Marra, Philip G. Pardey, and Ty Wyatt (2001) "A Meta-analysis of Rates of Return to Agricultural R&D," *Research Report 113.* Washington, D.C.: International Food Policy Research Institute. Available at www.ifpri.cgiar.org.

Alston, Julian M., and Jennifer James (2002) "The Incidence of Agricultural Policy." Pp. 1689–784 in *Handbook of Agricultural Economics, vol. 2B: Agricultural and Food Policy*, edited by Bruce Gardner and Gordon Rausser. Amsterdam: Elsevier, 2002.

Alston, Julian M., and Philip G. Pardey (1996) *Making Science Pay. The Economics of Agricultural R&D Policy.* Washington, D.C.: American Enterprise Institute.

Alston, Julian M., Philip G. Pardey and Johannes Roseboom (1998) "Financing Agricultural Research: International Investment Patterns and Policy Perspectives." *World Development* 26: 1057–71.

Alston, Julian M., Philip G. Pardey, and Michael Taylor (2001) "Changing Contexts for Agricultural Research Development." Pp. 3–12 in *Agricultural Science Policy. Changing*

Global Agendas, edited by Julian M. Alston, Philip G. Pardey, and Michael Taylor. Published for the International Food Policy Research Institute. Baltimore and London: Johns Hopkins University Press.

Alston, Lee J. (1983) "Farm Foreclosures in the United States during the Interwar Period." *Journal of Economic History* 43: 885–903.

Alston, Lee J., Wayne A. Grove, and David C. Wheelock (1994) "Why Do Banks Fail? Evidence from the 1920s." *Explorations in Economic History* 31: 409–31.

Alston, Lee J., and Robert Higgs (1982) "Contractual Mix in Southern Agriculture since the Civil War: Facts, Hypotheses and Tests." *Journal of Economic History* 42: 327–53.

Alston, Lee J., and Kauffman Kyle (1997) "Agricultural chutes and ladders: new estimates of sharecroppers and 'true tenants' in the South, 1900–1920" *Journal of Economic History* 57. 464–75.

Alston, Lee J., and Kauffman Kyle (1998) "Up, down and off the agricultural ladder: new evidence and implications of agricultural mobility for blacks in the Postbellum South." *Agricultural History* 72: 263–277.

Alston, Lee J., Gary Libecap, and Bernardo Mueller (1999) *Titles, Conflict, and Land Use: The Development of Property Rights and Land Reform on the Brazilian Amazon Frontier.* Ann Arbor: University of Michigan Press.

Ambrosoli, Mauro (1992) *Scienziati, Contadini e proprietari. Botanica e agricoltura nell' Europa Occidentale, 1350–1850.* Torino: Einaudi.

Amin, Shahid (1982) "Small Peasant Commodity Production and Rural Indebtedness: The Culture of Sugarcane, in Eastern U.P., c.1880–1920." Reprinted in *Credit Markets and the Agrarian Economy of Colonial India*, edited by Bose Sugata. Delhi: Oxford University Press, 1994, pp. 80–135.

Anderson, Jock R., and Peter Hazell (1989) "Synthesis and Needs in Agricultural Research and Policy." Pp. 339–56 in *Variability in Grain Yields. Implications for Agricultural Research and Policy in Developing Countries*, edited by Jock R. Anderson and Peter Hazell. IFPRI. Baltimore and London: Johns Hopkins University Press.

Anderson, Jock R., and Karla Hoff (1993) "Technological Change, Imperfect Markets and Agricultural Extension: An Overview." Pp. 471–77 in *The economic of rural organization. Theory, practice and policy*, edited by Karla Hoff, Avishay Braverman, and Joseph Stiglitz. Published for the World Bank. Oxford: Oxford University Press.

Anderson, Kym, and Yujiro Hayami (1986) *The Political Economy of Agricultural Protection: East Asia in International Perspective.* Sydney: Allen and Unwin.

Anderson, Kym, Yujiro Hayami, and Masayoshi Honma (1986) "The Growth of Agricultural Protection." Pp. 18–31 in *The Political Economy of Agricultural Protection: East Asia in International Perspective*, edited by Kym Anderson and Yujiro Hayami. Sydney: Allen and Unwin.

Anderson, Kym, and Rod Tyers (1992) "Japanese Rice Policy in the Interwar Period: Some Consequences of the Imperial Self-Sufficiency." *Japan and the World Economy* 4: 103–27.

Andreosso-O'Callaghan, Bernadette (2003) *The Economics of European Agriculture.* Basingstoke: Palgrave.

Antholt, Charles H. (1998) "Agricultural Extension in the Twenty-First Century." Pp. 354–69 in *International Agricultural Development*, edited by Carl K. Eicher and John M. Staatz. 3rd edition. Baltimore and London: Johns Hopkins University Press.

Anthony, Kenneth R., Bruce F. Johnston, William O. Jones, and Victor C. Uchendu (1979) *Agricultural Change in Tropical Africa.* Ithaca and London: Cornell University Press.

Antsiferov, Alexis (1930) *Russian Agriculture during the War*. New Haven: Yale University Press.

Arcari, Paola M. (1936) "Le variazioni dei salari agricoli in Italia dalla fondazione del Regno al 1933." *Annali di Statistica serie* VI, vol. 36. Rome: ISTAT.

Aricanli, Tosun (1986) "Agrarian Relations in Turkey: A Historical Sketch." Pp. 23–67 in *Food States and Peasants*, edited by Alan Richards. Boulder and London: Westview Press.

Arndt, H. W. (1963) *The Economic Lessons of the Nineteen-Thirties*. 2nd edition. London: Frank Cass.

Artadi, Elsa, and Xavier Sala-i-Martin (2003) "The Economic Tragedy of the Twentieth Century: Growth in Africa." NBER Working Paper n.9865.

Ash, Robert (1996) "Agricultural Development since 1978." Pp. 276–308 in *The Chinese Economy under Deng Xiao Ping*, edited by Robert Ash and Y. Y. Kueh. Oxford: Oxford University Press.

Ashton, Basil, Kenneth Hill, Alan Piazza, and Robin Zeitz (1984) "Famine in China, 1958–61." *Population and Development Review* 10: 613–45.

ASI (ad annum). *Annuario statistico italiano*, Rome: ISTAT.

Askari, Hossein, and John Thomas Cummings (1976) *Agricultural Supply Response*. New York: Praeger.

Atack, Jeremy, and Fred Bateman (1987) *To Their Own Soil: Agriculture in the Ante-bellum North*. Ames: Iowa State University Press.

Atack, Jeremy, Fred Bateman, and William N. Parker (2000a) "The Farm, the Farmer and the Market." Pp. 245–84 in *The Cambridge Economic History of the United States*, vol. 2: *The Long Nineteenth Century*, edited by Stanley Engerman and Robert Gallman. Cambridge: Cambridge University Press.

——— (2000b) "Northern Agriculture and the Westward Movement." Pp. 285-328 in *The Cambridge Economic History of the United States*, vol. 2: *The Long Nineteenth Century*, edited by Stanley Engerman and Robert Gallman. Cambridge: Cambridge University Press.

Atack, Jeremy, and Peter Passel (1994) *A New Economic View of American History*. 2nd edition. Norton: New York.

Attwood, Donald W. (1987) "Irrigation and Imperialism: The Causes and Consequences of a Shift from Subsistence to Cash Cropping." *Journal of Development Studies* 23: 342–64.

——— (1990) "Land registration in Africa: The impact on Agricultural Production." *World Development* 18: 659–71.

Austen, Roger B., and Michael B. Arnold (1989) "Variability in Wheat Yields in England: Analysis and Future Prospects." Pp. 100–106 in *Variability in Grain Yields: Implications for Agricultural Research and Policy in Developing Countries*, edited by Jock R. Anderson and Peter Hazell. IFPRI. Baltimore and London: Johns Hopkins University Press.

Aymard, Maurice (1982) "Autoconsommation et marches: Chayanov, Labrousse ou Le Roy Ladurie?" *Annales E.S.C.* 38: 1392–409.

Bachi, Riccardo (1926) *L'alimentazione e la politica annonaria in Italia*. Bari: Laterza.

Bairoch, Paul (1963) *Revolution industrielle et sous-développement*. Paris: Sedes.

——— (1989a) "European Trade Policy 1815–1914." Pp. 1–160 in *The Cambridge Economic History of Europe*, vol. 8, edited by P. Mathias and S. Pollard. Cambridge: Cambridge University Press.

——— (1989b) "Les trois révolutions agricoles du monde développé: rendements et productivité de 1800 à 1985." *Annales E.S.C.* 44: 317–55.

——— (1997) "New Estimates on Agricultural Productivity and Yields of Developed Countries 1800–1990." Pp. 45–64 in *Economic Development and Agricultural Productivity*, edited by A. Bhaduri and R. Skarstein. Cheltenham: Elgar.

——— (1999) *L'agriculture des pays développés. 1800 à nos jours*. Paris: Economica.

Bairoch, Paul, T. Deldycke, H. Gelders, and J. M. Limbor (1968) *La population active et sa structure*. Bruxelles: Universitè Libre de Bruxelles, Institut de Sociologie.

Balisacan, Arsenio M., and James A. Roumasset (1987) "Public Choice of Economic Policy: The Growth of Agricultural Protection." *Weltwirtschaftliches Archiv* 123: 232–47.

Ball, Eldon V., Jean-Christophe Bureau, Jean-Pierre Butault, and Richard Nehring (2001) "Levels of Farm Sector Productivity: An International Comparison." *Journal of Productivity Analysis* 15: 5–29.

Ban, Sung Hwan (1979) "Agricultural Growth in Korea, 1918–71." Pp. 90–116, 313–49 in *Agricultural Growth in Japan, Taiwan, Korea, and the Philippines*, edited by Yujiro Hayami, Vernon Ruttan, and H. Southworth. Honolulu: University Press of Hawai.

Banerjee, Abhijit (2000) "Prospect and Strategies for Land Reform." Pp. 253–73 in *Annual World Bank Conference on Development Economics 1999*, edited by Boris Pleskovic and Joseph E. Stiglitz. Washington: World Bank.

Banerjee, Abhijit, Paul J. Gertler, and Maitreesh Ghatak (2002) "Empowerment and Efficiency: Tenancy Reform in West Bengal." *Journal of Political Economy* 110: 239–80.

Banti, Alberto (1989) *Terra e denaro. Una borghesia padana dell'Ottocento*. Padua: Marsilio.

Barberi, Benedetto (1961) *I consumi nel primo secolo dopo l'Unità*. Milan: Giuffrè.

Barkaoui Ahmed, Jean-Christophe Bureau, and Jean-Pierre Butault (1997) "La mésure de la productivité par des functions de distance. *Economie et prévision* 127.

Barnett Margaret L. (1985) *British Food Policy during the First World War*. London: George Allen and Unwin.

Barral, Pierre (1979a) "Un secteur dominé: la terre." Pp. 351–400 in *Histoire économique et sociale de la France*, vol. 4.1, edited by Fernand Braudel and Ernst Labrousse. Paris: Presses Universitaires de France.

——— (1979b) "Les grandes épreuves: agriculture et paysannerie 1914–48." Pp. 823–60 in *Histoire économique et sociale de la France*, vol. 4.2, edited by Fernand Braudel and Ernst Labrousse. Paris: Presses Universitaires de France.

——— (1982) "Le secteur agricole dans la France industrialisée (1950–1974)." Pp. 1427–63 in *Histoire économique et sociale de la France*, vol. 4.3, edited by Fernand Braudel Ernst Labrousse and Jean Bouvier. Paris: Presses Universitaires de France.

Barro, Robert (1999) "Notes on Growth Accounting." *Journal of Economic Growth* 4: 119–37.

Barron, Hal (1984) *Those Who Stayed Behind*. Cambridge: Cambridge University Press.

Barry, Peter J., and Lindon J. Robinson (2001) "Agricultural Finance: Credit Constraints and Consequences." Pp. 514–71 in *Handbook of Agricultural Economics, vol. 1A: Agricultural Production*, edited by Bruce Gardner and Gordon Rausser. Amsterdam: Elsevier.

Barsky, Osvaldo (1991) "Social and technological transformations of the Argentine Pampa." Pp. 56–75 in *Modernization and stagnation. Latin American agriculture into the 1990s*, edited by Twomey Michael and Ann Helwege. New York and Westport: Greenwood Press.

Barzel, Yoram (1997) *Economic Analysis of Property Rights*. 2nd edition. Cambridge: Cambridge University Press.

Basset, Thomas J. (1993) "Introduction: The Land Question and Agricultural Transformation in Sub-Saharan Africa." Pp. 3–25 in *Land in African Agrarian Systems*, edited by Thomas J. Basset and Donald E. Crummey. Madison: University of Wisconsin Press.

Bateman, Fred (1968) "Improvement in American Dairy Farming, 1850–1910: A Quantitative Analysis." *Journal of Economic History* 28: 255–73.

——— (1969) "Labour Inputs and Productivity in American Dairy Agriculture, 1850–1910." *Journal of Economic History* 29: 206–29.

Bates, Robert H. (1981) *Markets and States in Tropical Africa*. Berkeley: University of California Press.

Bauer, Arnold J. (1975) *Chilean Rural Society from the Spanish Conquest to 1930*. Cambridge: Cambridge University Press.

Bautista, Romeo, and Alberto Valdes (1993) *The Bias against Agriculture. Trade and Macroeconomic Policies in Developing Countries*. San Francisco: ICS Press.

Becker, Jasper (1996) *Hungry Ghosts. China's Secret Famine*. London: John Murray.

Beckerman, Wilfred (2001) "A Skeptical View of Sustainable Development." Pp. 85–100 in *Agricultural Science Policy. Changing Global Agendas*, edited by Julian M. Alston, Philip Pardey, and Michael Taylor. Published for the International Food Policy Research Institute. Baltimore and London: Johns Hopkins University Press.

Behrman, Jere R., and Anil B. Deolikar (1988) "Health and nutrition." Pp. 631–711 in *Handbook of Development Economics*, vol. 1, edited by Hollis Chenery and T. N. Srinivasan. Amsterdam: North Holland.

Bekaert, Geer (1991) "Caloric Consumption in Industrializing Belgium." *Journal of Economic History* 51: 633–53.

Bellerby, J. R. (1968) "Distribution of Farm Income in the United Kingdom 1867–1938." Pp. 261–77 in *Essays in Agrarian History*, vol. 2, edited by Walter Minchinton. London: Newton Abbot.

Benedict, Murray R. (1953) *Farm Policies of the United States, 1790–1950*. New York: Twentieth Century Fund.

Benjamin, Dwayne (1995) "Can Unobserved Land Quality Explain the Inverse Productivity Relationship?" *Journal of Development Economics* 46: 151–84.

Berend, Ivan T. (1985) "Agriculture." Pp. 148–210 in *The Economic History of Eastern Europe 1919–1975*, vol. 1, edited by M. C. Kaser and E. A. Radice. Oxford: Clarendon Press.

Berend Ivan T., and Gyorgy Ranki (1974) *Economic Development in East-Central Europe in the 19th and 20th Centuries*. New York: Columbia University Press.

——— (1976) *Hungary: A Century of Economic Development*. London: Newton Abbot.

Berger, Helge, and Mark Spoerer (2001) "Economic Crises and the European Revolutions of 1848." *Journal of Economic History* 61: 293–326.

Bernard, Andrew B., and Charles Jones (1996) "Productivity across Industries and Countries: Time Series Theory and Evidence." *Review of Economics and Statistics* 78: 134–45.

Berry, Albert R., and William R. Cline (1979) *Agrarian Structure and Productivity in Developing Countries*. Baltimore and London: Johns Hopkins University Press.

Bértola, Luis (1998) *El PBI Uruguayo 1870–1936 y Otras Estimaciones*. Montevideo: Facultad de Ciencias Sociales..

Besley, Timothy (1995) "Saving, Credit and Insurance." Pp. 2123–201 in *Handbook of Development Economics*, vol. 3A, edited by Jere Behrman and T. N. Srinivasan. Amsterdam: Elsevier.

Besley, Timothy, and Robin Burgess (2000) "Land Reform, Poverty Reduction and Growth: Evidence from India." *Quarterly Journal of Economics* 115: 389–430.

Bevilacqua, Piero, and Manlio Rossi-Doria (1984) *Le bonifiche in Italia dal '700 ad oggi.* Bari: Laterza.

Biagioli, Giuliana (1980) "Agricoltura e sviluppo economico: una riconsiderazione del caso italiano." *Società e Storia*: 670–708.

——— (2000) *Il modello del proprietario imprenditore nella Toscana dell'Ottocento: Bettino Ricasoli.* Firenze: Olschki.

——— (2002) "La mezzadria poderale nell'Italia Centro-settentrionale in età moderna e contemporanea." *Rivista di storia dell'agricoltura* 42: 53–101.

Biagioli, Giuliana, Rossano Pazzagli, and Roberto Tolaini (2000) *Le Corse Agrarie. Lo sguardo del Giornale Agrario Toscano sulla Società rurale dell'Ottocento.* Pisa: Pacini.

Biggs, Stephen D., and Edward J. Clay (1981) "Sources on Innovation in Agricultural Technology." *World Development* 9: 321–36.

——— (1988) "Generation and Diffusion of Agricultural Technology: Theories and Experiences." Pp. 19–60 in *Generation and Diffusion of Agricultural Innovations: The Role of Institutional Factors,* edited by Ahmed Iftikhar and Vernon W. Ruttan. Aldershot: ILO and Gower.

Bill, Albert, and Adrian Graves (1984) "Introduction." Pp. 1–7 in *Crisis and Change in the International Sugar Economy 1860–1914,* edited by Albert Bill and Adrian Graves. Norwich and Edinburgh: ISC Press.

——— (1988) "Introduction." Pp. 1–23 in *The World Sugar Economy in War and Depression 1914–1940,* edited by Albert Bill and Adrian Graves. London: Routledge.

Binenbaum, Eran, Carol Nottemburg, Philip G. Pardey, Brian D. Wright, and Patricia Zambrano (2003) "South-North Trade, Intellectual Property Jurisdictions and Freedom to Operate in Agricultural Research on Staple Crops." *Economic Development and Cultural Change* 51: 309–35.

Binswanger, Hans P. (1995) "Predicting Institutional Change. What Building Blocks Does a Theory Need." Pp. 103–35 in *Induced Innovation Theory and International Agricultural Development: A Reassessment,* edited by Bruce M. Koppel. Baltimore and London: Johns Hopkins University Press.

Binswanger, Hans P., and Klaus Deininger (1997) "Explaining Agricultural and Agrarian Policies in Developing Countries." *Journal of Economic Literature* 35: 1958–2005.

Binswanger, Hans P., Klaus Deininger, and Feder Gershon (1997) "Power, Distortions, Revolt and Reform in Agricultural Land Relations." Pp. 2661–771 in *Handbook of Development Economics,* vol. 3B, edited by Jere Behrman and T. N. Srinivasan. Amsterdam: Elsevier.

Binswanger, Hans P., and Robert F. Townsend (2000) "The growth performance of agriculture in Sub-Saharan Africa." *American Journal of Agricultural Economics* 82: 1075–86.

Biraben, Jan Noel (1979) "Essai sur le nombre des hommes." *Population* 34: 13–24.

Birkhaeuser, Dean, Robert E. Evenson, and Feder Gershon (1991) "The Economic Impact of Agricultural Extension: A Review." *Economic Development and Cultural Change* 39: 607–50.

Blades, Derek W. (1975) *Non-monetary (Subsistence) Activities in the National Accounts of Developing Countries.* Paris: OECD Development Centre.

Blandford, David (1990) "The Cost of Agricultural Protection and the Difference Free Trade Would Make." Pp. 398–422 in *Agricultural Protectionism*, edited by Fred H. Sanderson. Washington, D.C.: Resources for the Future.

Blarel, Benoit, Peter Hazell, Frank Place, and John Quiggin (1992) "The Economics of Farm Fragmentation: Evidence from Ghana and Rwanda." *World Bank Economic Review* 6: 233–54.

Bleaney, Michael, and David Greenaway (1993) "Long-run Trends in the Relative Price of Primary Commodities and in the Terms of Trade of Developing Countries." *Oxford Economic Papers* 45: 349–63.

Block, Steven A. (1995) "The Recovery of Agricultural Productivity in Sub-Saharan Africa." *Food Policy* 20: 385–405.

Blomme, Jan (1992) "The Economic Development of Belgian Agriculture 1880–1980. A Qualitative and Quantitative Analysis." Brussels: Leuwen University Press.

Bloomfield, B. T. (1984) *New Zealand: A Handbook of Historical Statistics.* Boston: Hall.

Blum, Jerome (1978) *The End of the Old Order in Rural Europe.* Princeton: Princeton University Press.

Blyn, George (1966) *Agricultural Trends in India 1891–1947: Output Availability and Productivity.* Cambridge: Cambridge University Press.

Boahen, Albert Adu (1987) "L' Afrique face au défi colonial." Pp. 21–36 in *Histoire generale de l'Afrique, VII: L'Afrique sous domination coloniale 1880–1935.* Paris: UNESCO.

Bodenhorn, Howard (2000) *A History of Banking in Antebellum America.* Cambridge: Cambridge University Press.

Bogue, Allan G. (1963) *From Prairie to Corn Belt.* Chicago: University of Chicago Press.

Booth, Anne (1988) *Agricultural Development in Indonesia.* Sydney: Allen and Unwin.

——— (1991) "The economic development of South-East Asia 1870–1985." *Australian Economic History Review* 31: 20–52.

Booth, Anne, and R. M. Sundrum (1985) *Labour Absorption in Agriculture.* Oxford: Oxford University Press.

Bose, Sugata (1994) "Introduction." Pp. 1–22 in *Credit Markets and the Agrarian Economy of Colonial India*, edited by Sugata Bose. Delhi: Oxford University Press.

Boserup, Ester (1951) *The Conditions of Agricultural Growth: The Economics of Agrarian Change under Population Pressure.* Chicago: Aldine, reprinted 1966.

Bouis, Howarth (1994) "Consumption Effects of Commercialization of Agriculture." Pp. 65–78; *Agricultural Commercialisation, Economic Development and Nutrition*, edited by Joachim von Braun and Eileen Kennedy. Published for the IFPRI. Baltimore and London: Johns Hopkins University Press.

Bovard, James (1989) *The Farm Fiasco.* San Francisco: ICS Press.

Bowie, G.G.S. (1987) "New Sheep for Old—Changes in Sheep Farming in Hampshire, 1792–1879." *Agricultural History Review* 35: 15–24.

Boyd, Michael (1988) "The Performance of Private and Socialist Agriculture in Poland: The Effects of Policy and Organization." *Journal of Comparative Economics* 12: 61–73.

Brandão, Antonio Salazar, and Gershon Feder (1995) "Regulatory Policies and Reform: The Case of Land Markets." Pp. 191–209 in *Regulatory Policies and Reform: A Comparative Perspective*, edited by Claudio Frischtak. Washington, D.C.: World Bank.

Brandt, Loren (1989) *Commercialization and Agricultural Development.: Central and Eastern China 1870–1937.* New York: Cambridge University Press.

——— (1993) "Interwar Japanese Agriculture: Revisionist Views on the Impact of the Colonial Rice Policy and the Labour-Surplus Hypothesis." *Explorations in Economic History* 30: 259–93.

——— (1997) "Reflections on China's Late 19th and early 20th century economy." *China Quarterly* 150: 282–307.

Brandt, Loren, and Barbara Sands (1992) "Land Concentration and Income Distribution in Republican China." Pp. 179–206 in *Chinese History in Economic Perspective*, edited by Thomas Rawski and Lillian Li. Berkeley and Los Angeles: University of California Press.

Brassley, Paul (2000a) "Agriculture Science and Education." Pp. 594–49 in *The Agrarian History of England and Wales, vol. 7: 1850–1914*, part 1, edited by E.J.T. Collins. Cambridge: Cambridge University Press.

——— (2000b) "Farming Techniques D Land Drainage." Pp. 505–13 in *The Agrarian History of England and Wales, vol. 7: 1850–1914*, part 1, edited by E.J.T. Collins. Cambridge: Cambridge University Press.

——— (2000c) "Farming Techniques E Crop Varieties." Pp. 522–32 in *The Agrarian History of England and Wales, vol. 7: 1850–1914*, part 1, edited by E.J.T. Collins. Cambridge: Cambridge University Press.

——— (2000d) "Farming Techniques F Plant Nutrition." Pp. 533–47 in *The Agrarian History of England and Wales, vol. 7: 1850–1914*, part 1, edited by E.J.T. Collins. Cambridge: Cambridge University Press.

——— (2000e) "Farming Techniques G Weed and Pest Control." Pp. 548–54 in *The Agrarian History of England and Wales, vol. 7: 1850–1914*, part 1, edited by E.J.T. Collins. Cambridge: Cambridge University Press.

——— (2000f) "Farming Techniques I Animal Nutrition." Pp. 570–86 in *The Agrarian History of England and Wales, vol. 7: 1850–1914*, part 1, edited by E.J.T. Collins. Cambridge: Cambridge University Press.

——— (2000g) "Farming Techniques H Livestock Breeds." Pp. 563–69 in *The Agrarian History of England and Wales, vol. 7: 1850–1914*, part 1, edited by E.J.T. Collins Cambridge: Cambridge University Press.

——— (2000h) "Output and Technical Change in Twentieth-Century British agriculture." *Agricultural History Review* 48: 60–84.

Braun, Juan, Matias Braun, Ignacio Briones, Jose Diaz, Rolf Luders, and Gert Wagner (2000) *Economia Chilena 1810–1995: estadisticas historicas*, Documentos de trabajo n.187 Pontificia Universidad catolica de Chile Instituto de Economia Enero. Available at: http://volcan.facea.puc.cl/economia/publicaciones/documentos_trabajo.htm.

Bray, Francesca (1986) *The Rice Economies: Technology and Development in Asian Societies.* Berkeley: University of California Press.

Brenner, Robert, and Christopher Isett (2002) "England's Divergence from China's Yangzi Delta: Property Relations, Microeconomics and Patterns of Development." *Journal of Asian Studies* 61: 609–22.

Breusers, Mark (2001) "Searching for Livelihood: Land and Mobility in Burkina Faso." *Journal of Development Studies* 37: 49–80.

Brigden, Roy (2000) "Farming Techniques C: Equipment and Motive Power." Pp. 514–21 in *The Agrarian History of England and Wales, vol. 7: 1850–1914*, part 1, edited by E.J.T. Collins. Cambridge: Cambridge University Press.

Bringas Gutierrez, Miguel Angel (2000) *La productividad de los factores en la agricultura espanola.* Estudios de historia economica n 39. Madrid: Banco de Espana.

Brinkley, Gerald (1997) "The Decline of Southern Agricultural Output, 1860–1880." *Journal of Economic History* 57: 116–38.

Broadberry, Stephen (1998) "How Did the United States and Germany Overtake Britain? A Sectoral Analysis of Comparative Productivity Levels, 1870–1990." *Journal of Economic History* 58: 375–407.

Brooks, Karen, and Bruce Gardner (2004) "Russian Agriculture in the Transition to a Market Economy." *Economic Development and Cultural Change* 52: 571–86.

Brooks, Karen, and John Nash (2002) "The Rural Sector in Transition Economies." Pp. 1547–92 in *Handbook of Agricultural Economics, vol. 2A: Agricultural and Food Policy*, edited by Bruce Gardner and Gordon Rausser. Amsterdam: Elsevier.

Bruce, John W. (1988) "Indigenous Land Tenure and Land Concentration." Pp. 23–52 in *Land and Society in Contemporary Africa*, edited by S. P. Reyna and R. E. Downs. Hanover and London: University Press of New England.

——— (1993) "Do Indigenous Tenure Systems Constrain agricultural development?" Pp. 35–51 in *Land in African Agrarian Systems*, edited by Thomas J. Basset and Donald E. Crummey. Madison: University of Wisconsin Press.

Brunt, Liam (2003a) "Rehabilitating Arthur Young." *Economic History Review* 56: 265–99.

——— (2003b) "Mechanical Innovation in the Industrial Revolution: The Case of Plough Design." *Economic History Review* 56: 444–77.

——— (2004) "Nature or Nurture? Explaining English Wheat Yields in the Industrial Revolution, ca. 1770." *Journal of Economic History* 64: 192–225.

Buckley, Kenneth (1955) *Capital Formation in Canada 1896–1930.* Toronto: University of Toronto Press.

Bulmer-Thomas, Victor (1995) *The Economic History of Latin America since Independence.* Cambridge: Cambridge University Press.

Burnell, Colbert T. (2001) "Iowa Farmers and Mechanical Corn-pickers, 1900–1952." *Agricultural History* 74: 530–44.

Burrows, Geoff, and Ralph Shlomowitz (1992) "The Lag in the Mechanization of the Sugarcane Harvest: Some Comparative Perspectives." *Agricultural History* 66 (3): 61–75.

Butlin, Noel G. (1962) *Australian Domestic Product, Investment and Foreign Borrowing. 1861–1938/39.* Cambridge: Cambridge University Press.

——— (1994) *Forming a Colonial Economy: Australia 1810–1850.* Cambridge: Cambridge University Press.

Byerlee, Derek (1996) "Modern Varieties, Productivity and Sustainability: Recent Experience and Emerging Challenges." *World Development* 24: 697–718.

Campos, Mauro F., and Fabrizio Coricelli (2002) "Growth in transition: what we know, what we do not know and what we should." *Journal of Economic Literature* 40: 793–836.

Capalbo, Susan M., and Trang T. Vo (1988) "A review of the evidence on agricultural productivity and aggregate technology," Pp. 96–137 in *Agricultural productivity. Measurement and explanation*, edited by Susan Capalbo and John M. Antle. Washington: Resources for the Future.

Cardoso de Mello, Manoel João, and Maria da Conceição Tavares (1985) "The capitalist export economy in Brazil, 1884–1930." Pp. 82–136 in *The Latin American economies*, edited by Roberto Cortes-Conde and Shane J. Hunt. London: Holmes and Meier.

Cariola, Carmen, and Osvaldo Sunkel (1985) "The growth of the nitrate industry and socioeconomic change in Chile 1880–1930." Pp. 137–254 in *The Latin American Economies*, edited by Roberto Cortes-Conde and Shane J. Hunt. London: Holmes and Meier.

Carmona, Juan, and James Simpson (1999) "The 'rabassa morta' in Catalan viticulture: the rise and decline of a long-term sharecropping contract," *Journal of Economic History* 59: 290–315.

Caron, François (1979) *An Economic History of Modern France.* London: Methuen.

Cartier M. (1986) "Une Nouvelle Historiographie Chinoise. La formation d'un marché national vue par Wu Chengming." *Annales E.S.C.* 41: 1303–12.

Caselli, Francesco, and Wilbur John Coleman II (2001) "The U.S. Structural Transformation and Regional Convergence: A Reinterpretation." *Journal of Political Economy* 109: 584–616.

Chambers, J. D., and G. E. Mingay (1966) *The Agricultural Revolution, 1750–1880.* London and Sydney: Batsford.

Chandavarkar, A. G. (1983) "Money and Credit." Pp. 762–808 in *The Cambridge Economic History of India*, vol. 2: c. 1757–c. 1970, edited by Dharma Kumar. Cambridge: Cambridge University Press.

Chao, Kang (1982) "Tenure Systems in Traditional China." Pp. 269–91 in *Agricultural Development in China, Japan and Korea*, edited by Hou Chi-ming and Yu Tzong-shian. Taipei: Academia Sinica.

Chao, Kang (1986) *Man and Land in Chinese Economic History.* Stanford: Stanford University Press.

Charlesworth, Charles (1978) "Rich Peasants and Poor Peasants in Late Nineteenth-Century Maharashtra." Pp. 96–113 in *The Imperial Impact: Studies in the Economic History of Africa and India*, edited by Clive Dewey and A. G. Hopkins. London: Athlone Press.

Charlesworth, Neil (1985) *Peasants and Imperial Rule: Agriculture and Agrarian Society in the Bombay Presidency, 1850–1935.* Cambridge: Cambridge University Press.

Chaudhuri, B. B. (1984) "Rural Power Structure and Agricultural Productivity in Eastern India, 1757–1947." Pp. 100–70 in *Agrarian Power and Agricultural Productivity in South Asia*, edited by M. Desai, S. Hoebner Rudolph, and A. Rudra. Delhi: Oxford University Press.

Chavas, Jean Paul (2001) "Structural Change in Agricultural Production: Economics, Technology and Policy." Pp. 264–84 in *Handbook of Agricultural Economics, vol. 1A: Agricultural Production*, edited by Bruce Gardner and Gordon Rausser. Amsterdam: Elsevier.

Chayanov, A. V. (1966) *The Theory of Peasant Economy.* Homewood: Irwin. Originally published in 1925.

Chenery, Hollis, and Moshe Syrquin (1975) *Patterns of Development 1950–1975.* Oxford: Oxford University Press.

Cheng, Siok-Hwa (1968) *The Rice industry of Burma, 1852–1940.* Kuala Lumpur: University of Malaya Press.

Cheung, Steve (1969) *Theory of Sharecropping.* Chicago: Chicago University Press.

Chonchol, Jacques (1995) *Systemes agraires en Amérique Latine. Des agricultures préhispaniques à la modernisation conservatrice.* Paris: Editions de l'IHEAL.

Chorley, G.P.H. (1981) "The Agricultural Revolution in Northern Europe, 1750–1880: Nitrogen, Legumes and Crop Productivity." *Economic History Review* 34: 71–93.

Chowning, Margaret (1997) "Reassessing the Prospects for Profit in Nineteenth-Century Mexican Agriculture from a Regional Perspective: Michoacán, 1810–1860." Pp. 179–205 in *How Latin America Fell Behind: Essays on the Economic Histories of Brazil and Mexico, 1800–1914*, edited by Stephen Haber. Stanford: Stanford University Press.

Clarence-Smith, W. G. (1995) "Cocoa Plantations in the Third World, 1870s–1914." Pp. 157–71 in *The New Institutional Economics and Third World Development*, edited by John Harris, Janet Hunter, and Colin M. Lewis. London: Routledge.

Clark, Christopher (1990) *The Roots of Rural Capitalism: Western Massachusetts 1780–1860.* Ithaca and London: Cornell University Press.

Clark, Colin (1970) *The Economics of Irrigation.* 2nd edition. Oxford: Pergamon Press.

——— (1977) *Population Growth and Land Use.* 2nd edition. London: Macmillan.

Clark, Gregory (1987) "Productivity Growth without Technical Change in European Agriculture before 1850." *Journal of Economic History* 47: 419–32.

——— (1989) "Productivity Growth without Technical Change in European Agriculture: Reply to Komlos." *Journal of Economic History* 49: 979–91.

——— (1991) "Labor Productivity and Farm Size in English Agriculture: A note." *Explorations in Economic History* 28: 248–57.

——— (1993) "Agriculture and the Industrial Revolution, 1700–1850." Pp. 227–66 in *The British Industrial Revolution*, edited by J. Mokyr. Boulder: Westview.

——— (1998a) "Land Hunger: Land as a Commodity and a Status Good, England 1500–1910." *Explorations in Economic History* 35: 59–82.

——— (1998b) "Commons Sense: Common Property Rights, Efficiency and Institutional Change." *Journal of Economic History* 58: 73–102.

——— (2002a) "Land Rental Values and the Agrarian Economy: England and Wales, 1500–1914." *European Review of Economic History* 6: 281–308.

——— (2002b) "The Agricultural Revolution and the Industrial Revolution: England 1500–1912." Available at www.econ.ucdavis.edu/faculty/GClark/papers.

——— (2003) "The Great Escape: The Industrial Revolution in Theory and History." Available at www.econ.ucdavis.edu/faculty/GClark/papers.

Clark, Gregory, Michael, Huberman, and Peter Lindert (1995) "A British food puzzle, 1770–1850." *Economic History Review* 48: 215–37.

Clark, Gregory, and Anthony Clark (2001) "Common Rights to Land in England." *Journal of Economic History* 61: 1009–36.

Clark, Gregory, and Ysbrand van der Werf (1998) "Work in Progress? The Industrious Revolution." *Journal of Economic History* 58: 830–43.

Clarke, Roger A., and Dubravko J. I. Matko (1984) *Soviet Economic Facts, 1917–1981.* 3rd edition. New York: St. Martin's Press.

Clarke, Sally (1994) *Regulation and the Revolution in United States Farm Productivity.* Cambridge: Cambridge University Press.

Cleary, Mark, and Peter Eaton (1996) *Tradition and Reform: Land Tenure and Rural Development in South East Asia.* Kuala Lumpur: Oxford University Press.

Clout, Hugh (1980) *Agriculture in France on the Eve of the Railway Age.* London: Croom Helm.

Cock, James H. (1985) *Cassava: New Potential for a Neglected Crop.* Boulder and London: Westview Press.

Coelli, Tim, Sanzidumar Rahman, and Colin Thirtle (2002) "Technical, Allocative, Cost and Scale Efficiencies in Bangladesh Rice Cultivation: A Non-parametric Approach." *Journal of Agricultural Economics* 53: 607–26.

Cohen, Joel E. (1995) *How Many People Can the Earth Support?* New York: Norton.

Collins, E.T.J. (1972) "The Diffusion of the Threshing Machine in Britain 1790–1880." *Tools and tillage* 2: 16–33.

——— (1973) "Offerta e domanda di manodopera agricola in Europa dal 1800 al 1880." Pp. 89–131 in *Agricoltura e sviluppo economico*, edited by E. L. Jones and S.Woolf. Torino: Einaudi.

——— (1983) "The Farm Horse Economy of England and Wales in the Early Tractor Age, 1900–40." Pp. 73–101 in *Horses in European Economic History*, edited by F.M.L. Thompson. Reading: British Agricultural History Society.

Constance, Douglas H., Jere L. Gilles, and William D. Hefferman (1990) "Agrarian Policies and Agricultural Systems in the United States." Pp. 9–75 in *Agrarian Policies and Agricultural Systems*, edited by Alessandro Bonanno. Boulder: Westview Press.

Conway, Gordon (2001) "The Doubly Green Revolution: Balancing Food, Poverty and Environmental Needs in the 21st century." Pp. 17–34 in *Trade-offs or Synergies? Agricultural Intensification, Economic Development and the Environment*, edited by D. R. Lee and C. B. Barrett. Wallingford: CABI Publishing.

Conybeare, John (1987) *Trade Wars: The Theory and Practice of International Commercial Rivalry.* New York: Columbia University Press.

Coppola, Gauro (1981) *Il mais nell'economia agricola lombarda.* Bologna: Mulino.

Corni, Gustavo (1990) *Hitler's Peasants.* Oxford: Berg.

Corona Gabriella (2004) "Declino dei 'commons' ed equilibri ambientali: il caso italiano fra Otto e Novecento." *Società e Storia* 104: 357–83.

Correnti, Cesare, and Pietro Maestri (1864) *Annuario statistico italiano.* Torino: Tipografia Letteraria.

Cortes-Conde, Roberto (1997) *La economia argentina en el largo plazo: ensayos de historia economica.* Buenos Aires: Editorial Sudamericana-Universidad de San Andres.

Cote, Daniel, and Ginette Carre (1996) "1995 Profile: Agricultural Co-operation throughout the World: An Overview." *Review of International Co-operation* 89: 61–70.

Crafts, Nicholas F. R. (1985) *British Economic Growth during the Industrial Revolution.* Oxford: Clarendon Press.

Crafts Nicholas F. R., and Knick C. Harley (2004) "Precocious British Industrialization: A General Equilibrium Perspective." Pp. 86–107 in *Exceptionalism and Industrialisation. Britain and its European rivals, 1688–1815*, edited by Leandro Prados de la Escosura, Cambridge: Cambridge University Press.

Craig, Barbara J., and Philip Pardey (2001) "Inputs, Output and Productivity Developments in U.S. Agriculture." Pp. 37–55 in *Agricultural Science Policy. Changing Global Agendas*, edited by Julian M. Alston, Philip G. Pardey, and Michael J. Taylor. Baltimore and London: Johns Hopkins University Press.

Craig, Lee A., and Thomas Weiss (2000) "Hours at Work and Total Factor Productivity in 19th-Century U.S. Agriculture." *Advances in Agricultural Economic History* 1: 1–30.

Crisostomo-David, Cristina, and Randolph Barker (1979) "Agricultural Growth in Korea, 1918–71." Pp. 117–42, 351–88 in *Agricultural Growth in Japan, Taiwan, Korea and the Philippines*, edited by Y. Hayami, V. Ruttan, and H. Southworth. Honolulu: University Press of Hawai.

Crouzet, François (2003) "The Historiography of French Economic Growth in the Nineteenth Century." *Economic History Review* 56: 215–42.

Cuno, Kenneth M. (1992) *The Pasha's Peasants: Land, Society and Economy in Lower Egypt, 1740–1858.* Cambridge: Cambridge University Press.

Cuppari, Pietro (1870) *Manuale dell'agricoltore: Guida per conoscere, ordinare e dirigere le aziende rurali.* Firenze: Barbera.

Currie J. M. (1981) *The Economic Theory of Agricultural Land Tenure.* Cambridge: Cambridge University Press.

D'Antone, Lea (1991) " 'L'intelligenza' dell'agricoltura. Istruzione superiore profili intellettuali identità professionali." Pp. 391–426 in *Storia dell'agricoltura italiana*, vol. 3, edited by Piero Bevilacqua. Padua: Marsilio.

Dalton, George (1972) "Peasants in Anthropology and History." *Current Anthropology* 13: 385–406.

Daneo, Camillo (1980) *Breve storia dell'agricoltura italiana*, 1860–1970. Milano: Mondadori.

Danhof, Clarence (1979) "The Farm Enterprise: The Northern United States 1820–1860s." *Research in Economic History* 4: 127–91.

Darlymple, Dana G. (1979) "The Adoption of High-Yielding Grain Varieties in Developing Countries." *Agricultural History* 53: 704–26.

——— (1985) "The Development and Adoption of High-Yielding Varieties of Wheat and Rice in Developing Countries." *American Journal of Agricultural Economics* 67: 1067–73.

——— (1988) "Changes in Wheat Varieties and Yields in the United States, 1919–1984." *Agricultural History* 62: 20–36.

Dasgupta, Partha (1998) "The Economics of Food." Pp. 19–36 in *Feeding a World Population of More than Eight Billion People*, edited by J. C. Waterlow, D. G. Armstrong, L. Fowden, and R. Riley. New York: Oxford University Press.

David, Cristina C., and Keijiro Otsuka (1994) *Modern Rice Technology and Income Distribution in Asia.* Manila: IRRI.

David, Paul (1971) The Mechanization of Reaping in the Antebellum Midwest." Pp.210–38 in *The Reinterpretation of American Economic History*, edited by R. W. Fogel and S. Engerman. New York: Harper and Row.

——— (1974) "The Landscape and the Machine: Technical Interrelatedness, Land Tenure and the Mechanisation of the Corn Harvest in Victorian Britain." Pp. 233–88 in *Technical choice, innovations and economic growth*, edited by P. David. Cambridge: Cambridge University Press.

Davies, R. W. (1998) *Soviet Economic Development from Lenin to Khrushchev.* Cambridge: Cambridge University Press.

Deane, Phyllis (1967) *The First Industrial Revolution.* Cambridge: Cambridge University Press.

Deane, Phyllis, and William A. Cole (1969) *British Economic Growth 1688–1969: Trends and Structure.* Cambridge: Cambridge University Press.

De Bernardi, Alberto (1984) *Il mal della rosa.* Milan: Franco Angeli.

De Felice, Franco (1971) *L'agricoltura in Terra di Bari dal 1880 al 1914.* Milan: Banca Commerciale.

De Gorter, Harry, and Johan F. Swinnen (1994) "The Economic Polity of Farm Policy." *Journal of Agricultural Economics* 45: 312–26.

——— (2002) "Political Economy of Agricultural Policy." Pp. 1893–1943 in *Handbook of Agricultural Economics, vol. 2b: Agricultural and Food Policy*, edited by Bruce Gardner and Gordon Rausser. Amsterdam: Elsevier.

De Gorter, Harry, and Yacov Tsur (1991) "Explaining Price Policy Bias in Agriculture: The Calculus of Support-Maximizing Politician." *American Journal of Agricultural Economics* 73: 1244–54.

Deininger, Klaus, and Hans Binswanger (1999) "The Evolution of the World Bank's Land Policy: Principles, Experiences and Future Challenges." *World Bank Research Observer* 14: 247–76.

Deininger, Klaus, and Gershon Feder (2001) "Land Institutions and Land markets." Pp. 287–331 in *Handbook of Agricultural Economics, vol. 1A: Agricultural Production*, edited by Bruce Gardner and Gordon Rausser. Amsterdam: Elsevier.

De Janvry, Alain (1981) *The Agrarian Question and Reformism in Latin America.* Baltimore and London: Johns Hopkins University Press.

Delgado, Christopher, Jane Hopkins, and Valerie Kelly (1998) *Agricultural Growth Linkages in Sub-Saharan Africa.* Washington, D.C.: IFPRI.

Del Vita Anna, Elena Lombardi, Filomena Maggino, Edoardo Pardini, Alberto Rocchetti, Giovanna Stefania, and Gino Tesi (1998) "L'alta mortalità nel 1816–1817 e gli 'inverni del vulcano.'" *Bollettino di demografia storica.* 29: 71–89.

Deng, Kent G. (2000) "A Critical Survey of Recent Research in Chinese Economic History." *Economic History Review* 53: 1–28.

——— (2003) "Development and Its Deadlock in Imperial China, 221 B.C.–1840 A.D." *Economic Development and Cultural Change* 51: 479–522.

De Soto, Hernando (2000) *The Mystery of Capital. Why Capitalism Triumphs in the West and Fails Everywhere Else.* London and New York: Bantam Press.

De Vries, Jan (1984) *European Urbanization.* Cambridge: Harvard University Press.

——— (1994) "The Industrial Revolution and the Industrious Revolution." *Journal of Economic History* 54: 249–70.

De Vries, Jan, and Ad van der Woude (1997) *The First Modern Economy. Success, Failure and Perseverance of the Dutch Economy, 1500–1815.* Cambridge: Cambridge University Press.

Diamond, Jared (1997) *Guns, Germs and Steel.* New York: Norton.

Diaz Alejandro, Carlos (1970) *Essays on the Economic History of the Argentine Republic.* New Haven and London: Yale University Press.

Dong, Xiao-yuan, and Gregory K. Dow (1993) "Monitoring Costs in Chinese Agricultural Teams." *Journal of Political Economy 1993* 101: 539–53.

Dovring, Folke (1965) *Land and Labour in Europe in the Twentieth Century.* 3rd edition. The Hague: M. Nijhoff.

Duckham, A. N., and G. B. Masefield (1970) *Farming Systems of the World.* London: Chatto and Windus.

Duggan, William R. (1986) *An Economic Analysis of Southern African Agriculture.* New York: Praeger.

Dujon, Veronica (1997) "Communal Property and Land Markets: Agricultural Development Policy in St. Lucia." *World Development* 25: 1529–40.

Dye, A. (1998) *Cuban Sugar in the Age of Mass Production: Technology and the Economics of the Sugar Central.* Stanford: Stanford University Press.

Dyer, Graham (1997) *Class, State and Agricultural Productivity in Egypt: A Study of the Inverse Relationship between Farm Size and Land Productivity.* London: Frank Cass.

Earle, Carville (1992) "The Price of Precocity: Technical Choice and Ecological Constraint in the Cotton South 1840–1890." *Agricultural History* 66: 25–60.

Echevarria, Cristina (1997) "Changes in Sectoral Composition Associated with Economic Growth." *International Economic Review* 38: 431–51.

Eckstein, Alexander, Kang Chao, and John Chang (1974) "The Economic Development of Manchuria: The Rise of a Frontier Economy." *Journal of Economic History* 34: 239–64.

Eddie, Scott (1968) "Agricultural Production and Output per Worker in Hungary, 1870–1913." *Journal of Economic History* 28: 197–222.

——— (1971) "The Changing Pattern of Landownership in Hungary, 1867–1914." *Economic History Review* 20: 293–309.

Edelman, Marc (1992) *The Logic of the Latifundio. The Large Estates of Northwestern Costa Rica since the Late Nineteenth Century.* Stanford: Stanford University Press.

Eicher, Carl K., and C. Baker-Doyle (1992) "Agricultural Development in Sub-Saharan Africa: A Critical Survey." Pp. 1–328 in *A Survey of Agricultural Economics Literature, vol. 4: Agriculture in Economic Development, 1940s to 1990s*, edited by Lee R. Martin. Published for the American Agricultural Economics Association. Minneapolis: University of Minnesota Press.

Ellis, Frank (1988) *Peasant Economics.* Cambridge: Cambridge University Press.

Elson, R. E. (1990) "Peasant Poverty and Prosperity under the Cultivation System in Java." Pp. 24–48 in *Indonesian Economic History in the Dutch Colonial Era*, edited by Anne Booth, W. J. O'Malley, and Anna Weindemann. *Southeast Asia Studies Monographs* 35. New Haven: Yale University.

Engerman, Stanley (2003) "Great Disappointments: The Lessons from Nineteenth-Century Transition from Slavery to Free Labour." *Advances in Agricultural Economic History* 2: 1–20.

Engerman, Stanley, and Ken Sokoloff (1997) "Factor Endowment, Institutions and Differential Paths of Growth among New World Economies." Pp. 259–304 in *How Latin America Fell behind*, edited by Stephen Haber. Stanford: Stanford University Press.

Englander, Steven (1991) "International Technology Transfer and Agricultural Productivity." Pp. 291–311 in *Research and Productivity in Asian Agriculture*, edited by Robert Evenson and Carl Pray. Ithaca and London: Cornell University Press.

Epale, Simon Joseph (1985) *Plantations and Development in Western Cameroon, 1885–1975.* New York: Vantage Press.

Ercolani, Paolo (1969) "Documentazione statistica di base." Pp. 380–460 in *Lo sviluppo economico in Italia*, vol. 3, edited by G. Fuà. Milan: Franco Angeli.

Eschelbach Gregson, Mary (1993) "Rural Response to Increased Demand: Crop Choice in the Midwest, 1860–1880." *Journal of Economic History* 53: 332–45.

Esherick, Joseph (1981) "Number Games: A Note on Land Distribution in Prerevolutionary China." *Modern China* 7: 387–411.

Estey, Ralph H. (1988) "Publicly Sponsored Agricultural Research in Canada since 1887." *Agricultural History* 62: 51–63.

European Commission (2001) *The Agricultural Situation in the European Union, 2000 Report.* Bruxelles: European Comission.

Evans, B. L. (1969) *A History of Agricultural Production and Marketing in New Zealand.* Palmerston North: Kelling and Mundy.

Evenson, Robert E. (1991) 'IARC, NARC and Extension Investment, and Field Crop Productivity: An International Assessment." Pp. 314–28 in *Research and Productivity in Asian Agriculture*, edited by Robert Evenson and Carl Pray. Ithaca and London: Cornell University Press.

——— (2001) "Economic Impacts of Agricultural Research and Extension." Pp. 574–628 in *Handbook of Agricultural Economics, vol. 1A: Agricultural Production*, edited by Bruce Gardner and Gordon Rausser. Amsterdam: Elsevier.

——— (2002) "The Economic Contribution of Agricultural Extension to Agricultural and Rural Development." Paper available at www.fao.org/docrep/w5830e/w5830e06 .htm.

Evenson, Robert E., and Carl Pray (1994) "Food Production and Consumption: Measuring Food Production (with reference to South Asia)." *Journal of Development Economics* 44: 173–97.

Evenson, Robert E., and Larry E. Westphal (1995) "Technological Change and Technology Strategy." Pp. 2211–99 in *Handbook of Development Economics* vol 3, edited by J. Berhman and T. N. Srnivasan. Amsterdam: Elsevier.

Fafchamps, Marcel (1992) "Cash Crop Production, Food Price Volatility and Rural Market Integration in the Third World." *American Journal of Agricultural Economics* 74: 90–99.

Falkus, Malcolm (1991) "The Economic History of Thailand." *Australian Economic History Review* 31: 53–71.

Fan, Shenggen (1991) "Effects of Technological Change and Institutional Reform on Production Growth in Chinese Agriculture." *American Journal of Agricultural Economics* 73: 265–75.

Fan, Shenggen, and Xiaobo Zhang (2002) "Production and productivity Growth in Chinese Agriculture: New National and Regional Measures." *Economic Development and Cultural Change* 50: 819–37.

Fanfani, Roberto (1998) *Lo sviluppo della politica agricola comunitaria.* Rome: Carocci.

FAO book (ad annum) *Production Yearbook.* Rome: FAO

FAO (1952) *Engrais. Rapport sur la production et la consommation mondiales*, Rome: FAO.

——— (1961) *World Agricultural Structure.* Rome: FAO.

——— (1971) *Report on the 1960 World Census of Agriculture, vol. 5: Analysis and International Comparison of the Results.* Rome: FAO.

——— (1981) *1970 World Census of Agriculture. Analysis and International Comparison of the Results.* Rome: FAO.

——— (1990) World Agricultural Census. Available at: www.fao.org/waicent/faoinfo/economic/ees/census/wcav.htm.

——— (2001) "The State of Food Insecurity in the World." Rome: FAO. Available at: www.fao.org/Oocrep/003/Y1500.

——— (2003) "World Agriculture towards 2015/2030." Available at: www.fao.org/docrep/004/y3557e.

FAO AgriBankStat. Available at: www.fao.org/ag/ags/agsm/banks/invent.htm.

FAO Statistical Database. Available at: www.fao.org/.

Farrell M. J. (1957) "The Measurement of Productive Efficiency." *Journal of the Royal Statistical Society* (series A) 120: 253–81.

Farrell, Richard T. (1977) "Advice to Farmers: The Content of Agricultural Newspapers, 1860–1910." *Agricultural History* 51: 209–17.

Faure, David (1989) *The Rural Economy of Pre-liberation China: Trade Expansion and Peasant Livelihood in Jiangsu and Guangdong, 1870 to 1937.* Oxford: Oxford University Press.

Fearne, Andrew (1997) "The History and Development of the CAP 1945–1990." Pp. 11–55 in *The Common Agricultural Policy*, edited by Christopher Ritson and David Harvey. 2nd edition. Wallingford: CAB International.

Feder, Ernest, (1971) *The Rape of the Peasantry*. New York, 1971.

Feder, Gershon, and Roger Slade (1993) "Institutional Reform in India: The Case of Agricultural Extension." Pp. 530–42 in *The Economic of Rural Organization. Theory, Practice and Policy*, edited by Karla Hoff, Avishay Braverman and Joseph Stiglitz. Published for the World Bank. Oxford: Oxford University Press.

Feder, Gershon, Lawrence Lau, Justin Lin, and Xiaopeng Luo (1993) "The Determinants of Farm Investment and Residential Construction in Post-reform China." *Economic Development and Cultural Change* 41: 1–26.

Federico, Giovanni (1984a) "Commercio dei cereali e dazio sul grano in Italia (1863–1913). Una analisi quantitativa." *Nuova Rivista Storica* 68: 46–108.

——— (1984b) "Azienda contadina e autoconsumo fra antropologia ed econometria: considerazioni metodologiche." *Rivista di storia economica* (n.s.) 1: 78–124.

——— (1986) "Mercantilizzazione e sviluppo economico in Italia (1860–1940)." *Rivista di Storia economic* 3: 149–86. English translation in *The Economic Development of Italy since 1870*, edited by Giovanni Federico. Aldershot: Elgar 1994, pp. 305–55.

——— (1997) *An Economic History of the Silk Industry.* Cambridge: Cambridge University Press.

——— (2003) "A Capital Intensive Innovation in a Capital-Scarce World: Steam-threshing in 19th-Century Italy." in *Advances in Agricultural Economic History*, 2: 75–114.

——— (2004a) "The Growth of World Agricultural Production, 1800–1938." *Research Economic History* 22: 125–82.

——— (2004b) 'Not Guilty: Agricultural Overproduction in the 1920s and the Great Depression." Mimeo. Available at: www.iue.it/hec/people/faculty/profiles/federico.shtml

——— (2004c) "The Ultimate Causes of a Great Success Story: Agricultural Productivity Growth 1800–2000." Mimeo. Available at: www.inc.it/hec/people/faculty/profiles/federico.shtml

——— (2004d) "Protection and Italian Economic Development: Much Ado about Nothing?" In *Protection in 19th-Century Europe*, edited by Jean-Pierre Dormois and Pedro Lains. London: Routledge. Available at: www.iue.it/hec/people/faculty/profiles/federico.shtml

Federico, Giovanni, and Paolo Malanima (2004) "Progress, Decline Growth: Product and Productivity in Italian Agriculture 1000–2000." *Economic History Review.*

Federico, Giovanni, and Kevin O'Rourke (2000) "Much Ado about Nothing? The Italian Trade Policy in the 19th century." Pp. 269–96 in *The Mediterranean Response to Globalisation before 1950*, edited by J. Williamson and S. Pamuk. London: Routledge.

Feeny, David (1982) *The Political Economy of Productivity: Thai Agricultural Development, 1880–1975.* Vancouver: University of British Columbia Press.

Fegerler, Louis (1993) "Sharecropping Contracts in the Late-Nineteenth-Century South." *Agricultural History* 67: 31–46.

Fei, John C. H., and Gustav Ranis (1997) *Growth and Development from an Evolutionary Perspective.* Oxford: Blackwell.

Feinstein, Charles H. (1972). *National Income, Expenditure and Output of the United Kingdom, 1855–1965.* Cambridge: Cambridge University Press.

——— (1988a) "Agriculture." Pp. 267–80 in *Studies in Capital Formation in the United*

Kingdom 1750–1920, edited by Charles Feinstein and Sidney Pollard. Oxford: Clarendon.

——— (1988b) "Stock, Overseas Assets and land." Pp. 391–401 in *Studies in Capital Formation in the United Kingdom 1750–1920*, edited by Charles Feinstein and Sidney Pollard. Oxford: Clarendon.

——— (1998) "Pessimism Perpetuated: Real Wages and the Standard of Living in Britain during the Industrial Revolution." *Journal of Economic History* 58: 625–58.

Feinstein, Charles, Peter Temin, and Gianni Toniolo (1997) *The European Economy between the Wars*. Oxford: Oxford University Press.

Fennell, Rosemary (1997) *The Common Agricultural Policy*. Oxford: Clarendon.

Feuerwerker, Albert (1980) "Economic Trends in the Late Ch'ing Empire, 1870–1911." Pp.1–69 in *The Cambridge History of China*, vol. 11 part II, edited by John Fairbank and Kwang-Ching Liu. Cambridge: Cambridge University Press.

——— (1983) "Economic Trends 1912–1949." Pp. 28–127 in *The Cambridge History of China*, vol.12 part I, edited by John Fairbank. Cambridge: Cambridge University Press.

Finkelshtain, Israel, and James A. Chalfant. (1991) "Marketed Surplus under Risk: Do Peasants Agree with Sandmo?" *American Journal of Agricultural Economics* 73: 557–67.

Finlay, Mark R. (1988) "The German Agricultural Experiment Stations and the Beginnings of American Agricultural Research." In "Publicly Sponsored Agricultural Research in the United States: Past, Present and Future," edited by David D. Danbom. *Agricultural History* 62(4): 41–50.

Fitzpatrick, Sheila (1994) *Stalin's Peasants: Resistance and Survival in the Russian Village after Collectivisation*. Oxford: Oxford University Press.

Fogel, Robert W. (1989) *Without Consent or Contract*. New York and London: Norton.

——— (1991) "The Conquest of High Mortality and Hunger in Europe and America: Timing and Mechanisms." Pp. 33–71 in *Favorites of Fortune*, edited by Patricia Higonnet, David Landes, and Henry Rosovsky. Cambridge: Harvard University Press.

——— (2004). *The escape from hunger and premature death, 1700–2100*. Cambridge: Cambridge University Press.

Fogel, Robert W., and Stanley Engerman (1974) *Time on the Cross*. Boston: Little and Brown.

——— (1977) "Explaining the Relative Efficiency of Slave Agriculture in the Antebellum South." *American Economic Review* 67: 275–96.

Fornasari, Massimo, and Vera Zamagni. (1997) *Il movimento cooperativo in Italia*, Florence: Vallecchi.

Fortt, Jean M., and D. A. Hougham (1973) "Environment, Population and Economic History." Pp. 17–46 in *Subsistence to Commercial Farming in Present-day Buganda*, edited by Audrey Richards, Ford Sturrock, and Jean M. Fortt. Cambridge: Cambridge University Press.

Foster, Andrew, and Mark R. Rosenzweig (1996) "Technical Change and Human-Capital Returns and Investments: Evidence from the Green Revolution." *American Economic Review* 86: 931–53.

——— (2004) "Agricultural Productivity Growth, Rural Economic Diversity and Economic Reforms: India 1970–2000." *Economic Development and Cultural Change* 52: 509–42.

Fowler, Cary (1994) *Unnatural Science, Technology, Politics and Plant Evolution*. Chemin de la Sallaz: Gordon and Breach.

Framji, K. K., Garg B. C. and S.D.S. Luthra (1981) *Irrigation and drainage in the world. A global review* New Dehli: International commission on irrigation and drainage 3rd edition.

Francks, Penelope (1984) *Technology and Agricultural Development in Pre-war Japan.* New Haven and London: Yale University Press.

——— (1992) *Japanese Economic Development.* London: Routledge 1992.

——— (1996) "Mechanizing Small-scale Rice Cultivation in an Industrializing Economy: The Development of the Power-tiller in Japan." *World Development* 24: 781–96.

——— (1999) *Agriculture and Economic Development in East Asia.* London and New York: Routledge.

Fremdling, Rainer (1988) "German National Accounts for the 19th and Early 20th Centuries: A Critical assessment." *Vierteljahrsschrift fur Sozial und wirtschaftsgeschichte* 75: 339–55.

French, Michael (1997) *U.S. Economic History since 1945.* Manchester: Manchester University Press.

Frisvold, George B. (1994) "Does Supervision Matter? Some Hypotheses Using Indian Farm-level Data." *Journal of Development Economics* 43: 217–38.

Frisvold, George B., and Peter T. Condon (1998) "The Convention on Biological Diversity and Agriculture: Implications and Unresolved Debates." *World Development* 26: 551–70.

Fuijiki, Hiroshi (1999) "The Structure of Rice Production in Japan and Taiwan." *Economic Development and Cultural Change* 47: 387–400.

Fulginiti, Lilyan, and Richard K. Perrin (1997) "LDC Agriculture: Non-parametric Malmquist Productivity Indexes." *Journal of Development Economics* 53: 373–90.

——— (1999) "Have Price Policies Damaged LDC Agricultural Productivity?" *Contemporary Economic Policy* 17: 469–75.

Gaal L and P. Gunst (1977) *Animal husbandry in Hungary in the 19th and 20th centuries* Akademiai Kiadò Budapest.

Galassi, Francesco (1986) "Stasi e sviluppo nell'agricoltura toscana 1870–1914: primi risultati di uno studio aziendale." *Rivista di storia economica,* (n.s.)3: 304–37.

——— (1998) "Coordination and Monitoring in Lending Co-operatives: The Italian *Casse Rurali.*" In *Finance and the making of the modern capitalist world,* edited by P. L. Cottrell and J. Reis. London: Routledge.

——— (2001) "Measuring Social Capital: Culture as an Explanation of Italy's Economic Dualism." *European Review of Economic History* 5: 29–59.

Galassi, Francesco, and Jon Cohen (1994), "The Economics of Tenancy in Early Twentieth-Century Italy." *Economic History Review* 47: 585–600.

Galasso, Giuseppe (1986) "Gli anni della grande espansione e la crisi del sistema." Pp. 217–494 in *Storia del movimento cooperativo in Italia,* edited by Renato Zangheri, Giuseppe Galasso, and Valerio Castronovo. Turin: Einaudi.

Gallman, Robert E. (1970) "Self-sufficiency in the Cotton Economy of the Antebellum South." *Agricultural History* 44: 5–23.

——— (1986) "The United States Capital Stock in the Nineteenth Century." Pp. 165–213 in *Long-term Factors in American Economic Growth,* edited by S. Engerman and R. E. Gallman. Chicago: Chicago University Press.

Gallup John L., and Jeffrey Sachs (2000) "Agriculture, climate and technology: why are the Tropics falling behind?" *American Journal of Agricultural Economics* 82 pp. 731–737.

Galor, Oded, and Omer Moav (2002) "Natural Selection and the Origin of Economic Growth." *Quarterly Journal of Economics* 117: 1133–91.

Galor, Oded, and David N. Weil (2000) "Population, Technology and Growth: From Malthusian Stagnation to Demographic Transition and Beyond." *American Economic Review* 90: 806–28.

Gang, Deng (1993) *Development versus Stagnation: Technological Continuity and Agricultural Progress in Pre-modern China.* Westport: Greenwood.

Gardner, Bruce L. (1987) "Causes of U.S. Farm Commodity Programs." *Journal of Political Economy* 95: 290–310.

——— (1990) "The United States." Pp. 19–65 in *Agricultural Protectionism*, edited by Fred H. Sanderson. Washington, D.C.: Resources for the Future.

——— (1992) "Changing Economic Perspectives on the Farm Problem." *Journal of Economic Literature* 30: 62–101.

——— (1996) "The Political Economy of U.S. Export Subsidies to Wheat." Pp. 291–331 in *The Political Economy of American Trade Policy*, edited by Anne O. Krueger. Chicago: University of Chicago Press.

——— (2000) "Economic Growth and Low Incomes in Agriculture." *American Journal of Agricultural Economics* 82: 1059–74.

——— (2002) *American Agriculture in the Twentieth Century. How It Flourished and What It Cost.* Cambridge: Harvard University Press.

Gash, Norman (1972) *Sir Robert Peel. The Life of Sir Robert Peel after 1830.* Harlow: Longman.

Gattrell, Peter (1986) *The Tsarist Economy, 1850–1917.* London: Batsford.

Geertz, Clifford (1963) *Agricultural Involution. The Process of Agricultural Change in Indonesia.* Berkeley and Los Angeles: University of California Press.

Genung, Albert B. (1954) *The Agricultural Depression following World War I and Its Political Consequences.* Ithaca: Northeast Farm Foundation.

Gerdin, Anders (2002) "Productivity and Economic Growth in Kenyan Agriculture, 1964–1996." *Agricultural Economics* 27: 7–13.

Gerschenkron, Alexander (1966) "Agrarian Policies and Industrialization in Russia 1861–1917." Pp. 707–800 in *Cambridge Economic History of Europe*, vol. 6 Part 2, edited by H. J. Habbakuk and M. Postan. Cambridge: Cambridge University Press.

Ghatak, Maitreesh, and Priyanka Pandey (2000) "Contract Choice in Agriculture with Joint Moral Hazard in Effort and Risk." *Journal of Development Economics* 63: 303–26.

Ghatak, Subrata, and Ken Ingersent (1984) *Agriculture and Economic Development.* Brighton: Wheatsheaf.

Ghatak, Subrata, and James Seale (2001) "Supply Response and Risk in Chinese Agriculture." *Journal of Development Studies* 37: 141–50.

Gill, G. J. (1991) *Seasonality and Agriculture in the Developing World: A Problem of the Poor and the Powerless.* Cambridge: Cambridge University Press.

Giorgetti, Giorgio (1974) *Contadini e proprietari nell'Italia moderna.* Turin: Einaudi.

Glaeser, Berhard (1987) *The Green Revolution Revisited: Critique and Alternatives.* London: Unwin Hyman.

Gleave, M. B., and H. P. White (1969) "Population Density and Agricultural Systems in West Africa." Pp. 272–300 in *Environment and Land Use in Africa*, edited by M. Thomas and G. W. Whittington. London: Methuen.

Goddard, Nicholas (1983) "The Development and Influence of Agricultural Periodicals and Newspapers, 1780–1880." *Agricultural History Review* 31: 116–31.

——— (2000) "Agricultural Institutions, Societies, Associations and the Press." Pp. 651–90 in *The Agrarian History of England and Wales, vol. 7: 1850–1914*, Part 1, edited by E.J.T. Collins. Cambridge: Cambridge University Press.

Goetz, Stephan, Tanja Jaksch, and Rosemarie Siebert (2001) *Agricultural Transformation and Land Use in Central and Eastern Europe.* Aldershot: Ashgate.

Goldin, Claudia (1979) "N Kinds of Freedom." *Explorations in Economic History* 16: 8–30.

Goldsmith, Raymond (1955) *A Study of Savings in the United States.* Princeton: Princeton University Press.

——— (1969) *Financial Structure and Development.* New Haven and London: Yale University Press.

Gollin, Douglas, Stephen Parente, and Richard Rogerson (2002) "The Role of Agriculture in Development." *American Economic Review* 92: 160–64.

Good, David (1974) *The Economic Rise of the Habsburg Empire 1750–1914.* Berkeley: University of California Press.

Goodfriend, Marvin, and John McDermott (1995) "Early Development." *American Economic Review* 85: 116–33.

Goodwin, Barry (2001) "Problems with Market Insurance in Agriculture." *American Journal of Agricultural Economics* 83: 643–49.

Goosens, Martine (1992) *The Economic Development of Belgian Agriculture, 1812–1846: A Regional Perspective.* Leuwen: Leuwen University Press.

Gopinath, Munisamy, and Terry L. Roe (1997) "Sources of Sectoral Growth in an Economy-wide Context: The Case of U.S. Agriculture." *Journal of Productivity Analysis* 8: 293–310.

Gordon, Robert (2000) *Macro-economics.* Reading: Addison-Wesley.

Goswami, Omkar, and Aseem Shrivastava (1991) "Commercialisation of Indian Agriculture, 1900–1940: What Do Supply Response Functions Say?" *Indian Economic and Social History Review* 28: 229–61.

Gottschang, Thomas (1987) "Economic Change, Disasters and Migration: The Historical Case of Manchuria." *Economic Development and Cultural Change* 35: 461–90.

Grabowski, Richard (1995) "Induced Innovation. A Critical Perspective." Pp. 73–92 in *Induced Innovation Theory and International Agricultural Development. A Reassessment,* edited by B. M. Koppel. Baltimore and London: Johns Hopkins University Press.

Grant, Wyn (1997) *The Common Agricultural Policy.* Basingstoke and London: MacMillan.

Grantham, George (1975) "Scale and Organization in French Farming 1840–1880." Pp. 293–326 in *European Peasants and Their Markets,* edited by William Parker and Eric Jones. Princeton: Princeton University Press.

——— (1978) "The Diffusion of the New Husbandry in Northern France, 1815–1840." *Journal of Economic History* 38: 312–38.

——— (1980) "The Persistence of Open-field Farming in Nineteenth-Century France." *Journal of Economic History* 40: 515–30.

——— (1984) "The Shifting Locus of Agricultural Innovation in Nineteenth-Century Europe: The Case of the Agricultural Experiment Stations." In "Technique, Spirit, and Form in the Making of the Modern Economies: Essays in Honour of W. N. Parker," edited by Gary Saxonhouse and Calvin Wright. *Research in Economic History,* Supplement 3, pp. 191–214.

——— (1989a) "Agricultural Supply during the Industrial Revolution: French Evidence and European Implications." *Journal of Economic History* 49: 43–72.

——— (1989b) "Agrarian Organization in the Century of Industrialization: Europe Russia and North America." In "Agrarian Organization in the Century of Industrialization: Europe, Russia and North America," edited by George Grantham and Carole Leonard. *Research in Economic History,* Supplement 5 Part A, pp. 1–27.

——— (1991) "The Growth of Labour Productivity in the Production of Wheat in the *Cinq Grosses Fermes* of France 1750–1929." Pp. 341–63 in *Land, Labour and Livestock: Historical Studies in European Agricultural Productivity,* edited by Mark Overton and Bruce Campbell. Manchester: Manchester University Press.

——— (1996) "The French Agricultural Capital Stock, 1789–1914." *Research in Economic History* 16: 39–84.

——— (1997) "The French Cliometric Revolution: A Survey of Contributions to French Economic History." *European Review of Economic History* 1: 353–405.

Gray, Leslie C., and Michael Kevane (2001) "Evolving Tenure Rights and Intensification in Southwestern Burkina Faso." *World Development* 29: 573–87.

Green, Alan G. (2000) "Twentieth-Century Canadian Economic History." Pp. 191–247 in *The Cambridge Economic History of the United States, vol. 3: The Twentieth Century,* edited by Stanley Engerman and Robert Gallman. Cambridge: Cambridge University Press.

Green Ewen (2000) "No longer the farmers' friend? The Conservative party and agricultural protection, 1880–1914." Pp.149-177 in *Agriculture and politics in England, 1815–1939,* edited by J.R. Wordie. Basingstoke: Macmillan.

Greenland, D. J., P. J. Gregory, and P. H. Nye (1998) "Land Resources and Constraints to Crop Production." Pp. 39–55 in *Feeding a World Population of More than Eight Billion People,* edited by J. C. Waterlow, D. G. Armstrong, L. Fowden, and R. Riley. New York: Oxford University Press.

Gregory, Paul (1980) "Grain marketings and peasant consumption Russia 1885–1913," *Explorations in Economic History 17*: 135–64.

——— (1982) *Russian National Income 1885–1913.* Cambridge: Cambridge University Press.

——— (1994) *Before Command: An Economic History of Russia from Emancipation to the First Five-year Plan.* Princeton: Princeton University Press.

Gregory, Paul R., and Mokhtari Manochehr (1993) "State Grain Purchases, Relative Prices and the Soviet Grain Procurement Crisis." *Explorations in Economic History* 30: 182–94.

Gregory, Paul R., and Robert C. Stuart (1986) *Soviet Economic Structure and Performance,* 2nd edition. New York: Harper and Row.

Gregson, Mary E. (1996) "Long-term Trends in Agricultural Specialization in the United States." *Agricultural History* 70: 90–101.

Griffin, Keith (1979) *The political economy of agrarian change. An essay on the Green revolution* Basingstoke: MacMillan 2nd edition.

——— (1981) *Land Concentration and Rural Poverty.* 2nd edition. New York: Holmes and Meier.

Grigg, David (1974) *The Agricultural Systems of the World: An Evolutionary Approach.* Cambridge: Cambridge University Press.

——— (1982) *The Dynamics of Agricultural Change.* London and Melbourne: Hutchinson.

——— (1987) "Farm Size in England and Wales from Early Victorian Times to the Present." *Agricultural History Review* 35: 179–89.

——— (1994) *Storia dell'agricoltura in Occidente.* Bologna: II Mulino.

Grilli, Enzo, and Maw Cheng Yang (1988) "Primary Commodity Prices, Manufactured Goods Prices and the Terms of Trade of Developing Countries: What the Long Run Shows." *World Bank Economic Review* 2: 1–47.

Grilliches, Zvi (1957) "Hybrid Corn: An Exploration in the Economics of Technological Change." *Econometrica* 25: 501–22.

——— (1958) "Research Costs and Social Returns: Hybrid Corn and Related Innovation." *Journal of Political Economy* 66: 419–31.

Grove, Wayne, and Craig Heinecke (2003) "Better opportunities or worse? The demise of cotton harvest labor; 1949–1964." *Journal of Economic History* 63: 736–67.

Guinnane, Timothy (1994) "A Failed Institutional Transplant: Raffeisen's Credit Co-operatives in Ireland 1894–1914." *Explorations in Economic History* 31: 38–61.

——— (2001) "Co-operatives as Information Machines: German Rural Credit Co-operatives 1883–1914." *Journal of Economic History* 61: 366–89.

——— (2002) "Delegated Monitors, Large and Small: Germany's Banking System, 1800–1914." *Journal of Economic Literature* 40: 73–124.

Guinnane, Timothy, and Ronald I. Miller (1997) "The Limits to Land Reform: The Land Acts in Ireland 1870–1909." *Economic Development and Cultural Change* 45: 591–611.

Gunst, Peter (1996) *Agrarian Development and Social Change in Eastern Europe, 14–19th Centuries.* Aldershot: Variorum Ashgate.

Gupta, Bishnupriya (2001) "The International Tea Cartel during the Great Depression 1929–1933." *Journal of Economic History* 61: 144–59.

Gutierrez, L., and M. M. Gutierrez (2003) "International R&D spillovers and poductivity growth in the agricultural sector. A panel cointegration approach." *European Review of Agricultural Economics* 30: 281–303.

Hadass, Yeal S., and Jeffrey G. Williamson (2003) "Terms of Trade Shocks and Economic Performance 1870–1940: Prebisch and Singer Revisited." *Economic Development and Cultural Change* 51: 629–58.

Hallam, Arne (1993) "Empirical Studies of Size, Structure and Efficiency in Agriculture." Pp. 204–31 in *Size, Structure and the Changing Face of American Agriculture*, edited by Arne Hallam. Boulder: Westview.

Hamilton, David E. (1985) "The Causes of the Banking Panic of 1930: Another view." *Journal of Southern History* 51: 582–605.

Hanley, Susan B., and Kozo Yamamura (1977) *Economic and Demographic Change in Preindustrial Japan 1600–1868.* Princeton: Princeton University Press.

Hansen, Bent (1983) "Interest Rates and Foreign Capital in Egypt under British Occupation." *Journal of Economic History* 43: 867–84.

Hansen, Bent, and M. Wattleworth (1978) "Agricultural Output and Consumption of Basic Foods in Egypt, 1886/87–1967/68." *International Journal of Middle East Studies* 9: 449–69.

Hansen, Gary, and Edward C. Prescott (2002) "Malthus to Solow." *American Economic Review* 92: 1205–17.

Hansen, Svend A. (1974). *Økonomisk vækst i Danmark.* København: Københavns Universitet Institutut for Økonomisk Historie.

Hansen, Zeynep K., and Gary D. Libecap (2004) "The Allocation of Property Rights to Land: U.S. Land Policy and Farm Failure in the Northern Great Plains." *Explorations in Economic History* 41: 103–29.

Hansen Zeynep K. and Gary Libecap (2004a) "Small farms, externalities and the dust bowl of the 1930s." *Journal of Political Economy* 112: 665–694.

Hanson, James, and Richard Just (2001) "The Potential for Transition to Paid Extension: Some Guiding Principles." *American Journal of Agricultural Economics* 83: 777–84.

Hardach, Gerd (1977) *The First World War 1914–18.* London: Allen Lane.

Harley, C. Knick (1980) "Transportation, the World Wheat Trade and the Kuznets Cycle, 1850–1913." *Explorations in Economic History* 18: 218–50.

——— (1993) "Reassessing the Industrial Revolution." Pp. 170–226 in *The British Industrial Revolution*, edited by J. Mokyr. Boulder: Westview.

Harley, C. Knick, and N.F.R. Crafts (2000) "Simulating the Two Views of the British Industrial Revolution." *Journal of Economic History* 60: 819–41.

Harris, William J. (1994) "Crop Choices in the Piedmont before and after the Civil War." *Journal of Economic History* 54: 526–42.

Harrison, Mark (1996) "Soviet Agriculture and Industrialization." Pp. 192–207 in *The Nature of Industrialization, vol 4: Agriculture and Industrialization from the Eighteenth Century to the Present Day*, edited by Peter Mathias and John A. Davis. Oxford: Blackwell.

Hatton, Tim, and Jeffrey Williamson (1998) *The Age of Mass Migration.* Oxford: Oxford University Press.

Hawke, Gary R. (1985) *The Making of New Zealand.* Cambridge: Cambridge University Press.

Hayami, Yujiro (1986) "The Roots of Agricultural Protectionism." Pp. 32–39 in *The Political Economy of Agricultural Protection: East Asia in International Perspective*, edited by Kym Anderson and Yujiro Hayami. Sydney: Allen and Unwin.

Hayami, Yujiro, and Masao Kikuchi (2000) *A Rice Village Saga: Three Decades of Green Revolution in the Philippines.* Manila: International Rice Research Institute.

Hayami, Yujiro, and Keijiro Otsuka (1993) *The economics of contract choice* Oxford: Clarendon Press.

Hayami, Yujiro, and Vernon Ruttan (1985) *Agricultural Development.* 2nd edition. Baltimore and London: Johns Hopkins University Press.

Hayami, Yujiro, and Saburo Yamada (1975) "Agricultural Research Organization in Economic Development: A Review of the Japanese Experience." Pp. 224–49 in *Agriculture in Development Theory*, edited by Lloyd Reynolds. New Haven and London: Yale University Press.

——— (1991) *The Agricultural Development of Japan: A Century's Perspective.* Tokyo: University of Tokyo Press.

Hayter, Earl W. (1939) "Barbed Wire Fencing—a Prairie Invention." *Agricultural History* 13: 189–207.

Hazell, Peter (1989) "Changing Patterns of Variability in World Cereal Production." Pp. 13–34 in *Variability in Grain yields: Implications for Agricultural Research and Policy in Developing Countries*, edited by Jock R.Anderson and Peter Hazell. Baltimore and London: Johns Hopkins University Press.

——— (1992) "The appropriate Role of Agricultural Insurance in Developing Countries." *Journal of International Development* 4: 567–81.

Hazell, Peter, Carlos Pomareda, and Alberto Valdes (1986) "Introduction." Pp. 1–13 in *Crop Insurance for Agricultural Development: Issues and Experience*, edited by Peter Hazell, Carlos Pomareda, and Alberto Valdes. Baltimore and London: Johns Hopkins University Press.

Hazell, Peter, and C. Ramsay (1991) *The Green Revolution Reconsidered: The Impact of High-Yielding Rice Varieties in South India.* Baltimore and London: Johns Hopkins University Press.

Headrick, Daniel (1988) *The Tentacles of Progress: Technology Transfer in the Age of Imperialism, 1850–1940.* Oxford: Oxford University Press.

Heinecke, Craig (1994) "African-American Migration and Mechanized Cotton Harvesting, 1950–1960." *Explorations in Economic History* 31: 501–20.

Held, Joseph (1980) "The Interwar Years and Agrarian Change." Pp. 197–233 in *The Modernization of Agriculture: Rural Transformation in Hungary, 1848–1975*, edited by Joseph Held. Boulder: East European Monographs.

Helling, Gertrud (1966) "Zur entwicklung der Produktivität in der Deutschen landwirtschaft im 19. Jahrhundert." *Jahrbuch für Wirtshchaftsgeschichte*, 1: 129–41.

Hendriks, Gisela (1991) *Germany and European Integration.* New York and Oxford: Berg.

Henretta, James A. (1978) "Families and Farms *Mentalité* in Pre-industrial America." *William and Mary Quarterly* 35: 3–32.

Henriksen, Ingrid (1992) "The Transformation of Danish Agriculture 1870–1914." Pp. 153–77 in *The Economic Development of Denmark and Norway since 1870*, edited by Karl G. Perrson. Cheltenham: Elgar.

——— (1998) "Why Danish Credit Co-operatives Are So Unimportant." *Scandinavian Economic History Review* 46: 32–54.

——— (1999) "Avoiding Lock-in: Co-operative Creameries in Denmark, 1882–1913." *European Review of Economic History* 3: 57–78.

Henriksen, Ingrid (2003) "Freehold tenure in late eighteenth Denmark" *Advances in agricultural economic history* 2: 21–39.

Herrmann, Roland (1997) "Agricultural Policies, Macroeconomic Policies and Producer Price Incentives in Developing Countries: Cross-country Results for Major Crops." *Journal of Developing Areas* 31: 203–20.

Heston, Alan (1983). "National income." Pp. 376–462 in *The Cambridge Economic History of India. II. c.1757–c.1970*, edited by Dharma Kumar and Meghnad Desai. Cambridge: Cambridge University Press.

Heywood, Colin (1981) "The Role of Peasantry in French Industrialization." *Economic History Review* 34: 359–76.

——— (1992) *The Development of the French economy, 1750–1914.* Basingstoke and London: Macmillan.

Higgs, Edward (1996) "Occupational Census and the Agricultural Workforce in Victorian England and Wales." *Economic History Review* 49: 700–716.

Hill, R. D. (1977) *Rice in Malaya. A Study in Historical Geography.* Kuala Lumpur: Oxford University Press.

Historical Statistics Canada (1965) *Historical Statistics of Canada.* 1st edition. Edited by M. C. Urquhart and K.A.H. Buckley. Cambridge: Cambridge University Press.

Historical Statistics Canada (1983) *Historical Statistics of Canada* 2nd edition. Edited by F. H. Leacy. Ottawa: Statistics Canada.

Historical Statistics Japan (1987–1988) *Historical Statistics of Japan.* Tokyo: Japan Statistical Association.

Historical Statistics (1975) U.S. Department of Commerce. Bureau of the Census. *Historical Statistics of the Unites States: Colonial times to 1970.* Washington, D.C.: U.S. Government Printing Office.

Historik Statistikk (1978) *Norges offisielle Statistikk XII.* Oslo: Statistik Sentralbyra.

Hjerrpe, R. (1989) *The Finnish Economy, 1860–1985. Growth and Structural Change.* Helsinki: Bank of Finland.

Hobbs-Pruit, Betty (1984) "Self-sufficiency and the Agricultural Economy of Eighteenth-Century Massachusetts." *William and Mary Quarterly* 41: 333–64.

Hoebner Rudolph, Suzanne (1984) "Introduction." Pp. 9–21 in *Agrarian Power and Agricultural Productivity in South Asia*, edited by M. Desai, S. Hoebner Rudolph, and A. Rudra. Delhi: Oxford University Press.

Hoff, Karla, and Joseph Stiglitz (1993) "Imperfect Information and Rural Credit Markets: Puzzles and Policy Perspectives." Pp. 33–51 in *The Economics of Rural Organization: Theory, Practice and Policy*, edited by Karla Hoff, Avishay Braverman, and Joseph Stiglitz. Oxford: Oxford University Press for the World Bank.

Hoff, Karla, Avishay Braverman, and Joseph Stiglitz (1993) "Introduction." Pp. 1–29 in *The Economic of Rural Organization: Theory, Practice and Policy*, edited by Karla Hoff, Avishay Braverman, and Joseph Stiglitz. Oxford: Oxford University Press. Published for the World Bank.

Hoffmann, Walther (1965) *Das wachstum der Deutschen wirtschaft seit der mitte des 19.jahrhunderts*. Berlin: Springer Verlag.

Hohenberg, Paul (1972) "Change in Rural France in the Period of Industrialization 1830–1913." *Journal of Economic History* 32: 219–40.

Holderness, B. A. (1988) "Agriculture 1770–1860." Pp. 9–34 in *Studies in Capital Formation in the United Kingdom 1750–1920*, edited by Charles Feinstein and Sidney Pollard. Oxford: Clarendon.

——— (2000) "Investment, Accumulation and Agricultural Credit." Pp. 863–929 in *The Agrarian History of England and Wales, vol. 7:1850–1914, part.1*, edited by E.J.T. Collins. Cambridge: Cambridge University Press.

Holgersson, Bengt (1974) "Cultivated Land in Sweden and Its Growth, 1840–1939." *Economy and History* 17: 20–51.

Holley, Donald (2000) *The Second Great Emancipation. The Mechanical Corn Picker, Black Migration, and How They Shaped the Modern South.* Fayetteville: University of Arkansas Press.

Holley, Richard, and Vernon W. Ruttan (1969) "The Philippines." Pp. 215–60 in *Agricultural Development in Asia*, edited by R. T. Shand. Canberra: ANU Press.

Holloway, Thomas H. (1980) *Immigrants on the Land. Coffee and Society in São Paulo.* Chapel Hill: University of North Carolina Press.

Holtfrerich, Carl Ludwig (1986) *The German Inflation 1914–23.* Berlin and New York: de Gruyter.

Homer, Sydney, and Richard Sylla (1991) *A History of Interest Rates.* 3rd edition. New Brunswick and London: Rutgers University Press.

Honma, Masayoshi, and Yujiro Hayami (1986) "The Determinants of Agricultural Protection Levels: An Econometric Analysis." Pp. 39–49 in *The Political Economy of Agricultural Protection: East Asia in International Perspective*, edited by Kym Anderson and Yujiro Hayami. Sydney: Allen and Unwin.

Hopkins, A. G. (1973) *An Economic History of West Africa.* New York: Columbia University Press.

Huang, Philip (1985) *The Peasant Economy and Social Change in North China.* Stanford: Stanford University Press.

——— (1990) *The Peasant Family and Rural Development in the Yangzi Delta, 1350–1988.* Stanford: Stanford University Press.

——— (2002) "Development or Involution in Eighteenth-Century Britain and China? A Review of Kenneth Pomeranz's 'The Great Divergence.'" *Journal of Asian Studies* 61: 501–37.

Huber, J. R. (1970) "Effects on Prices of Japan's Entry into World Commerce after 1858." *Journal of Political Economy* 78: 614–28.

Huberman, Michael (2004) "Working hours of the world unite? New international evidence of worktime." *Journal of Economic History* 64: 964–1001.

Hudson, Pat (1992) *The Industrial Revolution.* London: Edward Arnold.

Hueckel, G. (1981) "Agriculture and the Industrial Revolution." Pp. 182–203 in *The Economic History of Britain since 1700, vol. I: 1700–1860*, edited by Roderick Floud and Donald McCloskey. Cambridge: Cambridge University Press.

Huffman, Wallace E. (2001) "Human Capital, Education and Agriculture." Pp. 333–81 in *Handbook of Agricultural Economics, vol. 1A: Agricultural production*, edited by Bruce Gardner and Gordon Rausser. Amsterdam: Elsevier.

Huffman, Wallace E., and Robert E. Evenson (1993) *Science for Agriculture. A Long-term Perspective.* Ames: Iowa State University Press.

——— (2001) "Structural and Productivity Change in U.S. Agriculture, 1950–1982." *Agricultural Economics* 24: 127–47.

Huffman, Wallace E., and Richard E. Just (1999) "The Organization of Agricultural Research in Western Developed Countries." *Agricultural Economics* 21: 1–18.

——— (2004) "Implications of Agency Theory for Optimal Land Tenure Contract." *Economic Development and Cultural Change* 52: 617–42.

Hunt, E. H., and S. J. Pam (1997) "Prices and Structural Response in English Agriculture, 1873–1896." *Economic History Review* 50: 477–505.

——— (2002) "Responding to Agricultural Depression, 1873–1896." *Agricultural History Review* 50: 225–52.

Hunt, Shane J., (1985) "Growth and Guano in Nineteenth-Century Peru." Pp. 255–318 in *The Latin American Economies*, edited by Roberto Cortes-Conde and Shane J. Hunt. London: Holmes and Meier.

Hunter, Holland (1988) "Soviet Agriculture with and without Collectivization, 1928–1940." *Slavic Review* 47: 203–16.

Hurt, Douglas R. (1982) *American Farm Tools: From Hand Power to Steam Power.* Manhattan: Sunflower University Press.

——— (1994) *American Agriculture. A Brief History.* Ames: Iowa State University Press.

Hwa, Erh-Cheng (1988) "The Contribution of Agriculture to Economic Growth: Some Empirical Evidence." *World Development* 16: 1329–339.

Hyden, Goran, Robert W. Kates, and B. L. Turner II (1993) "Beyond Intensification." Pp. 401–39 in *Population Growth and Agricultural Change in Africa*, edited by Robert W. Kates, Goran Hyden, and B. L. Turner II. Gainesville: University Press of Florida.

Iftikhar, Ahmed, and Vernon W. Ruttan (1988) "Introduction." Pp. 1–17 in *Generation and Diffusion of Agricultural Innovations: The Role of Institutional Factors*, edited by Ahmed Iftikhar and Vernon W. Ruttan. Published for ILO. Aldershot: Gower.

Ilbery, Brian W. (1985) *Agricultural Geography.* Oxford: Oxford University Press.

Iliffe, John (1979) *A Modern History of Tanganyka.* Cambridge: Cambridge University Press.

——— (1995) *Africans: The History of a Continent.* Cambridge: Cambridge University Press.

INEA (1951) *I tipi d'impresa nell'agricoltura italiana* Roma: INEA

Ingham, Barbara (1979) "Vent for surplus Reconsidered with Ghanaian Experience." *Journal of Development Studies* 15: 19–37.

Ingram, James C. (1971) *Economic Change in Thailand 1950–1970.* Stanford: Stanford University Press. First edition published in 1955.

Institut Internationale d'Agriculture (ad annum), *Annuarie International de Statistique Agricole.* Rome: Institut Internationale d'Agriculture.

Institute of Developing Economics (1969) *One Hundred Years of Agricultural Statistics in Japan.* Tokyo: Institute of Developing Economics.

Irwin, Douglas (1988) "Welfare Effects of British Free-Trade: Debate and Evidence from the 1840s." *Journal of Political Economy* 96: 1142–64.

Isern, Thomas D. (1981). *Custom Combining on the Great Plains: A History.* Norton: Oklahoma University Press.

Ishikawa, Shigeru (1967) *Economic Development in an Asian Perspective:* Economic Research Series no. 8. Tokyo: Hitotusbashi University Kinokuniya Bookstore.

——— (1981) *Essays on Technology, Employment and Institutions in Economic Development: Comparative Asian Experience.* Economic Research Series no. 19. Tokyo: Hitotusbashi University Kinokuniya Bookstore.

Islam, Mufakharul M. (1978) *Bengal Agriculture, 1920–1946: A Quantitative Study.* Cambridge: Cambridge University Press.

ISTAT (1934) *Censimento generale dell'agricoltura 19 marzo 1930, vol. 1: Censimento del bestiame.* Rome: ISTAT.

——— (1936) *Censimento generale dell'agricoltura 19 marzo 1930, vol. 2: Censimento delle aziende agricole.* Rome: ISTAT.

——— (1939) *Catasto agrario per il regno d'Italia,* vol. 2. Rome: ISTAT.

——— (1957) *Indagine statistica sullo sviluppo del reddito nazionale dell'Italia dal 1861 al 1956.* Annali di statistica serie VIII, vol. 9. Rome: ISTAT.

Jacoby, Hanan G., Guo Li, and Scott Rozelle (2002) "Hazard of Expropriation: Tenure Insecurity and Investment in Rural China." *American Economic Review* 92: 1420–47.

James, Harold (2001) *The End of Globalization.* Cambridge: Harvard University Press.

Jamison, Dean T., and Lawrence J. Lau (1982) *Farmer Education and Farm Efficiency.* Baltimore and London: Johns Hopkins University Press.

Janick, Jules, Robert Schery, Frank Woods, and Vernon Ruttan (1974) *Plant Science. An Introduction to World Crops.* San Francisco: Freeman.

Jensen, Einar (1937) *Danish Agriculture: Its Economic Development.* Copenhagen: J. H. Schultz.

Jin, Songqing, Jikun Huang, Ruifa Hu, and Scott Rozelle (2002) "The Creation and Spread of Technology and Total Factor Productivity in China's Agriculture." *American Journal of Agricultural Economics* 84: 916–30.

Johansen, Hans C. (1985) *Dansk historical statistics 1814–1990.* Copenhagen: Gyldenal.

Johnson, D. Gale (1973) *World Agriculture in Disarray.* 1st edition. London: McMillan.

——— (2000) "Population Food and knowledge." *American Economic Review* 90: 1–14.

Johnson, Nancy (2001) "Tierra y Libertad: Will Tenure Reform Improve Productivity in Mexico's Ejido Agriculture?" *Economic Development and Cultural Change* 49: 291–309.

Johnson, Nancy L., and Vernon Ruttan (1994) "Why Are Farms So Small?" *World Development* 22: 691–706.

Johnston, A. E. (1998) "The Importance of Fertilizers in Plant Nutrient Management." Pp. 35–49 in *Nutrient Management for Sustainable Crop Production in Asia,* edited by A. E. Johnston and J. K. Syers. Wallingford: CAB International.

Johnston, Bruce F., and John W. Mellor (1961) "The Role of Agriculture in Economic Development." *American Economic Review* 51: 566–93.

Jones, Charles I. (2001) "Was an Industrial Revolution Inevitable? Economic Growth over the Very Long Run." *Advances in Macroeconomics* 1: 1–43.

Jones, William O. (1980) "Agricultural Trade within Tropical Africa: Historical Background." Pp. 10–45 in *Agricultural Development in Africa*, edited by Robert H. Bates and Michael F. Lofchies. New York: Praeger.

Jorgenson, Dale W. (1997) "Agriculture and the Wealth of Nations." *American Economic Review* 87: 1–12.

Jorgenson, Dale W., and Frank M. Gollop (1992) "Productivity Growth in U.S. Agriculture: A Postwar Perspective." *American Journal of Agricultural Economics* 74: 745–50.

Judd, Ann M., James K. Boyce, and Robert E. Evenson (1986) "Investing in Agricultural Supply: The Determinants of Agricultural Research and Extension Investments." *Economic Development and Cultural Change* 35: 77–113.

Juma, Calestous (1989) *The Gene Hunters. Biotechnology and the Scramble for Seeds.* Princeton: Princeton University Press.

Jussaume, Raymond A. (1991) *Japanese Part-time Farming: Evolution and Impacts*, Ames: Iowa State University Press.

Jwaideh, Albertine (1984) "Aspects of land tenure and social change in lower Iraq during late Ottoman times." Pp. 333–56 in *Land tenure and social transformation in the Middle East,* edited by Tarif Khalidi. Beirut: American University of Beirut.

Kahan, Arcadius (1978) "Capital Formation during the Period of Early Industrialization in Russia 1890–1913." Pp. 265–307 in *The Cambridge Economic History of Europe*, vol. 7, part 2, edited by Peter Mathias and Michel Postan. Cambridge: Cambridge University Press.

Kalirajan, K. P., M. B. Obwona, and S. Zhao (1996) "Productivity Growth: The Case of Chinese Agricultural Growth before and after Reforms." *American Journal of Agricultural Economics* 78: 331–38.

Kalirajan, Kaliappa, and Bhide Shasanka (2003) *A disequilibrium macroeconomic model for the Indian economy.* Aldershot: Ashgate.

Kang, Kenneth, and Vijaya Ramachandran (1999) "Economic Growth in Korea: Rapid Growth without an Agricultural Revolution." *Economic Development and Cultural Change* 47: 783–801.

Kaniki, M.H.Y. (1995) "The Colonial Economy: The Former British Zones." Pp. 383–419 in *General History of Africa, vol.7: Africa under Colonial Domination*, edited by Adu Boahen. Paris: UNESCO.

Kantor, Shawn Everett (1998) *Politics and Property Rights: The Closing of the Open Range in the Postbellum South.* Chicago: University of Chicago Press.

Kanwar, Sunil (1998) *Wage Labor in Developing Agriculture.* Aldershot: Ashgate.

Karlova, Natalia, Irina Khramova, Eugenia Serova, and Tatiana Tikhonova (2003) "Institutional Reforms in the Agro-industrial Complex." Pp. 542–84 in *The Economics of Transition*, edited by Yegor Gaidar. Cambridge: MIT Press.

Karshenas, Massoud (1993) "Intersectoral Resource Flows and Development: Lessons of Past Experiences." Pp. 179–238 in *Economic Crisis and Third World Agriculture*, edited by Aijt Sing and Hamid Tabatabai. Cambridge: Cambridge University Press.

——— (2001) "Agriculture and Economic Development in Sub-Saharan Africa and Asia." *Cambridge Journal of Economics* 25: 315–42.

Kates, Robert W., Goran Hyden, and B. L. Turner II (1993) "Theory, Evidence, Study Design." Pp. 1–39 in *Population Growth and Agricultural Change in Africa*, edited by Robert W. Kates, Goran Hyden, and B. L. Turner II. Gainesville: University Press of Florida.

Kato, Yuzuro (1970) "Development of Long-term Agricultural Credit." Pp. 324–51 in *Agriculture and Economic Growth: Japan's Experience*, edited by Kazushi Ohkawa, Bruce F. Johnston and Hiromitsu Kaneda. Princeton: Princeton University Press.

Katz, Friederich (1974) "Labor Conditions on Haciendas in Porfirian Mexico: Some Trends and Tendencies." *Hispanic American Historical Review* 54: 1–47.

Kauffman, Kyle (1993) "Why Was the Mule Used in Southern Agriculture? Empirical Evidence of Principal-Agent Solutions." *Explorations in Economic History* 30: 336–51.

Kawagoe, Toshihiko, and Yujiro Hayami (1985) "An Intercountry Comparison of Agricultural Production Efficiency." *American Journal of Agricultural Economics* 67: 87–92.

Kendrick, John W. (1961) *Productivity Trends in the United States.* Princeton: Princeton University Press.

Key, Nigel, Elisabeth Sadoulet, and Alain de Janvry (2000) "Transaction Costs and Agricultural Household Supply Response." *American Journal of Agricultural Economics* 82: 245–59.

Keyder, Caglar (1991) "Introduction: Large-scale Commercial Agriculture in the Ottoman Empire?" Pp. 1–13 in *Landholding and Commercial Agriculture in the Middle East*, edited by Caglar Keyder and Faruk Tabak. Albany: State University of New York Press.

Kherallah, Mylene, Christopher Delgado. Eleni Gabre-Madhin, Nicholas Minot, and Michael Johnson (2002) *Reforming Agricultural Markets in Africa.* Published for IFPRI. Baltimore and London: Johns Hopkins University Press.

Khrishnamurty, J. (1983) "The Occupational Structure." Pp. 535–50 in *The Cambridge Economic History of India, vol. 2: c.1757–c. 1970*, edited by Dharma Kumar and Meghnad Desai. Cambridge: Cambridge University Press.

Kibreab, Gavin (2002) *State Intervention and the Environment in Sudan 1889–1989: The Demise of Communal Resource Management.* Levinston: Edwin Mellen.

King, Russel (1977) *Land Reform.* London: Bell.

Kislev, Yoav, and Willis Peterson (1996) "Economics of Scale in Agriculture: A Reexamination of the Evidence." Pp. 156–70 in *The Economics of Agriculture: Papers in honour of D. Gale Johnson 1996*, vol. 2, edited by J. M. Antle and D. A. Sumner. Chicago: Chicago University Press.

Kloppenburg, Jack R. (1994) *First the Seed: The Political Economy of Plant Biotechnology.* Cambridge: Cambridge University Press.

Knapp, Joseph G. (1969) *The Rise of the American Co-operative Enterprise 1620–1920.* Danville, Ill.: Interstate Printers and Publishers.

Knibbe, Merijn T. (1993) *Agriculture in the Netherlands 1851–1950.* Amsterdam: NEHA.

——— (2000) "Feed, Fertilizer, and Agricultural Productivity in the Netherlands, 1880–1930." *Agricultural History* 74: 39–57.

Knudsen, Odin, and John Nash (1993) "Agricultural Price Stabilization and Risk Reduction in Developing Countries." Pp. 265–85 in *The Bias against Agriculture. Trade and Macro-economic Policies in Developing Countries*, edited by Romeo Bautista and Alberto Valdes. San Francisco: ICS.

Kogel, Tomas, and Alexia Prskawetz (2001) "Agricultural Productivity Growth and Escape from the Malthusian Trap." *Journal of Economic Growth* 6: 337–57.

Komlos, John (1988) "Agricultural Productivity in America and Eastern Europe: A comment." *Journal of Economic History* 48: 655–64.

——— (1998) "Shrinking in a Growing Economy? The Mystery of Physical Stature during the Industrial Revolution." *Journal of Economic History* 58: 778–95.

Komlos, John, and Peter Coclanis (1997) "On the Puzzling Cycle in the Biological Standard of Living: The Case of Antebellum Georgia." *Explorations in Economic History* 34: 433–59.

Kongsamut, Piyabha, Rebelo Sergio, and Xie Danyang (2001) "Beyond Balanced Growth." *Review of Economic Studies* 68: 869–82.

Koning, Niek (1994) *The Failure of Agrarian Capitalism: Agrarian Politics in the UK, Germany, the Netherlands and the USA.* London and New York: Routledge.

Koppel, Bruce M. (1995) "Introduction." Pp. 3–15 in *Induced Innovation Theory and International Agricultural Development: A reassesment*, edited by B. M. Koppel. Baltimore and London: Johns Hopkins University Press.

Kosarek, Jennifer, Philip Garcia, and Michael Morris (2001) "Factors Explaining the Diffusion of Hybrid Maize in Latin America and the Caribbean Region." *Agricultural Economics* 26: 267–80.

Kostrowicki, Jerzy (1980) *Geografia dell'agricoltura.* Milan: Franco Angeli.

Kotsonis, Yanni (1999) *Making Peasants Backward. Agricultural Co-operatives and the Agrarian Question in Russia, 1861–1914.* Basingstoke: Macmillan.

Kranton, Rachel, and Anand V. Swamy (1999) "The Hazards of Piecemeal Reform: British Civil Courts and the Credit Market in Colonial India." *Journal of Development Economics* 58: 1–24.

Krueger, Anne O. (1992) *The Political Economy of Agricultural Pricing Policy, vol. 5A: Synthesis of the Political Economy in Developing Countries.* Published for the World Bank. Baltimore and London: Johns Hopkins University Press.

Kumar, Dharma (1983) "The fiscal system." Pp. 905–44 in *The Cambridge Economic History of India, vol. 2: c. 1757–c. 1970*, edited by Dharma Kumar and Meghnad Desai. Cambridge: Cambridge University Press.

Kurosaki, Takashi (2003) "Specialization and Diversification in Agricultural Transformation: The Case of West Punjab, 1903–1992." *American Journal of Agricultural Economics* 85: 372–86.

Kuznets, Simon (1966) *Modern Economic Growth: Rate, Structure and Spread.* New Haven and London: Yale University Press.

Kwiecinski, Andrzej, and Pescatore Natasha (2003) "Sectoral agricultural policies and estimates of PSEs for Russia in the transition period." Pp.111–127 in *Russia agro-food sector: towards truly functioning markets*, edited by P. Werheim, E.V. Serovam, K. Frohberg, and J. von Braun. Boston and Dordrecht: Kluwer.

Kyokawa, Yujiro (1984) "The Diffusion of the new technologies in the Japanese sericiculture industry: the case of hybrid silkworm." *Hitostubashi Journal of Economics* 25: 31–59.

Lains, Pedro (2003) "New Wine in Old Bottles: Output and Productivity Trends in Portuguese Agriculture." *European Review of Economic History* 7: 43–72.

Laitner, John (2000) "Structural Change and Economic growth." *Review of Economic Studies* 67: 545–61.

Lamb, Russel L. (2003) "Inverse Productivity, Labor Markets and Measurement Error." *Journal of Development Economics* 71: 71–95.

Lambert, David, and Elliot Parker (1998) "Productivity in Chinese Provincial Agriculture." *Journal of Agricultural Economics* 49: 378–92.

Lampe, John R., and Marvin R. Jackson (1982) *Balkan Economic history, 1550–1950.* Bloomington: Indiana University Press.

Lapenu, Cécile, and Manfred Zeller (2001) "Distribution, Growth and Performance of Micro-finance Institutions in Afria, Asia and Latin America." Food Consumption and Nutrition Division Discussion Paper no.114. Washington, D.C.: International Food Policy Research Institute.

Lardy, Nicholas R. (1983) *Agriculture in China's Modern Economic Development.* Cambridge: Cambridge University Press.

Larson, Donald, Rita Butzer, Yair Mundlak, and Al Crego (2000) A Cross-country Database for Sector Investment and Capital. *World Bank Economic Review* 14: 371–91.

Lastarria-Cornhiel, Susana (1997) "Impact of Privatization on Gender and Property Rights in Africa." *World Development* 25: 1317–33.

Lauck, Jon (2000) *American Agriculture and the Problem of Monopoly.* Lincoln and London: University of Nebraska Press.

Law, Marc T. (2003) "The origins of state pure food regulation." *Journal of Economic History* 63: 1103–30.

Lawkete, Angela (2003) *Inventing the Cotton Gin: Machine and Myth in Antebellum America.* Baltimore and London: Johns Hopkins University Press.

League of Nations (1931) *The Course and Phases of the World Depression* [II Economic and financial 1931 II.A.21]. Geneva: League of Nations.

——— (1943) *Agricultural Production in Continental Europe during the 1914–18 War and the Reconstruction Period* [II.Economic and financial II.A.7]. Geneva: League of Nations.

Lebergott, Stanley (1984) *The Americans: An Economic Record.* New York: Norton.

Lee, T., and Y. Chen (1979) "Agricultural Growth in Taiwan 1911–1972." Pp. 59–89 and 265–312 in *Agricultural Growth in Japan, Taiwan, Korea and the Philippines*, edited by Y. Hayami, V. Ruttan and H. Southworth. Honolulu: University Press of Hawai.

Lélé, Sharachandra (1991) "Sustainable Development: A Critical Review." *World Development* 19: 607–21.

Lerman, Zvi (2000) "From Common Heritage to Divergence: Why the Transition Countries Are Drifting Apart by Measures of Agricultural Performance." *American Journal of Agricultural Economics* 82: 1140–48.

——— (2001) "Agriculture in Transition Economies: From Common Heritage to Divergence." *Agricultural Economics* 26: 95–114.

Lerman Zvi, Yoav Kislev, David Biton, and Alon Kriss (2003) "Agricultural Output and Productivity in the Former Soviet Republics." *Economic Development and Cultural Change* 51: 999–1018.

Levine, Ross (1997) "Financial Development and Economic Growth: Views and Agenda." *Journal of Economic Literature* 35: 688–726.

Levy-Leboyer, Maurice (1970) "L'heritage de Simiand: Prix, Profit et termes d'échange au XIXe siècle." *Revue Historique* 243: 77–120.

Levy-Leboyer, Maurice, and François Bourguignon (1985) *L'économie française au XIXe siècle.* Paris: Economica.

Lew, Byron (2000) "The Diffusion of Tractors on the Canadian Prairies: The Threshold Model and the Problem of Uncertainty." *Explorations in Economic History* 37: 189–216.

Lewin, Moshe (1985) *The Making of the Soviet System.* London: Methuen.

Lewis, William A. (1949) *Economic Survey 1919–1939.* London: Allen and Unwin.

——— (1954) "Economic Development with Unlimited Supply of Labour." *Manchester School of Economics and Social Studies* 22: 139–91.

——— (1979) *Growth and Fluctuation*. London: Allen and Unwin.

——— (1981) "The Rate of Growth of World Trade 1870–1913." In *The World Economic Order: Past and Prospects*, edited by S. Grassman and E. Lundberg. London and Basingstoke: Macmillan.

Li, Bozhong (1998) *Agricultural Development in Jiangnan*. New York: St. Martin's Press.

Libecap, Gary D. (1998) "The Great Depression and the Regulating State: Federal Government Regulation of Agriculture, 1884–1970." Pp. 181–224 in *The Defining Moment: The Great Depression and the American Economy in the Twentieth Century*, edited by M. Bordo, C. Goldin, and E. N. White. Chicago and London: University of Chicago Press.

Libecap, Gary D., and Zeynep K. Hansen (2002) " 'Rain Follows the Plow' and Dry Farming Doctrine: The Climate Information Problem and Homestead Failure in the Upper Great Plains, 1890–1925." *Journal of Economic History* 62: 86–120.

Lichtenberg, Erik (2002) "Agriculture and the environment." Pp. 1249–313 in *Handbook of Agricultural Economics, vol. 2A: Agriculture and Its External Linkages*, edited by Bruce Gardner and Gordon Rausser. Amsterdam: Elsevier.

Liepmann, Heinrich (1938) *Tariff Levels and the Economic Unity of Europe*. London: Allen and Unwin.

Lin, Alfred H. Y. (1997) *The Rural Economy of Guangdong, 1870–1937. A Study of the Agrarian Crisis and Its Origins in Southernmost China*. Basingstoke: Macmillan.

Lin, Justin Yifu (1992) "Rural Reforms and Agricultural Growth in China." *American Economic Review* 82: 34–51.

——— (1994) "The Nature and Impact of Hybrid Rice in China." Pp. 375–408 in *Modern Rice Technology and Income Distribution in Asia*, edited by Cristina C. David and Keijiro Otsuka. Manila: IRRI.

Lin, Justin Yifu, and Dennis Tao Yang (2000) "Food availability, Entitlements and the Chinese Famine of 1959–61." *Economic Journal* 110: 136–58.

Linder, Marc, and Lawrence S. Zacharias (1999) *Of Cabbages and Kings County*. Iowa City: University of Iowa Press.

Lindert, Peter (1987) "Who Owned Victorian England: The Debate over Landed Wealth and Inequality." *Agricultural History* 61 (4): 25–51.

——— (1988) "Long-run Trends in American Farmland values." *Agricultural History* 62: 45–89.

——— (1991) "Historical Patterns of Agricultural Policy." Pp. 29–83 in *Agriculture and the State*, edited by P. Timmer. Ithaca and London: Cornell University Press.

——— (2000) *Shifting Ground: The Changing Agricultural Soils of China and Indonesia*. Cambridge and London: MIT Press.

Lipton, Michael (1974) "Towards a Theory of Land Reform." Pp. 269–315 in *Peasants, Landlords and Governments*, edited by David Lehmann. New York: Holmes and Meier.

Little, Daniel (1992) "New Perspectives on the Chinese Rural Economy: A Symposium." *Republican China* 18: 23–176.

Livi Bacci, Massimo (1993) "On the Human Costs of Collectivisation in the Soviet Union." *Population and Development Review* 19: 743–66.

Lomborg, Bjørg (1998) *The Sceptical Environmentalist*. Cambridge: Cambridge University Press.

Lorenzoni, Giovanni (1939) *Inchiesta sulla piccola proprietà coltivatrice formatasi nel primo dopoguerra.* Rome: INEA.

Loubère, Leo (1978) *The Red and the White: The History of Wine in France and Italy in the 19th Century.* Albany: SUNY Press.

——— (1990) *The Wine Revolution in France: The Twentieth Century.* Princeton: Princeton University Press.

Lupo, Salvatore (1990) *Il giardino degli aranci.* Padua: Marsilio.

Lusigi, Angela, and Colin Thirtle (1997) "Total Factor Productivity and the Effects of R&D in African agriculture." *Journal of International Development* 9: 529–38.

Lyaschenko, Peter I. (1970) *History of the National Economy of Russia.* New York: Octagon. Originally published in 1939.

Maddison, Angus (1985) "Alternative Estimates of the Real Product of India, 1900–1946." *The Indian Economic and Social History Review* 22: 201–10.

——— (1991) *Dynamic Forces in Capitalist Development.* Oxford: Oxford University Press.

——— (1995) *Monitoring the World Economy 1820–1992.* Paris: OECD.

——— (1998) *Chinese Economic Performance in the Long Run.* Paris: OECD.

——— (2001) *The World Economy: A Millennial Perspective.* Paris: OECD.

——— (2003) *The World Economy: Historical Statistics.* Paris: OECD.

Madsen, Jakob (2001) "Agricultural Crises and the International Transmission of the Great Depression." *Journal of Economic History* 61: 327–65.

Magnac, Thierry, and Gilles Postel-Vinay (1997) "Wage Competition between Agriculture and Industry in Mid-Nineteenth Century France." *Explorations in Economic History* 34: 1–26.

Mahdi, Kamil A. (2000) *State and Agriculture in Iraq* Reading: Ithaca Press.

MAIC (1891) *Ministero di agricoltura, industria e commercio, Direzione generale dell'agricoltura. I contratti agrari in Italia.* Rome: Bertero.

Malatesta, Maria (1989) *I signori della terra.* Milan: Franco Angeli.

——— (1999) *Le aristocrazie terriere nell'Europa contemporanea.* Bari: Laterza.

Malle, Silvana (1985) *The Economic Organization of War Communism, 1918–1921*, Cambridge: Cambridge University Press.

Manarungsan, Sompop (1989) *Economic Development of Thailand, 1850–1950: Response to the Challenge of the World Economy.* Institute of Asian Studies Monograph no. 042. Bangkok: Chulalongkorn University.

Marchand, O., and C. Thelot (1991) *Deux siècles de travail en France.* Paris: INSEE Etudes.

Marcours, Karen, and Johann Swinnen (2000) "Impact of Initial Conditions and Reform Polices on Agricultural Performance in Central and Eastern Europe, the former Soviet Union and East Asia." *American Journal of Agricultural Economics* 82: 1149–55.

Marcus, Alan I. (1988) "The Wisdom of the Body Politics: The Changing Nature of Publicly Sponsored American Agricultural Research since the 1830s." In "Publicly Sponsored Agricultural Research in the United States: Past, Present and future," edited by David D. Danbom. *Agricultural History* 62: 4–26.

Marks, Robert M. (1998) *Tigers, Rice, Silk and Silt: Environment and Economy in Late Imperial South China.* Cambridge: Cambridge University Press.

Marshall, Alfred (1920) *Principles of Economics.* 8th edition. London: Macmillan.

Martin, Lee R. (1992) *A Survey of Agricultural Economics Literature, vol. 4: Agriculture in Economic Development, 1940s to 1990s.* Published for the American Agricultural Economics Association. Minneapolis: University of Minnesota Press.

Martin, Susan (1987) "Boserup Revisited: Population and Technology in Tropical African Agriculture, 1900–1940." *Journal of Imperial and Commonwealth History* 16: 109–23.

Martin, Will, and Devashish Mitra (2001) "Productivity Growth and Convergence in Agriculture versus Manufacturing." *Economic Development and Cultural Change* 49: 403–23.

Masters, William A. and Margaret McMillan (2001) "Climate and Scale in Economic Growth." *Journal of Economic Growth* 6: 167–86.

Matsuyama, Kiminori (1992) "Agricultural Productivity, Comparative Advantage and Economic Growth." *Journal of Economic Theory* 58: 317–34.

Matthews, R.C.O., Charles H. Feinstein, and J. C. Odling Smee (1982) *British Economic Growth.* Oxford: Clarendon.

Mauldon, Roger (1990) "Price Policy." Pp. 310–28 in *Agriculture in the Australian Economy*, edited by D. B. Williams. Sydney: Sydney University Press.

McAlpine, Michelle Burge (1983) *Subject to Famine. Food Crises and Economic Change in Western India, 1860–1920.* Princeton: Princeton University Press.

McCann, James C. (1995) *People of the Plow: An Agricultural History of Ethiopia, 1800–1990.* Madison: University of Wisconsin Press.

McCauley, Martin (1976) *Khruschev and the Development of Soviet Agriculture. The Virgin Land Program 1953–1964.* New York: Holmes and Meier.

McClelland, Peter D. (1997) *Sowing Modernity: America's First Agricultural Revolution.* Ithaca and London: Cornell University Press.

McCloskey, Donald (1980) "Magnanimous Albion: Free Trade and British National Income." *Explorations in Economic History* 17: 303–20.

——— (1981) "The Industrial Revolution, 1780–1860." Pp. 103–27 in *The Economic History of Britain since 1700, vol. 1: 1700–1860*, edited by R. Floud and D. McCloskey. Cambridge: Cambridge University Press.

McCorvie, Mary, and Christopher Lant (1993) "Drainage District Formation and the Loss of Midwestern Westlands, 1850–1930." *Agricultural History* 67: 13–39.

McCunn, Alan, and Wallace E. Huffman (2000) "Convergence in U.S. Productivity Growth for Agriculture: Implications of Interstate Research Spillovers for Funding Agricultural Research." *American Journal of Agricultural Economics* 82: 370–88.

McEvedy, C., and R. Jones 1978. *Atlas of World Population History.* New York: Facts on File.

McGreevey, William Paul (1971) *An Economic History of Colombia 1845–1930.* Cambridge: Cambridge University Press.

McGuire, Robert A. (1980) "A Portfolio Analysis of Crop Diversification and Risk in the Cotton South." *Explorations in Economic History* 17: 342–71.

McInnis, Marvin (1986) "Output and Productivity in Canadian Agriculture, 1870–71 to 1926–27." Pp. 737–78 in *Long-term Factors in American Economic Growth*, edited by Stanley L. Engerman and Robert E. Gallman. Chicago and London: University of Chicago Press.

McMichael, Philip (1984) *Settlers and the Agrarian Question: Foundations of Capitalism in Colonial Australia.* Cambridge: Cambridge University Press.

McMillan, John, John Whalley, and Lijing Zhu (1989) "The Impact of China's Economic Reforms on Agricultural Productivity Growth." *Journal of Political Economy* 97: 782–807.

Medvedev, Zhores (1987) *Soviet Agriculture.* New York and London: Norton.

Merrick, Thomas W., and Douglas H. Graham (1979) *Population and Economic Development in Brazil 1800 to the Present.* Baltimore and London: Johns Hopkins University Press.

Miller, Simon (1995) *Landlords and Haciendas in Modernizing Mexico: Essays in Radical Reappraisal.* Amsterdam: CEDLA.

Milward, Alan S. (1984) *The Reconstruction of Western Europe 1945–51.* London: Routledge. Second edition published in 1992.

——— (1992) *The European Rescue of the Nation-state.* London: Routledge.

Minde, Isaac J., Peter T. Ewell, and James M. Teri (1999) "Contributions of Cassava and Sweet-potato to Food Security and Poverty Alleviation in the SADC Countries: Current Status and Future Prospects." Pp. 27–40 in *Food Security and Crop Diversification in SADC Countries: The Role of Cassava and Sweetpotato,* edited by M. O. Akoroda and J. M. Teri. Ibadan: SADC.

Miracle, Marvin P. (1967) *Agriculture in the Congo Basin: Tradition and Change in African Rural Economies.* Madison: University of Wisconsin Press.

Mironov, Boris (1999) "New Approaches to Old Problems: The Well-being of the Population of Russia from 1821 to 1910 as Measured by Physical Stature." *Slavic Review* 58: 1–26.

Mirri, Mario (2004) "Andare a scuola di agricoltura." Pp. 13–59 in *Agricoltura come manifattura,* edited by Giuliana Biagioli and Rossano Pazzagli. Florence: Olschki.

Mishra, Statis Chandra (1982) "Commercialization, Peasant Differentiation and Merchant Capital in Late Nineteenth-Century Bombay and Punjab." *Journal of Peasant Studies* 10: 3–51.

——— (1983) "On the Reliability of Pre-independence Agricultural Statistics in Bombay and Punjab." *The Indian Economic and Social History Review* 20: 171–90.

Mitchell, Brian (1988) *British Historical Statistics.* Cambridge: Cambridge University Press.

——— (1998a) *International Historical Statistics: Africa, Asia, Oceania 1750–1993.* 3rd edition. London and Basingstoke: MacMillan.

——— (1998b) *International Historical Statistics: The Americas 1750–1993.* 4th edition. London and Basingstoke: MacMillan.

——— (1998c) *International Historical Statistics: Europe 1750–1993.* 4th edition. London and Basingstoke: MacMillan.

Mitchell, Brian, and Phyllis Deane (1962) *Abstract of British Historical Statistics.* Cambridge: Cambridge University Press.

Mokyr, Joel (1981) "Irish History with the Potato." *Irish Economic and Social History* 8: 8–29.

——— (1983) *Why Ireland Starved: A Quantitative and Analytical History of the Irish Economy, 1800–1850.* London: Allen and Unwin.

——— (1987) "Has the Industrial Revolution Been Crowded Out? Some Reflections on Crafts and Williamson." *Explorations in Economic History* 24: 293–319.

——— (1990) *The Lever of Riches.* Oxford: Oxford University Press.

——— (1993) "Editor's Introduction: The New Economic History and the Industrial Revolution." Pp. 1–131 in *The British Industrial Revolution: An Economic Perspective,* edited by Joel Mokyr. Boulder: Westview.

——— (2002) *The Gifts of Athena.* Princeton: Princeton University Press.

Mokyr, Joel, and John Nye (1990) "La Grande quantification." *Journal of Economic History* 50: 172–76.

Moon, David (1999) *The Russian Peasantry 1600–1930.* London and New York: Longman.

Morgan, W. B. (1969) "Peasant Agriculture in Tropical Africa." Pp. 241–72 in *Environment and Land Use in Africa,* edited by M. F. Thomas and G. W. Whittington. London: Methuen.

Morris, Michael (1998) "Overview of the World Maize Economy." Pp. 13–33 in *Maize Seed Industries in Developing Countries*, edited by Michael Morris. Published for CIMMYT. Boulder: Lynne Rienner.

Morris, Michael, and Derek Byerlee (1998) "Maintaining Productivity Gains in Post-Green Revolution Asian Agriculture." Pp. 458–73 in *International Agricultural Development*, edited by Carl K. Eicher and John M. Staatz. 3rd edition. Baltimore and London: Johns Hopkins University Press.

Morrisson, Christian, and Erik Thorbecke (1990) "The Concept of the Agricultural Surplus." *World Development* 18: 1081–95.

Mortimore, Michael (1998) *Roots in the African Dust.* Cambridge: Cambridge University Press.

Mosley, Paul (1983) *The Settler Economies: Studies in the Economic History of Kenya and Southern Rhodesia 1900–1983.* Cambridge: Cambridge University Press.

——— (2002) "Agricultural Marketing in Africa since Berg and Bates." Pp. 176–90 in *Renewing Development in Sub-Saharan Africa: Policy, Performance and Prospects*, edited by Deryke Bleshaw and Ian Livingston. London: Routledge.

Moulin, Annie (1991) *Peasantry and Society in France since 1789.* Cambridge: Cambridge University Press; and Paris: Editions de la Maison des Science de l'Homme.

Moulton, Forest R. (1942) *Liebig and after Liebig.* Washington, D.C.: American Association for the Advancement of Science.

Moyer, Wyane, and Tim Mosling (2002) *Agriculture policy reform* Aldershot: Ashgate.

Mubarik, Ali, and Derek Byerlee (2002) "Productivity Growth and Resource Degradation in Pakistan's Punjab: A decomposition analysis." *Economic Development and Cultural Change* 50: 839–63.

Mukherjee, Mridula (1984) "Commercialization and Agrarian Change in Pre-independence Punjab." Pp. 51–104 in *Essays on the Commercialisation of Indian Agriculture*, edited by K. N. Raj, Neelandri Bhattacharya, Sumit Guha, and Sakti Padhi. Delhi: Oxford University Press.

Mulgan, Aurelia (2000) *The Politics of Agriculture in Japan.* London and New York: Routledge.

Mundlak, Yair (2001) "Production and Supply." Pp. 5–85 in *Handbook of Agricultural Economics, vol. 1A: Agricultural Production*, edited by Bruce Gardner and Gordon Rausser. Amsterdam: Elsevier.

Mundlak, Yair, and Donald F. Larson (1992) "On the Transmission of World Agricultural prices." *World Bank Economic Review* 6: 399–422.

Mundlak, Yair, Donald F. Larson, and Rita Butzer (1999) "Rethinking within and between Regressions: The Case of Agricultural Production Functions." *Annales d'économie et statistique* 55–56: 475–501.

Mundlak, Yair, Donald F. Larson, and Al Crego (1997) "Agricultural Development: Issues, Evidence and Consequences." World Bank Policy Research Working papers no. 1811. Washington, D.C.: World Bank.

Murdoch, Jonathan (1999) "The Micro-finance Promise." *Journal of Economic Literature* 37: 1569–614.

Myers, Ramon H. (1970) *The Chinese Peasant Economy. Agricultural Development in Hopei and Shantung, 1890–1949.* Cambridge: Harvard University Press.

Mynt, Hla (1975) "Agriculture and Economic Development in the Open Economy." Pp. 327–54 in *Agriculture in Development Theory*, edited by Lloyd Reynolds. New Haven and London: Yale University Press.

Nagaraj K. (1984) "Marketing Structures for Paddy and Aricanut in South Kanara: A Comparison of Markets in Backward Agricultural Districts." Pp. 247–72 in *Essays on the commercialisation of Indian agriculture*, edited by K. N. Raj, Neelandri Bhattacharya, Sumit Guha, and Sakti Padhi. Delhi: Oxford University Press.

Nakamura, James I. (1966) *Agricultural Production and the Economic Development of Japan 1873–1922.* Princeton: Princeton University Press.

National Research Council (1993) *Soil and Water Quality: An Agenda for Agriculture.* Washington, D.C.: National Academy Press.

Naylor, Rosamond (1994) "Herbicide Use in Asian Rice Production." *World Development* 22: 55–70.

Nerlove, Marc (1996) "Reflections on the Economic Organization of Agriculture: Traditional, Modern and Transitional." Pp. 9–30 in *Agricultural Markets: Mechanisms, Failures and Regulations*, edited by David Martimort. Amsterdam: Elsevier.

Newell, Andrew, Kiran Pandya, and James Symons (1997) "Farm Size and the Intensity of Land Use in Gujarat." *Oxford Economic Papers* 49: 307–15.

Nghiep, L., and Yujiro Hayami (1979) "Mobilizing Slack Resources for Economic Development: The Summer-Fall Rearing Technology of Sericiculture in Japan." *Explorations in Economic History* 19: 163–81.

Nguyen, D. T., and Harry X. Wu (1999) "The Impact of the Economic Reforms on Agricultural growth." Pp. 52–99 in *Productivity and Growth in Chinese Agriculture*, edited by Kali P. Kalirajan and Wu Yanrui. Basingstoke: Macmillan.

Nicolau, Roser (1989) "Poblacion." Pp. 53–90 in *Estadisticas Historicas de Espana siglos XIX–XX*, edited by Albert Carreras. Madrid: Fundacion Banco Exterior.

Nin, Alejandro, Channing Arndt, and Paul Preckel (2003a) "Is Agricultural Productivity in Developing Countries Really Shrinking? New Evidence Using a Modified Nonparametric Approach." *Journal of Development Economics* 71: 395–415.

Nin, Alejandro, Channing Arndt, Thomas W. Hertel, and Paul V. Preckel (2003b) "Bridging the Gap between Partial and Total Factor Productivity Measures Using Directional Distance Functions." *American Journal of Agricultural Economics* 85: 928–42.

Norrie, K. H. (1975) "The Rate of Settlement of the Canadian Prairies, 1870–1911." *Journal of Economic History* 35: 410–27.

Nove, Alec (1969) *An Economic History of the URSS.* Harmondsworth: Penguin.

Novello, Elisabetta (2003) *La bonifica in Italia. Legislazione, credito e lotta alla malaria dall'Unità al fascismo.* Milan: Franco Angeli.

NPSA (ad annum) Ministero di Agricoltura industria e commercio, Ufficio di Statistica agraria. *Notizie periodiche di statistica agraria.* Roma: Bertero.

Nurkse, Ragnar (1953) *Problems of Capital Formation in Under-developed Countries.* Oxford: Basil Blackwell.

Nutzenadel, Alexandr (2001) "Economic Crisis and Agriculture in Fascist Italy: Some New Considerations." *Rivista di Storia economica* (n.s. 17): 289–312.

O'Brien, Davis, Valeri Patsiorkoski, and Larry D. Dershem (2000) *Household Capital and the Agrarian Problem in Russia.* Aldershot: Ashgate.

O'Brien, Patrick (1968) "The Long-term Growth of Agricultural Production in Egypt: 1821–1962." Pp. 162–95 in *Political and Social Change in Modern Egypt*, edited by P. M. Holt. London: Oxford University Press.

——— (1996) "Path Dependency or Why Britain Became an Industrialized and Urbanized Economy Long before France." *Economic History Review* 49: 213–49.

O'Brien, Patrick, and Caglar Keyder (1978) *Economic Growth in Britain and France 1780–1914.* London: Allen and Unwin.

O'Brien, Patrick, and Leandro Prados de la Escosura (1992) "Agricultural Productivity and European industrialization." *Economic History Review* 45: 514–36.

OECD (1994) *Farm Employment and Economic Adjustment in OECD countries.* Paris: OECD.

OECD (2001) *Agricultural Policies in OECD Countries: Monitoring and Evaluation.* Paris: OECD.

OECD (2002) *Agriculture and Trade Liberalization: Extending the Uruguay Round Agreement.* Paris: OECD.

OECD Statistical Database available at www.newsourceoecd.org.

Offer, Avner (1989) *The First World War: An Agrarian Interpretation.* Oxford: Clarendon.

——— (1991) "Farm Tenure and Land Values in England, c.1750–1950." *Economic History Review* 44: 1–20.

O'Grada, Cormac (1977) "The Beginnings of the Irish Creamery System 1880–1914." *Economic History Review* 30: 284–301.

——— (1993) *Ireland before and after the Famine.* 2nd edition. Manchester: Manchester University Press.

——— (1994) "British Agriculture 1860–1914." Pp. 145–72 in *The Economic History of Britain since 1700,* vol. 2, edited by D. McCloskey and R. Floud. 2nd edition. Cambridge: Cambridge University Press.

——— (1995) *Ireland. A New Economic History 1780–1939.* Oxford: Oxford University Press.

O'Grada, Cormac (1999) *Black'47 and beyond. The Great Irish Famine* Princeton: Princeton University Press

Ojala, E. M. (1952) *Agriculture and Economic Progress.* Oxford: Oxford University Press.

Okhawa, Kazushi, et al. (1967) *Prices in Estimates of Long-term Economic Statistics of Japan since 1868*, vol 8. Tokyo: Toyo Keizai Shinposha.

Okhawa, Kazushi, and Miyohei Shinohara (1979) *Patterns of Japanese Economic Development.* New Haven and London: Yale University Press.

Olmstead, Alan (1975) "The Mechanisation of Reaping and Mowing in American Agriculture 1833–1870." *Journal of Economic History* 35: 327–52.

Olmstead, Alan, and Paul W. Rhode (1993a) "An Overview of California Agricultural Mechanization." Pp. 83–109 in *Quantitative Studies in Agrarian History,* edited by M. Rothstein and D. Field. Ames: Iowa University Press.

——— (1993b) "Induced Innovation in American Agriculture: a Reconsideration." *Journal of Political Economy* 101: 100–18.

——— (1998) "Induced Innovation in American Agriculture: An Econometric Analysis." *Research in Economic History* 18: 103–19.

——— (2000a) "The Transformation of Northern Agriculture, 1910–1990." Pp. 693–743 in *The Cambridge Economic History of the United States, vol.3: The Twentieth Century*, edited by Stanley Engerman and Robert Gallman. Cambridge: Cambridge University Press.

——— (2000b) "Biological Innovation and American Agricultural Development." Mimeo. Institute of Governmental Affairs, University of California Davis, April 2000.

——— (2001) "Reshaping the Landscape: The Impact and Diffusion of the Tractor in American Agriculture." *Journal of Economic History* 61: 663–98.

——— (2002) "The Red Queen and the Hard Reds: Productivity Growth in American Wheat, 1800–1940." *Journal of Economic History* 62: 929–66.

——— (2004) "An Impossible Undertaking: The Eradication of Bovine Tubercolosis in the United States." *Journal of Economic History* 64: 734–72.

Olper, Alessandro (2001) "Determinants of Agricultural Protection: the Role of Democracy and Institutional Setting." *Journal of Agricultural Economics* 52: 75–92.

Olson, Mancur (1971) *The Logic of Collective Action.* Cambridge: Harvard University Press.

Orlando, Giuseppe (1969) "Progressi e difficoltà dell'agricoltura." Pp. 181–95 in *Lo sviluppo economico in Italia*, vol. 3, edited by G.Fuà. Milan: Franco Angeli.

——— (1984) *Storia della politica agraria in Italia dal 1984 ad oggi.* Bari: Laterza.

O'Rourke, Kevin (1997) "The European Grain Invasion." *Journal of Economic History* 57: 775–801.

——— (2002) "Culture, Politics and Innovation: Evidence from the Creameries." CEPR Discussion Paper Series no.3235.

O'Rourke, Kevin, and Jeffrey Williamson (1999) *Globalization and History.* Cambridge (Mass.), and London: MIT Press.

Osband, Kent (1985) "The Boll Weevil versus King Cotton." *Journal of Economic History* 45: 627–43.

Oskam, Arie, and Spiro Stefanou (1997) "The CAP and Technical change." Pp. 191–224 in *The Common Agricultural Policy*, edited by Christopher Ritson and David Harvey. 2nd edition. Wallingford: CAB International.

Otsuka, Keijiro, Hiroyuki Chuma, and Yujiro Hayami (1992) "Land and Labor Contracts in Agrarian Economies: Theories and Facts." *Journal of Economic Literature* 30: 1965–2018.

Overton, John (1994) *Colonial Green revolution? Food, irrigation and the State in Colonial Malaya*, Wallingford:CAB International.

——— (1996) *Agricultural revolution in England*, Cambridge: Cambridge University Press.

Overton, Mark, and Bruce Campbell (1991) "Productivity changes in European agricultural development." Pp. 1–51 in *Land, Labour and Livestock: Historical Studies in European Agricultural Productivity*, edited by Bruce Campbell and Mark Overton. Manchester: Manchester University Press.

Owen, Roger (1986) "Large Landowners, Agricultural Progress and the State in Egypt, 1800–1970: An Overview with Many Questions." Pp. 69–95 in *Food, States and Peasants*, edited by A. Richards. Boulder and London: Westview.

Owen, Roger, and Sevket Pamuk (1998) *Middle East Economies in the Twentieth Century.* London: I. B. Taurus.

Owens, Trudy, John Oddinott, and Bill Kinsey (2003) "The Impact of Agricultural Extension on Farm Production in Resettlement Areas of Zimbabwe." *Economic Development and Cultural Change* 51: 338–57.

Paarlberg, Robert, and Don Paarlberg (2000) "Agricultural Policy in the Twentieth century." *Agricultural History* 74: 136–61.

Pagiola, Stefano (1999) "Economic Analysis of Rural Land Administration Policy." World Bank, Environment Department. Available at www.worldbank.org/networks/ESSD/icdb .nsf.

Palairet, Michael (1997) *The Balkan Economies c.1800–1914: Evolution without Development.* Cambridge: Cambridge University Press.

Palanisami, K. (1997) "Economics of Irrigation Technology Transfer and Adoption." In *Irrigation Technology Transfer in Support of Food Security.* FAO Water Reports no. 14. Rome: FAO. Available at www.FAO.org/docrep/W7314E.

Palladino, Paolo (1996) "Science, Technology and the Economy: Plant Breeding in Great Britain, 1920–1970." *Economic History Review* 49: 116–36.

Pallot Judith (1999) *Land reform in Russia 1906-1917. Peasant responses to Stolypin's project of rural transformation.* Oxford: Clarendon Press.

Pamuk, Sevket (1986) *The Ottoman Empire and European Capitalism, 1820–1913.* Cambridge: Cambridge University Press.

Pardey, Philip, and Nienke B. Beintema (2001) "Slow Magic: Agricultural R&D a Century after Mendel." Washington International Food Policy Research Institute. Available at www.ifpri.cgiar.org.

Parikh, Kirit S. (1994) "Agricultural and Food System Scenarios for the 21st century." Pp. 26–43 in *Agriculture, Environment and Health: Sustainable Development in the 21st Century*, edited by Vernon W. Ruttan. Minneapolis :University of Minnesota Press.

Parker, William N. (1972) "The Land, Minerals, Water and Forests." In *American Economic Growth: An Economist's History of the United States*, edited by Lance Davis et al. New York: Harper and Row.

Parker, William N., and Judith Klein (1966) "Productivity Growth in Grain Production in the United States, 1840–1860 and 1900–10." Pp. 523–80 in *Output, Employment and Productivity in the United States after 1800.* Published for NBER. New York: Columbia University Press.

Parsons, Neil, and Robin Palmer (1977) "The Roots of Rural Poverty: Historical Background." Pp. 1–13 in *The Roots of Rural Poverty in Central and Southern Africa*, edited by Robin Palmer and Neil Parsons. London: Heinemann.

Pavlovsky, George (1968) *Agricultural Russia on the Eve of the Revolution.* Reprint. New York: Fertig.

Peacock, A. T., and J. Wiseman (1967) *The Growth of Public Expenditure in the United Kingdom.* 2nd edition. London: Allen and Unwin.

Pearse, Andrew (1980) *Seeds of Plenty, Seeds of Want.* Oxford: Clarendon.

Pearson, C. J. (1992) *Field Crop Ecosystems*. Amsterdam: Elsevier.

Perdue, Peter C. (1987) *Exhausting the Earth. State and Peasants in Hunan, 1500–1850*, Cambridge: Harvard University Press.

Perkins, Dwight (1969) *Agricultural Development in China 1368–1968.* Edinburgh: Edinburgh University Press.

Perkins, J. A. (1981) "The Agricultural Revolution in Germany 1870–1914." *Journal of European Economic History* 10: 71–118.

Perkins, John (1997) *Geopolitics and the Green Revolution: Wheat, Genes, and the Cold War.* New York: Oxford University Press.

Perren, R. (1995) *Agriculture in Depression 1870–1940.* Cambridge: Cambridge University Press.

Peterson, Wesley, and Siva Rama, Khrisna Vallaru (2000) "Agricultural Comparative Advantage and Government Policy interventions." *Journal of Agricultural Economics* 51: 371–87.

Petrusewicz, Marta (1996) *Latifundium: Moral Economy and Material Life in an European Periphery.* Ann Arbor: University of Michigan Press.

Pezzati, Mario (1993) "Agricoltura ed industria: i concimi chimici." In *Studi sull'agricoltura italiana*, edited by P. D'Attorre and A. De Bernardi. *Annali Feltrinelli* 29: 373–401.

——— (1995) "La Federazione dei consorzi agrari ed il mercato dei concimi chimici (1892–1932)." Pp.133–63 in *La Federconsorzi fra stato liberale e fascismo*, edited by Severina Fontana. Bari: Laterza.

Pingali, Prabhu (1989) "Institutional and Environmental Constraints to Agricultural Intensification." In *Rural Development and Population: Institutions and Policy*, edited by Geoffrey McNicoll and Mead Cain. *Supplement to Population and Development Review* 15: 243–60.

——— (1998) "Confronting the Ecological Consequences of the Rice Green Revolution in Tropical Asia." Pp. 474–91 in *International Agricultural Development*, edited by Carl K. Eicher and John M. Staatz. 3rd edition. Baltimore and London: Johns Hopkins University Press.

Pingali, Prabhu, Yves Bigot, and Hans. P. Binswanger (1987) *Agricultural Mechanisation and the Evolution of Farming Systems in Sub-Saharan Africa.* Published for the World Bank. Baltimore and London: Johns Hopkins University Press.

Pingali, Prabhu, and Paul W. Heisey (2001) "Cereal-crop Productivity in Developing Countries." Pp. 56–81 in *Agricultural Science Policy: Changing Global Agendas*, edited by Julian M. Alston, Philip G. Pardey, and Michael J. Taylor. Baltimore and London: Johns Hopkins University Press.

Pingali, Prabhu, Hossain Mahabub, and Gerbacio Roberta (1997) *Asian Rice Bowls: The Returning Crisis.* Wallingford: CAB International in association with the IRRI.

Pinstrup-Andersen, Per, Rajul Pandya-Lorch, and Mark W. Rosegrant (1999) *World Food Prospects: Critical Issues for the Early Twenty-first Century.* Washington, D.C.: International Food Policy Research Institute.

Place, Frank, and Keijiro Otsuka (2001) "Tenure, Agricultural Investment and Productivity in the Customary Tenure Sector in Malawi." *Economic Development and Cultural Change* 50: 77–99.

Planck, Ulrich (1987) "The Family Farm in the Federal Republic of Germany." Pp. 155–91 in *Family Farming in Europe and America*, edited by Boguslaw Galeski and Eugene Wilkening. Boulder: Westview.

Polanyi, Karl (1944) *The Great Transformation: The Political and Economic Origins of Our Time.* New York: Holt, Rinehart, and Winston.

Poliakov, Iuri (2000) *Naselenie Rossii v.XX Veke* vol. 1. Moskva.

Pomeranz, Kenneth (1993) *The Making of a Hinterland: State, Society and Economy in Inland North China, 1853–1937.* Berkeley and Los Angeles: University of California Press.

——— (2000) *The Great Divergence.* Princeton: Princeton University Press.

——— (2002) "Beyond the East-West Binary: Resituating Development Paths in the Eighteenth-Century World." *Journal of Asian Studies* 61: 539–90.

Pomfret, Richard (2002) "State-Directed Diffusion of Technology: The Mechanization of Cotton Harvesting in Soviet Central Asia." *Journal of Economic History* 62: 170–88.

Porisini, Giorgio (1971) *Produttività ed agricoltura: i rendimenti del frumento in Italia dal 1815 al 1922.* Turin: UTET.

Postel-Vinay, Gilles, and Jean-Marc Robin (1994) "Eating, Working and Saving in an Unstable World: Consumers in Nineteenth-Century France." *Economic History Review* 45: 494–513.

Powelson, John P., and Richard Stock (1990) *The Peasant Betrayed: Agriculture and Land Reform in the Third World.* Washington, D.C.: Cato Institute.

Prados de la Escosura, Leandro (2004) *El progreso economico de Espana (1850–2000).* Bilbao: Fundacion BBVA.

Prados de la Escosura, Leandro, Teresa Daban Sanchez, and Jesus Sanz Oliva (1993) "De te fabula narratur." Ministero de Economia y Hacienda Documentos de trabajo D 93009.

Pray, Carl (1984) "Accuracy of Official Agricultural Statistics and the Sources of Growth in the Punjab 1907–47." *The Indian Economic and Social History Review* 21: 312–33.

Pray, Carl, and Robert Evenson (1991) "Research Effectiveness and the Support Base for National and International Agricultural Research and Extension Programs." Pp. 355–71 in *Research and productivity in Asian agriculture*, edited by Robert Evenson and Carl Pray. Ithaca and London: Cornell University Press.

Pray, Carl E., and Dina Umali-Deininger (1998) "The Private Sector in Agricultural Research System: Will It Fill the Gap?" *World Development* 26: 1127–48.

Pray, Carl, and Ahmed Zafar (1991) "Research and Agricultural Productivity Growth in Bangladesh." Pp. 114–32 in *Research and Productivity in Asian Agriculture*, edited by Robert Evenson and Carl Pray. Ithaca and London: Cornell University Press.

Pretty, Jules N. (1995) *Regenerating Agriculture.* London: Earthscan.

Price, Roger (1983) *The Modernization of Rural France.* London: Hutchinson.

Primack, Martin (1966) "Farm Capital Formation as a Use of Farm Labor in the United States, 1850–1910." *Journal of Economic History* 23: 348–62.

——— (1969) "Farm Fencing in the Nineteenth Century." *Journal of Economic History* 29: 287–89.

Pringle, Peter (2003) *Food, Inc. Mendel to Monsanto—The Promises and Perils of the Biotech Harvest.* New York and London: Simon and Schuster.

Pryor, Frederic L. (1992) *The Red and the Green. The Rise and Fall of Collectivized Agriculture in Marxist Regimes.* Princeton: Princeton University Press.

Puhle, Hans-Jurgen (1982) "Agrarian Movements in German politics." Pp.159–88 in *Trasformazioni delle società rurali nei paesi dell'Europa Occidentale e Mediterranea*, edited by Pasquale Villani. Naples: Guida.

Quataert, Donald (1980) "The commercialization of agriculture in Ottoman Turkey, 1800–1914." *International Journal of Turkish Studies* 1: 38–55.

——— (1981) "Agricultural trends and government policy in Ottoman Anatolia 1800–1914." *Asian and African Studies* 15: 69–84.

Quibria, M.G., and Salim Rashid (1984) "The puzzle of sharecropping: a survey of theories." *World development* 12: 103–44.

Raby, Geoff (1996) *Making Rural Australia: An Economic History of Technical and Institutional Creativity.* Melbourne: Oxford University Press.

Ransom, Elizabeth, Lawrence Busch, and Gerard Middendorf (1998) "Can Co-operatives Survive the Privatization of Biotechnology in U.S. Agriculture?" Pp. 75–93 in *Privatization of Information and Agricultural Industrialization*, edited by Steven Wolf. Boca Raton: CRC Press.

Ransom, Roger L., and Richard Sutch (1977) *One Kind of Freedom.* Cambridge: Cambridge University Press.

——— (2001) "One Kind of Freedom (and Turbo Charged)." *Explorations in Economic History* 38: 6–39.

Rao, C. H. Hanumantha (1994) *Agricultural Growth, Rural Poverty and Environmental Degradation in India.* Delhi: Oxford University Press.

Rao, Moran J., and José Maria Caballero (1990) "Agricultural Performance and Development Strategy: Retrospect and Prospect." *World Development* 18: 899–913.

Rao, Moran J. (1986) "Agriculture in Recent Development Theory." *Journal of Development Economics* 22: 41–86.

Rao, Prasada (1993) "Intercountry Comparisons of Agricultural Output and Productivity." FAO Economic and Social Development Paper no. 112. Rome.

Rasmussen, Wayne D. (1968) "Advances in American Agriculture: The Mechanical Tomato Harvester as a Case Study." *Technology and Culture* 9: 531–43.

Rastegari-Henneberry, Shida, Muahmmad Ehsan Khan, and Kullapapruk Piewthongngam (2000) "An Analysis of Industrial-Agricultural Interactions: A Case Study of Pakistan." *Agricultural Economics* 22: 17–27.

Rawal, Vikas (2001) "Agrarian Reform and Land Markets: A Study of Land Transactions in Two Villages of West Bengal, 1977–1995." *Economic Development and Cultural Change* 49: 611–29.

Rawski, Thomas (1989) *Economic Growth in Pre-war China*, University of Berkeley: California Press.

Ray, Debray (1998) *Development economics.* Princeton:Princeton University Press

Reichrath, Susanne (2004) "Les débuts des études agronomiques en Allemagne jusqú à la fin du 19ème siècle." Pp. 81–97 in *Agricoltura come manifattura*, edited by Giuliana Biagioli and Rossano Pazzagli. Florence: Olschki.

Reis, Jaime (1992) *O atraso economico português.* Lisbon: Imprensa Nacional Casa de Moeda.

Reyna, S. P., and R. E. Downs (1988) "Introduction." Pp. 1–22 in *Land and Society in Contemporary Africa*, edited by S. P. Reyna and R. E. Downs. Hanover and London: University Press of New England.

Reynolds, Clark W. (1970) *The Mexican Economy. Twentieth-Century Structure and Growth.* New Haven and London: Yale University Press.

Reynolds, Lloyds (1977) "Agriculture in Development Theory: An overview." Pp. 1–24 in *Agriculture in Development Theory*, edited by L. Reynolds. New Haven and London: Yale University Press.

——— (1985) *Economic Growth in the Thirld World.* New Haven and London: Yale University Press.

Rhode, Paul (1995) "Learning, Capital Accumulation and the Transformation of California Agriculture." *Journal of Economic History* 55: 773–800.

Richards, Alan (1982) *Egypt's Agricultural Development, 1800–1980: Technical and Social Change.* Boulder: Westview.

Richards, J. F., James R. Hagen, and Edward S. Haynes (1985) "Changing Land Use in Bihar, Punjab, and Haryana, 1850–1970." *Modern Asian Studies* 19: 699–732.

Richardson, Philip (1999) *Economic Change in China, c.1800–1950.* Cambridge: Cambridge University Press.

Ritson, Christopher (1997) "Introduction." Pp. 1–6 in *The Common Agricultural Policy*, edited by Christopher Ritson and Harvey David. 2nd edition: Wallingford: CAB International.

Rizov Marian, Dinu Gavrilescu, Hamish Gow, Erik Mathijs, and Johan Swinnen (2001) "Transition and Enterprise Restructuring: The Development of Individual Farming in Romania." *World Development* 29: 1257–74.

Rogari, Sandro (1999) *La Confagricoltura nella storia d'Italia.* Bologna: Mulino.

Rogowski, Ronald (1989) *Commerce and Coalition.* Princeton: Princeton University Press.

Rolfes, M. (1976a) "Landwirtschaft 1850–1914." Pp. 495–526 in *Handbuch der deutschen wirtschafts-und sozialgeschichte*, edited by H. Aubin and W. Zorn. Stuttgart: Ernst Klett.

——— (1976b) "Landwirtschaft 1914–1970." Pp. 741–95 in *Handbuch der deutschen wirtschafts-und sozialgeschichte*, edited by H. Aubin and W. Zorn. Stuttgart: Ernst Klett.

Rooth, Tim (1992) *British Protectionism and the International Economy: Overseas Commercial Policy in the 1930s.* Cambridge: Cambridge University Press.

Rosengrant, Mark W., and Peter Hazell (2000) *Transforming the Rural Asian Economy: The Unfinished Revolution.* Published for the Asian Development Bank. Oxford: Oxford University Press.

Rosenzweig, Mark R., and Hans P. Binswanger (1993) "Wealth, Weather Risk and the Composition and Profitability of Agricultural Investments." *Economic Journal* 103: 56–78.

Rossi, Nicola, Andrea Sorgato, and Gianni Toniolo (1993) "I conti economici italiani: una ricostruzione statistica." *Rivista di storia economica* (n.s.) 10: 1–47.

Rossini, Egidio, and Carlo Vanzetti (1986) *Storia della agricoltura italiana*, Bologna: Edagricole.

Rothenberg, Winifred (1992) *From market-places to a market economy*, Chicago-London: University of Chicago Press.

Roy, Tirthankar (2000) *The Economic History of India, 1857-1947*, Oxford: Oxford University Press.

Rozelle, Scott D. and Johan Swinnen (2004) "Transition and Agriculture." *Journal of Economic Literature* 42: 404–560.

Runge, Ford, Benjamin Senauer, Philip G. Pardey, and Mark W. Rosengrant (2003) *Ending Hunger in Our Lifetime: Food Security and Globalization.* Published for IFPRI. Baltimore and London: Johns Hopkins University Press.

Russell, Edmund P. (1999) "The Strange Career of DDT: Experts, Federal Capacity and Environmentalism in World War II." *Technology and Culture* 40: 770–96.

Russell, Walter (1973) *Soil Conditions and Plant Growth.* 10th edition. Longman: London.

Ruttan, Vernon W. (1978) "Structural Retardation and the Modernization of French Agriculture: A Skeptical View." *Journal of Economic History* 38: 714–28.

——— (1994) "Sustainable Agricultural Growth." Pp. 3–19 in *Agriculture, Environment and Health: Sustainable Development in the 21st century*, edited by Vernon W. Ruttan. Minneapolis: University of Minnesota Press.

——— (2002) "Productivity growth in world agriculture: Sources and constraints." *Journal of Economic Perspectives* 16: 161–84.

Ruttan, Vernon W., and Yuijiro Hayami (1995) "Induced Innovation Theory and Agricultural Development." Pp. 169–88 in *Induced Innovation Theory and International Agricultural Development: A Reassessment*, edited by Bruce M. Koppel. Baltimore and London: Johns Hopkins University Press.

Saito, Teruko, and Lee Kin Kiong (1999) *Statistics on the Burmese economy. The 19th and 20th Centuries.* Canberra: Institute of Southeast Asian studies.

Saker-Woeste, Victoria (1998) *The Farmer's Benevolent Trust. Law and Agricultural Cooperation in Industrial America 1865–1945.* Chapel Hill and London: University of North Carolina Press.

Salaman, Redcliffe (1949) *The History and Social Influence of the Potato.* Cambridge: Cambridge University Press.

Sandgruber, Roman (1978) *Osterreisches Agrarstatistik 1750–1918.* Vienna: Verlag fur Geschichte und Politik.

Sargen, Nicholas P (1979) *"Tractorization" in the United States and Its Relevance for the Developing Countries.* New York and London: Garland.

Schaefer, Donald (1983) "The Effect of the 1859 Crop Year upon Relative Productivity in the Antebellum Cotton South." *Journal of Economic History* 43: 851–65.

Schiff, Maurice, and Alberto Valdes (1992) *The Political Economy of Agricultural Pricing Policy. Volume 4: A Synthesis of the Economics in Developing Countries.* Published for the World Bank. Baltimore and London: Johns Hopkins University Press.

——— (2002) "Agriculture and the macro-economy, with Emphasis on Developing Countries." Pp. 1423–54 in *Handbook of agricultural economics, vol. 2A Agriculture and its external linkages*, edited by Bruce Gardner and Gordon Rausser. Amsterdam: Elsevier.

Schilcher, Linda (1991) "The Grain Economy of Late Ottoman Syria and the Issue of Large-Scale Commercialisation." Pp. 1173–95 in *Landholding and Commercial Agriculture in the Middle East*, edited by Caglar Keyder and Faruk Tabak. Albany: State University of New York Press.

Schimmelpfenning, David, and Colin Thirtle (1994) "Cointegration and Causality: Exploring the Relationship between Agricultural R&D and Productivity." *Journal of Agricultural Economics* 45: 220–31.

——— (1999) "The Internationalization of Agricultural Technology: Patents, R&D Spillovers and Their Effects on Productivity in the European Union and United States." *Contemporary Economic Policy* 17: 457–68.

Schimmelpfenning, David, Colin Thirtle, Johan van Zyl, Carlos Arnade, and Youngesh Khatri (2000) "Short and Long-run Returns to Agricultural R&D in South Africa, or Will the Real Rate of Return Please Stand Up?" *Agricultural Economics* 23: 1–15.

Schmitt, Gunther (1991) "Why Is the Agriculture of Advanced Western Economies Still Organized by Family Farms? Will This Continue To Be So in the Future?" *European Review of Agricultural Economics* 18: 443–58.

Schmookler, Jacob (1966) *Invention and Economic Growth.* Cambridge: Harvard University Press.

Schon, Lennart (1995) *Jordbruk med binäringar, 1800–1980*, vol. 24. Lund: Skrifter utgivna av ekonomisk historika föreningen i Lund.

Schultz, Theodor W. (1964) *Transforming Traditional Agriculture.* New Haven and London: Yale University Press.

Scott, Roy (1970) *The reluctant farmer. The rise of agricultural extension to 1914*, Urbana-Chicago-London: University of Illinois Press.

——— (1976) *The Moral Economy of the Peasant: Rebellion and Subsistence in South-East Asia.* New Haven and London: Yale University Press.

Sen, Amartya (1981) *Poverty and Famines: An Essay on Entitlement and Deprivation.* Oxford: Oxford University Press.

Sereni, Emilio (1968) *Il capitalismo nelle campagne.* Turin: Einaudi, First edition was published in 1949.

Serpieri, Arrigo (1910) *Il contratto agrario e le condizioni dei contadini dell'Alto Milanese.* Milan: Società Umanitaria.

Sexton, Richard, and Nathalie Lavoie (2002) "Food Processing and Distribution: An Industrial Organization Approach." Pp. 863–932 in *Handbook of Agricultural Economics, vol. 1B: Marketing. Distribution and Consumers*, edited by Bruce Gardner and Gordon Rausser. Amsterdam: Elsevier.

Shaban, Radwan Ali (1987) "Testing between Competing Models of Sharecropping." *Journal of Political Economy* 95: 895–920.

Shammas, Carol (1982) "How Self-sufficient Was Early America?" *Journal of Interdisciplinary History* 13: 247–72.

Shanin, Theodor (1973) "The Nature and Logic of Peasant Economy." *Journal of Peasant Studies* 1: 63–79, 186–206.

Sharma, Rita, and Thomas Poleman (1993) *The New Economics of India's Green Revolution: Income and Employment Diffusion in Uttar Pradesh.* Ithaca and London: Cornell University Press.

Shaw, Alan (1990) "Colonial Settlement 1788–1945." Pp. 1–18 in *Agriculture in the Australian Economy*, edited by D. B. Williams. Sydney: Sydney University Press.

Shaw-Taylor, Leigh (2001) "Parliamentary Enclosures and the Emergence of an English Agricultural Proletariat." *Journal of Economic History* 61: 640–62.

Sheingate, Adam (2001) *The Rise of the Agricultural Welfare State. Institutions and Interest Group Power in the United States, France and Japan.* Princeton: Princeton University Press.

Shiel, Robert S. (1991) "Improving Soil Productivity in the Pre-fertiliser era." Pp. 51–77 in *Land, Labour, and livestock: Historical Studies in European Agricultural Productivity*, edited by Bruce Campbell and Mark Overton. Manchester: Manchester University Press.

Shiva, Vandana (1991) *The Violence of the Green Revolution. Third World Agriculture, Ecology and Politics.* London: Zed Books.

Shukla, Tara (1965) *Capital Formation in Indian Agriculture.* Bombay: Vora and Co.

Sicsic, Pierre (1992) "City-Farm Wage Gaps in Late Nineteenth Century France." *Journal of Economic History* 52: 675–95.

Simms, James (1977) "The crisis in Russian agriculture at the end of the 19th century: a different view." *Slavic Review* 36: 377–98.

Simpson, James (1995) *Spanish Agriculture. The Long Siesta, 1765–1965.* Cambridge: Cambridge University Press.

——— (1997) "Did Tariffs Stifle Spanish Agriculture before 1936?" *European Review of Economic History* 1: 45–67.

——— (2000) "Co-operation and Co-operatives in Southern European Wine Production: The Nature of Successful Institutional Innovation." *Advances in Agricultural Economic History* 1: 95–126.

Singh A. J., and Derek Byerlee (1990) "Relative Variability in Wheat Yields across Countries and over Time." *Journal of Agricultural Economics* 41: 21–32.

Singh, Inderjit, Lyn Squire, and John Strauss (1986) *Agricultural Household Models. Extensions, Applications and Policy.* Baltimore and London: Johns Hopkins University Press.

Singh, Nirvikar (1991) "Theories of sharecropping." Pp.33–72 in *The economic theory of agrarian institutions*, edited by Pranab Bardhan. Oxford: Clarendon Press.

Sivasubramonian, S. (2000) *The National Income of India in the Twentieth Century.* New Delhi: Oxford University.

Sluglett, Peter, and Marion Farouk-Sluglett (1984) "The application of the 1858 Land code in greater Syria: some preliminary observations." Pp. 333–356 in *Land tenure and social transformation in the Middle East*, edited by Tarif Khalidi. Beirut: American University of Beirut.

Smale, Melinda (1997) "The Green Revolution and Wheat Genetic Diversity: Some Unfounded Assumptions." *World Development* 25: 1257–69.

Smethurst, Richard J. (1986) *Agricultural Development and Tenancy Disputes in Japan, 1870–1940.* Princeton: Princeton University Press.

Smil, Vaclav (2001) *Enriching the Earth: Fritz Haber, Carl Bosch and the Transformation of World Food Production.* Cambridge: MIT Press.

Smith, Louise (1961) *The Evolution of Agricultural Co-operation.* Oxford: Basil Blackwell.

Smith, Mark (1998) *Debating Slavery.* Cambridge: Cambridge University Press.

Sokoloff, Kenneth, and David Dollar (1997) "Agricultural Seasonality and the Organization of Manufacturing in Early Industrial Economies: The Contrast between England and the United States." *Journal of Economic History* 57: 288–321.

Solberg, Carl E. (1987) *The Prairies and the Pampas: Agrarian Policy in Canada and Argentina, 1880–1930.* Stanford: Stanford University Press.

Solomou, Solomos, and Weike Wu (1999) "Weather Effects on European Agricultural Output, 1850–1913." *European Review of Economic History* 3: 351–72.

——— (2003) "Weather Effects on European Agricultural Price Inflation, 1870–1913." *Advances in Agricultural Economic History* 2: 115–43.

Solow, Robert M. (1957) "Technical change and the Aggregate Production Function." *Review of Economics and Statistics* 39: 312–20.

Spulber, Nicholas (2003) *Russia's Economic Transitions: From Late Tsarism to the New Millennium.* Cambridge: Cambridge University Press.

Stanton, B. F. (1993) "Changes in Farm Size and Structure in American Agriculture in the Twentieth Century." Pp. 42–70 in *Size, Structure and the Changing Face of American Agriculture*, edited by Arne Hallam. Boulder: Westview.

Stead, David "Risk and risk management in English agriculture c 1750–1850." *Economic History Review* 57: 334–61.

Steckel, Richard H. (1995) "Stature and the Standard of Living." *Journal of Economic Literature* 33: 1903–40.

Stein, Burton (1992) "Introduction." Pp. 1–31 in *The Making of Agrarian Policy in British India*, edited by Burton Stein. Delhi: Oxford University Press.

Stillson, Richard T. (1971) "The Financing of Malayan Rubber, 1905–1923." *Economic History Review* 24: 589–98.

Stokes, Eric, B. Chaudhuri, H. Fukazawa, and Dharma Kumar (1983) "Agrarian Relations." Pp. 36–241 in *The Cambridge Economic History of India, vol 2: c.1757–c.1970*, edited by Dharma Kumar and Meghnad Desai. Cambridge: Cambridge University Press.

Stokes, Nancy L. (2001) "A Quantitative Model of the British Industrial Revolution, 1780–1850." *Carnegie-Rochester Conference Series on Public Policy* 55: 55–109.

Stone, Bruce (1988) "Developments in Agricultural Technology." In "Food and Agriculture in China during the Post-Mao Era," edited by K. Walker and Y. Y. Kueh. Special issue of *China Quarterly*, no. 116(December): 767–822.

Stone, Ian (1984) *Canal Irrigation in British India. Perspective on Technological Change in a Peasant Economy.* Cambridge: Cambridge University Press.

Strauss, Frederick, and Lois H. Bean (1940) "Gross Farm Income and Indices of Farm Production and Prices in the United States 1869–1937." *United States Department of Agriculture Technical Bulletin*, no. 703. Washington: US Government Printing Office.

Strauss, John, and Thomas Duncan (1995) "Human Resources: Empirical Modeling of Household and Family Decisions." Pp. 1883–2023 in *Handbook of Development*

Economics, vol. IIIA, edited by Jere Behrman and T. N. Srinivasan. Amsterdam: North Holland.

Sturmey, S. G. (1968) "Owner Farming in England and Wales, 1900–1950." In *Essays in Agrarian History*, vol. II, edited by Walter Minchinton. Newton Abbot: David and Charles.

Suharlyanto, Kecuk, Angela Lusigi, and Colin Thirtle (2001) "Productivity Growth and Convergence in Asian and African Agriculture." Pp. 258–73 in *Asia and Africa in Comparative Economic Perspective*, edited by P. Lawrence and C. Thirtle. London and Basingstoke: Palgrave.

Sullivan, Richard J. (1984) "Measurement of English Farming Technological Change, 1523–1900." *Explorations in Economic History* 21: 270–89.

Sumner, Daniel, and Stefan Tangermann (2002) "International Trade Policy and Negotiations." Pp. 1999–2055 in *Handbook of Agricultural Economics, vol. 2B: Agricultural and Food Policy*, edited by Bruce Gardner and Gordon Rausser. Amsterdam: Elsevier.

Sunding, David, and David Zilberman (2001) "The Agricultural Innovation Process: Research and Technology Adoption in a Changing Agricultural Sector." Pp. 207–61 in *Handbook of Agricultural Economics, vol. 1A: Agricultural Production*, edited by Bruce Gardner and Gordon Rausser. Amsterdam: Elsevier.

Svedberg, Peter (2003) "Undernutrition Overestimated." *Economic Development and Cultural Change* 51: 5–36.

Svennilson, Ingar (1954) *Growth and Stagnation in the World Economy.* Geneva: United Nations.

SVIMEZ (1961) *Un secolo di statistiche italiane.* Roma: SVIMEZ.

Swamy, Anand (1998) "Factor Markets and Resource Allocation in Colonial Punjab." *Journal of Development Studies* 34: 97–115.

Swinnen, Johan (1998) "Agricultural Reform in Central and Eastern Europe." Pp. 586–601 in *International Agricultural Development*, edited by Carl K. Eicher and John M. Staatz. 3rd edition. Baltimore and London: Johns Hopkins University Press.

Swinnen, Johan F., Anurag N. Banerjee, and Harry de Gorter (2001) "Economic Development, Institutional Change and the Political Economy of Agricultural Protection. An Econometric Study of Belgium since the 19th century." *Agricultural Economics* 26: 25–43.

Swinnen, Johan F., Harry de Gorter, Gordon Rausser, and Anurag N. Banerjee (2000) "The Political Economy of Public Research Investment and Commodity Policies in Agriculture: An Empirical Study." *Agricultural Economics* 22: 111–22.

Tamura, Robert (2002) "Human Capital and the Switch from Agriculture to Industry." *Journal of Economic Dynamics and Control* 27: 207–42.

Tan, Elaine S. (2002) "The bull is half the herd': Property Rights and Enclosures in England, 1750–1850." *Explorations in Economic History* 39: 470–89.

Tattara, Giuseppe (1973) "Cerealicoltura e politica agraria durante il fascismo." Pp. 373–404 in *Lo sviluppo economico italiano*, edited by Gianni Toniolo. Bari: Laterza.

Tauer, Loren W. (1998) "Productivity of New York Dairy Farms Measured by Nonparametric Malmquist Indices." *Journal of Agricultural Economics* 49: 234–49.

Taylor, Alan M. (1997) "Latifundia as Malefactor in Economic Development? Scale, Tenancy and Agriculture on the Pampas." *Research in Economic History* 17: 261–300.

Taylor, David (1987) "Growth and Structural Change in the English Dairy Industry, c.1860–1930." *Agricultural History Review* 35: 47–64.

Temin, Peter (1983) "Patterns of Cotton Agriculture in Antebellum Georgia." *Journal of Economic History* 43: 661–74.

——— (2002) "The Golden Age of European Growth Reconsidered," *European Review of Economic History* 6: 3–22.

Temple, Jonathan (1999) "The New Growth Evidence." *Journal of Economic Literature* 37: 112–56.

Thiam, Abdourahmane, Boris E. Bravo-Ureta, and Teodoro E. Rivas (2001) "Technical Efficiency in Developing Countries Agriculture: A Meta-analysis." *Agricultural Economics* 25: 235–43.

Thies, Clifford, and Daniel Gerlowski (1993) "Bank Capital and Bank Failure, 1921–1932: Testing the White Hypothesis." *Journal of Economic History* 53: 908–14.

Thiesenhusen, William (1996) "Mexican land reform, 1934–1991: success or failure?" Pp. 35–47 in *Reforming Mexico's agrarian reform*, edited by Laura Randall. Armonk: Sharpe.

Thirtle, Colin, and P. Bottomley (1992) "Total Factor Productivity in UK Agriculture, 1967–90." *Journal of Agricultural Economics* 43: 381–400.

Thirtle, Colin, David E. Schimmelpfenning, and Robert F. Townsend (2002) "Induced Innovation in United States Agriculture, 1880–1990: Time Series Tests and an Error Correction Model." *American Journal of Agricultural Economics* 84: 598–614.

Thompson, F.M.L. (1968) "The Second Agricultural Revolution." *Economic History Review* 21: 62–77.

Thorner, Daniel (1975) "Peasant Economy as a Category in Economic History." Pp. 202–17 in *Peasants and Peasant Societies*, edited by T. Shanin. Harmondsworth: Penguin.

Thornton, Russel (2000) "Population History of Native North Americas." Pp. 9–50 in *A Population History of North America*, edited by Michael Haines and Richard Steckel. Cambridge: Cambridge University Press.

Tiffen, Mary, and Michael Mortimore (1990) *Theory and Practice in Plantation Agriculture: An Economic Review.* London: Overseas Development Institute.

Tillack, Peter, and Eberhard Schulze (2000), "Decollectivization and restructuring of farms" Pp. 447–470 in *Russia agro-food sector: towards truly functioning markets*, edited by P. Werheim, E. V. Serovam, K. Frohberg and J. von Braun. Boston and Dordrecht: Kluwer.

Tilly, R. H. (1978) "Capital Formation in Germany in the Nineteenth Century." Pp. 383–441 in *The Cambridge Economic History of Europe*, vol. 7 part 1, edited by Peter Mathias and Michel Postan. Cambridge: Cambridge University Press, 1978.

Timmer, Peter C. (1969) "The Turnip, the New Husbandry and the English Agricultural Revolution." *Quarterly Journal of Economics* 83: 375–95.

——— (1988) "The Agricultural Transformation." Pp. 276–328 in *Handbook of Development Economics*, vol. 1, edited by H. Chenery and T. N. Snirvasan. Amsterdam: Elsevier.

——— (2002) "Agriculture and Economic Development." Pp. 1487–1546 in *Handbook of Agricultural Economics, vol. 2A: Agricultural and Food Policy*, edited by Bruce Gardner and Gordon Rausser. Amsterdam: Elsevier.

Tokarik, Stephen (2003) "Measuring the impact of distortions in agricultural trade in partial and general equilibrium." IMF Working paper WP/03/110.

Tomlinson, Brian R. (1993) *The Economy of Modern India 1860–1970.* Cambridge: Cambridge University Press.

Tonizzi, M. Elisabetta (2001) *L'industria dello zucchero.* Milan: Franco Angeli.

Topik, Steven (1987) *The Political Economy of the Brazilian State, 1889–1930.* Austin: University of Texas Press.

Tostlebe, Alvin (1957) *Capital in Agriculture: Its Formation and Financing since 1870.* Published for the NBER. Princeton: Princeton University Press.

Toutain, Jean-Claude (1961) "Le produit de l'agriculture francaise de 1700 à 1958." In "Histoire quantitative de l'economie francaise," parts 1 and 2. *Cahiers de l'Institut de science économique appliquee*, serie AF 1 and 2. Paris: ISEA.

——— (1997) "La Croissance Française 1789–1990. Nouvelles Estimations." *Economies et sociétés. Cahiers de l'ISMEA.* Serie Histoire quantitative de l'economie francaise. Serie HEQ n.1. Paris: PUF

Towne, Marvin W., and Wayne D. Rasmussen (1960) "Farm Gross Product and Gross Investment in the Nineteenth Century." *In Trends in the American Economy in the Nineteenth Century.* Studies in Income and Wealth, vol. 24. Princeton: Princeton University Press.

Tracy, Michael (1989) *Government and Agriculture in Western Europe 1880–1988.* 3rd edition. New York: New York University Press.

Traxler, Greg, and Derek Byerlee (2001) "Linking Technical Change to Research Effort: An Examination of Aggregation and Spillovers Effects." *Agricultural Economics* 24: 235–46.

Treadgold, Donald (1957) *The Great Siberian Migration.* Westport: Greenwood Press.

Tripp, Robert (2001) *Seed Provision and Agricultural Development.* London: Overseas Development Institute.

Trueblood, Michael, and Jay Coggins (2004) "Intercountry Agricultural Efficiency and Productivity: A Malmquist Index Approach." Mimeo.

Turner, Michael (1996) *After the Famine: Irish Agriculture 1850–1914.* Cambridge: Cambridge University Press.

——— "Agricultural Output, Income and Productivity." Pp. 224–320 in *The Agrarian History of England and Wales, vol. 7: 1850–1914*, edited by E.J.T. Collins. Cambridge: Cambridge University Press.

Turner, Michael, John V. Beckett, and Bethanie Afton (2001) *Farm Production in England 1700–1914.* Oxford: Oxford University Press.

Tyers, Rod, and Kym Anderson (1992) *Disarray in World Food Markets.* Cambridge: Cambridge University Press.

UMA (1968) *Quarant'anni di motorizzazione agricola in Italia.* Rome: UMA.

United Nations (ad annum) *Statistical Yearbook.* New York: United Nations.

United Nations (2001) *Development Report 2001.* New York: United Nations.

United Stated Department of Agriculture (1997) *Agricultural Census.* Available at www.ers.usda.gov/StateFacts/US.

——— (1998) *Rural Business—Cooperative Service Cooperative Information Report 1.* Washington, D.C.: U.S. Government Printing Office.

——— (1999) *Agricultural Income and Finance Situation.* Resource Economics Division. Available at usda.mannlib.cornell.edu/usda/.

——— (2000) *Agricultural Statistics.* Washington, D.C.: U.S. Government Printing Office.

——— (2002) *Agricultural Census 2002.* Available at www.nass.usda/gov/Census/Census 02/volume1.

——— (2004) *Agricultural Statistics 2004.* Available at www.usda.gov/nass/pubs/agr04/acro04.htm.

Unwin, Tim (1991) *Wine and the Vine. An Historical Geography of Viticulture and the Wine Trade.* London and New York: Routledge.

Urquhart, M. C. (1993) *Gross National Product, Canada 1870–1926: The Derivation of the Estimates.* Montreal and Kingston: McGill and Queens University Presses.

Valensi, Lucette (1985) *Tunisian Peasants in the Eighteenth and Nineteenth Century.* Paris and Cambridge: Editions de la Maison des Sciences de l'Homme and Cambridge University Press.

van der Eng, Pierre (1992) "The Real Domestic Product of Indonesia." *Explorations in Economic History* 29: 343–73.

——— (1996) *Agricultural Growth in Indonesia.* London and Basingstoke: Macmillan.

——— (2004) "Productivity and Comparative Advantage in Rice Agriculture in Southeast Asia." *Asian Economic Journal* 18.

van der Meer, Cornelis, and Saburo Yamada (1990) *Japanese Agriculture: A Comparative Economic Analysis.* London: Routledge.

Van Veldhuizen, Laurens, Ann Waters-Bayer, Ricardo Ramirez, Debra A. Johnson, and John Thompson (1997) *Farmers' Research in Practice: Lessons from the Field.* London: International Technology Publications.

Van Zanden, Jan Luiten (1988) "The First Green Revolution: The Growth of Production and Productivity in European Agriculture 1870–1914." *Research Memorandum* 42. Amsterdam: Vrije Universiteit Facultit der Economische Wetenschappen en Econometrie.

——— (1991) "The First Green Revolution: The Growth of Production and Productivity in European Agriculture, 1870–1914." *Economic History Review* 44: 215–39.

——— (1994) *The Transformation of European Agriculture in the 19th Century: The Case of the Netherlands.* Amsterdam: Vu Uitgeverij.

——— (2000) "Estimates of GNP." Available at http://nationalaccounts.niwi.knaw.nl.

Van Zanden, Jan Luiten, and Arthur van Riel (2004) *The strictures of inheritance. The Dutch economy in the nineteenth century.* Princeton: Princeton University Press.

Verdier, Daniel (1994) *Democracy and International Trade.* Princeton: Princeton University Press.

Vidal, Jean-François (1990) *Les fluctuations internationales de 1890 à nos jours.* Paris: Economica.

Virts, Nancy (1991) "The Efficiency of Southern Tenant Plantations, 1900–1945." *Journal of Economic History* 51: 385–95.

Visaria, Leela, and Pravin Visaria (1983) "Population (1757–1947)." Pp. 463–532 in *The Cambridge Economic History of India, vol 2: c.1757–c.1970,* edited by Dharma Kumar and Meghnad Desai. Cambridge: Cambridge University Press.

Vitali, Ornello (1968) *La popolazione agricola* Rome: Istituto di demografia.

Vitali, Ornello (1990) "I censimenti e la composizione sociale italiana." Pp. 377–414 in *Storia dell'agricultura italiana in età contemporanea, vol. 2: Uomini e classi,* edited by P. Bevilacqua. Padua: Marsilio.

Volin, Lazar (1970) *A Century of Russian Agriculture.* Cambridge: Harvard University Press.

Von Braun, Joachim (1996) "Production, Employment and Income Effects of Commercialization of Agriculture." Pp. 37–64 in *Agricultural Commercialisation, Economic Development and Nutrition,* edited by Joachim von Braun and Eileen Kennedy. Published for the IFPRI. Baltimore and London: Johns Hopkins University Press.

Von Braun, Joachim, Howarth Bouis, and Eileen Kennedy (1996) "Conceptual framework." Pp. 11–33 in *Agricultural Commercialisation, Economic Development and Nutrition,* edited by Joachim Von Braun and Eileen Kennedy. Published for the IFPRI. Baltimore and London: Johns Hopkins University Press.

Von Jankovich, Bela (1912) "Index-number von 45 Waaren in der oesterreisch-ungarischen Monarchie (1867–1909)." *Bulletin de l'Institut internationale de la statistique* 19(3): 136–56.

Von Oppen, M., B. K. Njehia, and Ijiami Abdelatif (1997) "The Impact of Market Access on Agricultural Productivity: Lessons from India, Kenya and the Sudan." *Journal of International Development* 9: 117–31.

Voros, Antal (1980) "The Age of Preparation: Hungarian Agrarian Conditions between 1848 and 1914." Pp. 21–129 in *The Modernization of Agriculture: Rural Transformation in Hungary, 1848–1975*, edited by Joseph Held. Boulder: East European Monographs.

Voth, Hans-Joachim (2001) "The Longest Years: New Estimates of Labour Input in England." *Journal of Economic History* 61: 1064–82.

Wade, William W. (1981) *Institutional Determinants of Technical Change and Agricultural Productivity Growth: Denmark, France and Great Britain, 1870–1965.* New York: Arno Press.

Wai, U. Tun (1977) "A Revisit to Interest Rates Outside the Organized Market of Underdeveloped Countries." *Banca Nazionale del Lavoro Quarterly Review* 30: 291–312.

Walker, Thomas, and N. S. Jodha (1986) "How Small Farm Households Adapt to Risk." Pp. 17–34 in *Crop Insurance for Agricultural Development: Issues and experience*, edited by Peter Hazell, Carlos Pomareda, and Alberto Valdes. Baltimore and London: Johns Hopkins University Press.

Walsh Sanderson, Susan R. (1984) *Land Reform in Mexico: 1910–1980.* New York: Academic Press.

Wan Guang H, and Ejiang Chen (2000) "A micro-empirical analysis of land fragmentation and scale economies in rural China." Pp. 133–147 in *China's agriculture at the crossroads.* edited by Yongzheng Yang and Weiming Tian. Basingstoke: Macmillan.

Wang, Yuru (1992) "Economic Development in China between the Two World Wars (1920–1936)." Pp. 58–77 in *The Chinese Economy in the Early Twentieth Century. Recent Chinese Studies.* edited by Tim Wright. New York: St Martin's Press.

Warman, Arturo (2004) *Corn and Capitalism: How a Botanic Bastard Grew to Global Dominance.* Chapel Hill and London: University of North Carolina Press, 2004.

Webb, Steven B. "Agricultural Protection in Wilhelminian Germany: Forging an Empire with Pork and Rye." *Journal of Economic History* 42: 309–26.

Whatley, Warren C. (1987) "Southern Agrarian Labor Contracts as Impediments to Cotton Mechanization." *Journal of Economic History* 47: 45–70.

Wheatcroft, Stephen G. (1991) "Crises and the Condition of the Peasantry in Late Imperial Russia." Pp. 128–71 in *Peasant Economy, Culture and Politics of European Russia, 1800–1921*, edited by Esther Kingston-Mann and Timothy Mixter. Princeton: Princeton University Press.

——— (1997) "Soviet Statistics of Nutrition and Mortality during Times of Famine, 1917–22 and 1931–33." *Cahiers du Monde Russe* 38: 525–58.

——— (1999) "The Great Leap Upwards: Anthropometric Data and Indicators of Crises and Secular Change in Soviet Welfare Levels, 1880–1960." *Slavic Review* 58: 27–60.

Wheatcroft, S. G., and R. W. Davies (1994a) "Population." Pp. 57–80 in *The Economic Transformation of the Soviet Union, 1913–1945*, edited by R. W. Davies, M. Harrison and S. G. Wheatcroft. Cambridge: Cambridge University Press.

——— (1994b) "Agriculture." Pp. 106–30 in *The Economic Transformation of the Soviet Union, 1913–1945*, edited by R. W. Davies, M. Harrison and S. G. Wheatcroft. Cambridge: Cambridge University Press.

Whitcombe, Elisabeth (1983) "Irrigation and Railways." Pp. 677–761 in *The Cambridge Economic History of India, vol. 2: c.1757–c.1970*, edited by Dharma Kumar and Meghnad Desai. Cambridge: Cambridge University Press.

White, Eugene (1984) "A Reinterpretation of the Banking Crisis of 1930." *Journal of Economic History* 44: 119–38.

White, William (2001) "An Unsung Hero: The farm Tractor's Contribution to Twentieth-Century United States Economic Growth." *Journal of Economic History* 61: 493–96.

Wicker, Elmus (1996) *The Banking Panics of the Great Depression.* Cambridge: Cambridge University Press.

Wiggins, Steve (2002) "Smallholder Farming in Africa: Stasis and Dynamics." Pp. 102–19 in *Renewing Development in Sub-Saharan Africa. Policy, Performance and Prospects*, edited by Deryke Bleshaw and Ian Livingston. London: Routledge.

Wik, R. M. (1953) *Steam Power on the American Farm.* Philadelphia: University of Pennsylvania Press.

Williamson, Jeffrey (1990) "The Impact of Corn Laws Just Prior to Repeal." *Explorations in Economic History* 27: 123–56.

——— (2002) "Land, Labor and Globalization, 1870–1940." *Journal of Economic History* 62: 55–85.

Wilson, Geoff, and Olivia J. Wilson (2001) *German Agriculture in Transition.* London: Palgrave.

Wines, Richard A (1985) *Fertilizers in America.* Philadelphia: Temple University Press.

Winters, Paul, Alain de Janvry, Elisabeth Sadoulet, and Kostas Stamoulis (1998) "The Role of Agriculture in Economic Development: Visible and Invisible Surplus Transfers." *Journal of Development Studies* 34: 71–97.

Wolf, Steven (1998) "Institutional Relations in Agricultural Information: Transitions and Consequences." Pp. 3–22 in *Privatization of Information and Agricultural Industrialization*, edited by Steven Wolf. Boca Raton: CRC Press.

Wong, Lung-Fai, and Vernon Ruttan (1990) "Comparative Analysis of Agricultural Productivity Trends in Centrally Planned Economies." Pp. 23–47 in *Soviet Agriculture: Comparative Perspectives*, edited by Kenneth R. Gray. Ames: Iowa State University Press.

Wood, Stanley, Kate Sebastian, and Sara J. Scherer (2000) "*Pilot Analysis of Global Ecosystems: Agroecosystems.*" International Food Policy Research Institute and World Resource Institute, Washington, D.C. Available at www.ifpri.org/pubs/books/page.htm.

World Bank (2002) *Building Institutions for Markets. World Development Report 2002.* Oxford: Oxford University Press.

——— (2003) *Land Policies for Growth and Poverty Reduction.* Washington, D.C., and Oxford: World Bank and Oxford University Press.

World trade analyzer Statistics Canada International Trade Division World trade analyzer Ottawa.

Wright, Gavin (1986) *Old South New South.* Baton Rouge: University of Louisiana Press.

——— (1988) "American Agriculture and the Labor Market: What Happened to Proletarisation?" *Agricultural History* 62: 182–209.

Wright, Gavin, and Howard Kunreuther (1975) "Cotton, Corn and Risk in the 19th Century." *Journal of Economic History* 35: 526–51.

Wright, Tim (1992) "Introduction." Pp. 1–19 in *The Chinese Economy in the Early Twentieth Century*, edited by Tim Wright. *Recent Chinese Studies.* New York: St Martin's Press.

Wrigley, C. C. (1970) *Crops and Wealth in Uganda: A short agrarian history.* Nairobi: Oxford University Press.

WTO Statistical Database. Available at www.wto.org.

Wu, Ziping, and Zhifang Wang (2000) "The accuracy of China's grain production data: evidence from two county-level case studies." Pp. 166–179 in *China's agriculture at the crossroads*, edited by Yongzheng Yang and Weiming Tian. Basingstoke: Macmillan.

Wu, Yanrui and Hong Yang (1999) "Productivity and Growth in China: A review." Pp. 29–51 in *Productivity and Growth in Chinese Agriculture*, edited by Kali P. Kalirajan and Wu Yanrui. Basingstoke: Macmillan.

Yaron, Jacon (1994) "What Makes Rural Finance Institutions Successful." *World Bank Research Observer* 9: 49–70.

Yates, Paul L. (1959) *Forty Years of Foreign Trade.* London: Allen and Unwin.

Yoong-Deok, Jeon, and Kim Young-Yong (2000) "Land Reform, Income Redistribution and Agricultural Production in Korea." *Economic Development and Cultural Change* 48: 253–68.

Yotopolous, Pan, and Jeffrey Nugent (1974) *Economics of Development: Empirical Investigations.* New York: Harper and Row.

Xu, Xinwu (1992) "The Process of the Disintegration of Modern China's Natural Economy." Pp. 113–33 in *The Chinese Economy in the Early Twentieth Century*, edited by Tim Wright. Recent Chinese Studies. New York: St Martin's Press.

Xu, Guohua, and L. J. Peel (1991) *The Agriculture of China.* Oxford: Oxford University Press.

Zamagni, Vera (1993) *An Economic History of Italy.* Oxford: Oxford University Press.

Zangheri, Renato (1986) "Nascita e primi sviluppi." Pp. 3–216 in *Storia del movimento cooperativo in Italia*, edited by Renato Zangheri, Giuseppe Galasso and Valerio Castronovo. Turin: Einaudi.

Zimmermann, Clemens (1999) "La modernisation des campagnes allemandes (19eme–20eme siècles). Les apports de l'historiographie récente en Allemagne." *Histoire and Sociétés Rurales* 11: 87–108.

Zong, Ping, and John Davis (1998) *Economics of Marketable Surplus Supply. A Theoretical and Empirical Analysis for China.* Aldershot: Ashgate.

Zweig, David (1989) *Agrarian Radicalism in China, 1968–1981.* Cambridge: Harvard University Press.

INDEX